Topics in Atmospheric and Oceanographic Sciences

Editors: Michael Ghil Robert Sadourny Jürgen Sündermann

To Fred A Franklin
with appreciation of a stimulating
friendship
Allan R. Robinson
July 1986.

Eddies in Marine Science

Edited by
Allan R. Robinson

With 268 Figures

Springer-Verlag
Berlin Heidelberg New York Tokyo 1983

Dr. ALLAN R. ROBINSON
Gordon McKay Professor of Geophysical Fluid Dynamics
Harvard University, Division of Applied Sciences
Pierce Hall, Cambridge, MA 02138, USA

Series Editors:

Professor Dr. MICHAEL GHIL
Courant Institute of Mathematical Sciences, New York University
251 Mercer Street, New York, NY 10012/USA

Professor Dr. ROBERT SADOURNY
Laboratoire de Météorologie Dynamique, Ecole Normale Supérieure
24 rue Lhomond, 75231 Paris Cedex 05/France

Professor Dr. JÜRGEN SÜNDERMANN
Universität Hamburg, Institut für Meereskunde
Heimhuder Straße 71, 2000 Hamburg 13/FRG

ISBN 3-540-12253-2 Springer-Verlag Berlin Heidelberg New York Tokyo
ISBN 0-387-12253-2 Springer-Verlag New York Heidelberg Berlin Tokyo

Library of Congress Cataloging in Publication Data. Main entry under title: Eddies in
marine science. (Topics in atmospheric and oceanographic sciences). Includes index.
1. Oceanic mixing. 2. Eddies. I. Robinson, Allan R. II. Series. GC 299.E 32
1983   551.47   83-4391.

Typesetting and printing: Zechnersche Buchdruckerei, Speyer.
Bookbinding: J. Schäffer OHG, Grünstadt
2132/3130-543210

Foreword

It is now well known that the mid-ocean flow is almost everywhere dominated by so-called synoptic or meso-scale eddies, rotating about nearly vertical axes and extending throughout the water column. A typical mid-ocean horizontal scale is 100 km and a time scale is 100 days: these meso-scale eddies have swirl speeds of order 10 cm s^{-1} which are usually considerably greater than the long-term average flow. Many types of eddies with somewhat different scales and characteristics have been identified.

The existence of such eddies was suspected by navigators more than a century ago and confirmed by the world of C.O'D. Iselin and V.B. Stockman in the 1930's. Measurements from R/V Aries in 1959/60, using the then newly developed neutrally buoyant floats, indicated the main characteristics of the eddies in the deep ocean of the NW Atlantic while a series of Soviet moored current-meter arrays culminated, in POLYGON-1970, in the explicit mapping of an energetic anticyclonic eddy in the tropical NE Atlantic. In 1973 a large collaborative (mainly U.S., U.K.) program, MODE-1, produced synoptic charts for an area of the NW Atlantic and confirmed the existence of an open ocean eddy field and established its characteristics. Meso-scale eddies are now known to be of interest and importance to marine chemists and biologists as well as to physical oceanographers and meteorologists. The geophysical fluid dynamics of meso-scale eddies is difficult, and their small scale and long lifetime combine to make statistically significant observations time-consuming and expensive. Cooperative work is necessary and sience 1975 scientists from the U.S. and U.S.S.R., together with groups sponsored by other nations, have been collaborating in such sprojects as POLYMODE and NEADS to make relevant theoretical studies and field observations in the N Atlantic.

It was with such joint studies in mind that the Scientific Committee on Oceanic Research created a Working Group, with Professor A.R. Robinson as Chairman, to "identify the critical scientific problems of the internal dynamics of the ocean and to suggest the most appropriate ways to study them; and to advise on the design of mid-ocean dynamics experiments."

This book is one result of their work: with the kind help of many experts in particular aspects the group has produced an account of the meso-scale eddies of the ocean, and of their influence in other fields of marine science, which will be of interest and value to a wide range of marine scientists. I welcome this volume, on behalf of SCOR, who are most grateful to its editor, Professor Robinson, and his colleagues for their efforts.

Readers may also like to know that SCOR has invited Professor A.S. Monin to act as Editor of a more specialized account of the theoretical aspects of eddy dynamics, which it is hoped will be published in due course.

E.S.W. SIMPSON
President SCOR

Preface

This book surveys the results of recent research in eddy science and explores its implications for ocean science and technology. It attempts a comprehensive review suitable for a wide audience of marine scientists. Much progress has occurred and the subject is rapidly advancing. However, the eddy-current phenomenon remains unexplored in vast areas of the world ocean and many of the most fundamental dynamical questions of eddy dynamics are unanswered or indeed as yet unasked. The intention of a survey now is to contribute to a global synthesis, to help facilitate researches in new regions and further reasearch into eddy dynamics, and to encourage applications. Eddy currents are energetically dominant and have the ability to transporte and to influence mixing. Thus knowledge of the physical science of the eddies has important implications for biological, chemical and geological oceanography; for modern interdisciplinary ocean science; and for practical activities in the sea including exploitation and management of the marine environment and its resources. Hopefully, this book will contribute to the communication between specialists in eddy dynamics and other marine scientists, engineers and managers, which is necessary for efficient utilization of the new physical knowledge.

The generic term "eddies" is used here without further qualification to encompass a large class of time and space variable ocean currents as discussed in the Introduction. The scales of variability range from tens to hundreds of kilometers and from weeks to months. These currents are usually approximately geostrophic, hence the name "geostrophic eddies" has been invoked. Many phenomena in the ocean are almost geostrophic and it is the nature of the small deviation from the geostrophic momentum balance which in fact subtly governs the dynamics of the flow. The general nature of this deviation for eddy currents is usually consistent with the assumption of the theory of quasigeostrophy. Thus the phenomena is sometimes referred to by the terminology "quasigeostrophic eddies". But eddies are usually, but not always, geostrophic and quasigeostrophic. Other qualifications in common usage refer to scales. Horizontal eddy scales are on the order of the so-called Rossby internal deformation radius (the product of the vertical scale and the ratio of the Brunt-Vaisala frequency to the Coriolis parameter). Motions on the scale of the atmosphere's internal deformation radius are referred to by meteorologists as "synoptic scale" motions, since they include the midlatitude cyclones and anticyclones which are major features of weather maps. By dynamical scale analogy many authors refer to oceanic eddies as "synoptic eddies". Another term, in very common usage since the discovery of evidence for eddies as a major oceanic phenomena and still prevalent, is "mesoscale eddies". This phrase was coined simply to indicate motions of less than oceanic gyre scale. Referring to the overall phenomena inclusively simply as "eddies" reserves the role of qualifiers for the identification of important general subclasses of eddy-currents.

Eddies are a most important physical phenomenon in the ocean and have been the subject of a difficult and vigorous research effort which has required a large scale of international cooperation to be carried out. The foundation for eddy science today was constructed by the dedicated research programs carried out during the 1970's. A major role in the coordination, planning and communication of eddy science was performed by the Working Group on Internal Dynamics of the Sea (WG-34) of SCOR (The Scientific Committee on Oceanic Research of the International Council of Scientific Unions). This multiauthored book was produced unter the auspices of WG-34 and constitutes the final scientific report of the Working Group (see SCOR Proceedings 1982 Vol 18 Annex 5).

The wealth of material presented here results from the scientific efforts and cooperation of numerous scientists, technicians, students and other around the world. Acknowledgements for support and help by the authors and the editor appear collectively in the Acknowledgement Section. In my introductory chapter I draw freely upon the material presented in the book without specific references. The advice, encouragement and efforts of Henry Stommel, Henry Charnock, and Curtis Collins were instrumental in making this book possible.

ALLAN R. ROBINSON

Acknowledgements

1. Overview and Summary of Eddy Science
A.R. ROBINSON

I wish to express my thanks to numerous colleagues for their stimulating scientific interactions during our pursuit of eddy dynamics. I am especially indebted to my graduate students and postdoctoral fellows. Mr. Robert Heinmiller helped in the initial organization of meetings and material and Ms. Marsha Cormier's assistance in the preparation and editing of the final manuscript was invaluable. I am grateful for the support given to this project by grants from the National Science Foundation to Harvard University and the Massachusetts Institute of Technology and by contracts from the Office of Naval Research to Harvard University.

2. Gulf Stream Rings
P.L. RICHARDSON

This is contribution number 4816 from the Woods Hole Oceanographic Institution and from the Mid-Ocean Dynamics Experiment (POLY-MODE). The manuscript was written with funds from the National Science Foundation (Grant OCE78-18017). Don Olson, Glenn Flierl, Tom Spence and Terry McKee made helpful comments on an earlier version of this paper.

3. Western North Atlantic Interior
C. WUNSCH

Supported in part by National Science Foundation Grant OCE78-19833. I am grateful to all those colleagues, technicians, seamen, students, and program directors, who have made possible the acquisition of these complex data sets. This chapter was largely written while I was a visitor at the Department of Applied Mathematics and Theoretical Physics, University of Cambridge, for whose hospitality I am grateful.

4. The Western North Atlantic – A Lagrangian Viewpoint
H.T. ROSSBY, S.C. RISER, and A.J. MARIANO

We are grateful to our numerous Mode and Polymode colleagues for the many contributions that they have made to this research over the past decade. We would especially like to thank Dr. James Price of Woods

Hole Oceanographic Institution for making some of his unpublished
float dispersion results available to us so that this article might be as
complete as possible.

We are thankful for the very generous level of support for this work
that has been provided over the years by a number of grants from the National Science Foundation and the Office of Naval Research. One of us
(S.C.R.) was supported by the Joint Institute for the Study of the Atmosphere and Ocean at the University of Washington during the preparation of this paper.

5. The Local Dynamics of Eddies in the Western North Atlantic

J.C. McWilliams, E.D. Brown, H.L. Bryden, C.C. Ebbesmeyer, B.A. Elliott,
R.H. Heinmiller, B. Lien Hua, K.D. Leaman, E.J. Lindstrom, J.R. Luyten,
S.E. McDowell, W. Breckner Owens, H. Perkins, J.F. Price, L. Regier,
S.C. Riser, H.T. Rossby, T.B. Sanford, C.Y. Shen, B.A. Taft, and J.C. Van Leer

The LDE has required the efforts of many people. We are grateful for the
financial support provided by the National Science Foundation (in particular, we thank the program manager, Dr. Curt Collins) and the Office
of Naval Research. We appreciate the counsel and organizational assistance from our colleagues in the POLYMODE Program (Drs. Allan Robinson and Henry Stommel, Co-Chairmen; Ms. Carol Ramey, Office
Manager). Engineering, technical, and operational contributions were
crucial to our experiment; in particular we thank Mr. Doug Webb and
Dr. Albert Bradley for their work on the SOFAR floats, Mr. Bob Williams who directed the CTD measurements made by the GEOSECS Operations Group, and Mr. Gerry Metcalf for his direction of the POLY-
MODE XBT program. We appreciate the assistance given us by the
masters and crews of the ships involved in the LDE: R.V. Atlantis II,
USNS Bartlett, R.V. Gillis, R.V. Gyre, R.V. Iselin, and R.V. Oceanus. Finally, we thank our scientific assistants, programmers, and secretaries for
their direct contributions to our work.

9. Subpolar Gyres and the Arctic Ocean

G.T. Needler

In many of the northern areas, particularly the Arctic Ocean, information
concerning eddies is very recent and/or unpublished. The author wished
to particularly thank those people who have provided him with material
not readily available.

12. The South Pacific Including the East Australian Current

A.F. Bennett

I am most grateful to the referees of this chapter and also to several of
the authors of the following references, who have assisted me in assem-

bling this material. The help of Dr. George Cresswell is especially to be acknowledged.

13. Eddies in the Southern Indian Ocean and Agulhas Current
M.L. GRÜNDLINGH

The author expresses his appreciation to his colleagues for their suggestions and gratefully acknowledges the help and constructive criticism provided by oceanographers of the Australian CSIRO in Cronulla, Australia.

14. The Southern Ocean
H.L. BRYDEN

This work was supported by the Office for the International Decade of Ocean Exploration of the National Science Foundation under grants OCE 77–22887 and OCE 77–19403. Discussions with Dan Georgi, Terry Joyce and Ray Schmitt enlarged the scope of this review. Comments by Jim Baker, Nan Bray, Adrian Gill, Arnold Gordon, Worth Nowlin, John Toole and Dan Wright helped clarify the text. This is Contribution Number 4773 from Woods Hole Oceanographic Institution.

15. Global Summaries and Intercomparisons: Flow Statistics from Long-Term Current Meter Moorings
R.R. DICKSON

Travel funds for this work were provided by the National Science Foundation through the POLYMODE Office and the author gratefully acknowledges their assistance.

17. Eddy-Resolving Numerical Models of Large-Scale Ocean Circulation
W.R. HOLLAND, D.E. HARRISON, and A.J. SEMTNER Jr.

The National Center for Atmospheric Research is sponsored by the National Science Foundation.

18. Periodic and Regional Models
D.B. HAIDVOGEL

This review was initiated while the author was a Visiting Senior Lecturer in the Department of Mathematics, Monash University. He thanks the members of that department and, in particular, Dr. A.F. Bennett, for their hospitality. This work has been supported by National Science Foundation Grant No. OCE 78–25700 (IDOE), whose support is gratefully acknowledged.

23. Eddies and Acoustics
R.C. SPINDEL and Y.J.-F. DESAUBIES

This work was performed under Office of Naval Research Contract N00014-79-C-0071. This is Woods Hole Oceanographic Institution Contribution Number 4795.

24. Instruments and Methods
R.H. HEINMILLER

The author wishes to acknowledge the help and support of many colleagues over the years in his involvement with oceanographic instrumentation. In the preparation of this chapter, conversations with the following scientists and engineers were of particular value: John Dahlen, Russ Davis, Charles Eriksen, David Halpern, James Hannon, James Luyten, James McCullough, James McWilliams, Peter Niiler, Lloyd Regier, Douglas Webb, and Carl Wunsch.

Contents

List of Contributors

ANGEL, MARTIN V., Institute of Oceanographic Sciences, Brook Road, Wormley, Godalming, Surrey GU8 5UB, Great Britain

BENNETT, ANDREW F., Institute of Ocean Sciences, P.O. Box 6000, Sidney, B.C., Canada V8L 4B2

BERNSTEIN, ROBERT L., University of California, Scripps Institution of Oceanology, La Jolla CA 92093, USA

BRECKNER OWENS, W., Woods Hole Oceanographic Institution, Woods Hole, MA 02543, USA

BROWN, E.D., Woods Hole Oceanographic Institution, Woods Hole, MA 02543, USA

BRYDEN, HARRY L., Department of Physical Oceanography, Woods Hole Oceanographic Institution, Woods Hole, MA 02543, USA

DESAUBIES, Y.J.-F., Woods Hole Ocenaographic Institution, Woods Hole, MA 02543, USA

DICKSON, ROBERT R., Ministry of Agriculture, Fisheries and Food, Directorate of Fisheries Research, Fisheries Laboratory, Lowestoft, Suffolk NR33 OHT, Great Britain

EBBESMEYER, C.C., Evans-Hamilton, Inc. 6306 21st Avenue, NE, Seattle, WA 98115, USA

ELLIOTT, B.A., Ocean Physics Department, Applied Physics Laboratory, University of Washington, Seattle, WA 98105, USA

EMERY, WILLIAM J., Department of Oceanography, University of British Columbia, 6270 University Boulevard, Vancouver, B.C., V6T 1W5, Canada

FASHAM, M.J.R., Institute of Oceanographic Sciences, Brook Road, Wormley, Godalming, Surrey GU8 5UB, Great Britain

GILL, ADRIAN E., Department of Applied Mathematics and Theoretical Physics, University of Cambridge, Silver Street, Cambridge, CB3 9EW, Great Britain

GOULD, W. JOHN, Institute of Oceanographic Sciences, Wormley, Godalming, Surrey GU8 5UB, Great Britain

GRÜNDLINGH, MARTEN L., NRIO, P.O. Box 17001, Congella 4013, South Africa

HAIDVOGEL, DALE B., Woods Hole Oceanographic Institution, Woods Hole, MA 02543, USA

HARRISON, D.E., Department of Meteorology and Physical Oceanography, Massachusetts Institute of Technology, Cambridge, MA 02139, USA

HEINMILLER, ROBERT H., OMET, 70 Tonawanda Street, Boston, MA 02124, USA

HOLLAND, WILLIAMS R., National Center for Atmospheric Research, P.O. Box 3000, Boulder, CO 80307, USA

LEAMAN, K.D., RSMAS, University of Miami, 4600 Rickenbacker Causeway, Miami, FL 33149, USA

VAN LEER, J.C., RSMAS, University of Miami, 4600 Rickenbacker Causeway, Miami, FL 33149, USA

LIEN HUA, B., Woods Hole Oceanographic Institution, Woods Hole, MA 02543, USA

LINDSTROM, E.J., Department of Oceanography, University of Washington, Seattle, WA 98195, USA

LUYTEN, J.R., Woods Hole Oceanographic Institution, Woods Hole, MA 02543, USA

MARIANO, ARTHUR J., Graduate School of Oceanography, University of Rhode Island, Narragansett, RI 02881, USA

McDOWELL, S.E., EG & G Environmental Consultants, 300 Bear Hill Road, Waltham, MA 02254, USA

McWILLIAMS, JAMES C., National Center for Atmospheric Research, P.O. Box 3000, Boulder, CO 80307, USA

NEEDLER, GEROLD T., Atlantic Oceanographic Laboratory, Bedford Institute of Oceanography, P.O. Box 1006, Dartmouth, Nova Scotia B2Y 4A2, Canada

PERKINS, H., NORDA, NSTL Station, MS 39529, USA

PRICE, J.F., Woods Hole Oceanographic Institution, Woods Hole, MA 02543, USA

REGIER, L., Scripps Institution of Oceanography, La Jolla, CA 92093, USA

RICHARDSON, PHILIP P., Woods Hole Oceanographic Institution, Woods Hole, MA 02543, USA

RISER, S.C., Department of Oceanography, University of Washington, Seattle, WA 98195, USA

ROBINSON, ALLAN R., Pierce Hall, 100-C, Harvard University, 29 Oxford Street, Cambridge, MA 02138, USA

ROOTH, CLAES G.H., RSMAS University of Miami, 4600 Rickenbacker Causeway, Miami, FL 33149, USA

ROSSBY, H. THOMAS, Graduate School of Oceanography, University of Rhode Island, Kingston, RI 02881, USA

SANFORD, T.B., Ocean Physics Department, Applied Physics Lab, University of Washington, Seattle, WA 98105, USA

SEMTNER JR., ALBERT J., National Center for Atmospheric Research, P.O. Box 3000, Boulder, CO 80307, USA

SHEN, C.Y., Department of Oceanography, University of Washington, Seattle, WA 98195, USA

SIEDLER, GEROLD, Institut für Meereskunde an der Universität Kiel, Abteilung für Meeresphysik, Düsternbrücker Weg 20, D-2300 Kiel 1, Fed. Rep. of Germany

SMITH, PETER C., Atlantic Oceanographic Laboratory, Bedford Institute of Oceanography, P.O. Box 1006, Dartmouth, Nova Scotia, Canada B2Y 4A2

SPINDEL, ROBERT C., Woods Hole Oceanographic Institution, Woods Hole, MA 02543, USA

SWALLOW, JOHN C., Institute of Oceanographic Sciences, Brook Road, Wormley, Godalming, Surrey GU8 5UB, Great Britain

TAFT, B.A., PMEL/NOAA, Seattle, WA 98195, USA

WATTS, D. RANDOLPH, Graduate School of Oceanography, Narragansett Bay Campus, University of Rhode Island, Kingston, RI 02881, USA

WUNSCH, CARL, Department of Earth and Planetary Sciences, M.I.T., 54-913, Cambridge, MA 02139, USA

Introduction

1. Overview and Summary of Eddy Science

A.R. Robinson

1.1 Eddy Currents in the Ocean

Ocean currents and their associated fields of pressure, temperature, and density vary energetically in both time and space throughout the ocean. Such variability in fact contains more energy than any other form of motion in the sea. Partly organized, yet highly irregular, these motions have dominant spatial scales in the range of tens to hundreds of kilometers and dominant temporal scales in the range of weeks to months. The variability is distributed unevenly with energy levels and dominant scales changing substantially from place to place in the ocean. These turbulent motions are the internal weather of the deep sea, and many types of synoptic events and peculiar marine internal storm systems are now known to occur. Types of variability which have been identified and studied include the meandering and filamenting of intense current systems, semi-attached and cast-off ring currents, advective vortices extending throughout the entire water column, lens vortices, planetary waves, topographic waves and wakes, etc. All of these types of variable flow are commonly referred to by physical oceanographers by the generic term "eddies".

Although energetically dominant and pervasive in the ocean, eddies have been definitively described and intensively studied only recently. The traditional descriptive picture of the ocean emerged essentially from a geographically sparse data set compositing years of discrete measurements widely spaced in the horizontal and the vertical. Thus the eddy signals were aliased in space and time and the conceptualization of currents and circulation was based upon smeared fields derived from incomplete and inadequate data sets. Except for the meandering of the strongest currents and a few ring observations, prior to the 1970's only suggestive evidence for the existence of eddies was available. Since that time, however, vigorous and dedicated investigations have been carried out with sensitive, continuously recording instruments from platforms and systems capable of providing adequate horizontal, vertical, and temporal resolutions. Variability types have been identified, kinematics described, and statistical characteristics quantified. Dynamical and energetic studies have been carried out, novel theories constructed and powerful new computer models which resolve eddies developed. However, vast areas of the oceans remain unexplored, most fundamental questions are partially or totally unanswered, important processes require accurate quantification and models need to be made more realistic. Experience indicates that as the resolution of measurements continues to increase, as observational records lengthen and as new geographical regions are opened up, totally new phenomena will be discovered. But the

Eddies in Marine Science
(ed. by A.R. Robinson)
© Springer-Verlag Berlin Heidelberg 1983

threshold has been crossed, the conceptualization of ocean currents and circulation is now more realistic, and scientific results of permanent value are being established.

1.1.1 Observing the Eddies

Ideally, the description of the eddy field would consist of an adequately resolved continuous synoptic time series of the three-dimensional physical (and chemical) fields throughout the global ocean. After a few accurate synoptic realizations local, regional, and global processes could be delineated. After a sufficient number of independent synoptic realizations ensemble statistics could be calculated. Then specific time series could be continued where and when necessary for special scientific or practical purposes or for studies of very long time scale phenomena such as climatic change. Meteorologists, whose synoptic phenomena are larger and faster, whose gaseous medium is much more accessible for observations and measurements, and who have been supported by society's interest in forecasts, have come a long way toward achieving such a description for the atmosphere with their daily weather maps. Nonetheless practicing forecasters and research atmospheric scientists alike feel hampered by data-sparse regions, e.g., over oceans. In 1978 and 1979, with an enormous international effort, meterologists carried out the First GARP Global Experiment. This was an attempt to observe the entire atmosphere in detail for the first time in order to understand better the motions, to develop forecast and general circulation models, and to assess the limits of predictability (GARP/WMO-ICSU 1973).

Research oceanographers have of course had to take a much more limited and focused approach to the description and investigation of their eddies, designing exploratory observations and physical experiments under the constraints of oceanic time and space scales and available resources. With great effort they have achieved a few three-dimensional synoptic realizations over one or several eddies in specially selected or interesting regions. Such data sets, although limited to regions of a few tens to a few hundreds of kilometers on a side and to durations of days to several months are invaluable. A small subset of such experiments is suitable for the detailed quantitative study of local dynamical processes, e.g., by mapping detailed budgets of energy, heat, and vorticity fluxes and balances. Some large-scale synoptic maps of surface indication of eddies and some large-scale maps of the geographical distribution of eddy statistics are now available from remotely sensed data. There is great future promise from satellite studies of eddies. The scales of eddy variability are particularly accessible to remote sensing techniques and the development of signal to noise separation schemes and interpretative models necessary to exploit fully dedicated new ocean satellite systems is progressing rapidly. Sections or swaths from ships of opportunity provide partial realizations of substantial value because of their extensiveness or repetitiveness. Time series from moored instruments provide kinematical and statistical information of the geographical distribution of eddy characteristics. A few records are long enough to provide the set of statistical moments necessary to describe mean fields and eddy transports, and moorings have been arrayed close enough together to provide the field gradients necessary for time series of local dynamical balances at the central location. Free floating devices set to drift with the water at various depths yield the Lagrangian dynamical and statistical description of the region they move through and their dispersion has provided important direct new estimates of eddy diffusivity in a few places. Accurate hydrographic and chemical sampling within eddy scales in now allowing water mass analyses to address questions of origin, transport, and mixing.

1.1.2 Modeling the Eddies

These measurements are complemented at present by two general types of eddy-resolving dynamical computer models. Indeed the development of these dynamical models has been guided by the field measurements, which in turn have been guided by model experimental results. Additionally a new type of statistical modeling, including objective analysis and optimal interpolation schemes similar to meteorologists', has been brought to bear on the design of field experiments and the analysis and interpretation of results. The first type of dynamical model deals with a block of ocean arbitrarily selected for study out of its larger-scale context. Such regional dynamical models relate directly to intensive data sets, address questions of local dynamics of eddies, are capable of providing adequate resolution both vertically and horizontally, and are suitable both for investigating turbulent aspects of the eddy field and for the direct assimilation of data. Important results have been abtained, but the physics of such models must be generalized, the consequences of assumptions and parameter choices evaluated, and many more particular and typical regions investigated.

The second type of dynamical model is the eddy-resolving general circulation model (EGCM). To date model studies have been carried out predominantly in large-scale but limited geometries idealizing more or less the mid-latitude North Atlantic with adequate horizontal resolution and coarse vertical resolution. Driven by smooth and steady mean surface wind fields, EGCM's can be run to statistical equilibrium after the spontaneous generation of eddies by internal processes arising from the conversion of mean kinetic and potential energies. EGCM experiments require large amounts of time on the fastest available computers and relate most naturally to the statistics of the field observations and their geographical distributions. Often necessary extensions of EGCM research include studies of generalized physics with active thermodynamics, of the effects of more realistic geometries and forcing functions, and the modeling of many other regions of the world's ocean. It is clear that a symbiosis between regional and general circulation model studies is possible and desirable, e.g., with an EGCM providing a large-scale environment for a regional study and regional process studies contributing to the interpretation of EGCM results. Moreover, both types of numerical dynamical model studies are of course naturally rooted in, interactive with, and influential upon analytic theories which form the fundamental abstract basis for the understanding of eddy-dynamics as a problem in geophysical fluid dynamics.

1.2 Status of Eddy Science

1.2.1 Distribution and Generation

From the synthesis of observations, experiments, models, and theories a general picture is emerging and some general results have been established. Eddies exist almost everywhere they have been looked for, but the distribution of synoptic types and eddy activity is spatially heterogeneous, intermittent, and eventful. Energy levels vary by orders of magnitude regionally and remarkable evidence exists which indicates that mean gradients of eddy statistics can occur on scales less than thought to be characteristic of the gyre-scale circulation. The most vigorous variability is associated with intense flow regions such as the Gulf Stream current, its extension and recirculation. There, eddies are produced by eddy-mean field interactions which are large amplitude processes related conceptually to the simple baroclinic instability of a strongly horizontally sheared current or a combination of the two processes. The baroclinic process is said to "convert" energy, which would otherwise be stored in the mean flow (available) potential energy field, to eddy energy. The barotropic process converts mean flow kinetic energy to eddy energy. Much of the eddy energy produced in intense current regions is exported by transport and radiative mechanisms to populate large regions of the open ocean with eddies. In locations far from major currents, such as portions of the Northwestern Pacific which are relatively quiescent, the variability can be wave-like, and could in some instances be wind-driven. Topographic effects certainly cause eddies. Frontal regions are generally unstable and water-mass formation processes may generate variability. But the major production mechanism identified to date is eddy-mean field interaction. Dedicated research in this area is of the utmost importance now. The processes are believed to be complicated and variable with energy conversions occurring also back from the eddies to the mean flow. The actual physics and location of interaction processes in the real ocean are essentially unknown. However, such processes are expected to occur in all western boundary current systems, as well as being suspected of occurring in several other strong current environments including eastern boundary, equatorial, and polar regions. The Antarctic Circumpolar Current is unstable to the baroclinic process in analogy to the midlatitude atmosphere. The dissipation mechanisms for eddies are not known; interactions with the bottom and with smaller scales, including internal waves, have been suggested.

1.2.2 Physics

Eddy currents are almost horizontal and almost geostrophic in their momentum balance, but to understand their evolution and local dynamics the small vertical velocity component and the small deviation from geostrophic balance must be taken into account. Both of these effects are typically about 1%, and in many instances may be accounted for in terms of a general theoretical model called quasigeostrophy. Quasigeostrophic motions are approximately hydrostatic and occur in a rotating stratified fluid when the horizontal scale of the motion is on the order of the Rossby internal deformation radius, (i. e., the vertical scale multiplied by the ratio of the Brunt-Vaisala buoyancy frequency to the Coriolis parameter). This is the case for ocean eddies as it is for the atmospheric weather systems for which the theoretical approximations were first developed. Quasigeostrophic motions are governed by a general equation (the so-called quasigeostrophic potential vorticity conservation) which represents a wealth of phenomena ranging from simple planetary waves to the complexities of geostrophic turbulence. Aspects of the theory are reasonably well advanced, but the surface has only been scratched and nothing like the general nonlinear solution is available or even conceivable.

Nonlinear interactions transfer energy and vorticity among quasigeotrophic scales of motion. Such transfers, called cascades, depend upon the strength of the flow, the initial scales in which the energy resides, the density stratification, the underlying topography, and the latitudinal variation of the Coriolis parameter (β-effect). The predominately horizontal geostrophic motions tend to transfer energy to larger scales while transferring vorticity (enstrophy) to smaller scales. This requires a straining and filamenting motion characteristic of two-dimensional turbulence. This process is dominant when the currents are depth-independent or when the motions at different depths are essentially uncoupled, which is the case for scales less than the internal deformation radius. The baroclinic instability mechanism can transfer energy from thermocline intensified currents larger in scale than the internal radius to shorter scales. When energy occurs in scales on the order of the internal deformating radius itself, the nonlinear interactions very efficiently transfer it vertically and tend to produce depth-independent currents (barotropification). The flow will remain vertically sheared only if it is weak enough not to be dominated by nonlinearity, or if compensating effects due to bottom topographic slopes are strong enough. The variable Coriolis parameter of course gives rise to the familiar planetary waves which are solutions of the linearized quasigeostrophic equation and which propagate phase westward. Empirically westward propagation is known to continue to characterize turbulent-like eddies into the nonlinear regime. The β-effect can inhibit the horizontal cascade of energy to larger scales. Planetary wave processes become effective when the energy reaches a scale at which the eddy turn-around time (advective time scale) is greater than the planetary wave period. (This scale is measured by the square root of the flow speed divided by the latitudinal gradient of the Coriolis parameter.) Then energy can become wave-like and dispersed, and zonally dominant anisotropic flows are favored.

Experimental determination of time and space scales of the variability and even of local dynamical balances in the ocean provides important evidence that much of the eddy field is quasigeostrophic. Away from production regions the nonlinear interactions govern the local statistical dynamical equilibrium for strong enough flows if the cascade rates are faster than the regional transit rate of the energy. Aspects of measured statistics rationalizable by the transfer processes and vorticity balances characteristic both of planetary waves and geostrophic turbulence have been measured, the latter taking the form of nonlinear advection of relative vorticity filaments by the larger energy-containing eddy scales.

Important circumstances are known, however, when the eddy currents are not quasigeostrophic. The contorted meandering and eddying of intense currents, rings, and the smaller long-lived isolated lenses often have strong centripetal acceleration (cyclostrophic effects) and thus are not geostrophic. Both warm and cold core Gulf Stream rings are now known to interact frequently with the current and with each other. Ring eddies, readsorption, multiple ring and ring plus current interactions are identifiable processes requiring a concerted research effort. Progress in understanding the movement of isolated rings has been achieved. The rings move westward due to the β-effect with meridional flow caused by intrusions and radial assymetry of the swirl motion. Rings tend to maintain an inner core of trapped water, exchange water around their edges and ultimately retain their identity only in the thermocline since planetary wave radiation can effectively disperse the energy of the weaker deeper flow. Some insight into the dynamics of rings and lenses is being sought in the framework of the general theory of coherent nonlinear structures (solitons). For such isolated stable features nonlinearities play a role essentially opposite to their role in geostrophic turbulence, i.e., by acting to overcome dispersive tendencies and to maintain the form of the object and therefore the energy in its initial scale range. Ageostrophic effects definitely occur over steep topography, near seamounts, islands, and continental boundaries. Such effects are usually studied in terms of a so-called primitive equation dynamical model. The physics is quite general except that the almost hydrostatic assumption is made for the vertical momentum balance. Studies with the eddy-resolving regional and general circulation computer dynamical models to date have mostly employed quasigeostrophic dynamics, but some primitive equation studies exist.

1.2.3 Role in the General Circulation

The existence of eddies has important dynamical consequences on physical phenomena on longer and shorter scales. Indeed much of the motivation for intensive research effort of the past several years has been to determine the role of the eddies in the general ocean circulation. Some progress has been made, but this question requires vigorous further pursuit and should provide a focus for research now and in the immediate future. Eddies influence the general circulation directly through eddy-mean field interaction and eddy – eddy interaction in the mean (general types of Reynolds stresses). As noted above, such interactions occur most strongly in intense current systems and the role of eddy momentum flux divergence as a driving mechanism for the Gulf Stream recirculation is noteworthy. Eddy heat transport by Southern Ocean eddies is important to the global heat budget and eddies transport heat significantly in the vicinity of the Gulf Stream. The effects of eddies can, however, be more subtle. Models exhibit deep permanent currents caused by eddies but not directly driven by local eddy flux divergence, and such effects must be expected to be occurring in the real ocean.

The parameterization of eddy-effects on the larger, slower scales of motion is an essentially unsolved problem of great interest and consequence. What is required is a simplified but accurate representation of the statistical effect of eddies so that they need not be dealt with explicitly in the consideration of longer time scale problems. Such a successful parameterization will probably not be entirely in terms of local mean field quantities and will certainly be spatially heterogeneous and generally nonisotropic. In certain circumstances the use of the traditional constant horizontal eddy diffusivities may be approximately correct but care must be exercised. This approximation could be appropriate in places for turbulent diffusion of one quantity of interest, e.g., heat, but not for another, e.g., momentum. Moreover, it is well-established that the turbulent fluxes can occur both down and up the gradients of mean field quantities. Direct measurements of the diffusivity by dispersion of floats in the Northwestern Atlantic yields values from less than 10^7 to greater than 10^8 cm^2 s^{-1} with an unexpected but important indication that the diffusivity is directly proportional to the eddy kinetic energy level. Theoretical studies and the analysis of some EGCM numerical experiments have identified the turbulent eddy flux of (potential) vorticity as a key to successful parameterization hypotheses and suggested the existence of interesting regions where eddy mixing has homogenized this quantity. Much more research with resolved eddy scales is required before the fundamental theoretical issues are settled and reliable engineering type formulas are established.

1.3 Influences of Eddies

1.3.1 Scientific Processes and Practical Consequences

It is apparent not only that eddies themselves are a dominant physical phenomenon requiring a major continuing research effort but that eddy-related effects profoundly influence physical oceanographic research on a very broad base. But the existence of the eddies as the dominant flow over much of the ocean has implications for ocean scientific research much more generally. Eddies transport, entrap, and disperse chemicals, dissolved substances, particulate matter, nutrients, small organisms, heat, etc. Thus, many fundamental questions in geochemistry, chemical and biological oceanography, and marine biology are influenced significantly. Aspects of the concentration distribution of chemical and biological variables will mirror the physical variability. This will be true both for snapshots, and for the geographical distribution of statistics. Factors of biological consequence that are already known to exhibit variability on eddy length scales include: mixed layer properties: productivity; the characteristics, concentrations, and structure of populations; the daily vertical migration within a species; the partitioning of the standing crop between species. Sampling strategies and experimental designs are obviously affected. Moreover, hypotheses and models for biological and chemical phenomena which are crucially dependent upon physical transport processes must be constructed or modified to take them into account in a contemporarily realistic manner. Analogously to recent advances in physical oceanography some basic ideas in biological and chemical oceanography are changing and new theories need to be constructed. Multidisciplinary ocean scientific research is essential now. Moreover, the eddies themselves provide a natural and compelling focus for the development of modern interdisciplinary ocean science, which could be one of the most profound impacts of the phenomenon on marine science.

The existence of eddy currents throughout the deep ocean regions with structures that reach out to various boundaries has important effects on fluxes and exchanges across those boundaries. Thus the variable currents have implications for large-scale air – sea exchanges influence atmospheric science, climate research, and global geochemistry. Climate changes from year to year and on longer time scales are thought to be governed substantially by oceanic processes. The role of eddy processes via both direct transports and indirect effects on the general ocean circulation must be understood. Eddy transport processes of geological scientific significance occur near the seabed and continental rises and slopes. Near coastal boundaries eddies are dissipated by the generation of topographic wave energy which is radiated to the rise, slope, and shelf region. Eddies can induce upwelling and eddy momentum flux can contribute to the maintenance of long-shore pressure gradients. Onshore-offshore eddy exchanges of heat, salt, and nutrients can contribute significantly to shelf-wide balances. Although these and possibly other eddy processes undoubtedly are crucial for important coastal ocean science research questions, little is known definitely as yet in this area.

Energetic current and temperature fluctuations, eddy transport mechanisms and the influence of eddies on interface fluxes and internal concentration distribution of dissolved and particulate matter have, of course, also a very significant impact on research in applied marine science and on activities relating to the management of the sea and the exploitation of its resources. Waste chemicals are released at the sea surface; containers filled with low-level radioactive wastes are dumped onto the sea bottom; industrially generated carbon-dioxide is continually accumulating through sea surface exchange with the atmosphere; the chemical residue from broken refrigerators, aerosol cans, and atmospheric nuclear bomb tests etc. is gradually moving down and throughout the world's ocean. Several nations are seriously considering seabed and subseabed disposal of high-level nuclear wastes. What is the ability of the ocean to disperse such pollutants into harmless dilutions, or to isolate toxic substances from humans and the components of the main marine ecosystems until they are no longer dangerous? What is the risk to health, what are the long-term damage to life-sustaining planetary resources and the esthetic consequences of introducing pollutants into ocean compared with alternative disposals? What can

ocean scientists learn of fundamental interest from such tracers already introduced in measurable quantities into the sea?

1.3.2 Dispersion and Mixing

Ring currents and strong thermocline eddies tend to maintain their internal water mass although some mixing and entrainment takes place. These kinds of advective eddies are effective transporters and provide distinct boundaries for biological variables and an internal environment which can support a locally anomalous ecosystem. The smaller-scale lenses often have very distinct water mass
characteristics indicative of distant origin, long lifetime, and negligible or limited mixing. These lenses obviously transport but their overall importance cannot yet be assessed because statistical estimates are not available. The dispersive mixing and transport properties of mid-ocean eddies are not so easily summarized. These effects and the underlying processes involved undoubtedly depend upon local environmental, kinematical, and dynamical factors. Direct
eddy effects plausibly come into play for blobs of tracer material ("dye")
which are large enough to be sheared by the gradients of the eddy currents, i. e.,
10 km or larger in extent. The horizontal gradients, or more accurately the gradients along constant density surfaces (isopycnals), then begin to shear out the
dye into filaments or streaks. This process is believed to occur more rapidly
and efficiently than smaller-scale mixing. Thus the blob is stirred by eddies
into elongated streaks. Essentially undiluted ribbons of dye first reach the eddy
scale and then begin to overlap and become intertwined. At this stage, volume
average concentrations can be quite dependent upon the volume chosen to average over. Even the choice of the most appropriate statistical descriptors for
this type of dispersion is a current research question. Later, when the dyed water has become larger in extent than a single eddy and some mixing has had
time to occur, a traditional eddy diffusivity becomes plausibly meaningful. The
smaller-scale mixing processes along densitiy surfaces (isopycnal mixing), the
"vertical" or across-density mixing processes (diapycnal mixing), the interaction of eddies with smaller scales and the eddy dissipation mechanisms are
linked research problems now that require merit investigation. Candidate phenomenon for the mixing processes include internal wave breaking, double diffusion, interleaving events and frontal instabilities on a hierarchy of scales all
in the presence of the vertical and horizontal shears of the horizontal eddy currents. Definite knowledge of the eddy and smaller-scale stirring and mixing
processes will have a substantial impact upon both fundamental scientific progress and technical capabilities for ocean management and marine resource exploitation.

1.3.3 Description and Prediction

A knowledge of the actual distribution of ocean currents and their evolution in
time on eddy scales is of considerable interest for a number of practical pur-

poses. This involves the problem of describing and predicting the synoptic realizations, i. e., of providing an oceanic version of the meterologists' weather
forecasting service. Ocean forecasting is useful for field operations related to
commercial fisheries, mineral and energy exploration and extraction and scientific research; for the transportation industry; and for ocean dumping management. Moreover, the temperature fluctuations caused by eddies significantly alter the sound speed in the ocean, producing a transient range-dependent
acoustic environment; the direct effect of the Doppler shifting by currents
could also be of some relevance. Turning the problem around, the alteration of
sound propagation by eddy currents can be used to infer the distribution of the
currents themselves if travel times and something about the structure of the
temperature field are known. This is, of course, the basis of the new acoustic
tomographic techniques.

It is generally apparent now that the practical ocean forecasting problem is
in many aspects closely tied to basic research problems in eddy kinematics and
dynamics. Global synoptic forecasts are not yet practical, but accurate regional
forecasts for (arbitrary) blocks of the ocean are feasible. Thus local dynamical
studies and regional models with open boundary conditions are implicated.
Such forecasts with reasonable accuracy requirements could be made efficiently if modern optimal estimation techniques which pool all available information from different sources are developed and utilized. Within such schemes, contemporary observations from different kinds of instruments are com-
posited, interpolations are carried out based on the past statistics of the regional flows (objective analysis), and the analyzed fields are melded with the
predicted fields from a dynamical computation, which is also continually pro-
jected forward in time. This approach draws upon the powerful techniques of
the dynamic-stochastic methods of modern engineering and four-dimensional
data assimilation methods of contemporary meteorology. The contribution and
relevance of remotely sensed data to such forecast methods and their related
fundamental dynamical studies cannot be overemphasized. Relevant data includes sea surface height, temperature, winds and the interrogation of drifting
instruments. Satellites provide a unique space-time coverage and the descrip-
tive-predictive system provides the requisite interpretative and assimilative
scheme.

1.4 Evolution and Outlook

1.4.1 A Time of Transition

From our assessment of the status of knowledge of eddy science and its implications, it is apparent that enough has been eccomplished to make the present
time ripe for scientific progress and for technical innovation. A wealth of information and ideas now exists, mostly from data gathered and models constructed recently. A partial synthesis has been achieved, some general physical
insights exist, the class of fundamental marine scientific issues involved is be-

ing clarified. Some important applications have been initiated and others are being identified. On the other hand, over much of the world's ocean the kinematics and dynamics of the eddy field are essentially unknown, understanding of the origin and effects of eddies is sketchy and their ultimate fate is generally obscure. The introduction of variability concepts in biological, chemical, and geological oceanography is embryonic, as is the initiation of truly interdisciplinary studies on eddy scales. Many practical activities in the marine environment of major societal concern where eddy effects matter do not yet take them adequately into account. The potential applicability of the new and evolving physical and dynamical knowledge is considerable and greater than presently being realized. Challenge and opportunity exist in this time of transition for eddy science. Previously eddies were unknown, ignored or singled out for special study. Now they are known to be commonplace and dominant currents which must be routinely taken into account in the planning of scientific investigations with broadly based scientific objectives. Central to this transition has been a recent series of dedicated research programs involving a very large commitment of resources and substantial international scientific cooperation among physical oceanographers. These dedicated programs were carried out during the 1970's when they were possible both to conceive and to execute because of the accumulation of prior suggestive evidence and the availability of adequate instrumentation and technology.

1.4.2 A Decade of Eddy Research

Historical remarks attributable to eddy-current effects can be found in scientific writings certainly over the past century and a half. Two interesting discussions of the historical evidence and the evolution of physical oceanography to the point of readiness for dedicated scientic eddy investigation are given by Brekhovskikh, Fedorov, Fomin and Yampolsky (1971) and Swallow (1976). Interesting events during the 1930's include Iselin's identification of a Gulf Stream eddy from hydrographic data and Stockman's initiation of direct current measurements via a 3-week anchor station time series in the Caspian. During the 1950's Gulf Stream meanders were delineated, the existence of rings was established and pioneering direct current observations were obtained from arrays of moored current meters and from deep acoustically tracked floats. The decade of the 1960's was characterized by relevant and essential instrumental and technical developments, the detection of some eddy signals in data and a growing awareness of the necessity of scientific pursuit of the variability phenomena (Stommel 1970). The setting of the stage for the vigorous programmatic research of the 1970's is presented by Stockman, Koshlyakov, Ozmidov, Fomin and Yampolsky (1969) and by Deacon (1971). The decade was opened by POLYGON-70 (Brekhovskikh, Fedorov, Fomin, Koshlyakov and Yampolsky 1971) which measured the eddy currents for several months in the North Atlantic Equatorial Current from moored current meters and by hydrography. The focus of the Mid-Ocean Dynamics Experiment (MODE) was an accurate four-dimensional mapping of a midocean eddy during the Spring of

1973. This program, centered southwest of Bermuda, intercompared a number of instruments and systems initiated by a series of long-term Eulerian and Lagrangian statistical measurements (The MODE Group 1978). Robinson (1975), Koshlyakov and Monin (1978) and McWilliams (1979) are a series of general reviews which survey progress in eddy science during the decade.

The largest, most intensive and extensive undertaking was POLYMODE (Robinson 1982), a joint U.S.-U.S.S.R. program which included: a synoptic dynamical experiment (SDE) (Academy of Sciences of the U.S.S.R. 1979, 1980; Nelepo, Bulgakov, Timtchenko et al. 1980), a Local Dynamics Experiment (LDE), and a statistical geographical experiment. The field phase of POLYMODE ended in 1979; much of the information in the very large data set acquired is still being analyzed. The center of the POLYMODE SDE which lasted longer than a year coincided with the MODE center making this few hundred kilometer block of ocean the most thoroughly mapped in the world. Typically two or three eddies were in the region and about four or five independent realizations occurred. The center of the overlapping LDE region was northward in the Gulf Stream recirculation and accurate enough measurements were obtained to directly infer balances. A dedicated intensive study of Northeastern Atlantic eddy currents similar to the SDE and LDE was recently completed by French und U.K. scientists (Groupe Tourbillon 1983). The POLYMODE statistical geographical experiment stretched out through the northwestern Atlantic to measure the distribution of eddy amplitudes and scales from moored and other instruments. A complementary and noncoherent array of moorings was maintained in the northeastern basin by the NEADS (Northeastern Atlantic Dynamics Study) program scientists (F.R.G., France, U.K.) for about 2 years to determine eddy energy levels. The distribution, movement structure, and dynamics of cold core rings was investigated cooperatively with field studies in 1976 and 1977 by the Ring Group (1981). The program contained an important biological and chemical scientific component. An intensive interdisciplinary cooperative study of warm core Gulf Stream rings (Warm Core Ring Executive Committee 1982) is currently in progress. The results of the dedicated programs of 1982 and the 1970's were largely responsible for the development of eddy science. In combination with the increasing amount of information on eddies elsewhere in the world's ocean and the results of a number of independent researchers, these results have brought about the present transitional and opportunistic scientific situation.

1.4.3 Eddies in Marine Science

This book is meant to serve as a timely and comprehensive review and summary of what is known about eddies in the world's ocean and to indicate the implications and applications of that knowledge to marine science broadly. A general survey at this time can expose similarities and differences in phenomena, help to identify important gaps in our knowledge and hopefully provide the basis for some further generalizations. Early overviews of application areas could help to construct the framework for further studies and to foster progress. The methodological basis of recent research is presented and unanswered questions posed as a possible guide to new undertakings. Chapters 2 through 14 present what is known about eddies on a regional basis throughout the world ocean. There is inevitably some unevenness of presentation due in a large part to the great difference in our knowledge from place to place as well as to differences in regional phenomena and author's emphasis. The various data-gathering techniques behind the results presented in these chapters are summarized in Chapter 24. The regional chapters contain kinematical and dynamical discussions. Authors have attempted to indicate important interrelationships among all the chapters but readers with special interests will certainly find more. Chapters 15 and 16 are attempts to overview results globally and include a substantial survey of current meter measurements. The two types of numerical dynamical models are the foci of the dynamical discussions of Chapters 17 and 18. The models are emphasized in this way because their importance to the design and analysis of experiments and because they serve as the natural vehicle for the application of our knowledge of the eddy currents to scientific and practical problems. The remaining Chapters, 19 through 23, deal directly with the implications and the applications of eddy science and treat several problems from marine science including climate, coastal effects, dispersion, biological processes, and acoustics. The rapid evolution of eddy science has created the possibility of new applications which at this time can be most effectively carried out by specialists in eddy dynamics working cooperatively with other marine scientists, engineers and managers. Hopefully this book will contribute to the communication required to facilitate such cooperation, as well as to the advancement of our knowledge of eddy dynamics itself.

Regional Kinematics, Dynamics, and Statistics

2. Gulf Stream Rings

P.L. Richardson

2.1 Introduction

Gulf Stream rings are a special type of eddy whose origin has been well documented; they form from cut-off Gulf Stream meanders (Fuglister 1972). Rings are the most energetic eddies in the ocean and their thermocline displacements, swirl speeds and volume transports are nearly equivalent to those of the Gulf Stream.

Recently, field studies of rings have concentrated on obtaining a description of their distribution, structure, biology, movement and life histories. Rings constantly and sometimes rapidly change their size, shape and position. They interact with the Gulf Stream and with other rings. Because of this complexity, accurate measurements of ring dynamics are difficult to obtain.

The descriptive data that we have accumulated are being used to develop and refine models of ocean eddies (see Chap. 18). Recent eddy-resolving general circulation models contain eddies that look and act very much like rings. Although these models are still very idealized, they have been used to identify important mechanisms that contribute to the dynamics of the Gulf Stream system. Model eddies (and by implication, rings) are key sites in which kinetic and potential energy is transformed and transmitted within the system. Eddies generate mean flow; they help drive the deep Gulf Stream, and they enhance the Stream's transport significantly. Eddies are vital in transporting water and its physical, chemical and biological components across the Stream.

The real role of rings in the dynamics of the Gulf Stream is still being assessed with field programs and model studies. The large number of highly energetic rings that coexist near the Gulf Stream, coupled with results of model studies, suggest that rings are a vital component of the Gulf Stream system. In order to understand the Gulf Stream's role in weather and climate we must understand the part played by rings.

2.2 History

Jonathan Williams, grandnephew of Benjamin Franklin, was the first to mention a warm-core ring (Williams 1793). In 1790 he measured surface temperatures, plus some surface velocities by lowering a cooking pot on the end of a long sounding line and concluded:

Eddies in Marine Science
(ed. by A.R. Robinson)
© Springer-Verlag Berlin Heidelberg 1983

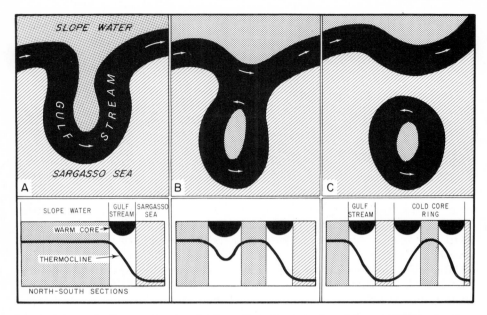

Fig. 1. Schematic diagram showing the formation of a cyclonic, cold-core, Gulf Stream ring. A cold-core ring consists of a closed segment *(ring)* of the Stream circulating around a mass of cold water detached from its former position in the Slope water region north of the Gulf Stream. (Adapted from Fuglister 1972)

"The evidence of this various current in so short a space, the heat of the water not being raised to the heat of the stream, and our situation to the Northward made me conclude that this to be the whirlpools of the eddy of the gulf stream just on the northern edge of it."

The "eddy" of the Gulf Stream refers to the westward flowing countercurrent and thus "whirlpools" is the reference to rings.

More evidence for rings was accumulated in the early 1930's. In the period 1929–1931, Church (1932, 1937) examined ship thermograph records and found warm eddies debouching from the Gulf Stream into the Slope water region. In 1932 on the ATLANTIS, Iselin (1936) began a series of deep temperature-salinity sections across the Gulf Stream which clearly show what today we would interpret as rings. Not knowing their cause, Iselin suggested that a warm-core ring was a "permanent eddy" and a cold-core ring could have been a large amplitude internal wave generated by a severe northeast storm. Iselin's (1940) later sections in 1937 and 1938 contained several additional ring observations, indicating that rings were relatively numerous.

Synoptic oceanography of the Gulf Stream began in 1946–47. LORAN and BT's were first used to circumnavigate a cold-core ring (Fuglister and Worthington 1947, Iselin and Fuglister 1948). These studies led to the discovery that rings are generated from cut-off Gulf Stream meanders. The data also showed that surface speeds in the Gulf Stream reach 200–250 cm/s and in rings

150 cm/s, higher than were commonly thought to occur. In 1950, Fuglister and Worthington (1951) were the first to observe in detail the formation of a cold-core ring from a meander, and they clearly circumnavigated the new ring. Again in 1960, a cold-core ring was observed forming at 60°W near the New England Seamounts (Fuglister 1963).

The first study to concentrate on cold-core rings was carried out in the mid-1960's by F.C. Fuglister. From September 1965 to February 1966 he dedicated seven cruises to following the evolution of two rings, and, from March to October 1967, nine cruises to measuring the life history of another ring. The data from these studies, plus some additional measurements, served as the basis for the first good description of the distribution, movement, and decay of cold-core rings (Fuglister 1972, 1977, Parker 1971, Barrett 1971).

In the 1970's many people became interested in both cold- and warm-core rings and began a series of individual and cooperative experiments.[1] New measurement techniques such as airborne XBT's, satellite infrared images, SO-FAR floats, satellite-tracked drifters, and vertical profilers enabled researchers to follow rings and measure their properties in new ways. Scientists from the United States Naval Oceanographic Office also began to study rings more actively during this time and publish data in the *Gulf Stream Monthly Summary*. This publication was replaced by *Gulf Stream* (1975–81) which has since been replaced by *Oceanographic Monthly Summary*. Recent analyses of satellite images and other data that show the Gulf Stream and rings are produced several times a week by NOAA and are called *Oceanographic Analysis*.

During 1976–1977 several investigators carried out a cooperative and interdisciplinary experiment during which two cold-core rings were followed over their lives and their physical, chemical, and biological properties and changes with time measured (Ring Group 1981). Several other rings were measured coincidently. Probably the most significant result of the recent experiments is a description of the complexity of a ring's life history. Rings split into pieces, merge, interact with the Gulf Stream, reform as modified rings, coalesce completely with the Stream. A recent summary of this work has been published by the Ring Group (1981).

2.3 Cold-Core Rings

2.3.1 Formation

Cold-core rings form from Gulf Stream meanders which pinch off to the south of the Stream (Fig. 1). The Gulf Stream loops to the right of its downstream direction and the two sides of the loop with currents flowing in opposite direc-

---

1 My own interest in rings began in 1967 during a chance encounter with an intense one off Cape Hatteras (Richardson and Knauss 1971) and deepened with the results of Fuglister's 1965–67 studies and a second close encounter with a powerful ring near Cape Hatteras (Richardson et al. 1973).

tions approach each other and merge trapping a central core of cold Slope water originally located north of the Stream (Fuglister 1972, Doblar and Cheney 1977). When the closed meander separates from the Stream, a ring is born. It consists of a closed segment of the Stream revolving cyclonically (counter-clockwise) around the cold water core (Fig. 2). Some of the five to eight cold-core rings which form each year (Fuglister 1972) are generated quite quickly, in a week, others more slowly, in several weeks. Evidence suggests that the near surface ring pinches off first, followed thereafter by the deeper structure (Fuglister and Worthington 1951, Fuglister 1963).

A new ring is often elliptical, but it usually becomes nearly circular as it moves away from the Stream. Typical overall diameters are 200–300 km and surface speeds 150 cm/s. Newly formed rings can be observed by satellite infrared images due to their initial surface temperature distribution and their sea surface depressions of approximately 0.5–1.0 m (Cheney and Marsh 1981a).

Rings form from 70°W eastward. Most rings have been observed in the region 60°–70°W with a maximum number north of Bermuda near 65°W, suggesting a preferred formation region. Ring generation is apparently common along 60°W near the New England Seamounts (Fuglister and Worthington 1951, Fuglister 1963), where the Stream makes particularly large amplitude meanders. The seamounts seem to be responsible for the large meanders and for a semi-permanent ring-meander located over the seamounts. Periodically rings pinch off from this structure and also coalesce with it. South and east of the Grand Banks the Stream seems to break down into several branches which also shed current rings.

Shedding of cold-core rings, their injection into the Sargasso Sea and subsequent decay there represents a significant transfer of heat across the Gulf Stream (Newton 1961, Cheney and Richardson 1976). The heat transfer includes (1) a ring of warm Gulf Stream water near the surface, and (2) a deep cold-core. The surface water can exchange heat directly with the atmosphere; the deeper layer represents an injection of heat deficit into the internal region of the Sargasso. Mintz (1979) has described the role of rings as interpreted from numerical model studies of ocean circulation. The large-scale Gulf Stream gyre carries heat from its lower latitude source to mid-latitudes where the heat is transferred to rings, both warm and cold, which carry the heat across the mean position of the Gulf Stream front.

2.3.2 Structure and Velocity

Cyclonic rings are characterized by a large raised dome in thermal, salinity and density fields (Figs. 2, 3). Deep hydrographic sections across a few new rings show that the dome structure extends down to the sea floor. The bell-shape has prompted the use of a Gaussian-shaped density field for the initial condition in many models of rings (Flierl 1977b, McWilliams and Flierl 1979, Mied and Lindemann 1979).

Frequently different water properties (temperature-salinity, temperature-oxygen, etc.) can be identified in the upper few hundred meters of a ring (Fuglister 1972, Hagan, Olson, Schmitz and Vastano 1978, Richardson, Maillard and Sanford 1979a, Vastano, Schmitz and Hagan 1982, Ring Group 1981). Initial water properties in cold-core rings vary from ring to ring. This is partly due to seasonal and geographical variations of the ingredient water properties and partly due to the specific details of the formation process for each ring. A ring consists of various amounts of Slope, Shelf, and Gulf Stream water (Fig. 4). Slope water is a mixture of water from northern sources and Shelf and Gulf Stream water. Gulf Stream water is a mixture of southern water plus water entrained along its path northward. West of 60°W the Gulf Stream entrains significant amounts of Shelf, Slope, and Sargasso water plus water from warm- and cold-core rings which have coalesced with the Stream. Some, or all, of these different water masses can be the initial ingredients of a ring.

Continuous profiles of T, S, O_2 show pronounced layering in the upper 500 m of rings (Lambert 1974, Ring Group 1981). Layers 25–100 m thick of Sargasso and Gulf Stream water penetrate into the ring core and core water penetrates into the Gulf Stream remnant. Continuous velocity profiles show strong inertial layers in the ring core as well as in the high velocity region (Richardson et al. 1979a). In the upper 1000 m the layers were 50–100 m thick, had amplitudes of 10 cm s^{-1} and oscillated with a period near the inertial. Given strong gradients of water properties, the inertial layers could provide a mechanism by which property gradients and mixing are enhanced.

Vastano and Hagan (1977) suggest that rings are sites where the Sargasso Sea water is produced by combining, mixing, and subsequent detrainment of Gulf Stream and Slope water. The formation of rings provides a mechanism by which the Gulf Stream front is convoluted and lengthened many times over the straight line mean path of the Stream and enables the different water masses on either side of the Stream an increased area in which to mix.

A young ring has surface swirl velocities of 150 cm s^{-1} (3 knots). Speed increases with distance from the center, reaches a maximum near a radius of 30–60 km and then drops off again towards the outer limits of the ring (Cheney and Richardson 1976, Fuglister 1977, Olson 1980). Figure 3 shows a velocity section through ring Bob at age 6 months (Olson 1980). A pronounced jet surrounds the core which is in nearly solid body rotation. A maximum in relative vorticity occurs just inside the radius of maximum current. The zone inside these maxima is the portion of fluid in the upper layers which translates for long periods of time (months – years) with the ring as it moves through the Sargasso Sea. At any one time the area of fluid translating with the ring extends beyond the radius of maximum swirl speed (Flierl 1976). The strength of the velocity jet is responsible for the maintenance of temperature-salinity and other anomalies in the ring core (Flierl 1976, Olson 1980).

The high speeds and high kinetic energy of rings are partly responsible for the large peak in Eddy Kinetic Energy (E_K) associated with the Gulf Stream region (Wyrtki, Magaard and Hager 1976). Buoys in rings yield an E_K of ~6000 cm^2 s^{-2} in the ring region south of the Stream; buoys in the same area but outside of rings yield values of ~1000 cm^2 s^{-1}, and those near 30°N, south of the ring region, ~200 cm^2 s^{-1} (Richardson 1982b). These E_K values near the Gulf Stream are significantly higher (2–10 times) than those given by Wyrtki et al.

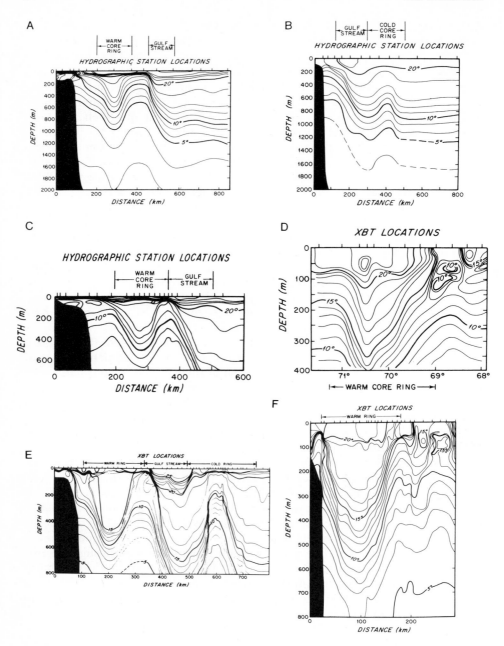

Fig. 2 A–F

Fig. 2. Vertical temperature sections through several rings. **A** After Iselin 1936, Fig. 14. **B** After Iselin 1936, Fig. 10. **C** After Iselin 1940, Fig. 13. **D** After Saunders 1971, Fig. 5. **E** After Richardson, Cheney and Worthington 1978, Fig. 4a. **F** KNORR 71, 1977. **G** After Richardson, Maillard and Sanford 1979a, Fig. 10. **H** After Richardson 1980, Fig. 5. **I** After Lai and

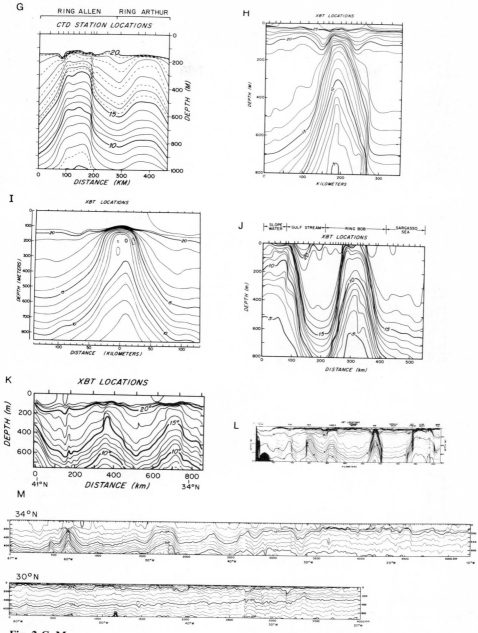

Fig. 2 G–M

Richardson 1977, Fig. 1. **J** After Richardson 1980, Fig. 16. **K** Big Babies along 58 °W after
McCartney, Worthington and Schmitz 1978, Fig. 3. **L** Section from Florida to New England
through the western Sargasso Sea after Richardson 1980, Fig. 5 b. **M** Sections along 34 °N
(top) and 30 °N (bottom) after the Mode I Atlas Group, Figs. 3.3 (2) and (3)

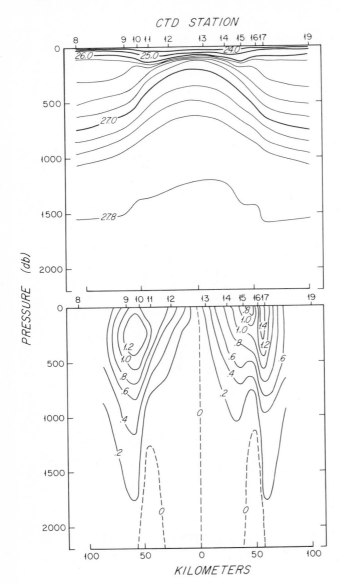

Fig. 3. *Top* Distribution of potential density σ_θ (kg/m³) for a CTD section through ring Bob. (After Olson 1980). *Bottom* Gradient currents (m/s) corresponding to the CTD section above and a 2500 db reference level. (After Olson 1980)

(1976) derived from ship drift data; the buoy E_K values near 30°N, are significantly lower (one half).

In the central part of rings, near-surface period of rotation ranges from about 2 days for a young ring to 5–10 days for older, less energetic rings (Richardson 1980). Occasionally, periods of rotation down to 1.5 days were measured in rings interacting with the Stream, corresponding to times that the central region of the ring was being spun up (Richardson, Maillard and Sanford 1979a).

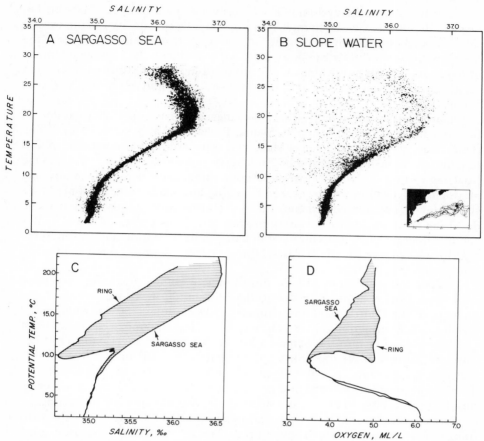

Fig. 4. A, B Temperature and salinity diagrams for the Gulf Stream meander region between 60° and 70 °W showing T-S anomalies across the Stream.
A Plot of Sargasso Sea stations to the south of the Gulf Stream center, stations whose temperature at 400 m was greater than 15 °C.
B Plot of Slope water stations located to the north of the Gulf Stream center, stations whose temperature at 400 m was less than 15 °C.
C, D Temperature-salinity and temperature-oxygen profiles from the center of a ring and the Sargasso Sea, outside of any rings. (Richardson, Cheney and Mantini 1977). The ring station has anomalously low salinity and high oxygen from the surface to at least 10° which is indicative of its Slope water origin

Swirl speeds and the rotation rates decrease with depth (Fig. 3). There is a controversy whether the cyclonic circulation extends coherently to the sea floor under a ring. Vertical shear extends to the bottom under young rings – shear in geostrophic velocity sections as well as shear in absolute current meter profiles. Richardson et al. (1979a) suggest that if the ring translation is subtracted from some current profiles, the cyclonic circulation of a ring then extended to the sea floor. McCartney, Worthington, and Schmitz (1978), however, analyzed current meter measurements from the time that a ring drifted past a mooring and concluded that the cyclonic velocity only extended to about 2000 m; below this depth was a weak counter-rotating eddy. Deep velocity fluctuations in the western boundary undercurrent large enough to occasionally reverse the deep southwestward flow can be visually correlated with the passage overhead of older rings (Mills and Rhines 1979), although the current fluctuations due to the rings do not seem to be a simple downward extension of the ring's surface velocity field.

Neutrally buoyant SOFAR floats have measured currents in rings at depths of 750–1300 m (Cheney, Gemmill, Shank, Richardson and Webb 1976). When three floats were launched in a ring, two were retained for periods of 2–3 months; the loops had a period of 20–40 days and speeds of 3–11 cm s^{-1}. One float came out quickly, in 2–3 weeks. Another float drifting outside of any ring at a depth of 1290 m was overtaken by a ring and made three cyclonic loops before being left behind (Cheney 1977a). During these loops (period of 40 days) the float speeded up from approximately 5 cm s^{-1} to 24 cm s^{-1}. Recently a float at 700 m was entrained in a ring at its formation and stayed in the ring for eleven months (Richardson, Price, Owens, Schmitz, Rossby, Bradley, Valdes and Webb 1981).

Conceptually, there are at least three kinds of volume transport associated with rings. First is a transport due to their formation. Each ring consists of a segment of the Gulf Stream approximately 500 km long, 100 km wide, and 2 km deep. The formation of 5–8 cold rings per year amounts to a transport of Gulf Stream water into the Sargasso Sea of 20×10^6 m^3 s^{-1} (Fuglister 1972). The Gulf Stream itself transports approximately 30×10^6 m^3 s^{-1} off Florida reaching a maximum of 150×10^6 m^3 s^{-1} north of Bermuda (Knauss 1969). Thus, rings constitute a significant part of the Gulf Stream's recirculation or return flow especially when one considers the added transport of warm rings.

Second, there is transport of water circling around the ring core – the swirl transport, estimates of which range up to 73×10^6 m^3 s^{-1} (Richardson et al. 1979a). Because of problems in determining how deep the swirl transport extends and the outer limits and appropriate reference station values, there is a large uncertainty in the calculated values of swirl transport.

Third, there is a transport of water carried by rings as they drift westward; the amount depends on whether they are advected or self-propelled. If rings are carried passively by a large-scale barotropic flow, then their whole volumes are transported with the flow. If, however, rings are self-propelled through an ocean at rest, only their upper part where the swirl speed exceeds the translation speed is trapped and carried along with the ring (Flierl 1976, Olson 1980). The trapped region resembles a cone with its base at the surface and its apex pointed downward. The area is largest near the surface where the surface swirl speed is largest; the area decreases with depth as the swirl speed approaches the translation speed of the ring and its center shifts northward toward the radius of maximum swirl speed. Evidence confirming a trapped region in rings includes surface drifters and anomalous water properties and biota which remain in the upper 500 m of rings for long periods of time (months to years).

2.3.3 Distribution and Number

Approximately ten cold-core rings co-exist at a single time (west of 55°W). This estimate comes from two attempts at making a ring census (Lai and Richardson 1977, Richardson, Cheney, and Worthington 1978; McCartney, Worthington, and Schmitz 1978) plus numerous ring sightings identified on single sections that have tended to confirm the number of rings observed on the censuses. In the spring of 1975 several XBT surveys plus satellite infrared images provided a nearly synoptic view of the number and distribution of rings (Fig. 5). The strongest rings, those with the highest raised dome in their centers, 500–600 m above normal background levels, were found north and northwest of Bermuda. Rings west of Bermuda and those east of 62°W had reduced displacements of the main thermocline. The ring census (Fig. 5) was based on data taken over four months. The effect of a three month gap in time between sur-

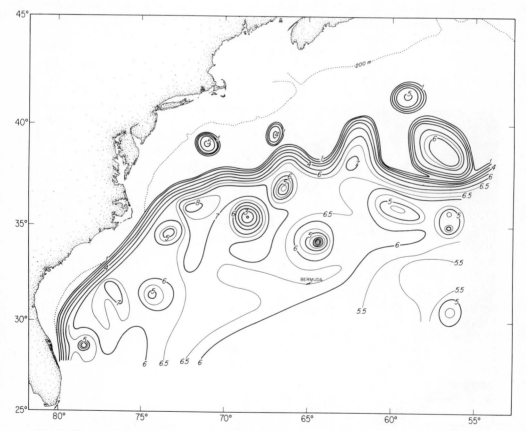

Fig. 5. Chart of the topography (hectometers) of the 15° isothermal surface showing the Gulf Stream, nine cold-core and three warm-core rings. Contours are based on XBT, CTD, hydrographic and satellite infrared data from the period March 16 to July 9, 1975. (Richardson, Cheney and Worthington 1978)

veys was to spread the rings out in an east-west direction near 62°W. It is estimated that if the region had been surveyed simultaneously approximately 10
rings would have been found.

An analysis of hydrographic stations, BT's, XBT's, and satellite infrared
images from the period 1932–1976 has resulted in the identification of 225
cold-core ring observations in the area west of 50°W (Parker 1971, Lai and
Richardson 1977). Rings were most frequently observed in the northwestern ʹ
Sargasso Sea (Fig. 6). This reflects the preferred ring formation and migration
patterns there plus the higher density of data in that region. The youngest
rings, those with the coldest temperature anomalies, are found between 64°–
70°W just south of the Stream. Many rings were observed in the western Sargasso, along a path 200 km offshore of the Gulf Stream axis where rings often
moved southwestward.

Between 50° and 60°W only 16 rings have been identified, all of them since
1970. This small number is due to the low density of data east of 65°W, and
due to the weaker near-surface temperature signal in the rings there that makes
identification more difficult especially with shallower mechanical BT's (used
by Parker 1971). Rings probably occur much more often east of 60°W than is
suggested by the present number of sightings there.

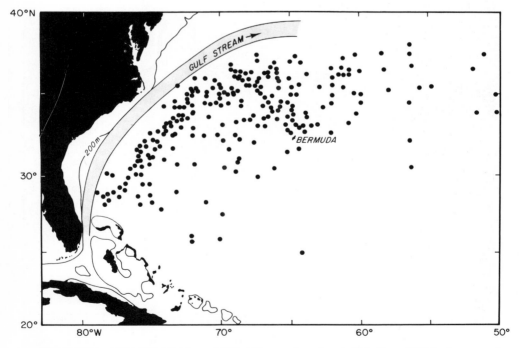

Fig. 6. Geographical distribution of 225 cold-core ring observations from 1932 to the present
from Parker (1971) and Lai and Richardson (1977) plus a few recent observations of rings in
the eastern area from Bulgakov, Djiganshin and Belous (1977) and Cheney and Marsh (1980)

South of 30°N the eastern area is devoid of ring observations. It is likely that rings do not enter this region or that if they do enter it they have decayed beyond the point of recognition. The few long XBT sections that exist in this region do not show ring-like deviations from large-scale variations in temperature (Fig. 2).

2.3.4 Movement

Rings usually move westward when they are not touching the Gulf Stream and eastward when they are attached to it (Fig. 7, Richardson 1980, Fuglister 1972, 1977). The westward movement, sometimes northwest, sometimes southwest, is characterized by a mean speed of 5 cm s^{-1}, although there are large variations about this, including periods when rings remain nearly stationary. Two hundred miles offshore of the Gulf Stream axis, between 28°–36°N, a ring corridor is located in which rings drift southwestward (Parker 1971; Richardson, Strong and Knauss 1973, Lai and Richardson 1977). At least 12 rings were observed to move down this path during the period 1970–1976. Recent ring trajectories, measured continuously with buoys, are more complicated than earlier

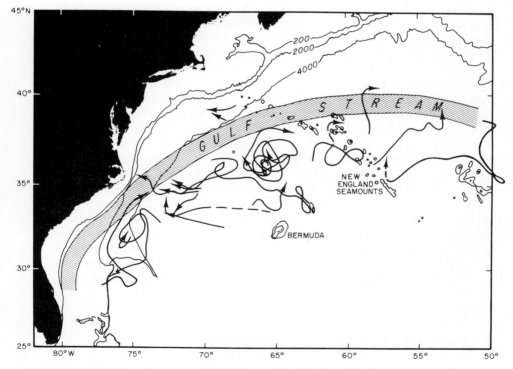

Fig. 7. Ring trajectories. *Dark lines* represent rings continuously tracked with free drifting buoys and SOFAR floats. (Fuglister 1977, Richardson 1980, Cheney and Marsh 1981a, Richardson et al. 1981)

observations based on repeated sightings (Richardson 1980). Although repeated sightings of rings suggest that they can drift into the southern Sargasso
Sea, no rings tracked with buoys moved into this region. There is a hint from
temperature anomaly observations that rings can penetrate southward near the
Bahama Islands to as far south as 25°N, but the anomalies in these rings were
quite weak and the observations tentative (Parker 1971). The limited data east
of 60°W suggest rings also drift westward there.

Frequently rings become attached to the Gulf Stream and are advected parallel to the Stream in a downstream direction. This eastward movement can be
quite fast; maximum daily speeds ranged from 25–75 cm s^{-1} for several rings.
Often rings split off from the Stream after an eastward advection and began a
westward movement. The result of successive ring attachments to the Gulf
Stream is that many rings moved in large clockwise loops with a characteristic
period of about 2.5 months and diameter of 175 km. These loops are prevalent
north and west of Bermuda (Fuglister 1977, Richardson et al. 1979a, Richardson 1980).

A semi-permanent ring-meander overlies the New England Seamount. It is
called a ring-meander because at times a ring seems to be separating from the
meander, but often complete separation does not occur and the ring remains
trapped near the seamounts. Several buoys moving eastward in the Stream became entrained for various lengths of time in this structure. Rings that appeared to have separated from the Stream and that were tracked with buoys
did not last long before they coalesced with the Stream.

The movement of rings is apparently controlled by a combination of advection by the large scale flows and the tendency for propagation to the west as a
packet of Rossby modes. The overall translation of rings is in the same direction as the mean flow which is to the west on either side of the Gulf Stream and
which has a mean speed of ~5 cm s^{-1} (Worthington 1976a, Luyten 1977,
Schmitz 1980). There are periods during which rings appear to propagate normal to the inferred large-scale flow. This motion, and possibly part of the regular translation, are probably due to the tendency for longer-scale disturbances
to propagate westward on a β-plane. The propagation of a wave-like vortex has
been treated by Adem (1956), Warren (1967) and Flierl (1977b). The nonlinearity of the ring circulation can also induce a northward (southward) component
of translation to a cyclonic (anticyclonic) ring (McWilliams and Flierl 1979,
Mied and Lindemann 1979). The models to date provide a set of basic physical
processes with which to understand ring movement. Further progress is
needed, however, before a clear understanding of ring trajectories such as
those in Fig. 7 is obtained. The combined effects of propagation and advection
are of particular interest at the present time.

2.3.5 Interaction and Coalescence with Gulf Stream

The ultimate fate of rings seems to be complete coalescence with the Stream
(Cheney et al. 1976, Richardson, Cheney, and Mantini 1977, Watts and Olson
1978, Richardson 1980). Although there is the possibility that rings can move

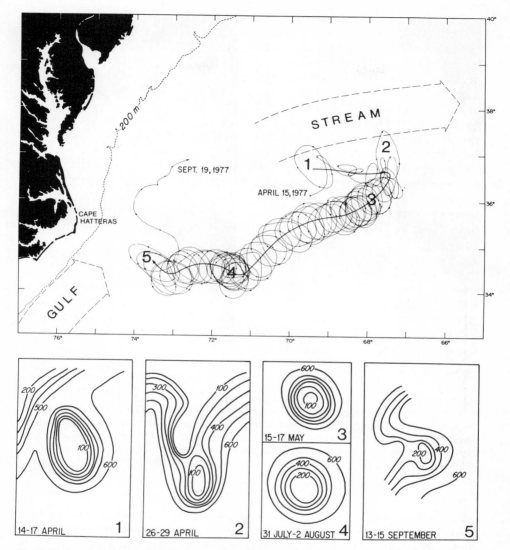

Fig. 8. Trajectory of ring Bob studied by the ring group in 1977. Top panel shows trajectory of free-drifting buoy looping around Bob's core. Bob moved eastward while interacting with the Gulf Stream, split off from the Stream, drifted southwestward through the Sargasso Sea, then coalesced with the Stream near Cape Hatteras after a lifetime of 7 months. Lower panels show the depth of the 15° isotherm in Bob at several times during its life. (After Ring Group 1981)

southward into the southern Sargasso Sea and slowly decay there, separately from the Stream, we have no evidence of this from rings tracked by buoys.

Frequently rings were observed to become attached to the Gulf Stream and interact with it. "Attached" means that ring contours – the topography of the

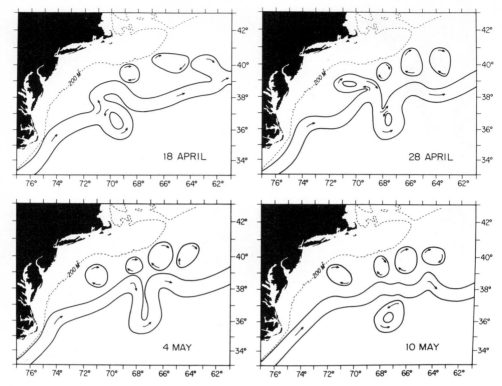

Fig. 9. Schematic representation of the interaction between ring Bob and the Gulf Stream. Bob became attached to the Stream in mid-April, 1977, was advected rapidly downstream, exchanged water and energy with the Stream, and reformed in early May. During this period of time the Gulf Stream formed two warm-core rings north of the Stream. Although the third panel shows an open meander, the evidence from a buoy in Bob suggests that closed circulation was maintained in the central region of the ring-meander. (After Richardson, 1980)

main thermocline, 15° isotherm for example – merge to some extent with the Stream's and that the two are exchanging water and energy, etc. (Figs. 8–10). Although satellite infrared images have shown this interaction since the early 1970's, only recently have we obtained drifting buoy and shipboard data to document it more fully (Watts and Olson 1978, Richardson et al. 1979a, Richardson 1980, Vastano, Schmitz, and Hagan 1980). A recent estimate suggests that on the average approximately six rings (cold- and warm-core) are in proximity to the Stream and that each of these interacts with the Stream every 2 months (Richardson 1980).

At least three modes of interaction have been identified in the available observations; the first two are fatal to the ring and result in its complete coalescence. The first mode occurs when a relatively intense ring becomes attached to the Stream, and the main Gulf Stream current is diverted and flows around the ring. The ring center opens to the north of the Stream which is in the form of

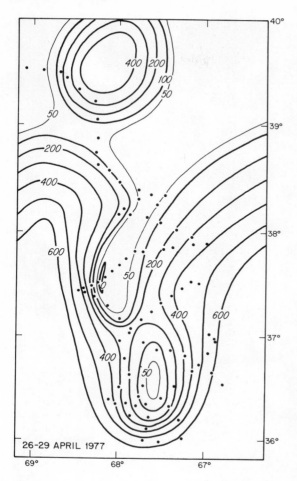

Fig. 10. Topography of the 15° isothermal surface in ring Bob while it was attached and interacting with the Gulf Stream 26–29 April based on an XBT survey, drifting buoys and satellite infrared images. A warm-core ring is located on the northern edge of the Stream. Vertically mixed, 12 °C water from the warm ring was observed to be entrained into the Gulf Stream; this water was possibly incorporated into Bob as it reformed in early May. (After Richardson 1980)

an open meander (Richardson, Cheney and Mantini 1977, Watts and Olson 1978). It is possible for the meander to then dissipate or to pinch off as another ring. The steps in this mode resemble the formation of a ring except that the steps are reversed in time. During the final coalescence, water from the ring center can be transported back across the Stream into the Slope water region.

The second mode occurs when a relatively weak ring becomes attached to the Stream, is advected downstream, and coalesces completely with the Stream (Richardson et al. 1979a, Richardson 1980). Rather than forming a meander as in the previous mode, the Stream seems to absorb the ring and to accelerate it downstream until it can no longer be identified. On two occasions the final coalescence of a ring may have been triggered by another ring which was advected downstream and collided with the ring subsequently lost.

The third and nonfatal mode occurs when a ring becomes attached to the Stream, moves downstream and reforms as a modified ring.

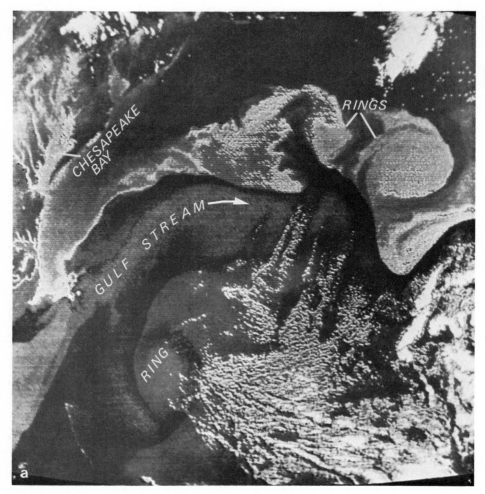

Fig. 11. Satellite infrared images of the Gulf Stream region showing the entrainment of warm Gulf Stream water into cold-core rings. Black and white photographs were made from color-enhanced images to emphasize certain features. **a** 28 April, 1974; **b** 13 April, 1977. (Color photos courtesy R. Legeckis and A. Strong, NOAA, NESS)

During the time the ring is connected to the Stream, significant amounts of energy and water can be exchanged between the two. Although this process can be easily observed in the near surface water with satellite images (Fig. 11) and with before and after XBT sections, it is much more difficult to observe in the deeper water of a ring (Vastano, Schmitz, and Hagan 1980). During an interaction, warm-core Gulf Stream water is entrained into a ring and advected cyclonically around its center. Frequently the warm, near-surface water entrained into rings is then mixed into surrounding areas and sometimes exchanged from ring to ring. Because of the rapid rotation and translation of rings warm Gulf Stream water can be spread throughout wide areas of the

Fig. 11b

Slope water region and Sargasso Sea in periods of weeks. This spreading of warm water may be an important step in maintaining the temperature structure of these areas despite their large wintertime loss of heat. Rings may also limit the size and location of areas in which 18° water forms and the amount of such water that does form.

During several interactions of rings and the Gulf Stream, the rotation rate of the central ring region increased. On two of these occasions the hydrographic properties were measured and the upper part of the ring core was vertically stretched (Richardson et al. 1979a, Vastano et al. 1980, Olson 1980, Olson and Watts 1982). On one of these occasions the 18° isotherm rose 250 m (relative to 10°C), 60% of the thickness before the interaction, and the rotation rate increased from a period of 3.5 days to 1.5 days (Richardson et al. 1979a). Thus, during an interaction, a ring can become spun up; the increase in rotation rate and the stretching of the upper layer are in agreement with the conservation of potential vorticity.

2.3.6 Decay

Rings exhibit both catastrophic decay and a slower spindown decay. Catastrophic decay occurs when a ring coalesces with the Stream and is lost. Rapid coalescence can span periods of time from a few days (Watts and Olson 1978) to a few weeks (Richardson 1980). A second type of catastrophic decay occurs when a ring bifurcates into separate "ringlets" as has been observed with two large rings (Cheney et al. 1976, Richardson et al. 1979a).[2] A third type of catastrophic decay occurs when a ring moves over the Blake Escarpment into depths of 800–1 000 m. Although this last process has not been observed in detail, one ring that was followed onto the Blake Plateau was significantly weaker afterward (Cheney and Richardson 1976, Schmitz and Vastano 1977). Another ring that was tracked with a satellite buoy had its normal ring circulation disrupted as it impinged on the escarpment; the buoy made increasing large loops in the ring and finally left it entirely.

In the slower spindown the dome of cooler, fresher water that forms the heart of the ring slowly subsides towards the Sargasso Sea background. This dome represents a reservoir of available potential energy (2–3 times the kinetic energy) which is slowly released during decay (Barrett 1971, Cheney and Richardson 1976, Olson 1980, Reid, Elliott and Olson 1981). The region of maximum available potential energy (compared to the Sargasso Sea) is centered near the 15° isotherm (Olson 1980). Regions of high swirl speed, volume transport, and kinetic energy, all of which are concentrated near the surface, are related to the cold-core and decay as the core collapses. The decay of ring Bob suggests a slow outward radial shift in the position of the ring front (Vastano et al. 1980, Olson 1980). The outward spread is consistent with inward radial velocity in the surface layers, compensated for by sinking in the ring core and outward radial velocity at depth. Calculations of variations in Bob's core (decreased layer thickness and decreased rotation rate) are consistent with the conservation of potential vorticity (Olson 1980).

Long term rates of subsidence of the core range from about 0.4 m d^{-1} to 1.3 m d^{-1} (Parker 1971, Fuglister 1972, Cheney and Richardson 1976, Richardson et al. 1979a, Vastano et al. 1982); however, the first value is for a moderately old ring which traversed the Blake Escarpment (Cheney and Richardson 1976) and the second is for a young ring that interacted with the Stream and may have bifurcated (Richardson et al. 1979a; Vastano et al. 1982). Considering all the complexities of a ring's life – interactions, infusions of new water, bifurcations, disruption due to topography, variations in background conditions – these values must be viewed as approximate; they imply overall ring lifetimes of 1 to 4 years. The numerous rings encountered in the southwestern Sargasso Sea also show a core subsidence with latitude and an implied life of 1 to 2 years in agreement with the estimated mean southwestward speeds in this region (Parker 1971, Lai and Richardson 1977). Recent evidence suggests that the most frequent finish of a ring is not through long-term decay but through coal-

---

2 Vastano et al. (1982) conclude that one of these rings did not bifurcate but that two separate rings partially merged, interacted, and then split apart.

escence with the Stream; thus the mean age at disappearance is approximately 1 year. One reason the longevity of a ring is difficult to calculate is due to the murky behavior of ring-Gulf Stream interactions. How many "new rings" are really recycled older rings?

Various mechanisms have been suggested as contributing to ring spindown and decay. The earliest of these is simple frictional decay leading to symmetric radial overturning which spins down the ring (Molinari 1970). While such a mechanism is certainly plausible, there are several alternatives which can accomplish the same end without resorting to large eddy viscosities. These include dispersion of the ring as a packet of Rossby waves, and a possible secondary hydrodynamic instability of the ring to the growth of subring scale perturbations. Wave dispersion, postulated as a decay mechanism for rings by Flierl (1977b), could be a fairly effective means for the release of ring energy and property distributions into the Sargasso Sea at the expense of the ring. The nonlinearity of the ring circulation apparently acts to overcome dispersion such that the ring behaves more like a solitary wave (McWilliams and Flierl 1979, Flierl 1979). The other mechanism involving either baroclinic, barotropic, or combined instability in the ring may also be of some importance at least in younger rings (Saunders 1973, Hart 1974, Olson 1976). Observations imply that the necessary conditions for instability are met in rings and that asymmetric disturbances of large enough amplitude to account for ring decay do occur (Olson 1980). Just as in the translation question, it is not certain to what extent each of the three mechanisms mentioned here plays a role in actual rings.

Some recent eddy-resolving general circulation models which develop quite realistic Gulf Stream rings (Holland 1978, Semtner and Mintz 1977, see also Chap. 18) begin to show the complex energy flows within the ocean. One of the dominant pathways of kinetic and potential energy within the system is through rings, where energy is transformed and transmitted both laterally and vertically to other parts of the ocean. Progress in understanding rings will be made as these models are refined, and the life histories of individual model rings are studied and understood.

2.4 Warm-Core Rings

Warm-core, anticyclonic rings form throughout the Slope water, a triangular region bounded on the south by the Gulf Stream and on the north by the continental slope (Saunders 1971, Gotthardt 1973). The largest rings, 200–300 km in diameter, consist of an annular ring of Gulf Stream surrounding a Sargasso Sea core; these are usually seen east of Georges Bank. Smaller rings, 100 km in diameter, form in the west; their formation resembles (on satellite infrared images) an aneurism during which the side of the Stream bulges out to the north and pinches off from the main current. Occasionally two rings pinch off from the same aneurism. This type of formation may explain why most warm-core rings have a smaller amplitude thermocline displacement in their centers (a few hundred meters) than that across the Stream – only part of the Gulf Stream, not its whole width, separates from the main current.

Typically, five warm rings form per year; their size (approximately 100 km) varies, but the average appears to be somewhat smaller in diameter than that of

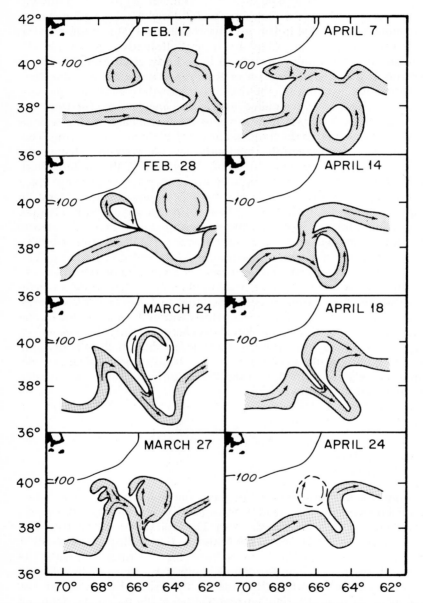

Fig. 12. Series of warm and cold ring interactions with the Gulf Stream from satellite images during 1976. (After Mizenko and Chamberlin, 1979). One hundred meter depth contour is shown in *upper left*

cold rings (Bisagni 1976, Lai and Richardson 1977, Halliwell and Mooers 1979). The proximity of the continental shelf to the mean Gulf Stream position may be a factor in limiting the size of northward meanders. Approximately three rings exist at a single time. Although warm rings form as far east as the Grand Banks (50°W), they are most frequently observed in the western region, due to more frequent satellite and ship coverage in the western region. Warm rings move westward with a mean speed of 5 cm s^{-1}, a speed similar to the mean flow in the Slope water region; this implies that these rings might be advected by the mean flow. As warm rings drift westward they gradually shrink in size. When they reach Cape Hatteras, after a lifetime of about 6 months, they coalesce with the Stream (Gotthardt and Potocsky 1974).

Warm rings exhibit considerable variation from the typical pattern described above. Frequently they interact with the Gulf Stream – sometimes separating again, sometimes coalescing completely. Many interactions may be seen with satellite infrared images which show warm Gulf Stream entrainments into rings and ring water entrainments into the Gulf Stream (Fig. 12). Rings also often entrain cold fresh Shelf water and advect it around their eastern sides into the Gulf Stream (Morgan and Bishop 1977). The spilled oil from the ARGO MERCHANT, tracked with a drifter, followed this path into the Stream. At times no warm rings can be identified in the Slope region and at other times the Slope region appears to be nearly filled with rings (May 1980).

During winter the warm-core of these rings is cooled, and deep (down to 500 m) vertically mixed isothermal layers are formed analogous to the 18° subtropical mode water south of the Gulf Stream (Fig. 2). The mixed layer in rings, however, can cool to a temperature of 11°–12°C. During subsequent summer heating, a seasonal thermocline forms over the top of the deeper mixed water. The mixed water is entrained by the Gulf Stream during ring-Gulf Stream interactions and final coalescence. This water can be incorporated into newly formed warm and cold rings.

An interdisciplinary and cooperative study of warm-core rings is presently being carried out. The experiment, scheduled to occur from 1981–1985, should provide significant new information on these rings.

2.5 Gulf of Mexico Rings

Large, anticyclonic, warm-core rings separate from the loop current in the eastern Gulf of Mexico. Elliott (1981) has prepared a summary of these rings and reinterpreted some older data. Early observations were made by Leipper (1970) and Nowlin, Hubertz and Reid (1968). These rings form at an average rate of one to two per year; a maximum of three rings were observed to form in one eight-month period. Their average diameter is 370 km, and they drift westward with a mean speed of 2.4 cm s^{-1}. One ring was tracked all the way to the western boundary of the Gulf. The estimated lifespan of these rings, as defined by an e-folding time, is one year.

2.6 Eastern Rings

"Rings" will be used here as a general term for eddies formed by the Gulf Stream system. This is a broader definition than that proposed by Fuglister (1972), who described a ring as being formed from a closed segment of the entire width of the Stream. The eastern rings seem to form much like the western rings with the exception that in the eastern region the Gulf Stream has become broader and less jet-like and often splits into separate filaments or branches. In general, eastern rings do not consist of a closed segment of current which includes the whole width of the Stream. The large, cold-core eastern rings have become popularly known as "Big Babies" (Worthington 1976b).

Ring-like eddies generated by the Gulf Stream system have been observed over a broad area of the North Atlantic. These ring observations coincide closely with a region of large thermocline displacements, high eddy potential energy density (Dantzler 1977) and high eddy kinetic energy (Wyrtki, Magaard and Hager 1976). The region of large thermocline displacements follows the Gulf Stream eastward and is bounded on the south by latitude 32°N and on the east by the mid-Atlantic Ridge; the northern limits are not well known. Two regions of high energy extend eastward across the mid-Atlantic Ridge near latitudes 33°N and 42°N.

Within this region are three sub-areas which contain rings of different characteristics. South of the Stream and extending eastward over the mid-Atlantic Ridge are large, cold-core eastern rings. Overlying the southeast Newfoundland Ridge are cold rings whose cores originate in the Labrador Current. East of the Grand Banks are smaller warm and cold-core North Atlantic Current rings.

Numerous large cold-core rings have been found south of the Gulf Stream and east of 60°W. Some of these rings appear to be larger than western rings and have overall diameters greater than 300 km. Some also have a smaller isotherm slope to their flanks, multiple bumps, a weak near surface temperature signal, and are often not very circular (Fig. 2, McCartney et al. 1978). A large (500 km) triangular-shaped ring was surveyed twice by Soviet scientists (Bulgakov et al. 1977). It was located near 32°N, 50°W, and moved westward with a speed of 4 km/d as did other rings in this area (McCartney et al. 1978). A large eastern ring, 450 km in diameter, was identified near 33°N, 35°W (Gould 1976).

Only twice has the formation of these rings been observed with supporting in situ data. The first was recorded by both satellite images (La Violette, pers. comm.) and a SOFAR float which was trapped in the ring as it pinched off from the Stream near 38°N, 50°W (Richardson et al. 1981). The ring which was approximately 300 km in diameter moved southward, then westward (4 cm s^{-1}) skirting the Corner seamounts and then probably coalesced with the Stream after a lifetime of 11 months, (Fig. 7). A second ring was observed by Gould (pers. comm.) to pinch off from a front lying south of the Azores near 33°N, 33°W. The ring had a diameter of 100–150 km, a thermocline displacement of 150 m, surface speeds of 25 cm s^{-1}, and was propagating westward with a speed of 2.9 cm s^{-1}.

The limited evidence suggests that rings can form from southward meanders of the Stream east of 60°W and also from meanders along the front at the
eastern terminus of the subtropical gyre. Some of the observed rings may have
formed from the cold water wedge associated with the Gulf Stream recirculation. This wedge or ridge of cold water lies along 35°N and extends from near
40°W to at least 55°W. A good synoptic depiction of this wedge was given by
Emery, Ebbesmeyer and Dugan (1980) and a map of the mean temperature
field of it at 450 m was shown by Richardson (1981). The front south of the
Azores is the eastern extension of the south side of this wedge.

A southeastward projecting ridge of cold water overlies the Southeast Newfoundland Ridge. (An early observation of this cold water and a large iceberg
50 m high, 2 km in diameter was made in a letter from F.D. Mason to J. Williams dated Clifton, England, 20 June 1810; see Furlong 1812). Often the surface manifestation of the cold ridge can be seen on satellite images as an extension of the cold Labrador Current into the warmer Gulf Stream water. Occasionally, water from the cold ridge pinches off to the southeast forming cold-
core rings (Mann 1967, Mountain and Shuhy 1980). A study of infrared images
shows that within a 5 year period 15 distinct rings formed, and that, at least initially, they drifted southeastward (LaViolette 1981). The location of the cold
ridge and subsequent formation of rings from it seem to be dynamically tied to
the sea floor ridge underneath. It is not known whether these rings are trapped
by the bottom topography or drift away from their source region. Since their
cores contain very cold and fresh water from the Labrador Current, it seems
clear that they are distinctly different from the large eastern rings described
above.

Southeastward of the Grand Banks and Flemish Cap region, North Atlantic
Current warm and cold-core rings, approximately 100 km in diameter, are frequently observed both with in situ data (Voorheis, Aagaard, and Coachman
1973, Schmitz 1981a) and with satellite images (La Violette 1981).

Rings are probably formed all along the northeastern extension of the
North Atlantic Current. Howe and Tait (1967) observed an elongated cold-core
ring 220 × 85 km in size, located near 53°N, 19°W. They suggested its core was
formed by an eastward intrusion of water originally located west of a branch of
the North Atlantic Current. Krauss and Meinke (1981) recently made an XBT
and hydrographic survey of the region north of the Azores between 38° and
46°N; they found four cold and three warm current rings with an average diameter of 150 km.

2.7 Current Rings from Other Currents[3]

Current rings have been seen wherever swift and narrow currents are found; in
the South Atlantic (Brazil Current), North Pacific (Kuroshio), South Pacific

---

3 See other chapters for a more complete description of these areas, the currents and the rings
  there.

(East Australian Current), Indian Ocean (Agulhas Current) and Southern Ocean (Circumpolar Current). The Kuroshio meanders and sheds warm and cold-core rings in much the manner of the Gulf Stream. Recently Kawai (1980) has identified 90 cold-core ring observations made since 1927. He also used a large, areal, synoptic survey of Kuroshio region consisting of 30 ships in the summer of 1939 to chart the coexistence of 13 cold rings and 2 warm rings. These data suggest that cold rings form at three or four specific sites, and that they subsequently drift southwestward through the western Pacific along several preferred paths. Some cold-core rings which formed to the east of the Izu Ogasawara Ridge apparently drifted through gaps in this Ridge south of Japan. A recent synoptic study found three warm rings and three cold rings (Cheney 1977b). One of the cold rings, 250 km in diameter, moved northwestward and coalesced with the Kuroshio after an estimated lifetime of 5 months (Cheney, Richardson, and Nagasaka 1980).

Warm-core Kuroshio rings are generally 150 km in diameter and have lifetimes of 2–10 months (Kitano 1975). One warm ring remained nearly stationary just east of Japan interacting occasionally with the Kuroshio. It finally coalesced with the Kuroshio forming a meander which spawned another warm-core ring (Hata 1969). Another ring moved northeastward with a mean speed of 1 cm s^{-1} over a 20-month period (Hata 1974). During two winters, deep (500 m) layers of isothermal water were formed in the core of this ring – 10° during the first winter, 6° during the second. In summer a seasonal thermocline developed over the isothermal layers. Occasionally two isothermal layers are seen in a single ring. Tomosada (1978), who reviewed warm Kuroshio ring observations, suggests that this occurs when Kuroshio water containing subtropical mode water overrides an older eddy which already contains an isothermal layer.

The Oyashio, a current located north of the Kuroshio, also sheds warm and cold current rings (Cheney 1977b).

Warm-core rings are generated from poleward meanders of the East Australian Current (Hamon 1965, Andrews and Scully-Power 1976, Nilsson and Cresswell 1981). These rings are typically 250 km in diameter and have lifetimes of about one year. Most rings coalesce with East Australian Current meanders, often pinching off again as modified rings. Deep (350 m) isothermal layers form in the core water during winter with temperatures down to 16°C. Recently two warm-core rings, each with an isothermal layer, were observed to coalesce resulting in a ring with two different isothermal layers in it (Cresswell 1981). Observations of cold-core East Australian Current rings have not been reported although there is the possibility they exist.

Infrared images have been used to observe Brazil Current rings (Legeckis and Gordon 1982). Forty-three warm-core ring observations were made in the area bounded by 38°–48°S, 48°–56°W. These include 20 individual warm rings seen during the period September 1975–April 1976. The implication is that at least 20 rings per year can form. However, only three rings were observed during 4 months of 1978, so the formation rate could be much less than 20 on the average. Warm-core rings are usually elliptical with a mean diameter of 150 km. The preferred drift direction was southward. Only four cold-core rings were identified; they were also elliptical and their size ranged from 100–300 km.

South of Cape Town, South Africa, the Agulhas turns abruptly eastward and forms a westward-projecting wedge of warm surface water. At times the tip of the wedge separates from the main current and forms warm-core rings 150–200 km in diameter (Duncan 1968, Gründlingh 1978). The Agulhas return current flows eastward near 40°S in a series of semi-permanent warm and cold meander-rings, some of which probably separate from the main current (Gründlingh 1978, Harris, Legeckis, and van Foreest 1978, Harris and van Foreest 1978). The location of the meander-rings appears to be connected dynamically with sea floor topographic features such as the Agulhas Plateau, which is much like the Gulf Stream where it overlies the New England Seamount Chain and Southeast Newfoundland Ridge.

Cold-core cyclonic rings have been observed to form from the Antarctic Polar Front, a transition zone between Antarctic and Subantarctic water which coincides with a core of high velocity. Two rings were reported in the Drake Passage (Joyce, Patterson and Millard 1981, Peterson, Nowlin and Whitworth 1981); they had diameters of 100 km and they drifted northeastward in the Antarctic Circumpolar Current at speeds of 5–10 cm s^{-1}. The available potential energy of these was on the order of 1/20 of Gulf Stream rings; this is due to smaller vertical displacements and weaker stratification than that which is found in Gulf Stream rings (Joyce et al. 1981). Another cyclonic, cold-core, Polar Front ring was observed south of Australia (Savchenko, Emery, and Vladimirov 1978). This ring was south of and combined with the Subantarctic Front and was moving north northeastward at a speed of 3 cm s^{-1}. The authors suggest it contained a deep anticyclonic eddy beneath the cyclonic ring in the surface layer (<2 000 m).

3. Western North Atlantic Interior

C. Wunsch

3.1 Introduction

The western North Atlantic Ocean is the best-studied of the oceans both in terms of its mean flows and of its variability. A dominant feature, and perhaps the major characteristic, of mesoscale variability in this ocean basin is its very strong qualitative dependence upon proximity to the Gulf Stream system (see Schmitz 1978). The focus of this chapter is upon studies made during the POLYMODE program of the character of mesoscale variability at great distances from the Gulf Stream system.

Because of the vast increase in eddy energy toward the Gulf Stream, and because the Stream thus remains the primary candidate as the source of most, if not all, the eddy energy in the North Atlantic Ocean, it is not readily apparent that a geographical separation of the eddy field into the "interior western North Atlantic" is useful. Nonetheless, we will attempt to describe here the major features of the measurements made in the years 1977–79 as a part of POLYMODE. Much of the data have been described more fully by Wunsch (1981) and Fu, Keffer, Niiler and Wunsch (1982) to which the reader is referred for more detailed information. The rationale for the measurements came largely from questions which were raised in the MODE-1 program (The MODE Group 1978). This latter experiment had seen one mappable feature for a brief period. MODE-1 had shown the possible dynamical and kinematical importance of mesoscale variability in a particular region. But there remained a series of nearly qualitative questions: (a) Whether such eddies were present all the time or only intermittently; (b) whether larger and smaller scale features with longer lifetimes were present simultaneously but not visibly so because of the particular observational strategy that had been used; (c) whether eddies tend to be isolated features, or occurred more or less uniformly and "tightly packed". Generally speaking, these first three questions can be related, at least in part, to the statistical problem of finding the average spectrum of motion in the MODE area as a function of depth; (d) whether in regions where the energy levels were known to be different there were other changes in the qualitative characteristics of the eddies. (Schmitz 1976 had described an early exploration of the variability along 70°W using single mooring sites, and had shown a very large increase in energy as the Gulf Stream was approached.) It was not known whether the energy changes that did occur were linked primarily with proximity to the Stream, to changes in topography, to distance to the western boundary, or to changes in atmospheric forcing, amongst other possibilities.

Eddies in Marine Science
(ed. by A.R. Robinson)
© Springer-Verlag Berlin Heidelberg 1983

To answer these and other questions, the POLYMODE program deployed three clusters A, B, C, collectively known as Array 3, as shown in Fig. 1. Small clusters of moorings were used in order to obtain rudimentary information concerning the spatial scales of the motions. Each mooring contained several instruments (current meters and temperature-pressure recorders) as tabulated by Koblinsky, Keffer and Niller (1979) and Fu and Wunsch (1979). Fu et al. (1982) describe the specific motivation behind the different choices of position. The major points were these: POLYMODE arrays 1 and 2 (see Schmitz 1978, 1980, Richman, Wunsch and Hogg 1977) had begun the exploration of the changes of mesoscale variability between the Gulf Stream and the MODE-1 area and of the region just to the east of MODE-1. Array 2 had focused specifically upon the so-called Gulf Stream recirculation region (e.g., Worthington 1976) a dynamically interesting region of high energy both in the mean and in the fluctuations (Schmitz 1980). Cluster A was intended to examine the structure of the eddy field on the western flanks of the mid-Atlantic Ridge, but well clear of the intense recirculation of the Stream. Cluster B began the exploration of the eastern basin of the North Atlantic and sought any indication of a deep eddy "western intensification" on the flanks of the Ridge. Both Clusters A and B are in regions of exceedingly rough, three-dimensional, topography. A dominant motive for the deployment of Cluster C was to examine the North Equa-

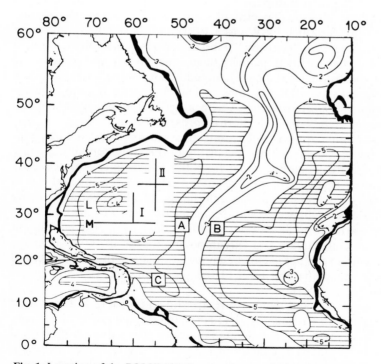

Fig. 1. Location of the POLYMODE moored arrays and general bathymetry. *L* Local Dynamics Experiment; *M* MODE; *I* and *II* Array 1 and 2; *A, B,* and *C* Array 3 Clusters A, B, and C

torial Current region. Calculations by Gill, Green and Simmons (1974) had suggested that a region like this one might be one of baroclinic instability, but less intense than in the Gulf Stream, and hence accessible by conventional mooring techniques. Cluster C was deployed over nearly flat terrain.

In addition, all three clusters were in regions where there was little extant knowledge of the large-scale time average flow field. Any such determination was to be a useful bonus. The mean field is discussed by Fu et al. (1982) and Wunsch and Grant (1982).

3.2 Observations

The basic cluster deployment period was one year; this interval sets the maximum accessible time scale and limits the statistical reliability of the results in a major way. A single mooring was deployed for a second year in Clusters A and B in order to improve the data base.

3.2.1 Spectrum and Time Scales

Figures 2 and 3 display temperature and stick velocity plots for not-untypical moorings in the three clusters. Spectra of representative moorings are displayed in Fig. 4–7 (as described by Wunsch 1981) and these tend to show the same qualitative features: An "isotropic" band at periods shorter than about 50 days in which both components of velocity have equal energy and some of the characteristics of geostrophic turbulence; an "eddy-containing" band at about 50–100 days much as in the MODE-1 area, and an "interannual band" ("secular band" in Schmitz' 1978 terminology) with a definite tendency to zonality (not readily apparent here with such short records).

The roughly -2 slope appears to be a universal feature of the records with an excess of energy above the -2 line occurring in the eddy-containing band. Additional such spectra, including some from the Pacific and Antarctic, are displayed in Wunsch (1981). It is argued there that the three spectral bands seem to be a universal phenomenon although the spectral level varies by many orders of magnitude (see Chap. 15), and the center frequency of the eddy-containing band also varies from an extreme of 10–20 days in intense currents of high energy like the Gulf Stream and Circumpolar Current, to 100 to 150 days in some of the records seen here. This behavior contrasts with that of internal waves where both the shape and level of the spectrum tend to be universal (e.g., Garrett and Munk 1979).

Time scales of velocity and temperature are shown in Table 1. These were computed as described in Richman et al. (1978) from the integral of the square autocorrelation function. A bracketed value indicates that values at individual moorings differed by as much as a factor of 2. The temperature time scale is the most consistent among the moorings in all clusters. The mid-water minimum in temperature time scale in Clusters A and B is related to the presence of Medi-

Table 1. Time scales of variability at the three clusters compared to that at MODE Center. Values were computed from integral of square of correlation function. Parenthetical values indicate that individual records differed by a factor of 2 or more from each other

Depth range	POLYMODE – 3A (28°N, 48°W)			POLYMODE – 3B (27°N, 41°W)			POLYMODE – 3C (16°N, 54°W)			MODE Center (28°N, 70'W)		
m	u	v	T	u	v	T	u	v	T			
120– 215	26	30	42	(62)	(44)	26	(34)	21	30			
230– 260									43			
320– 340							(45)	20	41			
480– 540			(41)			(62)	(51)	19	33	70	23	36
660– 850									20			
1 420–1 530	32	35	50	(52)	68	23			39	20	28	32
2 440–2 830							52ᵃ	33ᵃ	35ᵃ			
3 400–4 040	(26)	(23)	(47)	(20)	(14)	37ᵃ	(34)	29	(33)	21	26	23

ᵃ Two record average.

terranean water. In general, the velocity time scales are shorter than the temperature scales. At the upper levels, the time scales are longer in B than in A and in Cluster C there is a pronounced anisotropy of time scale. We see a general confirmation of the suggestion by Richman et al. (1978) that time scales tend to lengthen as one proceeds eastward along 28°N away from the MODE-1 region (whose time scales are also shown in Table 1). But at Cluster C, the time scales seem, if anything, shorter than in the MODE-1 area.

3.2.2 Spatial Scales

Study of the spatial correlation scales, in a similar manner, and with results described in Fu et al. 1982, suggest that in all three clusters, the horizontal scales have grown somewhat as compared to MODE-1. The lagged spatial correlations show that in Clusters A and B propagation is to the west, with B displaying a distinct southward movement in the meridional velocity as well. For meridional velocity, the estimated phase speed is about 3 km/day. In Cluster C, phase propagation is dominantly westward also at 3–5 km day^{-1}. In the MODE-1 area the propagation was southwestward at about 4 km day^{-1} (The MODE Group 1978).

In some ways, the most interesting inter-cluster differences appear in the vertical structure. We describe it here in terms of the empirical orthogonal functions (EOF's) of the vertical covariance matrix (Lumley 1970. For the uninitiated, the EOF's are related to what is known as principal component analysis in statistics. The functions are orthogonal in space and time and capture in the simplest way the structure of correlated changes between instruments). The EOF's are shown in (Fig. 8 and 9 for the vertical displacement, $\zeta \propto -T/(\partial T/\partial Z)$) and horizontal velocity, $\underline{v}$ respectively.[1] The velocity functions are drawn

---

[1] The pressure recorders showed that there was no significant mooring motion that would contaminate these results.

C. Wunsch

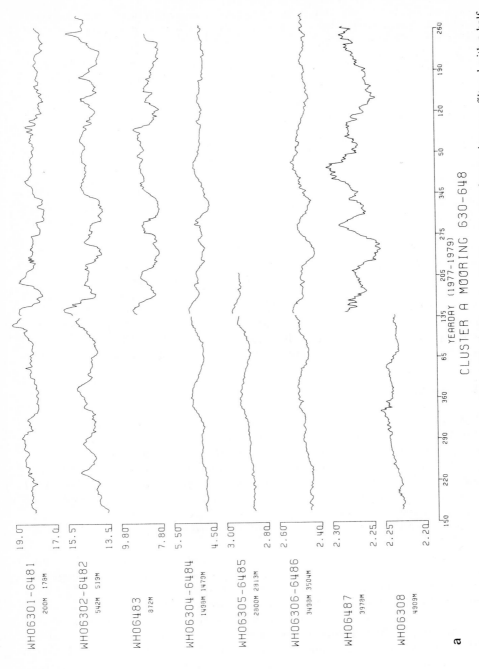

Fig. 2. Temperature as a function of time at moorings **a** 630, **b** 623, and **c** 81. The raw records were low-pass filtered with a half-power cutoff of two days for moorings 630 and 623 and four days for mooring 81, then subsampled daily

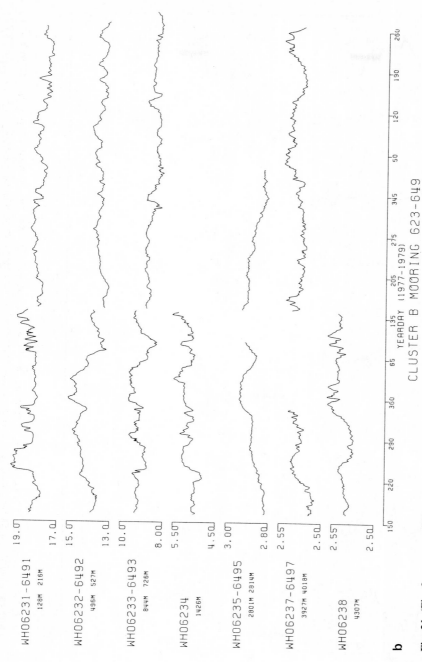

b

Fig. 2b (Fig. 2c see page 52)

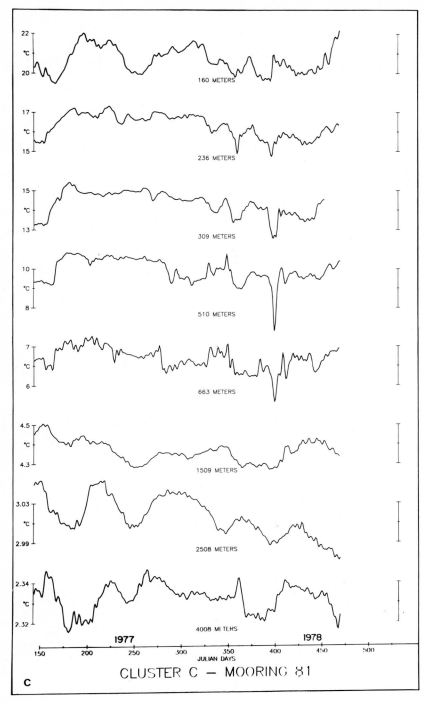

Fig. 2c

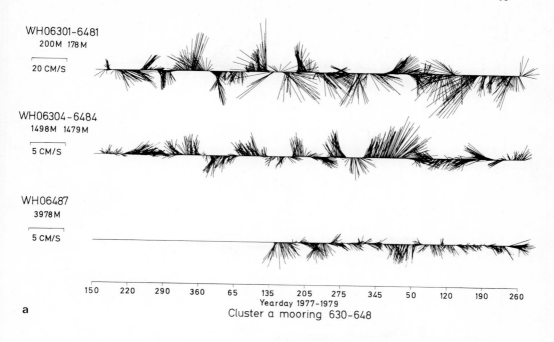

WH06301-6481
200M 178M

20 CM/S

WH06304-6484
1498M 1479M

5 CM/S

WH06487
3978M

5 CM/S

150   220   290   360   65   135   205   275   345   50   120   190   260
Yearday 1977-1979

a

Cluster a mooring 630-648

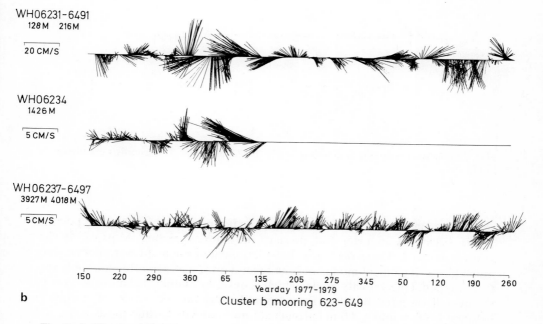

WH06231-6491
128M 216M

20 CM/S

WH06234
1426M

5 CM/S

WH06237-6497
3927M 4018M

5 CM/S

150   220   290   360   65   135   205   275   345   50   120   190   260
Yearday 1977-1979

b

Cluster b mooring 623-649

Fig. 3a, b (Fig. 3c see page 54)

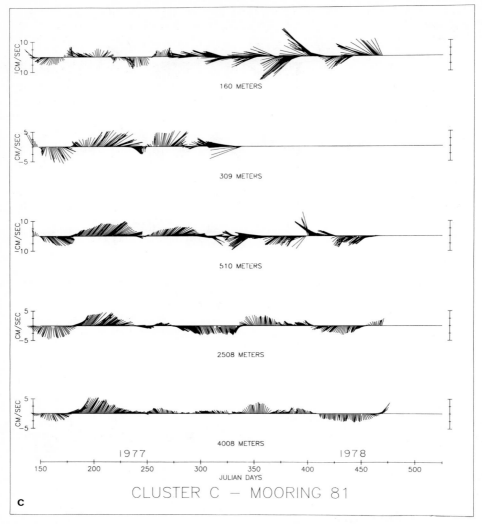

Fig. 3. Daily stick diagrams at moorings **a** 629, **b** 623, and **c** 81. The records were filtered as in Fig. 2

so that the maximum vector is unity and the bottom vector is along the positive x axis. Cluster A, mooring 630, presents the simplest picture. Most of the variance is associated with the vertical displacement of the main thermocline, very much like a dynamical mode, and the second most energetic fluctuation is surface trapped. One velocity mode, which is quite baroclinic, accounts for 96% of the overall variance, but does not account for the variance in detail at 4000 m because 1500 m and 4000 m currents are not correlated while those at the 200 m and 1500 m levels are. This cluster is much like the MODE-1 area which has often been summarized by saying that the motion is dominated by a single baroclinic mode – i.e., a bulk vertical displacement of the thermocline.

In Cluster B, both at moorings 623 and 625, 96% of the variance field of velocity is also representative of a single velocity mode which looks identical to that at Cluster A. However, to represent 98% of the variance in the vertical displacement structure requires three modes: One is bottom trapped (61%); the second most energetic mode (24%) resembles the most energetic mode in Cluster A. We also computed the displacement EOF's at the southernmost mooring of Cluster B number 627, and here two modes (64% and 30%) are required to account for 94% of the variance, with significantly different shapes than in 623. Thus, despite their close proximity, clusters A and B are in some ways quite different.

Cluster C has the most complex vertical structure. At mooring 81 (and also 79) three modes are required to represent 86% of the displacement variance (47%, 23%, 16% at mooring 81 und 65%, 21%, and 9.2% at mooring 79) although the dominant mode again closely resembles a bulk vertical displacement of the

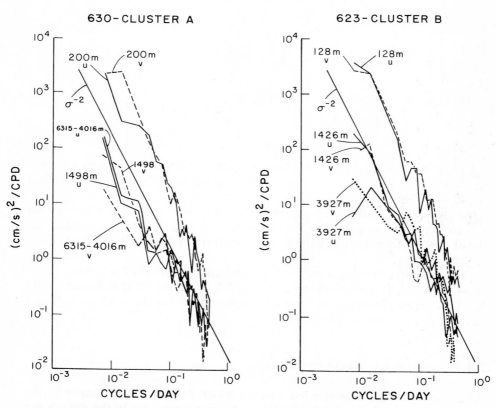

Fig. 4. Zonal (u) and meridional (v) kinetic energy density spectra in Cluster A for mooring 630

Fig. 5. Same as Fig. 4 except for mooring 623 of Cluster B

82-CLUSTER C

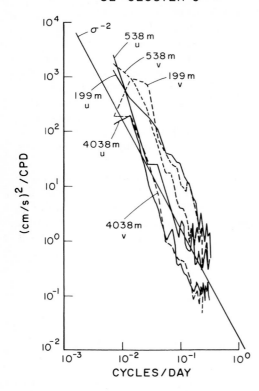

Fig. 6. Same as Fig. 4 except for mooring of Cluster C

thermocline. Two modes are required to represent 95% of the velocity variance at both 81 and 82. The first velocity modes are similar with nodes at a nominal depth of 300 m.

Recalculation of the displacement EOF's without the records where spatial inhomogeneity is suspected (1500 m record at B and the 750 m record at C), does not significantly change the vertical structure or energy distribution of the modes.

In summary, the apparent vertical coherence of energy-containing variations through the water column in Cluster A is similar to that found in the MODE area, whereas the variations in the vertical observed in Clusters B and C contain more complex features.

3.2.3 Kinetic and Potential Energy

The velocity modes display what would perhaps best be shown simply by the total energy at each depth, the suppression of kinetic energy in the deep water over the very rough topography of the mid-Atlantic ridge in Clusters A and B. Table 2 provides the eddy kinetic and potential energies (defined about the re-

Table 2. Eddy kinetic and potential energies. Second year record averages from site moorings are shown in parenthesis

Depth range m	POLYMODE -3A (28°N, 48°W)	POLYMODE -3B (27°N, 41°W)	POLYMODE -3C (16°N, 54°W)	POLY-MODE -1 (28°N, 55°W)	MODE -Center (28°N, 67.7°W)	MODE -East (28°N, 67.7°N)	Array 2 (36°N, 53.8°W)	(31.5°N, 55°W)
120– 215	54.9 (74.0)	73.2 21.8	31.7 72.6					
230– 260			40.7					
320– 340			21.3 65.5	9.0 10.0	39.5 32.0	33.0 22.5		
480– 540	35.7	35.9* 55.6	26.5 39.4				269.0	49.4
600– 850		26.5+	39.7 5.4		7.4	3.1	61.0	10.8
1420–1530	1.8	4.0 (15.2)	5.3 3.7					
2440–2830	1.0	(11.4)		0.5 0.8	8.6	7.4		
3400–4040	4.8	(0.9) (2.2)	(4.0)* 1.6*		11.0	5.2	84.2	9.8

POLYMODE -3B: * #624 only + #623 only

POLYMODE -3C: (4.0)* *with #79   1.6* *with #79   (6.6)   (2.3)

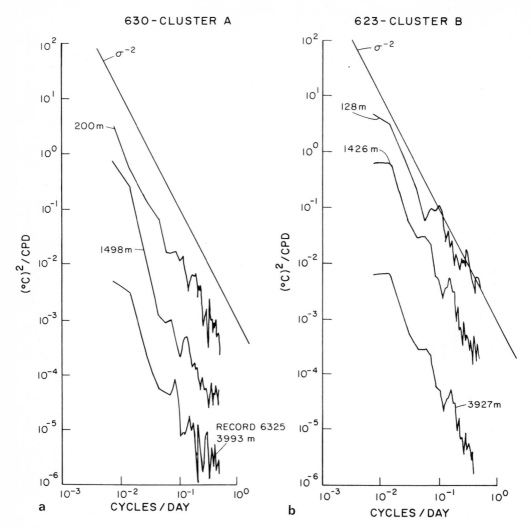

Fig. 7 a, b

cord means in the obvious way). From the moored records, the kinetic and potential eddy energy at 500 m along nominal 28°N latitude first decreases eastward from MODE Center to 55°W, then increases to Cluster A; the potential energy continues to increase across the ridge to Cluster B. Cluster C K.E.′ and P.E.′ at 500 m are of intermediate magnitude compared to the 28°N values. At 1 500 m, K.E.′ at Cluster A and B is less than at MODE Center and is comparable to MODE East; however, the apparent P.E.′ at A and B is three times larger than the value at MODE east. The Array II values reflect the very large kinetic energies at all depths approaching the Gulf Stream system (Schmitz 1978).

The low temperature variance at 1 500 m in Cluster A is consistent with the smoother low-frequency signal in the records and, because the water mass var-

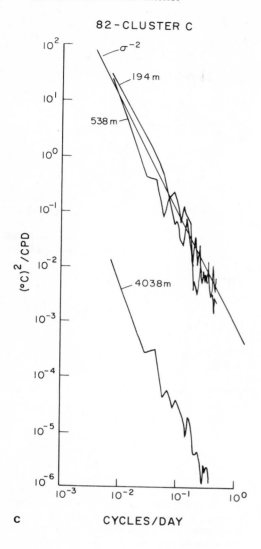

82-CLUSTER C

σ^{-2}

194 m

538 m

4038m

(°C)^2/CPD

CYCLES/DAY

c

Fig. 7 a, b, c. Temperature spectra from same moorings as in Figs. 4–6

iability is not apparent in the CTD traces there, we interpret our computed P.E.' as a good measure of the actual eddy potential energy. However, at Cluster B, all five 1 500 m records show a ragged, high-frequency temperature variability. How much of this temperature variance is due to vertical motions and hence potential energy variability, and how much is due to the horizontal advection of elements of Mediterranean Water with heterogeneous T-S relations is not clear. Note that at 1 500 m, K.E.' at Cluster B is double that of A. This increase probably reflects at least some general eddy energy increase in the upper layers on the eastern flank of the Mid-Atlantic Ridge rather than being wholly due to the K.E.' of Mediterranean water mass elements. Finally, at 4 000 m the K.E.' and P.E.' seem to be controlled both by bottom roughness and the upper

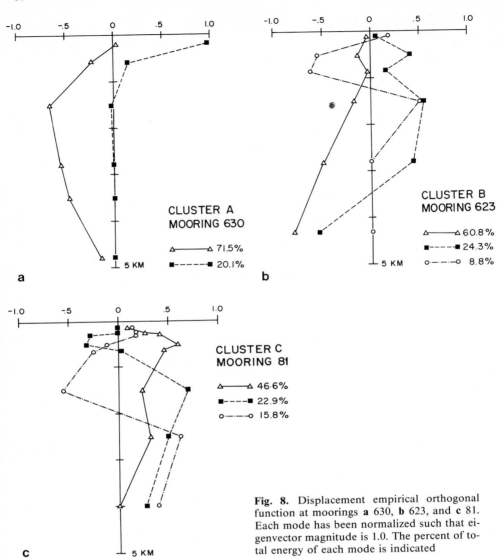

Fig. 8. Displacement empirical orthogonal function at moorings **a** 630, **b** 623, and **c** 81. Each mode has been normalized such that eigenvector magnitude is 1.0. The percent of total energy of each mode is indicated

level variability. MODE east and Cluster C are in slightly rough bottom areas and the bottom energy levels are comparable. Clusters A and B exhibit a more drastic vertical decay of energy to the rough bottom than occurs at MODE east or Cluster C. Evidence for increase of P.E.′ near the bottom is seen at Cluster A and, most dramatically of both K.E.′ and P.E.′ at mooring 79, Cluster C.

In Array 3 we obtained, for the first time in POLYMODE, long records at the top of the main thermocline, between 120–215 m. In each cluster three long current records exist and there is at most a 10% variation in K.E.′ among the separate moorings in each cluster. At all three clusters, K.E.′ decreases with in-

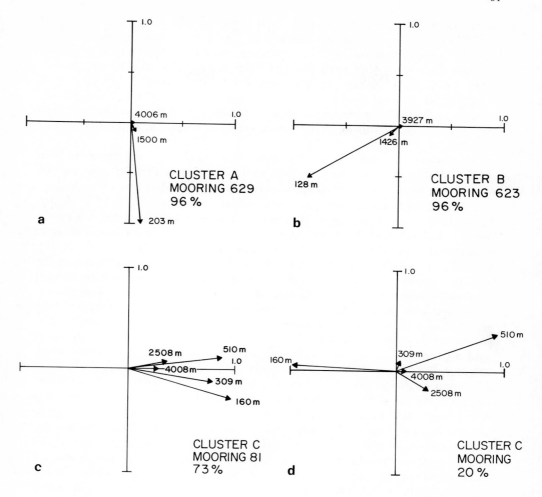

Fig. 9. The most energetic velocity (complex) empirical orthogonal modes at moorings **a** 629, **b** 623, and **c** 81. The vectors have been rotated so that the bottom vector lies along the x-axis and have been normalized so that the top vector has a magnitude of 1.0. **d** is the second mode at mooring 81

creasing depth through the main thermocline although this is less dramatic at C. At each upper level instrument in Clusters A and C, $\overline{v'^2} > \overline{u'^2}$ and in Cluster B, $\overline{u'^2} > \overline{v'^2}$. In Clusters A and C the spatial inhomogeneity of P.E.' within the cluster is larger than at B. Clusters A and C show a maximum value of P.E.' at the surface. At all moorings in C, P.E.' decreases with increasing depth and then increases again to a maximum between 320 and 540 m, but the exact verti-cal distribution is somewhat different at each mooring. Keffer (1981) computed the fluctuation potential energy distribution as a function of depth from the historical hydrographic data in the Cluster C area. The data show a similar minimum at 500 m depth as is found in the moored data in Table 2. Except for the 300–500 m level on 81, in Cluster C, $\overline{v'^2} > \overline{u'^2}$. In A, there appears to be a uniform vertical decrease of P.E.'. However, because of the vertical sampling, we do not know whether a relative maximum occurs between 500 and 200 m as there is in C. In B, the near surface value of P.E.' is less than at 500 m level. At A and C the thermal field variance in the upper portion of the main thermocline is horizontally more inhomogeneous than at B, while the thermal field variance at B is horizontally more inhomogeneous below the main thermocline than it is at A and C.

In summary, in Array 3 the mid-water K.E.' and P.E.' are comparable to central MODE regions values. There is a relative increase of K.E.' from the 500 m to 200 m level, and, over rough topography, a sharp decrease of both K.E.' and P.E.' to the ocean bottom. The horizontal inhomogeneity of the P.E.' distribution in and above the main thermocline is observed where there is a north-south polarization of the low-frequency currents; in some moorings there is bottom trapping of P.E.' and K.E.'. As noted earlier by Schmitz (1976, 1978), Richman et al. (1977) and Wunsch (1981), the gross energy variations of the mid- and upper-ocean remain consistent with a general intensification toward the Gulf Stream system, and with secondary maxima toward other boundaries. Consistent with Dantzler's (1977) potential energy diagram, there seems to be a broad minimum of kinetic and potential energy in the thermocline centered at around 28°N stretching at least from the Hatteras abyssal plain to Cluster B (and all the way to Africa according to Dantzler). There is no justification, however, for calling this region an "eddy desert", as there is only a quantitative, not a qualitative change, in energy levels over the North Atlantic. Generally speaking the differences between kinetic energy spectra of different parts of the North Atlantic are subtle. All seem to be red, with a -2 form from periods of a day or so through an eddy-containing band at 50–100 days, with a decline in energy, or at least a levelling off in the rate of rise, at longer periods. As Schmitz (1978) notes there does tend to be a shift toward higher frequencies (a "blue-shift") in the more energetic regions near the Gulf Stream recirculation, or equivalently, a "stretching-out" of the dominant time scales as one moves eastward along 28°N.

The most striking difference between the MODE area and the mid-Atlantic ridge is the loss of energy in the deep water over the rough topography. The potential energy over the entire western North Atlantic is much less variable than is the kinetic energy; Richman et al. (1977) note that this reduced variability is consistent with Schmitz' (1978) result that in the high kinetic energy regions, the motions have a much greater tendency to be barotropic.

3.3 Other Data

The POLYMODE program produced "statistical/geographical" data on the variability of the western North Atlantic in other ways as well. Ebbesmeyer and Taft (1979) have summarized the potential energy variability of the region, somewhat along the same lines as Dantzler (1977), but using hydrographic data instead. This data base permitted them to extend the analysis to greater depths than Dantzler could, and avoid bias problems intrinsic to the XBT.

The Ebbesmeyer and Taft (1979) study is confined to the region between 67° and 73°W and 23° and 35°N. Qualitatively, their results are much like Dantzler's with an increase in potential energy toward the Gulf Stream as one moves northward through the area, although the gradient is somewhat weaker than in Dantzler's study. At deeper levels, the potential energy increase is very small, consistent with the moored results of Richman et al. (1978). A considerably weaker westward energy increase is apparent in the region in the upper levels of the water column.

The other exploration tool of POLYMODE, the SOFAR floats, has provided data summarized by Riser and Rossby (1982). Their data, which is somewhat spatially inhomogeneous in its coverage, perhaps inevitable with a quasi-Lagrangian measurement method, is confined to the region between 58° and 76°W and 18° and 38°N, with most of it in the southwest triangle of the box. This data penetrates further west than any from previous velocity measurements and shows quite clearly a marked westward intensification of kinetic energy west of 70°W, in addition to the northward increase expected from the moored results. In some ways, the most marked character of these measurements is the pronounced zonal dominance in the fluctuations over most of the region (away from the immediate vicinity of the Blake-Bahama ridge) and which only becomes apparent in the moored results at very low frequencies (see discussion above). The Riser and Rossby (1982) results are in general agreement with the moored results, both those reported here, and previously (Richman et al. 1978), that the apparent eddy heat flux is quite weak, and perhaps not significant in the open western North Atlantic. (But it could still be very important in and near the Gulf Stream itself). A more extensive discussion of this data set can be found in Chapter 4.

A number of moored current meter records exists in the interior of the Atlantic in addition to those obtained in POLYMODE. Most of these records are now understood to be far too short in duration to properly define the statistics of the mesoscale variability. But within the context of the new long measurements, one can see for example, in the data of Tucholke, Wright and Hollister (1973) at about 24°N, 69°W a variability consistent with that nearby (i.e., the MODE area) albeit in some cases influenced by the Antilles Outer Ridge. Hogg and Schmitz (1980) have described 9-month long observations in the Charlie-Gibbs fracture zone at about 53°N, 35°W in rough topography somewhat similar to that in Clusters A and B. They interpret the motions as linear bottom trapped waves, reminiscent of those seen in the EOF's of Cluster B. At Bermuda, Hogg, Sanford, and Katz (1978) suggest from their 7-month-mooring deployment, that the velocity field is dominated by the passage of Gulf Stream rings. But Frankignoul's (1981) analysis of 24 years of Panulirus hydrographic data at this island seems to confirm the earlier conclusions that the spectrum is much like that found in the MODE area, although the Panulirus station is made on the south shore of the island and may to some extent be masked from interloping Gulf Stream rings.

The results, from both moored and drifting instrumentation, of the intensive Polymode Local Dynamics Experiment (LDE) are discussed in Chapter 5. There are some qualitative differences between those results and the ones described here, differences that are presumably the result of the much higher energy levels encountered in the near-Gulf Stream site. In the context of this chapter, the LDE results are dominated by proximity to a dynamical boundary, rather than being characteristic of the ocean interior.

3.4 Some Simple Conclusions

In other chapters of this volume, one will find descriptions of the eddy field in the eastern North Atlantic Ocean and elsewhere (see especially Chap. 15). The reader seriously interested in the structure of the North Atlantic eddy field should consult also the papers of Schmitz (1976, 1978, 1980), who describes in considerable detail the variability in the near-Gulf Stream region. A full discussion of his results is beyond the scope of a chapter on the ocean "interior", but a few remarks ought to be made. The most obvious large-scale inhomogeneities in the North Atlantic interior occur as one approaches and enters the Gulf Stream and its recirculation. The energy levels rise dramatically, the depth dependence of the velocity field decreases, and consistent with that, the potential energy in the variability is proportionally decreased (Richman et al. 1977). By presumption, the rather complex structure of the apparent mean flow in the near Gulf Stream is intimately tied up with the existence of this very strong variability. It is premature however, to claim that full understanding of the interactions is at hand. Chapter 17 discusses some of the results of numerical models that deal with this problem. We will thus confine ourselves to some conclusions directed at the overall, limited, goals of Array 3.

The variance charts of Dantzler (1977) had suggested a mid-ocean minimum – extending clear across to Africa at about the latitude of MODE-1 – in variability potential energy. Generally speaking, Array 3, the float, and the hydrographic data confirm this picture although the distinct minimum, at around $50°-55°$W within this zonal feature that Dantzler shows is not apparent (he had very little data there). Cluster A just to the west of the mid-Atlantic ridge has many of the characteristics of the MODE-1 region itself. As anticipated from array 1, there is a tendency for the dominant energies to occur at longer periods in the interior compared to the western and northern boundaries. Richman et al. (1977) note that the lengthened periods push the energy-dominant period toward the range where linear baroclinic waves may be applicable. Thus far, there has been no direct attempt to test linear dynamics in Clusters A and B, partly because of uncertainties about how to handle the finite amplitude three-dimensional bottom topography. The Hogg and Schmitz (1980) results suggest at least some applicability of linear dispersion relations in part of the frequency band. There is a marked suppression of eddy energy in the deep water over the very rough topography of the Mid-Atlantic ridge.

Cluster B shows some subtle, but nonetheless very real, differences from A even though the spatial separation between these clusters is only 400 km. The EOF's show some evidence of bottom trapping in the variability, and a much more pronounced effect of the Mediterranean Water superimposed upon the general decline in energy with depth.

Form observations over an abyssal plane, Cluster C has a surprisingly complex vertical structure. This array was set for the express purpose, in addition to the purely descriptive one, of determining if clear evidence for baroclinic instability of the North Equatorial Current as a generator of eddy motion could be found. Fu et al. (1982) describe the results in detail – but the answer is very simple. No clear evidence is found. The vertical phase propagation tests and the heat flux direction tests for simple baroclinic instability (Gill et al. 1974) both yield great intermooring differences, no stable result, and hence no substantial evidence for eddy generation here.

In a somewhat wider context, we have confirmed that mesoscale variability appears to be a universal feature of the North Atlantic ocean. Wherever we have seriously looked, the eddy energy exceeds that of the mean, and we see space and time scales of order 100 km and 100 days. As far as the interior western North Atlantic is concerned there are several formidable challenges that remain:

1. To understand the interaction of the ocean and its variability, with the finite amplitude, three-dimensional topography of the mid-Atlantic ridge. This interaction seems to have resulted in the "spinning-down" of the deep water, while leaving the thermocline energies largely untouched. The mechanisms and consequences of this interaction need to be understood.

2. The Gulf Stream region still remains as the only clearly plausible strong source of eddy energy. The process of radiation (whether similar to that described by Pedlosky 1977 or otherwise) needs to be understood.

3. The interaction between the weak mean flows of the ocean interior and the variability remains to be elucidated. In particular, a key question of the next few years will be to determine the role, if any, of the variability in transporting heat, momentum, mass, etc. through the interior ocean. At the present time, it appears that in the interior ocean, that the eddy contribution to these transports is small and perhaps negligible.

4. The Western North Atlantic – A Lagrangian Viewpoint

H.T. Rossby, S.C. Riser, and A.J. Mariano

4.1 Introduction

This article is an overview of what we have learned in recent years about the circulation of the western North Atlantic by direct observation with Sofar floats. These instruments, like their progenitor the Swallow float, are in essence tagged water parcels which allow us to obtain an explicit description of fluid motion over a wide range of space and time scales. On time scales of months to years the wandering of floats from one region to another yields information on the large-scale motions of water masses. In this sense floats may be viewed as a branch of modern hydrography. On time scales of days and weeks the floats are very effective tracers of mesoscale oceanic motion.

The float data provide valid descriptions of the large scale circulation, but it is recognized that the limited number of realizations to date, and particularly their nonuniform distribution, render some conclusions tentative. There is an additional uncertainty which arises from the fact that a float is an isobaric rather than isentropic device. Thus a float and a fluid parcel, together at some initial instant, will eventually move apart. In regions of moderate baroclinicity this does not appear to be a problem over times of up to several months (Riser 1982a), but the implications of this quasi-Lagrangian approximation in highly baroclinic systems such as the Gulf Stream have yet to be examined in datail. In the limit of motions at small spatial scales (such as internal waves), floats, which are about 5 m in length, clearly cannot represent particle motion.

This overview is structured in the following manner. We first summarize the studies that have been conducted and follow with a Lagrangian and an equivalent Eulerian summary of the total data base. We then discuss specific signatures and processes: mixing and dispersion, topographic influences, the evidence for discrete eddies, zonal flows in the gyre interior, and exchange processes across the Gulf Stream. We conclude with a summary of some of the most important results we feel have emerged from this program.

4.2 The Data Base

The data base is large. Between the fall of 1972 and November 1979 approximately 73 float years of data were collected. During this period the float program evolved from being a highly experimental technique into a well-estab-

Eddies in Marine Science
(ed. by A.R. Robinson)
© Springer-Verlag Berlin Heidelberg 1983

lished operational capability. Three experimental programs shaped this evolutionary process: the MODE program, the early POLYMODE experiments, and the POLYMODE Local Dynamics Experiment (LDE).

For MODE, centered at 28°N, 69°40'W, 20 floats were built and deployed at 1500 m depth. During the 4-month intensive experiment (1973), many floats were recovered and reset for technical reasons and in order to maintain a coherent array. At the end of the field experiment, the floats were left in the water and allowed to disperse, and several were trackable through 1975. Nearly all of the data was from south of 30°N and west of 68°W. Rossby et al. (1975) and Freeland et al. (1975) discuss this experiment in detail, and Dow et al. (1977) provide a complete catalogue of the data set.

During the early POLYMODE years (1975–1977) a series of exploratory experiments in different dynamical regimes and geographic locales were undertaken. These include (1) the gyre interior in the Nares Abyssal Plain area, northeast of Puerto Rico; (2) the Mediterranean eddy or "meddy" study near the Bahamas Escarpment; and (3) the waters of the South Hatteras Abyssal Plain. These data provide a description of the processes in the western North Atlantic outside the Gulf Stream recirculation region (Riser and Rossby 1983). O'Gara et al. (1982) provide a detailed data report.

The objectives of the third experiment, the POLYMODE LDE, were to examine in quantitative detail the dynamical structure of the mesoscale eddy field closer to the Gulf Stream and its recirculation in an area more energetic than that of MODE. The analyses of this program, which took place in 1978, are presently underway (see Chap. 5). Price and Rossby (1982), and Rossby et al. (1983) discuss aspects of the SOFAR float experiment in the LDE. Spain et al. (1980) have compiled a complete data report.

4.3 The Mean Field

The Lagrangian description of fluid motion seeks to determine the motional history of a water parcel and its properties such as dissolved oxygen or salinity as a function of time. It is a natural framework for examining the origin and spreading of different water masses, which are specified by their physical and chemical properties. Such is the classical technique of water mass analysis, which in recent years has gained tremendous impetus with the addition of other chemical elements and transient tracers. For the case of Sofar floats it is the float itself which is the tag or label by which we ascertain the history of different water parcels. Although the number of parcels that can be tagged is very limited, the technique has a clear advantage in that the space-time history of the parcel is well documented. We emphasize the spatial information that is presented throughout this overview, for the horizontal dimension is normally very difficult to explore.

In the past we have frequently used the spaghetti chart, which is the superposition of all float tracks without regard to time, to summarize this Lagrangian information. However, as the data base grows, these charts get very busy

and difficult to interpret. To simplify the task we shall summarize the low-frequency or mean field from both the Lagrangian and equivalent Eulerian points
of view. We hasten to add, however, that neither approach is truly rigorous. It
is our hope that the results are informative and that the concerns they raise will
stimulate more intensive research into how this Lagrangian information is best
ingested and presented.

4.3.1 The Lagrangian View

To get a clear overview of the fate of the floats over long times we have constructed two schematic spaghetti charts, one of the float tracks at 700 m (Fig. 1)
and one of all float tracks below the main thermocline (from 1300, 1500, and
2000 m; Fig. 2). We have only included those trajectories which span at least 6
months. Further, in order to emphasize the longer time scales we have subjectively filtered out the fast eddy time scales by limiting each trajectory to simple
curves with one or two inflection points. The arrows along the track are 100
days apart and are meant to indicate the time a float has spent in that portion
of the track.

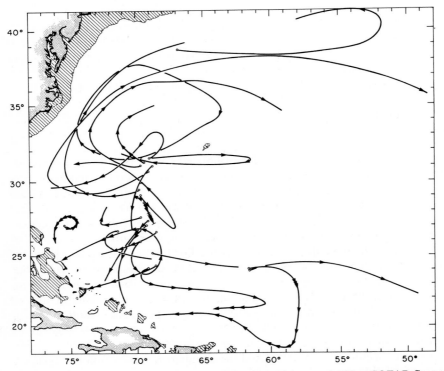

Fig. 1. A spaghetti plot of subjectively low-pass filtered (see text) 700 m SOFAR float trajectories. *Arrows* are 100 days apart. Note the high velocity of floats caught in the Gulf Stream

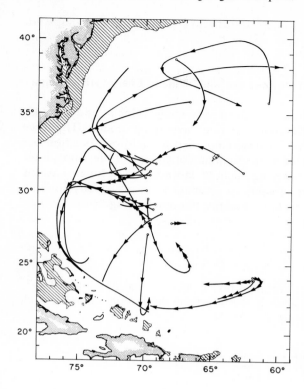

Fig. 2. As for Fig. 1, but for SO-FAR floats below the main thermocline. Note in general the slower velocity of the deep floats compared to the thermocline floats in Fig. 1 and the lack of a well-defined recirculation

At 700 m there is general motion to the west in the region immediately to the south of the Gulf Stream. Some of the floats become entrained and swept rapidly off to the east in a manner that is qualitatively reminiscent of Worthington's (1976) recirculation model. South of 25°N we find conspicuous displacement of the floats to the east, a subject we discuss below in more detail. It is noteworthy that between 26°N and 34°N along 70°W all trajectories tend to the west, yet the destinies of floats south and north of about 28°N are widely different. South of this latitude the motion is generally southwest with a tendency to curve to the southeast while floats to the north move north and west towards the Gulf Stream. Thus the latitude of about 28°N appears to the southern limit to the Gulf Stream recirculation system in the thermocline.

The deep water picture, Fig. 2, differs significantly. The entrainment of floats into the Gulf Stream is not as evident, and, as we show later, floats often cross the instantaneous path of the Gulf Stream. Unlike the situation at 700 m, the mean field of the Gulf Stream and its recirculation are weak and difficult to identify in the presence of the eddy activity. While some deep floats originating at 31°N (the LDE region) do migrate towards the northwest, others tend to the west and southwest. This southwest trend was even more evident during MODE, which was centered at 28°N. Thus the trajectories suggest a bifurcation of the mean field with floats north of approximately 31°N tending towards the Gulf Stream and floats south of there heading west and south. The Gulf

Stream recirculation in the deep water appears to be smaller and perhaps not as well defined as that of the upper ocean.

4.3.2 The Equivalent Eulerian View, and the Eddy Kinetic Energy Distribution

We have used this float data base to estimate the mean velocity and eddy kinetic energy field from an "equivalent Eulerian" averaging scheme. The three position fixes per day for each float were smoothed to yield one position per day, and velocities were computed as the time rate of change of these positions. The velocities from all floats entering each 1° latitude by 1° longitude box were then averaged, regardless of time, to estimate a mean velocity for that box. The resulting mean field was then spatially smoothed using a boxcar filter, that is,

$$\tilde{U}_{i,j} = \tfrac{1}{16}[4\,U_{i,j} + 2\,\{U_{i+1,j} + U_{i-1,j} + U_{i,j+1} + U_{i,j-1}\}$$
$$+ U_{i+1,j+1} + U_{i+1,j-1} + U_{i-1,j+1} + U_{i-1,j-1}]$$

where $U_{i,j}$ is the mean velocity vector in box i,j before smoothing and $\tilde{U}_{i,j}$ is the smoothed value in the box. Thus in the following discussion a number such as 25°30'N refers to the area from 25°–26°N.

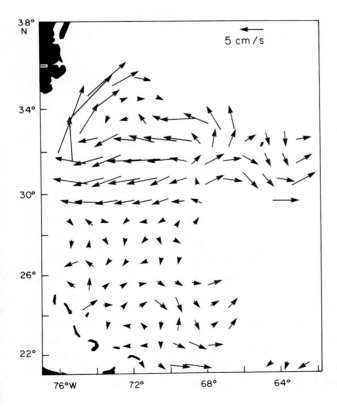

Fig. 3. The mean equivalent Eulerian circulation estimated from 700 m SOFAR float data

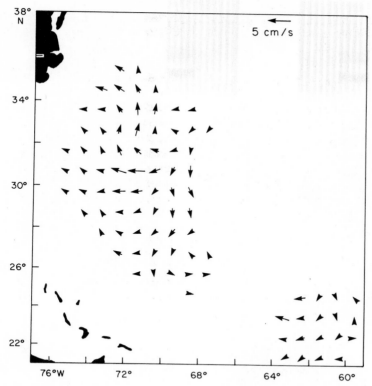

Fig. 4. The mean equivalent Eulerian circulation estimated from 1 300, 1 500, and 2 000 m SO-FAR float data

Assuming that there are a total of N velocity measurements from all floats in a given box, the eddy kinetic energy per unit mass, K_E, for the box i,j becomes

$$K_{Ei,j} = \frac{1}{2N} \sum_k |\underline{U}_{ki,j} - \tilde{\underline{U}}_{i,j}|^2$$

where the sum is taken over all N velocity measurements in the box. The K_E field computed by this method was then smoothed, using the boxcar scheme given above. In some boxes, especially in the Gulf Stream and along the perimeter of the sampled area, there are only a limited number of velocity observations and hence we cannot claim a high degree of statistical confidence in the K_E estimates.

The mean velocity fields at 700 m and below the thermocline are shown in Fig. 3 and 4. Though the data base used is somewhat different, these figures agree well with the results given by Riser and Rossby 1983), as well as with the strictly Eulerian estimates of Schmitz (1978). There are caveats, however. The striking divergence of the mean velocity field at 700 m near 31°30'N, 69°30'W is due to disparate advective states of the local eddy field at the time of the two

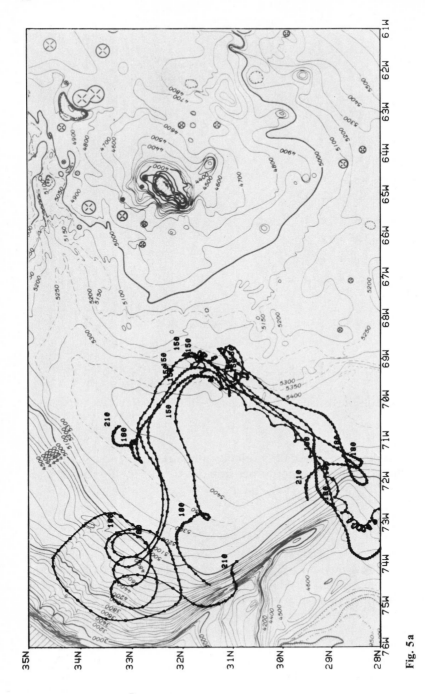

Fig. 5a

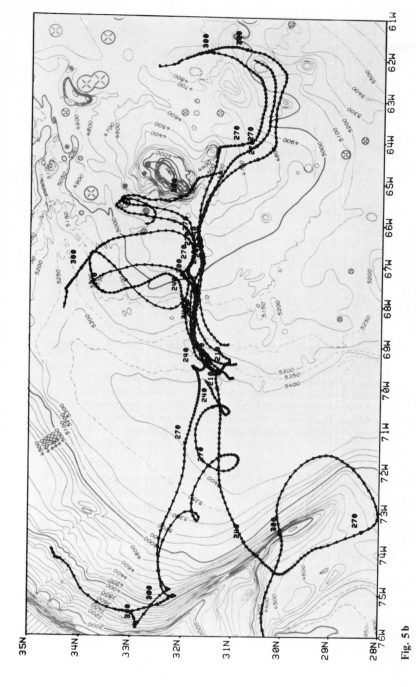

Fig. 5 b

Fig. 5. a A group of 700 m SOFAR floats launched in May 1978. Note how these drift rapidly to the west. Time is in year days.
b A group of 700 m SOFAR floats launched in July 1978, two months later. These drift to the east instead

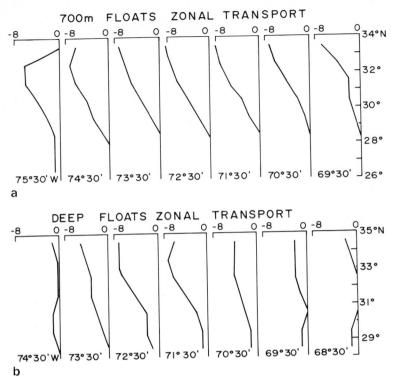

Fig. 6. a The cumulative zonal transport (in 10^6 m^3 s^{-1}) estimated for the 700 m floats from 26°N to the given latitude for each given longitude. A layer depth of 250 m and the average velocities of Fig. 3 were assumed. **b** The cumulative zonal transport estimated from SOFAR floats below the main thermocline from 28°N to the given latitude for each given longitude. A layer depth of 600 m and the average velocities of Fig. 4 were assumed

LDE float cluster settings, the first one in May and the second one in July 1978 (Fig. 5). Another difficulty associated with the inference of a Eulerian mean flow from float velocities is that strong K_E gradients, which imply eddy diffusivity gradients, may bias the mean flow calculation, since the center-of-mass of a float cluster will move up a gradient of eddy diffusivity. This bias was first derived in Freeland et al. (1975) and is discussed in more detail in the next section.

Despite these difficulties, Fig. 3 shows good agreement with the classical picture of the mean circulation of the upper thermocline waters in the western North Atlantic. A strong anticyclonic circulation gyre with its southern boundary at approximately 29°N as hypothesized by Worthington (1976) is clearly visible. The Gulf Stream recirculation region shown here is consistent with the dynamic height calculations of Ebbesmeyer and Taft (1979). The transition between the Gulf Stream recirculation and the eastward flow to the south is also evident. The cumulative zonal transports estimated from the float data for the region bounded by 26°30'N–33°30'N for each degree longitude between 69°W

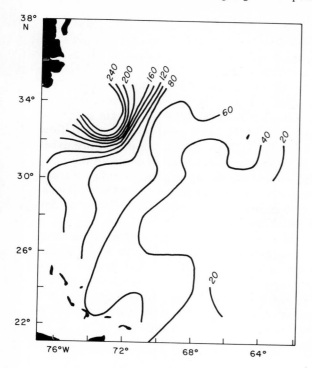

Fig. 7. The eddy kinetic energy in cm^2 s^{-2} estimated from 700 m SOFAR float data. Note in general the southeast to northwest increase and the high gradient region of the field north of 30°N

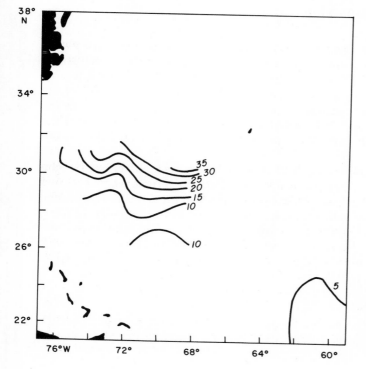

Fig. 8. The eddy kinetic energy in cm^2 s^{-2} estimated from SOFAR floats below the main thermocline. Note the lower values relative to Fig. 7

and 76°W as a function of latitude north of 26°N are shown in Fig. 6a. They are qualitatively consistent with Worthington's (1976) model of the upper thermocline layer. Note that the transport integral decreases north of 33° as one approaches the Gulf Stream along 75°W.

The estimated mean velocity field from the deep floats, shown in Fig. 4, exhibits four distinct regions: (1) a slow (<1 cm s^{-1}) westward flow centered at 23°N, 63°W; (2) a southerly flow centered at 28°N, 70°W of 1–2 cm s^{-1}; (3) a stronger flow of 3–4 cm s^{-1} centered at 30°N, 72°W, and (4) a northerly flow of comparable strength centered at 33°N, 72°W. Again, the possibility exists that some of the structure near 31°N may in part reflect changes in the eddy field between the two LDE settings a few months apart. The cumulative zonal deep transport north of 28°N, shown in Fig. 6b, is to the west and is somewhat less than but still comparable to that given by Worthington's (1976) model. There is little transport south of 30°N and east of 70°W.

The shallow K_E field shown in Fig. 7 exhibits a general increase from southeast to northwest with a strong gradient region just north of 31°N and west of 71°W. The strong gradient in the vicinity of the Gulf Stream return flow is also a striking feature of eddy-resolving gyre-scale models (see Rhines 1977). The deep K_E field shown in Fig. 8 exhibits a steady increase to the north at a given longitude; again, some of the detailed structure in the contours is probably associated with the space-time distribution of the data.

4.4 Mixing and Dispersion

In a seminal work, G. I. Taylor (1921) provided a direct relationship between the life history of a collection of fluid parcels and the effective eddy diffusivity of the motions of the fluid. For a better understanding of large-scale oceanic mixing and for developing better parameterizations of eddies in models of the general circulation, a knowledge of the effective diffusivity of eddies in the ocean is necessary. In addition, many chemical and biological studies of the ocean ultimately depend on some knowledge of bulk diffusivity due to eddies.

Consider a collection of parcels of fluid, in which each parcel has Lagrangian horizontal velocity components $u_i(t)$ as a function of time, where $i = 1$ or 2 for zonal or meridional velocity. It is assumed that the velocities $u_i(t)$ represent the residual fluid parcel velocities after the regional ensemble mean velocity, such as estimated above, has been removed. It is of some interest to know how the parcel motion at some time t is related to the motion at some later time $t + \tau$. As a useful measure of this, we can compute the Lagrangian autocorrelation function R_i (Taylor 1921) as

$$R_i(\tau) = \langle u_i(t) u_i(t + \tau) \rangle / \langle u_i^2 \rangle ,$$

where the brackets $\langle \rangle$ denote that in computing R_i we have averaged over the ensemble of marked parcels. It is assumed here that the statistics of the motion are stationary in time.

Essentially the function $R_i(\tau)$ is a measure of how long parcels of fluid remember their previous states. For very small τ, R_i has a value of very nearly 1; for very large τ, it is expected that R_i approaches zero. Taylor noted that there is an intimate connection between R_i and the mean-square dispersion $\langle x_i^2 \rangle$ of parcels from their release point,

$$\langle x_i^2 \rangle = 2 \langle u_i^2 \rangle \int_0^t dT \int_0^T R_i(\tau) d\tau \,,$$

where T is the time over which the fluid is observed. For times T long enough that the second integral has reached a steady value, Taylor showed that

$$(1/2) \, d/dt \langle x_i^2 \rangle = \langle u_i^2 \rangle I_i = \kappa_i \,,$$

where $I_i = \int_0^T R_i(\tau) d\tau$ is the Lagrangian integral time scale and κ_i is the eddy diffusivity along the direction given by the subscript. This remarkably simple relationship allows the computation of the diffusivities from the statistics of a collection of particle trajectories. When there exists a cluster of instruments, as with the LDE, the diffusivity κ_i may be computed directly from a knowledge of the dispersion $\langle x_i^2 \rangle$ of the cluster. In other cases, when the data consists of repeated settings of individual instruments in a region over a period of time (sometimes years), κ_i can be estimated from a knowledge of $\langle u_i^2 \rangle$ for the region (computed from the ensemble) and the integral time scale I_i computed from an estimate of $R_i(\tau)$ for the ensemble. In computing κ_i estimates from the total data set it has been necessary to use a combination of these approaches.

The values of κ_i for different regions and depths vary by only an order of magnitude, and all are consistent to within an order of magnitude with the horizontal eddy diffusivities computed some 30 years ago by Riley (1951). Yet what is most striking about the computed eddy diffusivities is their relationship to the local eddy kinetic energy level, as shown in Fig. 9. For most of the experiments in the western North Atlantic, at a variety of depths, the estimated diffusivity appears to be nearly linearly related to the eddy kinetic energy level (Price 1982), suggesting that the integral scale I_i has similar values over the explored portions of the western basin. The physical basis for this result is not yet clear; further studies must address this and related problems. It is results such as this, however, which will eventually lead to better parameterizations of mesoscale eddies in models of ocean circulation and mixing.

Spatially varying diffusivities, such as those shown in Fig. 9, can lead to somewhat unexpected effects in the Lagrangian frame. For example, it has been suggested (Freeland et al. 1975) that some caution is required in comparing Lagrangian statistics to their equivalent Eulerian counterparts, since a cluster of fluid parcels will always appear to diffuse up a gradient of eddy diffusivity, giving a component to the Lagrangian mean flow in that direction. It is possible to derive the result that

$$\hat{u}_i = \frac{d}{dt} \langle x_i \rangle = \overline{\nabla \kappa_i} \,,$$

where $\hat{u}_i$ is the induced mean velocity of the center of mass of the cluster and $\overline{\nabla \kappa_i}$ represents the mean Eulerian eddy diffusivity gradient. If we assume that the values in Fig. 9 give a good representation of the north-south gradient of

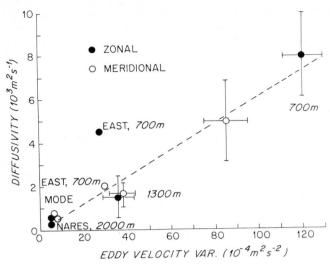

Fig. 9. Diffusivity κ_i as a function of the velocity variance, $\langle u_i^2 \rangle$. The MODE data points are from Freeland et al. (1975). The points labelled *East* are from the 700 m trajectories shown in Fig. 13. This figure is originally from Price (1982), with the East and Nares diffusivities computed in Riser and Rossby (1983) added here

eddy diffusivity along 70°W from 22°N to 32°N, and that this diffusivity varies by a maximum amount of 10^8 cm^2 s^{-1} over this interval, then we estimate that $|\hat{u}| \sim 1$ cm s^{-1}. As many of the equivalent Eulerian mean velocities estimated from the float data in Section 4.3.B above and shown in Fig. 3 and 4 are not appreciably larger than this value, some care is required in the interpretation of float data presented in this manner. For large values of the computed mean velocities, however, as are found farther west and north, this apparent induced mean velocity does not appear to be a problem.

4.5 Topographic Influences

If the shape of the bottom of the ocean influences the motion of water parcels, then it may also influence stirring and mixing of physical, chemical and biological properties in the ocean. The areal and depth dependence of such an influence on fluid motion thus needs to be better understood. In order to examine how topography might affect fluid motion, it is useful to begin with the concept of the conservation of potential vorticity, which can be derived from the Navier-Stokes equations. For an homogeneous ocean, this is

$$d/dt[(\zeta + f)/H] = 0 ,$$

where f is the Coriolis parameter, which varies with latitude; ζ is the relative vorticity; and H is the depth of the ocean. Simply stated, this implies that when following the motion of a marked parcel of fluid, the bracketed quantity is conserved. As an application of this concept, generally it is assumed that the mean circulation of the ocean abyss occurs along contours of f/H, though this need not be true for eddy-type motions where ζ may be large.

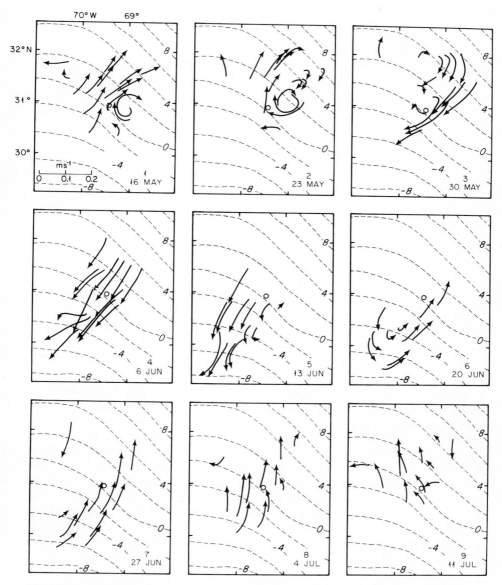

Fig. 10. Week-long segments of 1300 m float trajectories from the LDE superimposed over f/H contours. *Open circle* at 31°N, 69.5°W marks the location of the LDE central mooring. The panel number and the beginning date are noted at *lower right of each panel*

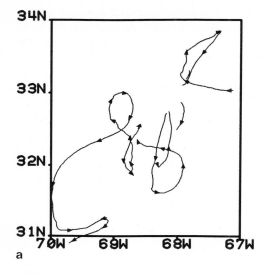

a

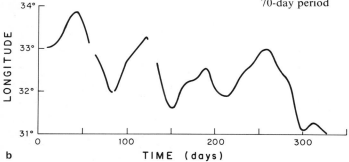

b

Fig. 11. a Float #1 of the pre-LDE group shown here from February 1976 at about 33°N, 67°W to December 1976 when it left this area. Note the similar oscillatory motion with that of the floats from Fig. 10. Gaps in the trajectory are due to tracking difficulties. **b** Latitude plotted as a function of time for the float shown in Fig. 11a. Note the long duration of this oscillation of approximately 70-day period

In this section we present some observational evidence, based on float measurements, for several different types of topographic influences on mesoscale variability in the western North Atlantic. The best documented case comes from the POLYMODE LDE (see also Chap. 5) when a group of 18 floats was set at 1300 m with an initial spacing between instruments of about 30 km, in the vicinity of 31°N 69°W. Twelve of the floats moved as a remarkably coherent cluster for two cycles of a wavelike oscillation having a period of about two months. This motion has been analyzed in detail by Price and Rossby (1982), who show that the Lagrangian vorticity balance computed from the float cluster is consistent with a topographically modified planetary wave solution to the Navier-Stokes equations for an homogeneous ocean. The wave which was most consistent with the observations had a period of 61 days, a wavelength of 340 km, and, perhaps most importantly, a group velocity from the west, suggesting the possibility of the western boundary region as a generation region. The floats did not follow contours of f/H and indeed often appeared to move normal to these contours (Fig. 10). It is useful to note the simi-

larity between the POLYMODE observations and the trajectory of a float
(POLYMODE float number 1) from several years earlier, shown in Fig. 11 a. Over
much of its path the motion of Float 1 was of a simple translational character
to the (south)west. However, between 34°N and 31°30'N, and 67°W and 69°W
the motion was of a much more oscillatory nature with Lagrangian periods of
40–80 days (Fig. 11 b). This is also essentially the same region where the Aries
eddies were observed 16 years earlier (Swallow 1971). Thus there are clear similarities between the LDE floats, Float 1, and the Aries observations, which
Rhines (1971 b) had previously suggested were strongly topographically related. This type of topographic influence can also apparently extend into the
main thermocline. For example, Riser (1982 b) has analyzed POLYMODE trajectory data from 700 m near the Blake-Bahama Outer Ridge, and finds topographically controlled wavelike motions not unlike those at 1300 m from the
LDE.

Bottom topography appears to influence parcel motion in other ways as
well. For example, during the post-MODE period, two floats at 1800 m depth
and some 2500 m above the bottom began a coherent, anticlockwise motion as
they moved westward over the Blake-Bahama Outer Ridge near 29°30'N. This

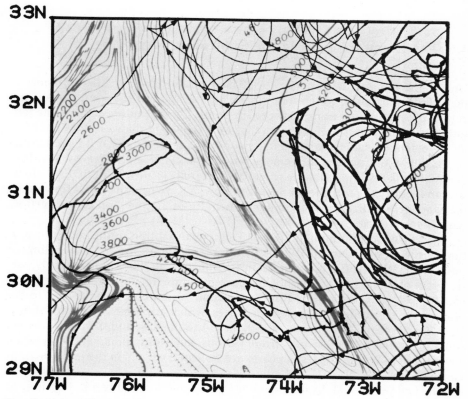

Fig. 12. Float trajectories near the Blake-Bahama Outer Ridge. Note that, while floats come
near to the steep flanks of the ridge, they rarely cross over the bathymetric contours

cyclonic motion persisted for nearly 5 months as the floats continued to drift west until they reached the Blake Escarpment. Due to the lack of any associated measurements of currents or hydrography, it is impossible to know whether such strong eddy-like particle motion resulted from the fluid moving directly across the top of the ridge or whether the floats were simply entrained into a previously existing feature (Riser et al. 1978). In general, however, from many tracks it has become clear that the Blake-Bahama Outer Ridge has a strong influence on fluid motion, as is shown in Fig. 12. The steep flank on the north side appears to be especially important in determining the character of the nearby flow. Note the very simple translation of the deep floats along contours of constant f/H. The paucity of float tracks which cross the ridge at both 700 and 2000 m is readily apparent.

Finally, we note that Richardson (1982) has reported observations from surface drifters near the Corner Rise Seamounts in the western North Atlantic which indicate that the influence of topography can extend even to the surface waters, suggesting that it is likely that fluid parcels throughout the water column are sometimes subject to topographic effects; to the extent that bottom topography introduces a net motion of fluid parcels, it may play an important role in stirring and mixing in the ocean.

4.6 The Eastward Flow in the Subtropical Gyre

One of the dynamical cornerstones of large-scale ocean circulation theory is the Sverdrup balance, which states that where the curl of the wind stress is negative, the meridional transport is to the south. Furthermore, south of the latitude where the curl is an extremum the zonal flow is to the west. While the latter statement depends upon boundary condition assumptions at the meridional walls, this is the picture that emerges from the classical studies by Sverdrup (1947), Munk (1950), and more recently by Veronis (1973). With SOFAR floats we can identify the meridional Sverdrup flow, but we find the zonal flow in the thermocline south of the extremum often to be in the opposite direction, namely to the east.

As part of several pilot studies of Lagrangian motion in the southern portion of the western North Atlantic (south of Bermuda), we have set a limited number of SOFAR floats in the main thermocline at 700 m. Surprisingly, virtually all observations of thermocline currents in the region 20°–30° north and east of 70°W suggest a weak flow to the south and a conspicuous flow to the east, as shown in Fig. 13. These trajectories, which are summarized in Table 1, yield ensemble mean velocities of 5.3 cm s^{-1} to the east and 0.6 cm s^{-1} to the south. Current meter records from various moorings along 28°N between 70°W and 55°W also suggest an eastward flow of 2 cm s^{-1} in the main thermocline (Richman et al. 1977, Schmitz 1978).

The net meridional displacement of each of the floats shown in Fig. 13 is to the south; moreover, the ensemble mean southerly velocity of 0.6 cm s^{-1} of these floats may be a direct manifestation of the Sverdrup circulation. Using a

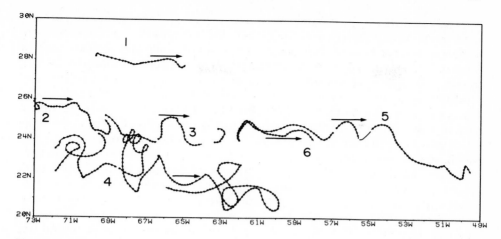

Fig. 13. A group of SOFAR floats in the main thermocline south of Bermuda over a number of years that show a net displacement to the east. See Table 1 for a summary of trajectories

wind stress curl of -1.3×10^{-8} dynes cm^{-2} (estimated from Fig. 6 of Leetmaa et al. (1978), which is also consistent with Evenson et al. (1975)), and a value for the planetary beta effect of 2×10^{-13} cm^{-1}s^{-1}, then the meridional (southward) Sverdrup transport per unit length is simply the quotient of these quantities, 7×10^4 cm^{2}s^{-1}. If we assume that this is distributed uniformly over the upper kilometer of the ocean, then this transport corresponds to a southward velocity at any level in this depth range of 0.7 cm s^{-1}. The remarkable numerical agreement may be fortuitous, but the uniformly southward displacement of all the floats lends some confidence to this result. In contrast, a float at 2 000 m depth which was tracked for over 1 000 days in the same region showed a mean southerly velocity of only 0.25 cm s^{-1}, suggesting that the meridional flow is substantially baroclinic (that is, limited to the upper ocean).

Table 1. A summary of thermocline floats with a net eastward displacement

Example number	Float number	Depth (m)	Duration (days)	Net east displacement (km)	Net north displacement (km)	$\bar{u}$	$\bar{v}$
						(cm s^{-1})	
1	4M	900	30	200	− 10	7.7	− 0.4
2	22P	700	78	400	− 66	5.9	− 1.0
3	13P	700	180	600	− 66	3.9	− 0.4
4	11P	700	615	600	− 242	1.1	− 0.5
5	6P	700	235	1 300	− 170	6.4	− 0.8
6	9P	700	70	400	− 50	6.6	− 0.8

$\langle u \rangle = 5.3$ cm s^{-1}
$\langle v \rangle = -0.6$ cm s^{-1}
M: MODE. P: Pre-LDE POLYMODE. The overbar denotes the mean over the trajectory. Brackets denote the ensemble mean.

The zonal flow is an enigma. With over 3 years of float data, and perhaps even more current meter data along 28°N, the quantity of direct observations of currents is not insignificant, and, as noted above, indicates the existence of an eastward mean flow in the thermocline. Moreover, there is extensive hydrographic evidence (indirect observations of currents) for an eastward flow, as Reid (1978) discusses in some detail. From a careful examination of the available density data, he shows that there should be an eastward flow in the upper waters south of 30°N in the western North Atlantic. For example, between 31°N and 24°N the surface geopotential anomaly increases from 1.5 to 1.7 dynamic meters, which corresponds to a surface geostrophic velocity of 4 cm s^{-1}. If such a speed were characteristic of the upper 1000 m, a transport of the order of 25×10^6 m^{3}s^{-1} would result – a number comparable to the basic meridional circulation. Reid (1978) emphasized how this zonal flow appears to be a continuation of the westward flow south of the Gulf Stream after it turns south and east near 30°N 75°W. Such a flow pattern can also be found in several barotropic numerical models, such as Bryan (1963) and Veronis (1966), suggesting the possibility that it may result from the inherent nonlinear properties of the Navier-Stokes equations.

Another notable aspect of this eastward flow is that it substantially overlaps the Mediterranean salt tongue, which is usually thought of as being maintained by advective processes which transport waters of high salinity from the eastern Atlantic to the west (Richardson and Mooney 1975). If the eastward flow discussed above is indeed real, then it appears that large-scale advection cannot be the only mechanism by which salt is carried west. We offer as another possibility the transport by small discrete eddies which actively translate westward through the surrounding waters, which will be discussed in the following section.

4.7 Discrete Eddies

One of the most striking phenomena to come to light in the POLYMODE program is the widespread existence of a class of apparently very stable, isolated, baroclinic eddies. Occurring over a wide range of scales, they have been found in and above the main thermocline as well as in the deeper water.

The first of these features to be documented, both dynamically and hydrographically was the Mediterranean eddy ("meddy") discovered in the extreme western Atlantic near the Bahamas Bank (McDowell and Rossby 1978). The eddy was lens-like (confined both horizontally and vertically) and had a positive salinity anomaly of 0.46 parts per thousand (about 17 standard deviations above the mean salinity of the surrounding waters). It was clear from the salinity that this water originated in the eastern Atlantic, perhaps from near 22°W. If the eddy did originate a distance of some 4000 km from where it was observed it may have been at least 3 years old, assuming a westward translation rate of 3 cm s^{-1}.

Other isolated eddies of a seemingly similar nature, though having varying property anomalies, have been observed in recent years in the western North Atlantic (see Chap. 6, for example). In addition, a number of float trajectories (Fig. 14) have suggested the presence of persistent, isolated eddies, though most of these examples have no accompanying hydrographic data. A summary of the eddy properties inferred from these float trajectories is given in Table 2. What is especially striking in these trajectories is the apparent westward drift of the eddies; it is only those examples close to the Gulf Stream or the continental margin which behave differently.

The orbital periods given in Table 2 may in many cases actually be an underestimate of the maximum rotation rate, which is presumably at the eddy center. Thus some of these features may have very high orbital velocities near their center. This strength and the small size of these features may give us a clue as to why they are long-lived: if the horizontal shear of the background flow were strong enough, a large eddy could be stretched apart in a cascade to smaller and smaller scales. Thus the size and ultimate fate of an eddy may not be governed as much by its internal properties as by the background mesoscale

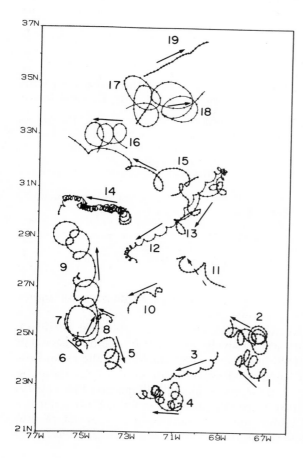

Fig. 14. All the observed trajectories of floats presumed to be caught in isolated, discrete eddies. Except for eddies near the western boundary and Gulf Stream, note that all are apparently displaced to the west. See Table 2 for a summary

Table 2. A summary of float observations near suspected isolated eddies

Example number	Float number	Depth (m)	Duration (days)	Rotation period (days)	Eastward translation rate (cm s^{-1})
1	11P	700	80	-20	-1.8
2	29P	2000	188	20	-0.8
3	12P	700	31	-7	-8.0
4	14P	700			
5	12P	700	45	-20	0.3
6	27P	2000	13	-6	3.0
7	59L	700	75	21	0.0
8	21P	700	13	9	-7.0
9	21P	700	69	4	0.0
10	22P	700	40	14	-5.0
11	30P	700	—	24	-1.0
12	52L	700	75	-3.5	-4.7
13	58L	1500	86	-15	-2.9
14	39M	1500	150	4	-2.1
15	84L	700	75	20	-8.2
16	61L	700	27	-10	-6.8
17	26P	700	41	18	0.0
18	55L	700	33	12	0.0
19	54L	700	33	-7	10.5

M: MODE. P: Pre-LDE POLYMODE. L: LDE.
Total: 14 shallow eddies; 4 deep eddies; 8 anticyclonic ($-$) eddies; 10 cyclonic ($+$) eddies.
Remarks:

3, 4   The Mediterranean eddy (McDowell and Rossby 1978; McDowell 1982); the water properties of this eddy suggest it came from the region 32°N 23°W (McDowell 1982).

11   Small intense eddy documented by Soviet POLYMODE XBT surveys. The eddy is moving north-northwest while the float passes it in the opposite direction.

12   Small eddy, possibly from the low oxygen tongue west of northwest Africa.

13   Deep anticyclonic eddy, probably formed near the Gulf Stream. Numbers 12 and 13 were the subject of thorough studies in the LDE and will be reported on in detail at a future time.

14   Only discrete eddy observed in MODE. It may have been formed by vortex stretching over the Blake-Bahama Outer Ridge (Riser et al. 1978).

17, 18 An intriguing aspect of these are the filaments of Gulf Stream water terminating over these eddies with cyclonic cusps as shown in Fig. 15. However, there are no reports of cold-core rings that might correspond to these cyclonic features.

19   A weak, anticyclonic motion advected along the edge of the Gulf Stream.

activity. This may help to explain why no very large Mediterranean eddies have been observed: they suffer a violent death!

Granted that the majority of these eddies are indeed stable and longlived, how are they propelled? If they are passively advected, it would appear that they are going in the wrong direction since we have shown that the mean flow at the level of the shallow eddies in the southeast corner (eddies 1, 2, 3) is to the east. If they are actively driven as in the model of McWilliams and Flierl (1979) then their speed must significantly exceed the local mean field. If the latter is 3 cm s^{-1} to the east, then the eddies must be translating at a speed of 0(6) cm s^{-1}. This is greatly in excess of the fastest baroclinic planetary wave, 0(3)

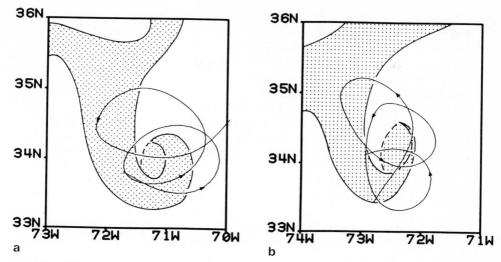

Fig. 15. a The trajectory of Float 54 during June 1978. *Hatched area* is a filament of Gulf Stream water which was shown in the monthly summary "Gulf Stream" for June 1978, published by the U.S. Department of Commerce. **b** The trajectory of Float 55 during May 1979. *Hatched area* is a filament of Gulf Stream water which was shown in the monthly summary "Gulf Stream" for May 1979. Compare this to the trajectory shown in **a** which is one year earlier

cm s^{-1}, which we shall assume is a measure of how fast these lenses can translate (some evidence for this limiting velocity is given by McWilliams and Flierl (1979).

A possible resolution to this observational dilemma may be found by noting that at the depths of the shallow eddies the effective potential vorticity gradient is significantly greater than the planetary beta-effect (that is, the change in the Earth's rotation that a particle feels as it moves north) due to the northward pinching of the isopycnal surfaces. McDowell has estimated that the ∇(f/h) field (where now h denotes the thickness of the layer of water from 12°–17°C) in the lower part of the North Atlantic thermocline is at least a factor two greater than the beta effect (Rhines and Young 1982). Since the scale of this field is much larger than that of the eddies, to them it appears as a mean field. Thus the effective speed of the shallow lenses might be doubled. This problem does not exist in the deep waters since the mean flow there is of 0(.3) cm s^{-1} to the west.

The observation of an eddy in the lower thermocline with a very strong Mediterranean salt anomaly suggests the possibility that such features may play an important transport role (Rossby 1982). The approximate coincidence of the ∇(f/h) maximum with the westwardly spreading salt anomaly suggests that it may be a preferential path for the westward motion of such features.

4.8 The Path of the Gulf Stream

Between the Straits of Florida and the region south of Nova Scotia the Gulf Stream with its core of Gulf of Mexico water is thought to increase in transport from 30 to as much as $150 \times 10^6 \, \mathrm{m^3 s^{-1}}$ (Knauss 1969). According to Worthington (1976), and as discussed earlier, this increase is largely due to entrainment of waters to the south of the Gulf Stream, which are supplied by a tight recirculation system between 45° and 75°W. This concept of advection of the warm Gulf Stream waters is well known from the literature and is substantiated by many satellite infrared photographs, which clearly show the core of warm water extending far to the east of Cape Hatteras. Vertical coherence of the Gulf Stream is presumed from the alignment and similar slope of the density field throughout the water column. Thus, as additional waters are entrained from the Sargasso Sea at all depths, they are added to and embed the waters from farther south.

It is against this backdrop of a Stream entraining waters into a continuously traceable baroclinic structure that the properties of a number of Sofar float trajectories in and near the Gulf Stream become of particular interest, as shown in Figs. 16 and 17. We see the entrainment of 700 m floats along the bathymetric contours of the Blake Spur and the Blake-Bahama Outer Ridge, whereas between 33° and 36°N floats can make frequent contact with the Gulf Stream without getting entrained (Fig. 16). Once floats reach 71°W, they appear to stay in the Gulf Stream, although in one case at 38°N, 70°W a 700 m float crossed into the slope water region for 50 days before becoming re-entrained. At 1300 m the Gulf Stream path is not as evident, as discussed earlier, and east of 70°W floats appear to have complete freedom in crossing the path of the Stream. Although the constant pressure properties of the floats make the interpretation of these crossings difficult, it seems to us that these paths may reflect Lagrangian cross-stream motion (Rossby 1982).

4.9 Concluding Remarks

In this overview we have sought to convey the richness and variety of information that has been obtained with SOFAR floats. Further, we believe a major strength of the Lagrangian method in oceanic studies has been the diversity of horizontal information which is obtained. This dimension appears naturally in all discussions above, whether it be the horizontal circulation, horizontal diffusivity, the role of bottom topography, eddies, or cross-frontal mixing and transport processes.

Of course, this horizontal information is not pure, but is instead intrinsically intertwined with the history or temporal evolution of the dynamical state of the fluid. At present our ability to interpret this space-time information is limited and effectively depends upon the existence of Eulerian models from which the Lagrangian signatures can be extracted and compared with data (Price and Rossby 1983, Riser 1982b, Flierl 1981); we believe that understanding the Lagrangian properties of simple Eulerian models continues to be a fruitful area for future research.

The fact that position is an integral of velocity means that it is well suited for low-frequency and mean flow studies. Dispersion studies are intrinsically

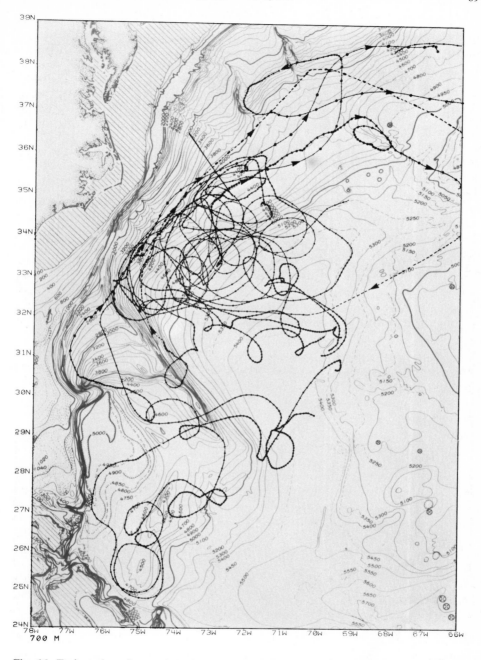

700 M

Fig. 16. Trajectories of several 700-m floats that have been in the Gulf Stream. Note that floats between 33°N and 36°N make frequent contact with the Stream but do not get entrained

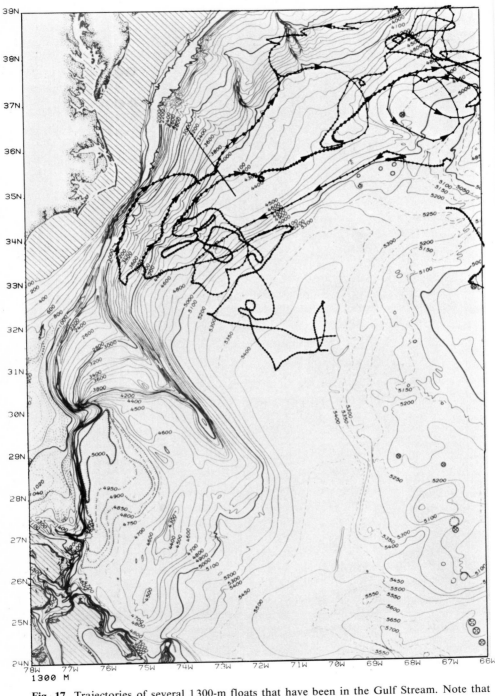

Fig. 17. Trajectories of several 1300-m floats that have been in the Gulf Stream. Note that floats east of 70°W often appear to cross the stream at this depth

Lagrangian, and information on where material spreads is obtained explicitly without resort to assumptions about the relative importance of advection and diffusion. In the future, the coupling of the history of tagged parcels of water to a knowledge of the large-scale hydrography in which they are embedded should lead us to a better understanding of how these distributions are maintained and to better predictions of the rates at which material is transported by the oceans.

5. The Local Dynamics of Eddies in the Western North Atlantic

J.C. McWilliams, E.D. Brown, H.L. Bryden, C.C. Ebbesmeyer,
B.A. Elliott, R.H. Heinmiller, B. Lien Hua, K.D. Leaman, E.J. Lindstrom,
J.R. Luyten, S.E. McDowell, W. Breckner Owens, H. Perkins, J.F. Price,
L. Regier, S.C. Riser, H.T. Rossby, T.B. Sanford, C.Y. Shen, B.A. Taft,
and J.C. Van Leer

5.1 Introduction

The local dynamics of oceanic mesoscale eddies is a subject of enormous scope
and detail. The scope of this paper, on the other hand, is restricted to a particular experiment, the POLYMODE Local Dynamics Experiment (LDE). The
LDE is unique among mesoscale experiments to date in its high sampling density and variety of measurements, and thus is well suited to considerations of
eddy dynamics in one locale. This uniqueness is, of course, one of degree not
kind. Other mesoscale mapping experiments are POLYGON (USSR), MODE,
Tourbillon (France), and the USSR component of POLYMODE. We, the authors, are the experimentalists and analysts of the LDE.

   The LDE measurements were made during 1978–79 near 31°N, 69.5°W.
This site is on the southern edge of the Gulf Stream recirculation gyre, which is
a northwestern intensification of the wind-driven subtropical gyre in the North
Atlantic Ocean (Worthington 1976). This location was chosen as a compromise
among a desire to work well within the gyre, with its associated large mean and
eddy currents, a conflicting desire to avoid the peculiar measurement problems
associated with especially intense currents, such as those of Gulf Stream Rings,
and a desire to use the familiar MODE results (from 28°N, 70°W) as a basis
for LDE experimental design; see McWilliams and Heinmiller (1978) for an
extensive discussion of the site selection. The observations extend through the
water column vertically and across a few hundred kilometers horizontally,
though the sampling distributions are nonuniform in both dimensions.

   Because the measurements were only recently made, our analyses are incomplete and, in most cases, unpublished. Consequently, the present article is
as much a preview as a summary of LDE results. Furthermore, we do not in-
tend this as the primary scientific report of the experiment since brevity pre-
cludes detailed documentation and breadth masks individual contributations.
The reader is referred to the present and future scientific literature for more
substantial presentations.

Eddies in Marine Science
(ed. by A.R. Robinson)
© Springer-Verlag Berlin Heidelberg 1983

5.2 Scientific Objectives

The experiment was designed to meet five scientific objectives. As will be shown in Sect. 5.4, all of them have been or are likely to be met, though not always quite in the manner envisioned. There have been, as any experienced oceanographer would expect, some phenomenological surprises.

5.2.1 Dynamical Balances

Several of the sampling schemes in the LDE, in particular that for the moored array in the thermocline, were designed to yield estimates of various components in the dynamical balance equations for mesoscale eddies. The primary balance of interest is that of potential vorticity, q. Under the approximation of quasigeostrophic and non-dissipative flow, q should be conserved following the geostrophic motion, and all terms in this balance can be estimated from observations except the contribution to the Eulerian flux divergence, $\nabla \cdot (\hat{v}\hat{q})$, from the sub-energetic scale (and sub-array scale) currents, denoted here by carets. The objective is to demonstrate the degree of q balance and to diagnose dynamical processes from the manner of the balance (an example of this type of diagnosis for MODE is McWilliams, 1976). For example, cascading turbulence and Rossby waves have quite distinct q balances. An unanticipated piquancy has been added to this inquiry by recent theoretical predictions of constant q regimes in a mid-depth region of the ocean, most plausibly applicable in the LDE to the upper thermocline, 500–750 m depth (Holland 1983, McWilliams and Chow 1981, Rhines and Young 1982). In such a regime the enstrophy $\overline{q'^2}$ should be smaller than a level determined by the sub-mesoscale diffusivities of momentum and heat. (An overbar denotes the time mean and a prime the instantaneous departure from it). Other informative balances which are to be estimated are those of energy and heat. Some balances such as momentum and continuity are usually ill-conditioned for direct estimation beyond confirming that mesoscale eddies are nearly geostrophic and horizontally non-divergent, although an exception has been found for one of the small-scale eddies discussed in Section 5.4.

5.2.2 Synoptic Structures

The LDE was also intended to provide a high resolution sampling of the energetic eddy scales in all four dimensions. These scales of variability may be defined, somewhat arbitrarily, as 50 km (horizontally), 500 m (vertically) and 15 days (temporally). The price of such high resolution, of course, is a limited extent in each of these dimensions; these are respectively 200 km, (the upper) 3 000 m, and 60 days. The value of such a complete, synoptic data set lies in the relatively unprejudiced descriptive analyses which can be made. Measurements were made of both the flow field, geostrophically constructed from geopotential and direct velocity measurements, and the (nearly) passive tracers, salinity and dissolved oxygen referenced to surfaces of constant potential density.

5.2.3 Geostrophic Fine Scales

In addition to a synoptic description of the energy-containing eddies, at least a statistical description was sought for the low-frequency, hence plausibly geostrophic, variablity on finer scales, both in flow variables and passive tracers. This is both as an exploration of a relatively unmeasured physical regime and for comparison with inertial range predictions from turbulence theory (e.g., Kraichnan 1967). We were quite surprised by the coherent vortex nature of some of the fine scale variability we encountered (see Sect. 5.4).

5.2.4 Surface Layer Mesoscale Eddies

The oceanic surface layer, which usually occupies the top 100 m or less, is one of significant atmospheric forcing and energetic ageostrophic currents (e.g., the Ekman layer). It is, however, contiguous to the also energetic mesoscale currents of the interior. By various means (see Sect. 5.3), LDE measurements were extended to near the ocean surface to allow a description of mesoscale variability in the surface layer during a time of general warming and formation of the seasonal thermocline, April through July.

5.2.5 The General Circulation

Much research effort, including the greater part of POLYMODE, has been expended in the last decade in beginning to map the means, variances, and covariances of currents, density, and other observables in the North Atlantic Subtropical Gyre. By being located in a relatively unexplored region, the LDE makes a contribution to this geographical documentation. Furthermore, in the Gulf Stream recirculation gyre, the magnitudes of the mean currents and eddy energy were anticipated to be larger than typical, and eddy-mean interaction processes (i.e., ones with non-trivial Reynolds stress, heat flux, and energy conversion) were thought likely to be occurring.

5.3 The Experiment

The LDE experimental design is presented in McWilliams and Heinmiller (1978). Vicissitudes in implementing this design have led to a somewhat, though not greatly, different distribution and quantity of observations. On the whole the field phase of the LDE was successful, and the rate of data recovery was high. A summary of the LDE measurements is presented in Table 1. Persons interested in examining the data may either request them from the responsible scientist or wait for their delivery to the National Oceanographic Data Center.

Table 1. Data from the LDE

Type	Area (N.lat, W.long)/ Depth	Number of Observations	Time (day-month-year)	Responsible scientist (Reference)
Current meters				
	30 36–31 23, 69 10–69 50/ 250–5 330 m	9 moorings 13 T/Ps 24 CM's with T	III 78–VII 79	Owens (Bradley 1981)
CTD/STD				
	29 57–32 00, 68 26–70 37/ 0–3 000 m	553 stations incl. oxygen	20 V 78– –14 VII 78	Taft
	28 02–32 05, 67 50–72 38/ 0–1 500 m	53 stations	23 VII 78– –13 VIII 78	Rossby
	30 08–31 32, 68 56–70 31/ 0–1 500 m	42 stations	6–18 VI 78	Sanford
	30 08–31 12, 69 04–69 35/ 0–1 000 m	105 stations	23 V 78– –21 VI 78	Leaman (Leaman and Chang 1980)
	30 34–31 26, 69 07–69 52/ 0–5 000 m	10 stations	20–24 VII 79	Millard (Bryden and Millard 1980)
Expendible profilers				
XBT (T-7)	30 00–31 30, 69 00–70 31/ 0 to 750 m	77 stations	8–18 VI 78	Sanford
	30 00–32 01, 68 26–70 37/ 0 to 750 m	532 stations	20 V 78– –14 VII 78	Taft
	30 00–32 00, 68 00–70 00/ 0 to 750 m	Approx. 200 stations	15 V 78– –14 VII 78	Heinmiller
XTVP	30 09–31 59, 68 56–70 27/ 0–1 800 m	52 stations	9–19 VI 78	Sanford

J.C. McWilliams et al.

Table 1. (continued)

Type	Area (N.lat, W.long)/ Depth	Number of Observations	Time (day-month-year)	Responsible scientist (Reference)
SOFAR Floats	30 20–31 47, 69 00–70 00 (Launch Locations)/		Launched between 15 V 78– –2 VII 78	Rossby (Spain et al. 1980)
			Data days through X 79 for floats approximately within 29 30–32 30, 68 00–71 00:	
	590–901 m,	20 floats	790 float-days	
	1 264–1 501 m	23 floats	1 670 float-days	
E-M Velocity Profiler	30 06–31 32, 68 56–70 30/ full depth	47 stations	6–18 VI 78	Sanford
Cyclesonde	30 49–31 08, 69 08–69 30/ 30–400 m	3 moorings, 118 profiles	26 V 78– –21 VI 78	Van Leer (Val Leer et al. 1980)
Doppler Currents	30 18–32 00, 68 27–70 34/ 35.5–87.5 m	237 stations	20 V 78– –6 VII 78	Regier

Microstructure observations were also made in the LDE area during early VI 1978 (T-11 XBTs and Microstructure Profiler). Responsible scientists are Sanford (T-11's) and Gregg (Microstructure Profiler).

5.4 The Phenomena

The mesoscale realization during the most intersively sampled phase (May–July, 1978) is dominated on the larger and more energetic scales by two features, a deep wave and a thermocline jet.

The deep wave is illustrated in Fig. 1 by the coherent movement of a cluster of floats near 1 300 m depth in an oscillatory and approximately co-linear manner. This is kinematically consistent with geostrophic wave motion where particles move perpendicular to the propagation direction. This oscillation can

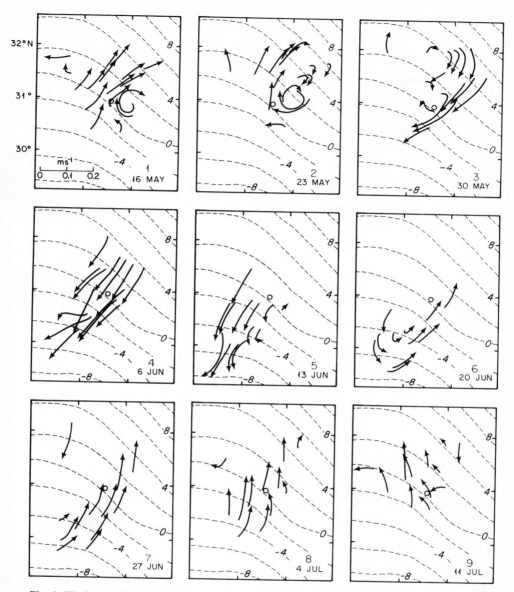

Fig. 1. Week-long float trajectories at 1300 m superimposed upon f/H contours (i.e., Coriolis frequency/fluid depth). *Open circle* at 31°N, 69.5°W marks the location of the LDE central mooring. (Price and Rossby 1982)

also be seen in the LDE central mooring current meter records for all depths beneath the thermocline (Fig. 2). During the period mid-May to early August, two full cycles occur. An interpretation of this phenomenon as a nearly barotropic, topographic Rossby wave has been made (Price and Rossby 1982). The

J.C. McWilliams et al.

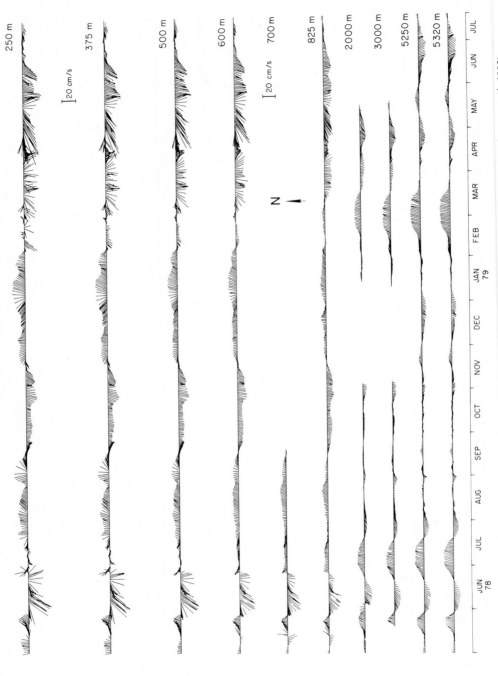

250 m

375 m

[20 cm/s

500 m

600 m

700 m

[20 cm/s

825 m

2000 m

3000 m

5250 m

5320 m

N

JUN 78   JUL   AUG   SEP   OCT   NOV   DEC   JAN 79   FEB   MAR   APR   MAY   JUN   JUL

Fig. 2. Vector time series of the low-frequency currents from the LDE central mooring at 31°N, 69.5°W (Owens et al. 1982)

phase velocity is 6 cm s^{-1} towards 300°T, the wavelength is 340 km, and the period is 61 days.

The Lagrangian q balance averaged over the float cluster has also been estimated (Price and Rossby 1982). Consistent with the wave interpretation, changes in the environmental potential vorticity, f/H, caused by cluster movement, are balanced by changes in the relative vorticity, $\underline{k} \cdot \underline{\nabla} x \underline{v}$, to within the uncertainty of the estimate (i.e., 10%–20%), where f is the Coriolis frequency, H is the fluid depth, $\underline{k}$ is a unit vertical vector, and $\underline{v}$ is the horizontal velocity vector. This wave behavior is a remarkably orderly and simple oceanic event; it is not yet clear, however, whether it has counterparts during the 15 month mooring time series (Fig. 2), and thus we must reserve judgment about its typicality.

The thermocline jet arises during June 1978, and lasts for perhaps a month. Its onset can be seen as strong, southwestward flow in and above the thermocline at the central mooring (Fig. 2). The dynamic height fields in Fig. 3 also show the jet, most clearly in the latter four surveys. This jet seems unusual among mid-ocean mesoscale phenomena in that it has a very large longitudinal extent, although its tranverse scale, perhaps 100 km in quarter-wavelength, is a familiar one. The longitudinal scale is essentially unresolved from the synoptic, 200 km diameter density array (Fig. 3), and some float trajectories asynoptically indicate a much longer scale (Fig. 9).

Another feature of the jet is its recurrence. The second most energetic feature in the mooring record is also a southwestward thermocline flow, which occurs during May 1979 (Fig. 2). Furthermore, a quite similar feature was observed from density, tracer, and float measurements during April 1977; in this case there is also an indication that its longitudinal scale might be 500 km or longer. This analysis is being led by Ebbesmeyer.

Our interpretation of this jet is presently quite tentative. It perhaps should be viewed as a transient aspect of the Gulf Stream recirculation gyre; this would be consistent with its recurrence, its longitudinal extent, and its not insignificant transport [we estimate 40×10^6 m^3 s^{-1} as the incremental (baroclinic) transport above 3000 m for both the 1977 and 1978 jets]. This characterization is further supported by the circuit which a 700 m float made of the recirculation gyre in a period of a little more than a year; it was launched slightly south of the LDE site in April 1977 and was subsequently carried through the LDE region by the 1978 jet (float 26, Riser and Rossby 1982). We know, from larger area XBT surveys in April 1978, that thermocline isotherm displacements were present in the same orientation and with as large an amplitude as those across the jet 2 months later (Gent et al. 1978). Initially, though, the horizontal separation between peak displacements was much larger, perhaps 350 km, and the associated currents were accordingly much weaker. Jet formation by some sort of local frontogenetic (i.e., gradient enhancing) mechanism is indicated, although no specific mechanism has yet been identified. All three present examples of the jet occur in spring or early summer, suggesting a quasi-annual periodicity. An intriguing question is whether it is only a coincidence that Gulf Stream transport through the Florida Straits also has a broad annual maximum at the same time (Niiler and Richardson 1973; their results are reproduced as Fig. 8 in Chap. 6 of this book).

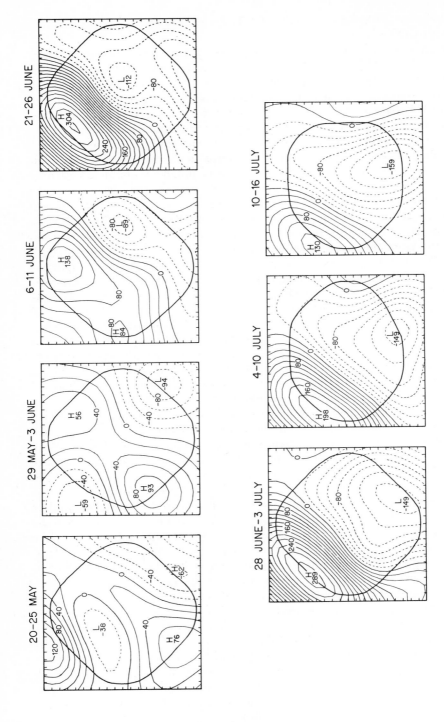

Fig. 3. Dynamic heights (0/3 000 db) for seven time periods. The contour interval is 2 cm, and negative contours are *dashed*. Extrema are labeled in mm. *Heavy solid line* marks the edge of the region where the relative error variance is less than 10%. The box dimensions are (250 km)^2, and the center is at 31°N, 69.5°W. A LDE mean value of 237.0 cm has been subtracted in these maps. (Figure provided by Shen)

Mesoscale structures extend into the oceanic surface layer. In the LDE the vertical shear patterns, represented by maps of the presure of various isotherms (Fig. 4), are generally coherent with depth in and below the 18°C thermostad. The thermostad, of course, marks the limit of surface layer penetration during wintertime. Shallower than this thermostad, the seasonal thermocline structures are partially coherent with deeper structures. This coherence includes the sea surface temperature, although it appears to diminish as the seasonal warming progresses (even when the mean warming is removed).

A first baroclinic vertical mode calculated from linear wave theory has p(T) variations in phase at all depths. The LDE observations are crudely consistent with this, except for the partial decoupling of the surface layer, due most probably to its exposure to atmospheric forcing.

The influence of the deep structures on the surface layer is clearly evident. The early sea surface temperature patterns are consistent with advection by mesoscale flows within the mean meridional temperature gradient. Deformation of sea surface salinity signals by mesoscale eddies is also observed. Direct velocity measurements near the surface indicate relatively weak vertical shears for the low-frequency (seen from moored cyclesondes; these data are being analyzed by Leaman, Perkins, and Van Leer) or mesoscale currents (seen from shipboard Doppler logs; Regier 1982). No departures from geostrophic flow were found on these scales at a resolution level of a few cm s^{-1}.

The means and variances of velocity for the LDE indicate that its site is at least partially within the return flow of the Gulf Stream recirculation gyre. The mean velocities are shown in Fig. 5. In the upper kilometer flow is to the southwest at 2.4 to 4.4 cm s^{-1}, consistent with Worthington's (1976) description of the mean circulation of the North Atlantic. The northeastward flow below the thermocline, however, has not been anticipated in previous observational descriptions of the general circulation, but it is consistent with the eastward flow of the deep counter-rotating gyre found in model solutions (Holland 1978). The velocity variances are approximately three times greater than those 300 km to the south during the MODE Experiment, ranging from 70 cm^2 s^{-2} near the bottom to 300 cm^2 s^{-2} at 250 m, the shallowest current meter (Owens et al. 1982). In addition, for periods between 2 and 880 days, the vertical structure of the kinetic energy spectrum is independent of frequency; this is similar to other measurements in the return flow (Schmitz 1978) and is a useful distinguishing characteristic compared to more typically mid-ocean mesoscale eddies (i.e., those outside the recirculation gyre) whose vertical structure is strongly frequency dependent.

The LDE eddies are also not typically mid-ocean in their significant local contributions to the maintenance of the general circulation. For example, in the thermocline there is an eddy heat flux, $\overline{y'T'}$, directed to the south and east, which is approximately opposite to the mean temperature gradient, $\nabla\overline{T}$ (Bryden 1982). This implies a conversion of potential energy from the mean to the eddy fields at a rate

$$\frac{-g\alpha}{\partial\overline{T}/\partial z}\,\overline{y'T'}\cdot\nabla\overline{T}\approx3\times10^{-5}\text{ erg cm}^{-3}\text{ s}^{-1}\ (0.3\times10^{-5}\text{ Jm}^{-3}\text{ s}^{-1}),$$

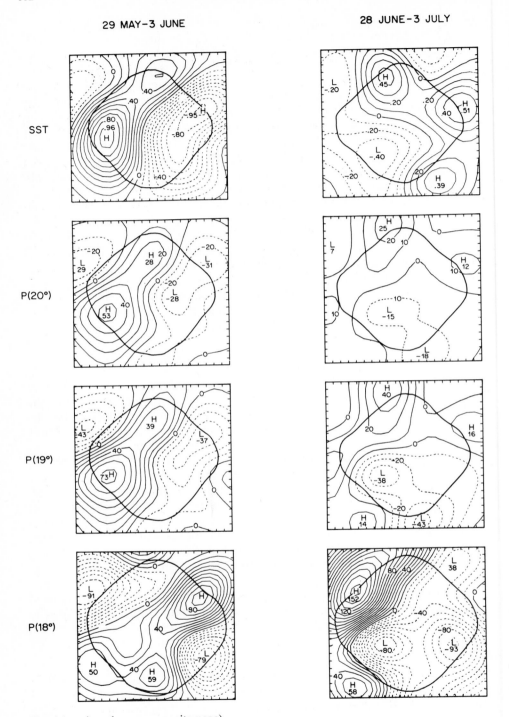

Fig. 4 (continuation see opposite page)

29 MAY–3 JUNE

28 JUNE–3 JULY

P(13°)

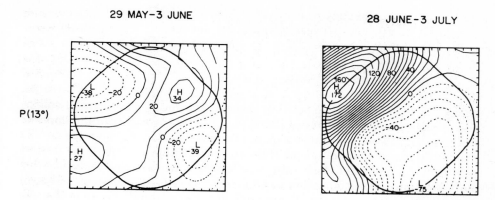

Fig. 4. Sea surface temperature (SST) and pressure of various isotherms, mapped with the same conventions as in Fig. 3. Contour intervals are 0.1 °C for the former and 10 db for the latter. These quantities represent departures from LDE mean values and, additionally for SST, a domain average for each survey period plus a further zonal and time average for each latitude. (Figure provided by Shen)

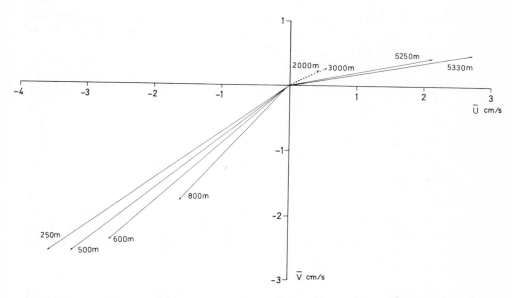

Fig. 5. Mean, record averaged velocities for the LDE central mooring (Owens et al. 1982)

which is larger than any other conversion yet measured in the North Atlantic Ocean (n.b., a much larger rate of 2×10^{-4} erg cm^{-3} s^{-1} was measured in the Antarctic Circumpolar Current; Bryden 1979). In this expression, g is the gravitational acceleration, α the coefficient of the thermal expansion, and z the vertical coordinate. The eddy heat flux is a relatively efficient process in that the correlation coefficients between v' and T' are not small; for example, at 600 m at the LDE center,

$$\overline{v'T'} \Big/ \sqrt{\overline{T'^2}\,\overline{v'^2}} = 0.5\,,$$

where v' is the northward velocity component. This local heat flux is, of course, a gyre-scale phenomenon, not a global one; that is, the flux is in the opposite direction (equatorward) to that required for global balance (Vonder Haar and Oort 1973). Preliminary estimates of the empirical orthogonal modes in the vertical indicate significant phase shifts with height, consistent with eddy heat flux and baroclinic conversion (work in progress by Owens). Furthermore, some of the theoretical modes for small amplitude departures about $\overline{y}(z)$ and the mean Brunt-Vaisalla profile $N(z)$ are baroclinically unstable ones (work in progress by McWilliams).

The Reynolds stress tensor, $\overline{y'y'}$, is also significant in the LDE. It, however, implies a barotropic conversion of mean energy at a rate

$$-(\overline{y'y'}) \cdot \cdot \nabla \overline{y} \approx -1.5 \times 10^{-5} \text{ erg cm}^{-3}\,\text{s}^{-1}\,.$$

Because the estimate is negative, this term represents a net loss of eddy kinetic energy following the mean flow. However, the energy budgets for the mean and eddy fields cannot be locally closed in the LDE. This is to be expected for regions without natural boundaries where energy fluxes are important. An indication of the latter is the discrepancy between the term above and the related mean kinetic energy gain rate, $-\overline{y}\nabla \cdot \cdot (\overline{y'y'})$: their sum differs from zero by more than the uncertainty of the estimates, which implies a net energy flux through the LDE thermocline. An additional energy loss rate from the eddies to internal waves of about 1×10^{-5} erg cm^{-3} s^{-1} has also been estimated (Bryden 1982); the eddy viscosity associated with this is discussed at the end of this section.

Much of the behavior of mesoscale eddies can plausibly be interpreted as realizations of geostrophic turbulence; this interpretation is argued in general terms in, for example, Rhines (1979). A well-known property of turbulent flow is diffusive transport of material. The spreading of float clusters in the LDE provides a direct measure of this process. Taylor (1921) defined the horizontal diffusivity κ_i as

$$\kappa_i = \frac{\mathrm{d}}{\mathrm{d}t}\langle x_i'^2 \rangle\,,$$

where x_i' is displacement of a particle from its mean position in the ith coordinate direction, and the brackets denote an average over all particles. The estimated κ_i from the LDE are listed in Table 2. These values are large compared to both the values estimated at 1500 m depth in MODE (i.e., less than 10^7 cm^2 s^{-1}; Freeland et al. 1975) and the values traditionally used in numerical models of the oceanic general circulation (e.g., Bryan and Lewis 1979). As a

Table 2. Taylor diffusivities in the LDE[a]

z [m]	κ_x [10^7 cm^2 s^{-1}]	κ_y [10^7 cm^2 s^{-1}]
700	8	5
1300	1.5	1.6

[a] Calculated by Price.

crude summary of the MODE and LDE results, κ_i can be related linearly to velocity variance by

$$\kappa_i = \overline{v_i'^2}\, \tau_I\,,$$

where the integral time scale τ_I is 8×10^5 s. A constant integral time scale is a somewhat unexpected result. Simple scaling arguments would suggest alternatively that, for energetic eddies of nearly constant horizontal scale, τ_I should be proportional to $(\overline{v_i'^2})^{-1/2}$, hence κ_i proportional to $(\overline{v_i'^2})^{1/2}$, instead of the preceding empirical result. Nevertheless, if further diffusion measurements support constant τ_I, then we will have a powerful and simple parameterization of mesoscale eddy mixing. We note that the mesoscale κ_i are neither isotropic nor homogeneous, which is again different from current modeling applications.

An alternative diffusivity, κ_*, can be defined in a manner analogous to κ_i from the separation of pairs of particles, δr, rather than the single particle displacements, x_i'. Within a turbulent inertial range, where the horizontal kinetic

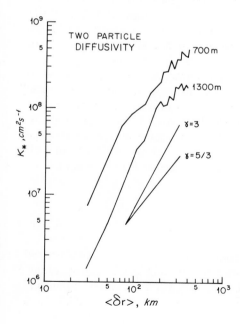

Fig. 6. Two particle diffusivities, κ_*, in the LDE, together with reference curves based upon power law energy spectra as described in the text. (Figure provided by Price)

energy spectrum has a power law dependence upon wavenumber, one would predict

$$\kappa_* \propto \langle \delta r \rangle^{(\gamma + 1)/2}$$

(Morel and Larchevec 1974). Here $\langle \partial r \rangle$ is the average pair separation distance at a particular time, and $-\gamma$ is the spectrum power law exponent. Figure 6 shows κ_* as a function of $\langle \delta r \rangle$ for the LDE. The dependence is as predicted until $\langle \delta r \rangle$ is sufficiently large as to be representing scales larger than the inertial range, and the exponent corresponds to $\gamma = 3$, also as predicted by simple theoretical arguments (Kraichnan 1967, Charney 1971). Since it would be extremely expensive to obtain a low-frequency wavenumber spectrum directly from an extensive mooring array, indirect inferences of the spectrum from $\kappa_*(\langle \delta r \rangle)$ are likely to be our most useful source of such information.

The Eulerian q belance in the thermocline has, as yet, been only crudely estimated, although with further processing of the data we anticipate being able to do so quite accurately (n.b., Sect. 5.2.1). The balance has an advection-dominated character, which is consistent with geostrophic turbulence and resembles the MODE result in this respect (McWilliams 1976). There also appears to be a tendency for time changes in relative vorticity, $\underline{k} \cdot \nabla \underline{x} \underline{v}$, and stretching vorticity, $-f \partial \eta / \partial z$, to be compensating ($\eta$ is the elevation of an isopycnal surface). This is suggestive of the constant q regime mentioned above, although we are presently uncertain about the degree of compensation (i.e., how small $\overline{q'^2}$ is). (This analysis is being carried out by Hua, McWilliams and Owens).

Far and away the greatest phenomological surprise of the LDE was found on spatial scales smaller than the most energetic ones. Since most of the authors are participating in the small-scale analyses, we have not attempted to accurately credit individual contributions here. Several intense, nearly axisymmetric circulations were found with relatively small horizontal and vertical extents. They have large associated tracer anomalies, indicating an efficient material transport and a long lifetime (in order to have transported the tracer concentration from the distant location where, on average, it is not anomalous). Their aspect ratio (vertical/horizontal) is of the same order as f/N, consistent with their being in approximate geostrophic adjustment. However, their Rossby numbers, R, are in the range 0.1–0.3, which is larger than those characterizing the energetic-scale currents, perhaps 0.05 or less. We are here using an R based upon velocity magnitude and streamfunction horizontal scale, $R = V_0 / fL_0$. Other definitions, such as those based upon the relative vorticity magnitude, typically yield higher R values. This large a value for R indicates that the radial momentum balance has a significant correction to geostrophy from the centrifugal force; i.e., the balance is

$$\frac{\partial \phi}{\partial r} = fv + v^2 / r + \ldots,$$

where ϕ is the geopotential, r is the radial coordinate, v is the azimuthal velocity, and the dots indicate contributions from friction, acceleration, and azimuthal or vertical fluxes. None of these latter terms has yet been shown to be significant.

Geostrophic turbulence theory has as yet no place for such large R, long-lived, small-scale eddies in the inertial range. A geostrophic turbulence inertial

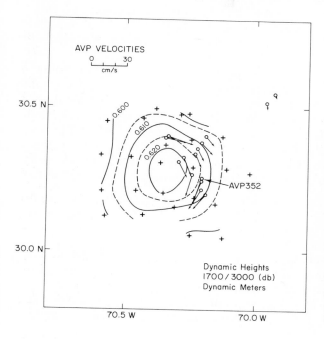

Fig. 7. Contours of dynamic heights measured at the crosses and AVP (Absolute Velocity Profiler) velocities at the *circles* (with the large-scale velocity and the vertically averaged velocity below 3 000 db removed), with measurement locations transformed into a reference frame moving with the large-scale flow (13.9 cm s^{-1} towards 219°T). These measurements were made during 11–12 June 1978. (Figure provided by Elliott and Sanford)

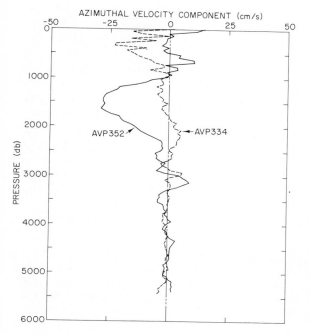

Fig. 8. Two profiles of the azimuthal velocity component in the eddy shown in Fig. 7. Profile AVP 334 was taken near the center of this feature during an earlier survey. The position of profile AVP 352 is located in Fig. 7. In each case the background flow and the vertically-averaged velocity below 3 000 db is removed. (Figure provided by Elliott and Sanford)

range with spectrum exponent $\gamma = 3$ is one with R independent of scale from the energy containing range downwards. How abundant they are, and therefore how much of a contribution they make to the horizontal kinetic energy spectrum, for example, is unknown, but the LDE shows that in at least one location they are not rare. Because they carry mass, they might provide an important and distinctly non-diffusive transport and mixing mechanism, if sufficiently abundant.

One example of such a circulation is shown in Figs. 7–8. They show a small, anticyclonic eddy whose vertical range is between 1 000 and 3 000 db, and whose maximum velocity of 29 cm s^{-1} occurs at 1 650 db and a radius of 14 km from the eddy center. These imply R = 0.3. Thus, in this anticyclone (v < 0), the azimuthal speeds exceed the geostrophic speeds by 40%. Of particular interest is whether there is a detectable angle in Fig. 7 between the velocity vectors and the dynamic height contours, which would indicate a finite radial velocity, u. Such a flow must satisfy the following azimuthal momentum balance:

$$\left[f + \frac{1}{r} \frac{\partial}{\partial r} (rv) \right] u = \ldots,$$

where any azimuthal pressure gradient is also buried in the dots. We have not yet decided whether a finite u is defensible. In particular, it would work against the maintenance of property anomalies within the eddy. At the eddy center there are anomalies of -0.05%₀ and 18 μmol kg^{-1} (0.2 ml l^{-1}) in salinity and oxygen, respectively; the former is quite large compared to the natural variability of the LDE region, the latter is not.

Another example is shown in Fig. 9. By good fortune, a SOFAR float was deployed at 700 m in a small-scale anticyclonic circulation, where it remained for more than 2 months until recovered. During this period the eddy appears to passivly move with the energetic mesoscale flow through several reversals of direction, as indicated by the nearly parallel track of a neighboring float not within the small eddy. In the middle of the period, both floats are in the thermocline jet discussed above; the bulk motion is rectilinear for nearly 650 km, which is a further indication of the large longitudinal scale of the jet. This eddy is a local pycnostad with anomalously low static stability between 550 and 850 db. Dugan et al. (1982) report many examples of locally low stability regions in Sargasso Sea XBT profiles. Its maximum azimuthal velocity, as best we can estimate it, is 25 cm s^{-1} at 750 db depth and a radius of 10 km (i. e., R = 0.35). Its tracer anomalies are -0.12%₀ and -24 μmol kg^{-1} (-0.6 ml l^{-1}) in salinity and oxygen. No significant evidence has been found for shape changes or decay over this 2-month period, which is a direct indication of a long lifetime.

Fig. 9. a Trajectories of two floats at 700 m during 14 May–23 July 1978, with labels referring ▶ to year dates. One float is trapped within a small-scale eddy, the other is not. b An objective demodulation of the eddy center trajectory and particle circuit moving with the eddy, as calculated from the trapped float in a. The circuit is plotted every 42 h, which corresponds to an average half rotation period for the float. (Figure provided by Rossby and Owens; a first appeared in Hartline 1979)

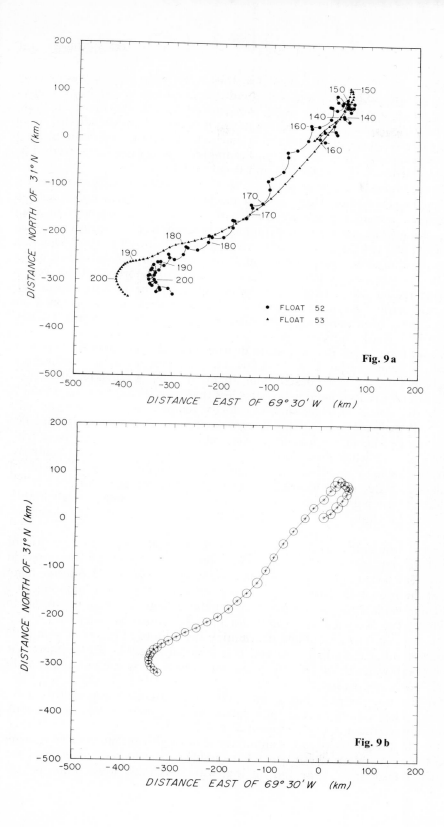

Fig. 9 a

Fig. 9 b

Indirect inferences of lifetimes of at least a few years can be made by assuming that the eddy tracer anomalies indicate the site of origin. For example, by such an argument the eddy in Fig. 9 originated at least 2000 km to the south and east of the LDE location. (This analysis is being led by McDowell). At an average migration speed of 3 cm s^{-1}, a time of 2 years would be required for it to arrive. There are also a few previous examples of long distances traversed by coherent vortices (e.g., McDowell and Rossby 1978).

We can only speculate at present about the mechanism of generation for these small-scale eddies. Three possible mechanisms are vortex shedding from topography, meander pinch-off from narrow jets (in the manner of Gulf Stream Ring formation), and geostrophic adjustment after penetrative convection. The latter is a particularly attractive possibility since it is consistent with the observed predominance of both small horizontal scale (characteristic of the scale of a convective "chimney" and a small radius of deformation in a weakly stratified region where penetrative convection could be forced) and anticyclonic circulation. Examples of such convection at sites not too distant from the LDE have been found in the Mediterranean and Labrador Seas (MEDOC Group 1970, Clark and Gascard 1982a).

We also measured the mesoscale distributions of some passive tracers on potential density surfaces in the LDE. Our analyses are quite preliminary, though, and the following descriptions are provisional. A much larger fraction of the variance is associated with scales less than, say, 25 km for tracers than is true for dynamical variables such as velocity, temperature (on a pressure surface), and dynamic height. This is consistent with a geostrophic turbulence inertial range where the horizontal wavenumber spectra of the latter are at least as steep as k^{-3}, while those of the former are proportional to k^{-1}. Similarly, vertical correlation lengths are shorter for tracers and do not extend outside of the classically described water masses (e.g., the Mediterranean water layer beneath the thermocline). Some, but by no means all, of the small-scale variance is associated with tracer anomalies in the coherent circulations described above.

Tracer probability distributions are also distinctive. They exhibit significant skewnesses and rather broad tails. These features are partially illustrated in Fig. 10. The broad tails appear as outliers not tightly connected with the main part of the probability distribution. In a given range of σ_θ, corresponding to a particular water mass, the outliers have a preferred sign, and thus contribute to skewness. The two small-scale eddies of Figs. 7–9 contribute some of the outliers in Fig. 10. The probability distribution for salinity in the 18° thermostad exhibits another remarkable feature. It is bimodal (Fig. 11). Furthermore, the two types of water involved have boundaries which are smooth on the scale of the energetic mesoscale eddies.

We are not yet prepared to interpret the mesoscale tracer distributions. However, they clearly contain information which is independent of and complementary to the mesoscale dynamical fields.

The final result we report is an inference of significant momentum transfer from the mesoscale to the internal wave field. A correlation between the low frequency horizontal shear and internal wave Reynolds stress, both as esti-

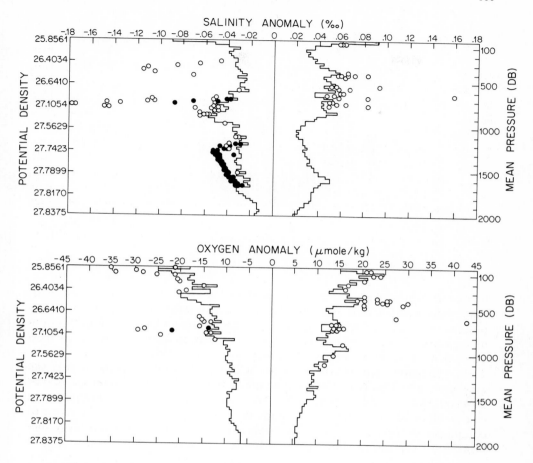

Fig. 10. Outliers in the frequency distributions for salinity and oxygen on σ_θ surfaces, featuring those anomalies outside envelopes determined by constancy of the cumulative frequency function over intervals of 0.006‰ and 1.5 μgm kg^{-1}, respectively. Anomalies from within the two eddies of Figs. 7–9 are *black circles,* and other anomalies are *white circles.* The data base here is CTDO profiles on a grid with spacing 25 km and diameter 200 km, surveyed seven times (n. b., Fig. 3); additional measurements, not included here, were made in local surveys of some of the anomalous eddies. The resolution in σ_θ corresponds to approximately 25 db in the mean pressure of σ_θ surfaces. (Figure provided by Taft and Lindstrom)

mated from the moored array, implies a horizontal eddy viscosity, ν, of order 10^6 cm^2 s^{-1} (Brown and Owens 1981). A viscosity of this magnitude would significantly deplete the mesoscale field on a time scale

$$\tau = L^2/\nu ,$$

where L is a mesoscale horizontal length. For the thermocline jet (L ≈ 100 km), a viscous decay time of 1000 days is indicated. For the small scale eddy in Fig. 9 (L ≈ 10 km), however, τ is 10 days, which is inconsistent with the documented

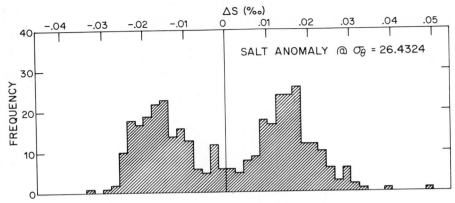

Fig. 11. Frequency distribution of salinity anomalies on $\sigma_\theta = 26.4324$, which lies within the 18° thermostad. The data base is as in Fig. 10. (Figure provided by Taft and Lindstrom)

much longer lifetime of the eddy. Thus, the eddy viscosity inference is misleading for the small-scale eddies, perhaps because of an insufficient scale separation from the internal waves. However, we cannot reject its applicability for the more energetic mesoscale currents (since the indicated τ is much longer than we have been able to observe an individual feature). Other processes, such as wave dispersion or turbulent cascades, will in any event be likely to alter the energetic mesoscale features in a time much shorter than τ. As reported above, however, the associated energy transfer to the internal waves is a not insignificant component of the mesoscale eddy energy budget. It will be of interest to compare the observed $\overline{q'^2}$ with the theoretical predictions, referred to in Section 5.2, of what upper bound this ν should set for it.

5.5 Summary

When our present analyses are completed, the LDE will have both accomplished its scientific objectives and yielded some surprises. Most prominent among the latter, perhaps, are the elongated thermocline jet, the coherent, small-scale vortices, and the peculiar tracer distributions.

A comparison of Sections 5.2 and 5.4 indicates the manner in which the objectives are being met. A Lagrangian q balance has been estimated for the deep wave, and an Eulerian q balance will be estimated in the thermocline, with possible extensions to other depths. Several conversion terms in the eddy and mean energy budgets have been estimated. A centrifugal force contribution to a primarily geostrophic momentum balance has been shown for one of the small-scale vortices. A relatively complete and accurate synoptic description has been obtained for the deep wave, the thermocline jet, and the mesoscale eddies in the surface layer. The geostrophic fine scales have been shown to contain coherent, long-lived eddies as well as a significant fraction of the total tracer variance. Finally, the LDE has been shown to lie near the edge but somewhat within the Gulf Stream recirculation gyre by virtue of its relatively large mean velocity, eddy energy, Reynolds stress, and heat flux.

We encourage the reader to examine the scientific publications from the LDE as they appear, since they will both complete what is presented here and add results we have judged too preliminary to include.

6. Gulf Stream Variability

D.R. Watts

6.1 Introduction

The current structure, transport and water mass properties of the Gulf Stream[1] vary geographically and temporally, with seasonal, mesoscale and interannual changes found whenever an appropriate investigation has sought such variability. The Gulf Stream path meanders with a broad spectrum of variability, particularly downstream of Cape Hatteras. These lateral translations range from interannual and seasonal shifts to the north or south, through substantial mesoscale 20–60 days variations, 4–14 day periodicities predominant along the continental margin, to significant rapidly translating and growing meanders with periodicities as short as 4 days downstream of Cape Hatteras.

There have been many recent developments which extend our observational and theoretical understanding of the Gulf Stream. The following sections discuss the path, current structure and transport, water mass variability, and basic dynamical balances in the Gulf Stream. In each section, historical studies are reviewed, and our knowledge of the "mean" Gulf Stream is presented as a necessary framework within which the variability itself may then be discussed. Studies of the meandering path of the Gulf Stream have been greatly advanced by satellite infrared imagery and inverted echo sounders, from which wave number-frequency spectra and an observational dispersion relationship for meanders have now been produced. Comparisons are made with theoretical models of meander characteristics, with promising similarities noted when spatial growth as well as temporal growth is included. The section on current structure and transport derives its newest advances from long-term current meter moorings near and under the Stream, which tend to confirm a tight recirculation gyre just south of the Stream. The strong eddy field near the Stream and the recirculation show an interestingly weak depth dependence. Recent observations on water mass properties in the Gulf Stream, with the advantage of continuous profiles of salinity, temperature, and oxygen content as a function of depth, reveal spatial and temporal variability, which different investigations have related to fluctuations in the circulation through the Florida Straits, changes in water mass formation processes associated with severe winters, and different water properties entrained into the Stream from either side.

---

1 I will use the term Gulf Stream to include the western boundary current from the Florida Straits to Cape Hatteras and its continuation to the south of the Grand Banks, thus including Iselin's "Florida Current" and "Gulf Stream".

Eddies in Marine Science
(ed. by A.R. Robinson)
© Springer-Verlag Berlin Heidelberg 1983

The dynamics section discusses the distribution of potential vorticity across the Stream, in relation to quasigeostrophic model results, and reviews observations on energy exchanges between the eddy fluctuations and the mean flow.

Because this chapter is to appear shortly after publication of Fofonoff's (1981) review article on the Gulf Stream, my purpose will be to complement that work with different emphases and only overlap it where essential. Other chapters within this book by Richardson (Chap. 2), Rossby et al. (Chap. 4), and Holland et al. (Chap. 17) are relevant to the subject of Gulf Stream variability. Time and practical limitations dictate that I choose only some subset of topics under the broad heading of Gulf Stream variability. Among several specialized aspects of Gulf Stream research which have been omitted are the following topics for which the reader may find these references a useful introduction: wave-like fluctuations along the continental margin (Brooks and Mooers 1977, Bane, Brooks and Lorenson 1981), inshore spin-off eddies from Florida to the Carolinas (Lee, Atkinson and Legeckis 1981), the Western Boundary Undercurrent (Richardson and Knauss 1971), and the North Atlantic Current beyond the Grand Banks (Clarke et al. 1980).

A major research advance during the last decade has been the construction of eddy resolving general circulation models and their application to studies of the subtropical and subpolar gyre circulation (Holland et al. Chap. 17). These studies, together with observation of the geographical distribution of low-order eddy statistics (Dickson Chap. 15 and Emery Chap. 16) have identified the Gulf-Stream meander, extension and recirculation regions as regions of important eddy-mean field interaction. This chapter is concerned primarily with details of the local dynamics of the Gulf-Stream current and the reader is referred to those chapters for study of this topic.

6.2 The Gulf Stream Path

The near-surface currents in the western central North Atlantic are schematically summarized in Fig. 1, which also indicates principal topographic features by the 1 000 m depth contour. Surface current speeds are indicated by arrow thickness. The broad North Equatorial Current and more intense Guiana Current flow relatively unimpeded by the Lesser Antilles into the Caribbean, since they are shallow currents. Additional flow proceeds northwestward intermittently as an Antilles Current (Gunn and Watts 1982), and some of this diverts to enter the Caribbean through the Windward Passage between Cuba and Hispaniola. The combined flow in the Caribbean enters the Straits of Yucatan, makes a time-varying loop in the Gulf of Mexico (Maul 1977, Chew 1974), and enters the Florida Straits.

The Gulf Stream emerges from the Florida Straits bounded to the west by the continental margin (200 m isobath) and flows northward to the Carolinas. It is 70 to 100 km wide. The central part of the current typically remains over water depths of 500–800 m. Off Onslow Bay, (near 34°N), the Blake Plateau ends and the continental slope descends steeply to 2 000–4 000 m depth. Down-

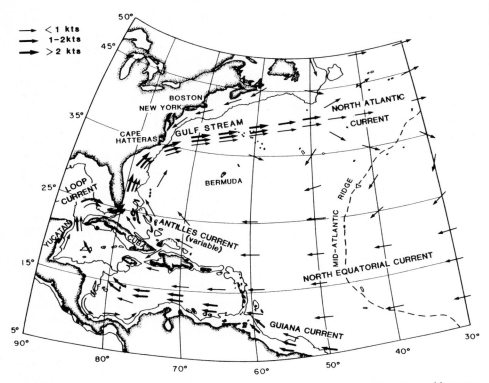

Fig. 1. Surface currents in the western North Atlantic. The current speed is indicated by *arrow thickness* according to key at *upper left-hand corner*. A *light line* at the 1000 m depth contour indicates principal topographic features

stream of Cape Hatteras the Stream no longer follows the shelf break, but instead the mean path continues to the northeast, into ever-deepening waters (Fig. 3b). Within 300 km downstream of Cape Hatteras, the mean water depth below the Stream is greater than 4000 m, and the current is bounded on the northwest by a widening wedge of colder, relatively stationary Slope Water rather than a solid coastal boundary. Near 60°–65°W the Stream encounters the New England Seamount chain, downstream of which the current appears to weaken and become more variable. Richardson (1981) discusses the trajectories of 35 satellite-tracked surface drifters in the Gulf Stream region and finds a quasi-permanent 100 km southeastward deflection of the Stream over the seamounts with the frequent formation of a ring meander. Part of the Stream recirculates in the subtropical gyre, and part supplies a North Atlantic drift current, which is presently insufficiently described.

Worthington (1976) has proposed a circulation system for the North Atlantic in which the recirculation from subsurface Gulf Stream waters differs greatly from the near-surface subtropical gyre: in his representation of thermocline, intermediate and deep water flow, the return flow becomes more concentrated to the north just offshore of the Gulf Stream for successively deeper flow patterns. For waters colder than 4°C (below ~1500 m), his recirculation gyre is confined within 500–800 km south of the Gulf Stream. Long-term current meter records obtained recently (Schmitz 1977, 1978, 1980) at locations along 55°W longitude show a mean westward flow in this recirculation region which covers a zonal band ~500 km wide at 600 m depth which becomes narrower at 1000 and 1500 m depths to a thin (~100 km) westward jet (u ~ −9 cm s^{-1}) at 4000 m. This flow is termed weakly depth-dependent in the sense that its amplitude and direction are approximately equal at thermocline and abyssal depths. The flow at 55°W under the Stream axis at 4000 m is also westward in the mean (−5 cm s^{-1}), yet between this and the aforementioned return flow is another weakly depth-dependent eastward flow (~6 cm s^{-1}). These stable averages such as at 55°W required long deployments because fluctuation and eddy kinetic energies exceed the mean below the thermocline. Hence this information comes principally from Schmitz's 55°W moorings, and could reflect some special flow pattern influenced by the New England Seamount Chain and Grand Banks of Newfoundland. Tracks from SOFAR floats launched along 55°W (Richardson et al. 1981) at 700 to 2000 m depth tend to confirm the narrow swift recirculation south of the Stream, and also suggest a northeastward flow into the Newfoundland Basin. The deep mean recirculation could be different west of the seamount chain.

The instantaneous path of the Stream differs from the foregoing mean description particularly downstream of Cape Hatteras. The motion is characterized by meanders, or lateral shifts of the entire baroclinic structure associated with the Gulf Stream. Figure 2 shows several instantaneous paths (Knauss 1967) of the "north wall" of the Stream as indicated by the 200 m depth contour of the 15°C isotherm. The meanders typically have along-stream wavelengths of approximately 300–500 km, and the amplitude can be large compared to the width of the current. Frequently the path is even more contorted, forming large elongated loops east of 70°W, particularly in the neighborhood of the New England Seamount chain (60°–65°W).

The Stream also shifts laterally in the region south of Cape Hatteras. The root-mean-square lateral displacements increase slowly from 5–10 km off Florida, to a local maximum of 25 km downstream of a topographic feature off Charleston, S.C. (the "Charleston Bump," 31°–32°N), in the lee of which they decay back to 10 km r.m.s. The r.m.s. displacement amplitudes in Fig. 3 upstream of Cape Hatteras are from Bane and Brooks (1979). The meander amplitudes along the Carolina shelf have been determined by satellite (Legeckis 1979), and by AXBT and moored current meters (Brooks and Bane 1981; Bane, Brooks and Lorenson 1981). These meanders have wavelengths of 150–200 km and are observed to propagate downstream at approximately 30–40 km day^{-1}.

As the Stream continues into deeper water downstream of Cape Hatteras, the envelope and r.m.s. amplitude of meandering broaden greatly (Fig. 3a). In the region 100 to 200 km beyond Cape Hatteras (1400–1500 km on Fig. 3), where the water depth is approximately 3000 m, the variance doubles in each 50 km step downstream as the r.m.s. amplitude increases from 15 to 20 to 30 km

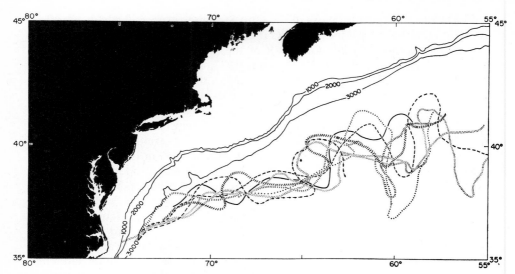

Fig. 2. Several "north wall" paths of the Gulf Stream as determined from the 15 °C/200 m isotherm contours in a sequence of monthly surveys. (With permission from Knauss 1967). These
paths discussed in detail by Hansen (1970)

(Watts and Johns 1982). Further along the Stream path, Halliwell and Mooers
(1979) have analyzed satellite imagery to find that the r.m.s. amplitude
continues to grow to ~80 km near 65°W and the seamount chain. The overall
envelope of meandering broadens to 200–300 km width, as can also be seen in
Fig. 2. In this region the Stream can shift through more than its own width and
the meander amplitude can be comparable to the wavelength. Surface drifter
trajectories (Richardson 1981) suggest that very large amplitude meanders begin at the seamounts, extending eastward. Thus all along the Gulf Stream path
the meander amplitude grows downstream and shows evidence of topographic
influence. Harrison and Robinson (1979) suggest that Gulf Stream meanders
may drive eddy variability in mid-gyre.

Regional characteristics of the path variability are summarized in Table 1.
In the early shipboard surveys (Fuglister and Worthington 1951, Fuglister
1963) the path of the Stream south of Nova Scotia was dominated by long wavelength (~400 km), large-amplitude meanders. Those which were studied in detail were found to be quasi-stationary, and were characterized by a slowly
evolving baroclinic field suggesting temporal growth. Near 72°W, however,
downstream propagation was observed; one meander crest moved at a rate of
approximately 17 km day^{-1} for 3 days (Fuglister and Worthington 1951).
Hansen (1970) later conducted monthly path surveys between 75°–55°W for
one year, with weekly surveys over a 2-month period, using a towed thermistor.
He found meanders on appreciably shorter scales, with wavelengths ranging
from 200–400 km. Mean downstream propagation rates of 5–10 km day^{-1} and
spatial growth rates $\kappa = 2 \times 10^{-3}$ km^{-1} were estimated by constructing lines of
constant phase between features on successive (superposed) path sequences.

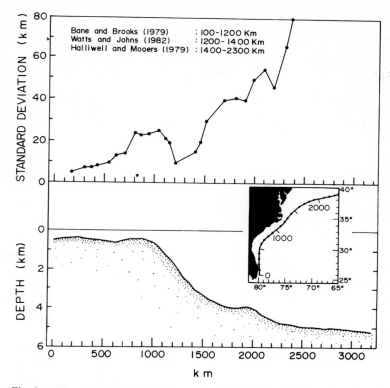

Fig. 3. a The r.m.s. lateral displacement amplitudes for the Gulf Stream path from the Florida Straits downstream 2000 km along the mean path indicated in the *inset*. Data sources for various segments are summarized in the key at *upper left*, with permission from the respective authors. **b** The bottom depth below the mean path, as a function of downstream distance. (With permission from Knauss (1967)

The rapid evolution between monthly paths, however, produced considerable ambiguity in identifying and tracking events. Weekly paths evolved more continuously and appeared not to be aliased by higher frequency/wavenumber phenomena; however later studies make even this conclusion tenuous.

In a subsequent study, Robinson et al. (1974) constructed repeated path segments at 26-h intervals, using XBT data in a two-degree square near 70°W, and they also delineated surface paths by airborne radiometer (ART) over more extensive areas. They found fluctuations within a wide range of time scales. ART paths at intervals of a few days revealed a pattern of meanders of length scale ∼350 km migrating eastward at mean phase velocities of 5–8 cm s^{-1}, consistent with Hansen's (1970) estimate. In addition, they observed one fluctuation with period ∼12 days and wavelength ∼250 km, indicating a rapid downstream propagation speed ∼20 km day^{-1}. Periodicities as long as 80 days were estimated in association with large-scale secular shifts of the path.

Table 1. Gulf Stream path variability

	Florida Straits	Carolina Capes	Cape Hatteras 73°–75° W	~65°–70° W
Periodicities (days)	9–12 (d) 10–13 (k) 7–14 (c)	3– 8 (b) 4– 7 (m) 30–36 (i)	2– 60 (l)	50–60 (f) 45 (i) {20– 80 and ~ 12} (j)
Wavelengths (km)	170 (k) 1000 (c)	200–250 (b) 90–260 (h)	150–600 (l)	320–360 (f) {350–1000 and ~ 250} (j) 200– 400 (g)
Propagation speeds (km day^{-1})	17S (k) 100S (c)	30–45N (b) 40N (h)	18–36NE (l)	6– 7E (f) {5– 8E and ~20E} (j) 5–10E (g) ~17E (e)
Amplitude r.m.s. (km)	5 (a)	10–25 (a) 10 (m)	15–30 (l)	50–80 (f)
Growth and decay characteristics	Slow growth compared to advective time scale (a)	Perturbation at "Charleston Bump" and decay downstream (a, h)	Rapid growth, doubling variance each 50 km (l)	Moderate growth, doubling variance in ~400 km (f)
Comments	Seasonal variation in transport	—	—	Wavelength decreases as amplitude increases downstream (f) Interannual shifts >50 km (f)

(a) Bane and Brooks (1979); (b) Brooks and Bane (1981); (c) Brooks and Mooers (1977); (d) Duing et al. (1977); (e) Fuglister and Worthington (1951); (f) Halliwell and Mooers (1979); (g) Hansen (1970); (h) Legeckis (1979); (i) Maul et al. (1980); (j) Robinson et al. (1974); (k) Schott and Düing (1976); (l) Watts and Johns (1982); (m) Webster (1961)

Recently Watts and Johns (1982) have presented an observational dispersion relationship for Gulf Stream meanders in deep water just downstream of Cape Hatteras. Their year-long measurements were from a moored array of inverted echo sounders, which continuously tracked the position of the "north wall" of the Gulf Stream at three cross-stream sections located 100, 150 and 200 km downstream of Cape Hatteras. These path-displacement records exhibit high coherence, stable phase delays, and rapid spectral growth downstream. The phase speed and spatial growth rate shown in Fig. 4 were determined from these observations. Path displacements in this region exist on a continuum of time scales between 2 and 60 days with over 96% of the variance at periods longer than 4 days. Propagation is downstream, ranging from 40 km day^{-1} at short periods and wavelengths (4 days, 160 km) to 20 km day^{-1} associated with (30 day, 600 km) meanders. These speeds are intermediate between values measured upstream along the Carolina coast (35–40 km day^{-1}, Brooks and Bane 1981) and those estimated further downstream (5–10 km day^{-1}, Hansen 1970, Robinson et al. 1974).

Rapid downstream meander growth was observed, doubling the total variance in each 50 km interval downstream (Watts and Johns 1982). The growth was entirely associated with periods longer than 4 days, most rapid for periodicities of 14 days and longer, and associated with wavelengths as short as 150 km and as long as 600 km. The observed growth can be described by a spatial growth rate κ or by a temporal growth rate σ in a frame translating downstream with the meander phase speed. Briefly summarized in Table 2 are the propaga-

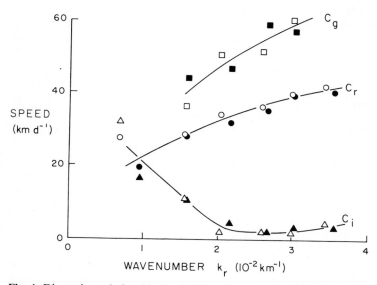

Fig. 4. Dispersion relationship for Gulf Stream meanders near 73°–75°W just downstream of Cape Hatteras. Phase speed components (C_r, C_i) and group speed C_g plotted vs. wave number k_r. *Circles, triangles,* and *squares* represent C_r, C_i, and C_g respectively, calculated at 50 and 100 km array spacings. (Watts and Johns 1982)

tion and growth characteristics in (a) a growth regime associated with short scales $(T, \lambda) \sim (4$ days, 160 km); (b) an intermediate minimum growth band of $(T, \lambda) \sim (9$ days, 250 km); and (c) the energetic rapid growth band for $(T, \lambda) \gtrsim (14$ days, 350 km). Growth rates observed by Hansen (1970) farther downstream were smaller, $\kappa \sim (2 \pm 1) \times 10^{-3}$ km^{-1} (for $\lambda \simeq 300$ km). Rates predicted theoretically by Tareev (1965) and Orlanski (1969) were somewhat larger, $\kappa = 8 \times 10^{-3}$ km^{-1} ($\lambda = 250$ km) and $\kappa = 12 \times 10^{-3}$ km^{-1} ($\lambda = 365$ km), respectively. A time-dependent thin jet model by Robinson et al. (1975) found for (31 days, 560 km) perturbations the corresponding growth rate was $\kappa = 5 \times 10^{-3}$ km^{-1}, in good agreement with these observations.

Table 2. Meander characteristics

Growth regime	T(days)	λ(km)	$k(10^{-2}$ km$^{-1})$	$\kappa(10^{-3}$ km$^{-1})$	$\sigma(d^{-1})$	C_r, C_i(km d$^{-1})$
a) Spatial	4.3	160	4.0	4.0	0.14	37,3.6
b) Minimum	9	250	2.4	2.5	0.07	29,2
c) Temporal	14	350	1.8	7.0	0.20	25,10

The propagation and growth curves for this region, shown in Fig. 4, closely resemble the theoretical results of Hogg (1976) in which he indicates a transition at $\kappa = 2.4 \times 10^{-2}$ km^{-1} from temporal growth for longer (T, λ), to spatial growth for shorter (T, λ). The wave number at which this transition occurs corresponds closely to the high wave number growth cutoff in Phillips' model (see Pedlosky 1979), for an appropriate choice of length scale of about twice the internal Rossby deformation radius. Pedlosky treats several other baroclinic instability models, in all of which the fluctuations propagate as exp ik$(x - ct)$ with k real and c complex, and in which the temporal growth $\propto \exp(kc_i t)$ exhibits a cutoff at high wave numbers. By contrast, as shown by Hogg, instabilities which include complex k can exhibit spatial growth at wave numbers greater than the temporal growth cutoff. Although these new measurements cannot indicate a clear distinction between temporal and spatial growth, and Hogg's model was for a weak baroclinic flow, the behavior closely resembles that predicted by Hogg. Further investigations with spatial growth should produce interesting results.

Several processes may be influential in the generation of meanders. Brooks (1978) has indicated that variable winds could force continental shelf waves. Where the current extends to the bottom, the topography can steer the path by vertical stretching: Warren (1963) and Niiler and Robinson (1967) have modeled this downstream of Cape Hatteras, concluding that this mechanism is insufficient by itself to generate the large curvatures which are frequently observed. Nevertheless, observationally, the r.m.s. amplitude of path deflections increases in the lee of a topographic feature off Charleston, S.C., and then decays further northward along the North Carolina coast (Bane and Brooks 1979). Further downstream, in the region of the New England Seamount chain, the current path and tracks of surface drifters become more contorted (Richardson 1981, Maul et al. 1978). Additionally, interaction with nearby Gulf Stream rings can strongly perturb the path of the Stream (Watts and Olson 1978, Olson and Watts 1983). Models mentioned above suggest that baroclinic and barotropic instability processes should be operative in the Gulf Stream, and the similarities with observations on rapidly growing meanders in deep water just downstream of Cape Hatteras indicate that these processes must be important there.

By contrast, fluctuations of the Stream path along the continental margin may well be stable (although inshore spin-off eddies can grow, and it is not clear that they should be treated distinctly from the Stream). As noted earlier the r.m.s. meander displacement decays downstream along the North Carolina coast, indicating that perturbations introduced at the "Charleston Bump" do not grow due to instability. Eddy/mean flow interactions have been investigated at several sections from Miami to Onslow Bay, N.C. (Brooks and Niiler 1977, Schmitz and Niiler 1969). Fofonoff (1981) carefully summarizes these interactions. While some portions in a transect of the current have positive or negative energy flux from the mean current to the fluctuations, they generally conclude that the total energy flux integrated across the entire width of the current is not significantly different from zero anywhere along the continental margin.

Given the propagation and growth characteristics (the complex dispersion relationship) for meanders downstream of Cape Hatteras and the observed high downstream coherence of displacements, work is underway to test whether the downstream path $x(y,t)$ may be predictable from "inlet" observations off Cape Hatteras. However, not all fluctuations grow as they move downstream, and other perturbations could develop downstream of the "inlet", in either case reducing the predictive accuracy.

The Gulf Stream path also exhibits variability on seasonal and interannual time scales northeast of Cape Hatteras, but at present there is only limited information available. Iselin (1940) suggested on the basis of 15 transects from Montauk to Bermuda during 2½ years that the Stream tends to flow further south in the spring when the transport is higher. Fuglister (1972) presents evidence of a similar trend for a north-south path displacement from data for a spatial average over every degree of longitude in the region south of New England during one year (1965–66). In Fig. 5, Worthington (1976) has compared this trend for the spatial average displacement in one year to a collection of geostrophic transport measurements south of New England taken over many years, which adds corroborating evidence for Iselin's suggestion. The trend is qualitatively similar to a separation mechanism shown by Veronis (1973, 1981): in a two-layer system with stationary lower layer the baroclinic transport is $T = (g^*/2f)(H_B^2 - H_A^2)$, where $g^* = g\Delta\rho/\rho_2$, $\Delta\rho = \rho_1 - \rho_2$, the subscripts 1,2 refer to the upper and lower layers respectively, and H_A, H_B are the interface depths on either side of the current. If the interface surfaces and the central gyre thermocline depth H_B is given (for example by thermohaline processes or other controlling mechanism), then $fT = g^* H_B^2/2$. For an upper layer transport, T, this specifies the highest f, and hence latitude, to which the current could pro-

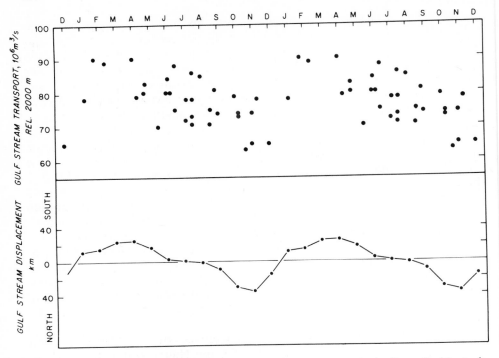

Fig. 5. *Top* Gulf Stream geostrophic transports relative to 2000 m principally on the Montauk-Bermuda section, plotted vs. month. In order to emphasize seasonal trends, the year runs through two cycles and all points are plotted twice. *Bottom* The along-stream average displacement of the Gulf Stream during 1965–66 plotted vs. month, as above. (With permission from Worthington 1976)

gress before it separates from the coast or reduces its transport. Quantitatively however, observations do not indicate that the product (fT) remains constant, since T changes by $\sim \pm 8\%$ whereas f tends to compensate by only $\sim \mp 4\%$. Evidence which tends to contradict this pattern comes from Halliwell and Mooers (1979), who find from satellite data that the Gulf Stream mean path shifted 30–60 km shoreward off New England in 1977 compared to the previous year, while Worthington (1977) states that the transport is high after the severe winter of 1977. On balance this seasonal trend is not well established, but the Stream path clearly varies on seasonal to interannual time scales.

6.3 Current Structure and Transport

Much of our knowledge about the Gulf Stream comes from the many hydrographic sections and relatively few direct velocity profiles which have been taken across the Stream. Current meter records have been limited to moorings within the bottom 1 000 m under the Gulf Stream, but technological developments are now being tested on the practicality of mooring up to thermocline levels within the Gulf Stream. While present-day research is studying the geographic and temporal *variability* in the structure and transport of the Gulf Stream, in fact the relative constancy of the structure from the Florida Straits to south of Nova Scotia is noteworthy. A strongly baroclinic frontal structure is always present and in most sections is found to extend to the ocean bottom, whether the mean depth is 500 or 5 000 m. The baroclinic structure maintains approximately the same width, of the same scale as the internal Rossby radius of deformation characterizing the main thermocline ($\lambda = (g^*H)^{1/2}/f \sim 35$ km), along this entire portion of its path. Stern (1975) finds the characteristic width δ of an inertial western boundary current as follows: for transport $T \sim Vh\delta$ of current V in depth h, and relative vorticity $\zeta \sim V/\delta \sim -\Delta f$ compensating for the change in planetary vorticity Δf as the water changes latitude, this requires $-\Delta f \sim T/h\delta^2$ or $\delta \sim (T/h\Delta f)^{1/2}$. For the Gulf Stream with $T \sim 60 \times 10^6$ m^3 s^{-1}, $h \sim 600$ m, $\Delta f \sim 0.7 \times 10^{-4}$ s^{-1}, this gives $\delta \sim 38$ km in good agreement with observations. The spatial similarity has been illustrated, for instance, in a sequence of five deep sections across the Stream from the Florida Straits to south of Nova Scotia shown by Worthington (1976). The temporal similarity has been illustrated by a set of 90 XBT temperature sections between New York and Bermuda made by the U.S. Naval Oceanographic Office (Gulf Stream Monthly Summary 1970, 1971, 1974). We use these N.Y. to Bermuda XBT sections below to examine small seasonal changes in the average Gulf Stream temperature cross-section and transport.

In the absence of direct current measurements, the downstream component of the velocity structure may be determined from the density structure by assuming geostrophic and hydrostatic balance in the cross-stream and vertical directions respectively. A reference velocity or reference transport is required to provide a constant of integration for this "dynamic method". We examine later such a section with reference velocities from deep Swallow floats by Warren and Volkmann (1968). Other sections using deep floats have been made by Barrett (1965), and Fuglister (1963), or using transport floats (Knauss 1969).

Directly measured velocity profiles provide a better description of the velocity field and also give information on cross-stream flow. Richardson, Schmitz and Niiler (1969) have measured the velocity structure across several sections along the continental margin from the Florida Straits to just south of Cape Hatteras, using transport floats to several intermediate depths at 100–200 m intervals (discussed below). Rossby (pers. comm.) presently has a program underway to repeatedly measure the current structure in an area 150 km downstream of Cape Hatteras with a free-drifting current profiler. This new technique can be expected to substantially improve our description of the current structure and transport variability in the deep offshore region.

The following salient points emerge regarding the current structure determined by the foregoing measurements:

a) The velocity field in the upper 800 m remains remarkably similar geographically and temporally; however, it appears to be somewhat more variable than is obvious in the density field. This variability can be attributed partially to the fact that measurements to geostrophically estimate the velocity field depend upon the horizontal derivative of the density field (see Fig. 7), and as such are subject to measurement error and erroneous influence by (ageostrophic) internal wave displacements. Similarly, the direct velocity or transport profiles are themselves somewhat aliased by short-term internal motion, especially inertial oscillations. These effects add to the inherent eddy variability in the structure of the Gulf Stream current itself.

b) There is a slight broadening and deepening of the current downstream of the constraining Florida Straits, as shown by the isotachs in Fig. 6 and 7, but the downstream similarity of the current structure is clearly displayed. Figure 6, excerpted from Richardson, Schmitz and Niiler (1969), shows three sections averaged from several transects with multi-depth transport floats. The Ft. Pierce section is at the northern limit of the Florida Straits, and the Jacksonville and Cape Fear sections are respectively about 300 and 600 km further downstream. Progressing downstream, the current neither slows down appreciably nor broadens, although the transport increases greatly. Figure 7 is taken from the Warren and Volkmann (1968) section near 38°N 69°W, about 1600 km downstream of the Ft. Pierce section. The current structure in the upper 900 m is remarkably similar to that shown in Fig. 6. The velocities in this transect have been recomputed using the gradient wind relationship to take into account the path curvature observed. Isotachs and isotherm contours have been superposed in this figure. The maximum velocity core (~ 200 cm s^{-1}) at the surface carries the warmest water northward from low latitudes. As a result of the warm core, the near surface currents increase downward before decreasing, and the region of swiftest subsurface currents is tilted downward toward the right edge, so that the maximum current at 200–400 m depth is 20 km offshore of the surface maximum.[2] Currents of 40–60 cm s^{-1} extend to 800 m depth.

c) The deep waters under the Stream exhibit significant baroclinicity usually down to the ocean bottom. Whether or not this indicates that the current itself extends to the bottom has been a question of considerable interest and importance due to dynamical implications regarding topographic steering of the current path. Several investigators have made observations relevant to this question (Barrett 1965, Fuglister 1963, Warren and Volkmann 1968, Barrett and Schmitz 1971, Richardson and Knauss 1971, Luyten 1977, Hendry 1981, Watts and Johns 1982). In these deep transects there is generally a region of downstream flow (~ 10–20 cm s^{-1}) extending to the bottom, usually somewhat offshore of the near-surface current maximum, flanked by regions of equally strong counterflow. Those deep transects, which include reference velocities, are too few in number to characterize the evidently strong variability in the deep currents. Two lines of deep current meters across the Gulf Stream near 70°W (Luyten 1977) showed *mean* currents (~ 1–2 cm s^{-1}) to the east in waters south of the 4000 m isobath and to the west (~ 3–8 cm s^{-1}) nearer the conti-

2 It is conventional to speak of the right (offshore) and left (shoreward) sides of the current as seen looking downstream.

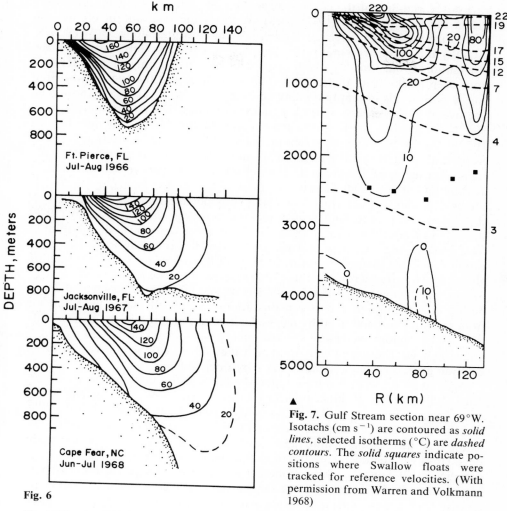

Fig. 6

Fig. 7. Gulf Stream section near 69°W. Isotachs (cm s^{-1}) are contoured as *solid lines,* selected isotherms (°C) are *dashed contours.* The *solid squares* indicate positions where Swallow floats were tracked for reference velocities. (With permission from Warren and Volkmann 1968)

Fig. 6. The Gulf Stream current structure on three sections along the coast. Average isotach (cm s^{-1}) contours are plotted from repeated sections where velocity profiles were obtained by dropping transport floats to several depths. (With permission from Richardson, Schmitz, and Niiler 1969)

nental margin. The variability, however, is large (~ 8–12 cm s^{-1} r.m.s.) compared to the mean currents and characterized by bursts of activity. Luyten reported surprisingly short downstream spatial scales, <50 km, and longer meridional scales, ~ 150 km. Hendry (1981) moored deep current meters under the mean path of the Gulf Stream near 55°W in a closely spaced zonal array in order to further examine these short zonal scales. In contrast to Luyten's observations, he reports good zonal coherence at all separations from 20 to 90 km.

Eddy variability of the currents near the Gulf Stream is much higher than within the gyre. Schmitz (1977, 1978) describes the distribution of low frequency kinetic energy determined on long-term moorings along 55°W (POLY-MODE) and other deep moorings along 70°W. Near the Gulf Stream the KE distribution is only weakly depth-dependent, increasing only by a factor of 2 from deep (4000 m) to thermocline (600 m) waters, and the spectral distribution at both levels is simply dominated by mesoscale (50–100 day) variability. The overall KE decreases smoothly southward along 55°W from a mooring at 37.5°N, ≈200 km south of the typical axis of the Stream, to moorings at 31.5°N and 28°N, which represent the interior of the gyre. At thermocline levels (600 m) the decrease is a factor of 30, and at abyssal depths (4000 m) the decrease is a factor of 100. Schmitz et al. (1981) confirm this vertical and lateral distribution of KE with Lagrangian float observations. Dantzler (1977) has shown a similar increase (factor of 16) in eddy potential energy in the thermocline from mid-gyre to Gulf Stream regions; however Kim and Rossby (1979) show that most of the change could arise from the geographical distribution of Gulf Stream rings. Cheney and Marsh (1981b) illustrate the increased eddy variability of sea surface height near the Gulf Stream as measured by satellite altimeter. They find a roughly threefold increase in r.m.s. surface height (~ tenfold energy change) between the Gulf Stream and mid-gyre. The spectral distribution of KE in deep waters remains approximately the same near and far from the Stream, dominated by mesoscale fluctuations, but in the thermocline of mid-gyre regions the largest spectral energy is associated with periods longer than 100 days. The strong eddy field near the Gulf Stream may drive the recirculation gyre, both with similar weak depth dependence.

Watts and Johns (1982) have tested the relationship between the deep currents and meanders of the near-surface Stream near 73°W. The current meters were positioned 1000 m off the bottom in 3000–3700 m water depths. The deep temperature records (T) are coherent with the Stream displacements, indicating that the deep baroclinic field translates laterally together with the meanders. However, the deep currents (u,v) associated with the most energetic periodicities (> 10 days) are neither coherent with the lateral near-surface displacements nor with T measured on the same meters. In fact, they are too energetic to arise from a simple mean-flow-plus-lateral-translation description. Kinematically this requires that the currents flow mainly along isothermal surfaces, slightly inclined to the horizontal, so that temperature and velocity are not coherent. Slightly greater or lesser inclination of the currents then serves to translate the deep baroclinic field laterally, and this can be associated with less energetic fluctuations on different time scales. The energetic along-isotherm flow in this case conforms well with the kinematics of topographic Rossby waves. Thus the existence of these energetic fluctuations effectively masks the current response to lateral translations of the deep baroclinic field, which are therefore evident only in the temperature records. The relationship to Schmitz's observations of a strong barotropic eddy field in the vicinity of the Gulf Stream is not clear at this time.

The volume transport of the Gulf Stream increases significantly downstream. Knauss (1969) examines 14 volume transport measurements, using either transport floats or geostrophic measurements relative to Swallow floats, extending from the Florida Straits to the transport maximum at about 65°W. He finds a rather smooth increase in transport of 7% per 100 km, roughly doubling the transport from 30×10^6 m³ s⁻¹ off Miami to 60×10^6 m³ s⁻¹ off the Carolinas and again to over 120×10^6 m³ s⁻¹ near 65°W. Volume transport values reported by Worthington (1976) are in agreement with these measurements. Superimposed on this general geographic variation are substantial temporal variations, which have been characterized best in the Florida Straits by a sequence of transport float transects as shown in Fig. 8 (Niiler and Richardson, 1973). They are arranged in this figure to emphasize the seasonal variability, with a maximum of 34 Sv in the summer and a minimum of 25 Sv in the winter. However, note that the seasonal trend only accounts for 45% of the variability. Short-term fluctuations, such as can be seen in the group of April 1968 measurements, account for the remaining 55% of the variability. It is interesting to note that the seasonal transport variation is out of phase with the wind curl forcing and the associated southward Sverdrup transport across the rest of the ocean at that latitude. Gunn and Watts (1982), in a study of the Antilles Current, have suggested as one possible controlling mechanism that the inflow through the Windward Passage to the Caribbean and thence through the Florida Straits is seasonally modulated. They indicate that there is inflow at sill depth through the Windward Passage in the summer when the Sverdrup transport is weak and the Antilles Current is absent, but that this inflow is absent in the winter when the Sverdrup transport is high and the Antilles Current is present. The evidence is incomplete and the subject requires further study.

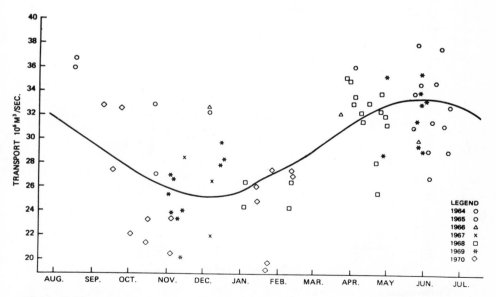

Fig. 8. Gulf Stream transport in the Florida Straits, plotted vs. month. (With permission from Niiler 1975)

Satellite altimetry observations can determine the surface height change, Δh, across the Gulf Stream (Cheney and Marsh 1981a, Kao and Cheney 1982). They discuss several altimeter passes across the Gulf Stream in the region 65°W–70°W. Interesting changes are found in Δh, which is proportional to geostrophic surface transport, varying by $\sim 30\%$ with time scales of a few days. Furthermore in a simple two-layer sense the transport above the main thermocline varies as $(\Delta h)^2$, and this has fluctuations as large as a factor of two during the 3-month study period in this region. Because such variations are not reflected in the hydrographic data, it is believed that they are associated with the barotropic component of the flow.

Downstream of Cape Hatteras near 70°W, a seasonal trend in the transport has already been noted (Iselin 1940, Worthington 1976), with the highest transport (~ 85 Sv) in the spring and the lowest (~ 70 Sv) in the fall (Fig. 5). These are all geostrophic calculations relative to 2000 m, mostly on the Montauk-Bermuda transect. Worthington has suggested the formation of 18° water by wintertime deep convective cooling as a possible mechanism to deepen the thermocline in the Sargasso Sea and thus increase the Gulf Stream baroclinic transport. The set of U.S. Naval Oceanographic Office XBT transects from New York to Bermuda (Gulf Stream Summary 1970, 1971, 1974) consists of 90 tem-

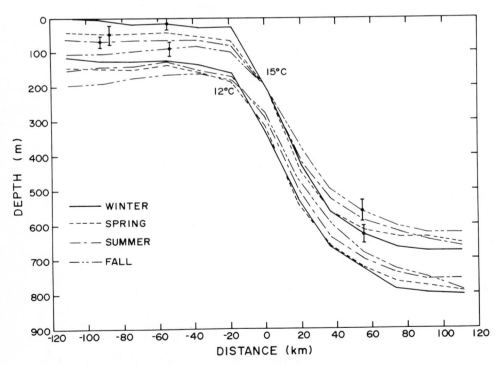

Fig. 9. Cross-stream profiles of the 12°C and 15°C isotherms in the Gulf Stream, averaged by season from three years of U.S. Naval Oceanographic Office XBT sections from New York to Bermuda

perature sections during 3 years. They have been categorized by season, shifted to align the "north wall" position, and then averaged together[3] to produce the four seasonal-mean sections in Fig. 9. The 15 °C and 12 °C isotherm depths across the Stream are shown with the four seasons superimposed. The section was almost perpendicular (80°) to the average path of the Stream near 70°W, but individual sections may have been somewhat broadened if the instantaneous current crossed the transect more obliquely. This average section must therefore be biased to be somewhat broader than the average *normal* section, but of course estimates of the baroclinic transport are unaffected. The winter grouping (January–April) clearly has the greatest change in thermocline depth across the Stream with progressively less in spring (May–June), summer (July–September), and fall (October–December). The overall differences between the profiles exceed the r.m.s. range of error bars which apply to individual seasons. For an appropriate choice of g* (3.2×10^{-2} m s^{-2}) the transport corresponding to $H_B^2 - H_A^2$ ranges from 75 Sv in the fall to 90 Sv in the winter. This transport cycle is qualitatively and quantitatively very similar to Worthington's (1976) observations. Such a seasonal variation in transport could be either wind-driven, or a result of a thermohaline forcing mechanism.

An interannual variation in transport at 70°W may also be noted by grouping the above transects by year. The corresponding annual averages (biased somewhat toward summer conditions when there were more frequent transects) are:

$$1970: <T> = 83 \text{ Sv}$$
$$1971: <T> = 83 \text{ Sv}$$
$$1973: <T> = 99 \text{ Sv}$$

This is of course only a preliminary indication of the range of variation in annual average transport, but the 1973 data set may be biased *low* if anything by the absence of winter measurements.

6.4 Water Masses

Gulf Stream water properties are vertically stratified, exhibiting great similarity along density surfaces which themselves are baroclinically inclined acrossstream like the isotherms plotted in Fig. 7. This similarity in water mass properties is best illustrated in terms of standard [TS], [σ_t O$_2$], or other characteristic diagrams, in which the rapid vertical variations play no explicit role; hydrographic stations from both cross-stream and along-stream locations conform to the same standard curves in this manner.

In order to study *variability* (either cross-stream, geographical or temporal) it is a useful technique to subtract the "standard" characteristics from a profile or section and thus display the residual "anomalies" more prominently. The standard curve need not have any particular physical significance nor be an ac-

---

3 Before averaging, sections were discarded that had a Gulf Stream ring near enough to perturb the thermocline shape across the Gulf Stream.

curate average in order to elucidate the anomalies; but of course the *same* "standard" must be removed in order to compare the variability from one survey to the next. In view of Worthington's (1976) recent monograph discussing the water mass structure in the Gulf Stream system, the purpose here is to be illustrative rather than attempt to be comprehensive. The discussion will proceed from the surface waters downward using the standard [TS] curve, shown as the heavy line in Fig. 10, as the primary means of identifying the various water masses. In the same manner as in Worthington's (1976, footnote 3, p. 44) monograph, this [TS] curve is constructed from the following segments: between 18 °C and 4 °C the [TS] curve of Iselin (1936, Fig. 53) has been used, and below 4 °C the [θS] curve of Worthington and Metcalf (1961); above 18 °C our own average of stations northeast of Cape Hatteras has been used.

The surface waters are subject to exchange of heat and moisture with the atmosphere, and hence exhibit seasonal as well as longer and shorter period variations. Below 100 m this variability is essentially absent. While there is a general tendency for the warmer surface waters to be more saline, in fact the maximum salinity is found at a sub-surface depth of 100–200 m in association with a core of Subtropical Underwater (SUW). This water mass is formed in the tropical Trade Wind zone of the North Atlantic where evaporation exceeds precipitation (Worthington 1976), and it spreads westward below the surface towards the Antilles and Caribbean Sea. Fluctuations in precipitation, evaporation, and mixing with other water masses can vary the amount of SUW produced as well as the strength of the maximum salinity by ~0.5‰. Despite the variability, a core of SUW (~36.8‰) is consistently present in the waters emerging from the Florida Straits, whereas further offshore out of the Stream no subsurface salinity maximum is reliably present at this latitude (Wennekens 1959). South of New England, in the middle of the Stream, a core of SUW (~36.8‰) is commonly observed, interleaved with less saline Sargasso Sea waters, as indicated by the stippled region in Fig. 10.

In a depth range which includes the SUW (100–200 m) and extends to about 600 m, Gulf Stream waters of Florida Straits and Caribbean origin have distinctly lower oxygen values than exist at the same density surfaces in the Sargasso Sea. Figure 10 shows the mean Sargasso [$\sigma_t O_2$] curve from Richards and Redfield (1955). The oxygen anomalies relative to that standard curve are -0.5 to -2.0 ml 1^{-1}. Richards and Redfield (1955) show several cross-sections of the Stream, from the Yucatan Channel to south of Nova Scotia (~62 °W), in which these anomalies are clearly present from 100 to 200 m and sometimes as deep as 600 m directly within and offshore of the high velocity core of the Stream.

Below the SUW, the vertical gradients of temperature and salinity decrease in the Subtropical Mode Water (STMW), as discussed by Masuzawa (1969) and Worthington (1959). In the North Atlantic this water type is associated with a thermostad close to 18 °C and has also been known as "18° Water". Continuous [TS] profiles through the STMW (17.9 ± 0.3 °C and 36.5 ± 0.10‰) generally exhibit a somewhat more pronounced halostad near 18 °C such as appears in Figure 10. Below this, the relatively straight portion of the [TS] curve from 18 °C to about 9 °C is called the North Atlantic Central Water (NACW).

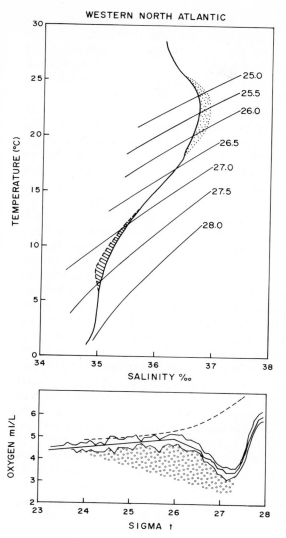

Fig. 10. Characteristic diagrams for Gulf Stream waters. *(top): Solid curve* is the [TS] standard curve described in text. *Dotted area* indicates high salinity anomalies associated with Subtropical Underwater. *Hatched area* indicates low salinity anomalies associated with Antarctic Intermediate Water. Contours of σ_t are labeled. *Bottom* The [$\sigma_t O_2$] relationship of Richards and Redfield (1955) for Sargasso Sea waters is shown on a *solid curve* with standard deviation lines above and below. Gulf Stream waters typically have lower oxygen levels in the region indicated by *open circles*. *Dashed line* is the oxygen concentration in saturated waters having the [TS] characteristics of Sargasso Sea water

Oxygen values below the SUW usually decrease through the STMW and the NACW and reach a minimum of less than 3.0 ml 1^{-1} above the oxygen-rich deep water. The presence of this oxygen minimum is a common feature in the deep main thermocline of the world's oceans. In this region the minimum typically occurs at approximately 800 m depth and consistently at the same sigma-t surface, 27.1 ± 0.1, or corresponding temperature ~ 10 °C. Richards and Redfield (1955) show an accentuated oxygen minimum in waters emerging from the Florida Straits. In fact this strengthened oxygen minimum persists in some of the sections out to 68 °W, but others exhibit only the standard oxygen minimum. It appears that the negative oxygen anomaly signature is somewhat more persistent in the more swiftly moving σ_t ~ 26.0 to 26.5 surfaces.

Below the oxygen minimum there is a salinity minimum (and an increased oxygen concentration) that is caused by the contribution of Antarctic Intermediate Water (AAIW). This water mass is present in all parts of the South Atlantic and as far north as 20°N in the North Atlantic, beyond which its distribution becomes fragmented. Throughout the Caribbean there is a well established salinity minimum which is strongest at temperatures near 6°C. (There are colder waters in the Caribbean, down to $\theta \sim 3$°C, which are influenced by NACW but not directly by Mediterranean Water; however, the limited depth of the Florida Straits prevents these from flowing out.) Waters as cold as 7°C (S ~ 34.88‰) and occasionally 6°C (S ~ 34.84‰) flow with the Gulf Stream out of the Florida Straits carrying the AAIW low salinity signature (Iselin 1934, Parr 1937). The sill depth of 570 m occurs in the northern end of the Florida Straits; however the density layers rise approximately 400 m against the left edge of the channel. Hence the 6°C water which is about 800 to 900 m deep in the Caribbean is able to flow out the Straits. Other offshore waters which join the Gulf Stream north of the Bahamas have distinctly higher salinities at these temperatures (~ 35.07 and 35.03‰ at 7° and 6°C respectively). The corresponding [TS] curve for Sargasso Sea water has no local minimum, only an inflection point, caused by the contribution of AAIW. Figure 10 has a hatched region between the standard Sargasso Sea [TS] curve and the somewhat fresher ($\sim .07$‰) waters between ~ 6° to 10°C which originate from the Caribbean Sea and Yucatan Straits. Downstream of Cape Hatteras one typically finds these water types interleaved, as will be illustrated below.

Further downstream, northern fresh water sources can influence the [TS] relationship below 10°C. Worthington (1976) shows a tongue of cold fresh water extending from the Grand Banks westward to south of Nova Scotia. Centered at approximately 6°C, Subarctic Intermediate Water contributes low salinities from 10° to between 5° and 4°C, and Labrador Sea Water (centered near 3.4° with S ~ 34.96‰) contributes low salinities in the 3° to 4°C range. Any fresh water influence in deep Gulf Stream or Slope Water regions south of New England at temperatures colder than 6°C certainly has a northern origin. We return to this subject in the discussion of variability.

In the Western North Atlantic the influence of high salinity Mediterranean Water is most pronounced in the 4° to 5°C temperature range at a depth of about 1300 m. It causes only a small inflection near 4.5°C in the standard [TS] curve of this region and hence for the waters which have joined the Gulf Stream south of New England.

Deeper yet, below 4°C, there is the North Atlantic Deep Water (NADW), with its complex constituency as described in Worthington (1976). Gulf Stream, Sargasso and Slope Waters all exhibit the same [TS] characteristics in these deep waters, and consequently cannot be distinguished by these properties. The bottom potential temperature gets as cold as 1.8°C, and is thought to have a 5% to 10% contribution of Antarctic Bottom Water (AABW), based upon [TS] and dissolved silicate considerations (Needell 1980, Amos et al. 1971).

In the above discussion of the "standard" characteristics some treatment of geographic variations has been implicit, akin to the larger-scale gradations shown for different North Atlantic regions in Wright and Worthington (1970). The properties associated with SUW, the O_2 minimum, and AAIW which characterize the Gulf Stream as it leaves the Florida Straits are still observed as the Stream enters deep water off Cape Hatteras, but variability in its temporal and spatial characteristics can only be preliminarily defined at present. Figure 11 illustrates this with a section near 73°W showing the salinity anomalies relative to the above standard curve. A core of SUW is clearly defined in the center and right-hand side of the current, and AAIW is observed as a continuous band between 7° and 10°C. This is probably an unusually continuous example of these anomalies, which are more commonly discontinuous or patchy in appearance, particularly as one proceeds downstream. Accordingly, the presence or absence of these anomalies in a section becomes less definite. Rossby (1936) suggested that the negative oxygen anomalies at the oxygen minimum seem to disappear within a few hundred kilometers downstream of Cape Hatteras in contrast to the Richards and Redfield results, but this contradiction is probably due to the relative scarcity of such data.

The temporal variability of the water mass properties in the Gulf Stream has been extensively studied only in the Florida Straits (Parr 1937, Wennekens

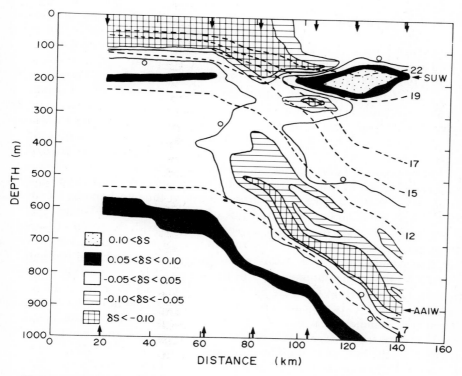

Fig. 11. Section of salinity anomaly on Chesapeake-Bermuda section. Selected isotherm contours are *dashed*

1959, Broida 1962, 1963, 1964, 1969, Niiler and Richardson 1973). Off Charleston, S.C., additionally Hazelworth (1976) reports on a sequence of biweekly hydrographic surveys across the Gulf Stream to characterize the variability. There are some indications of seasonal and longer-term variations, as well as fluctuations on time scales of a day to a week induced by tidal currents, wind forcing, and other fluctuations in the current strength or position. In other regions a similar range of variability is observed, but the associated time scales are not known; it is unlikely, however, that tidal and wind forcing are as important in the deep water regions.

Across the Florida Straits, Parr (1937) made a series of five anchor stations from Miami to Bimini with observations every 2 h at each station for 24 h. He presents rather convincing evidence of tidal modulation of the strength of the SUW core. It is carried within the central-to-right-hand part of the current. On the left side, intrusions of SUW and fresher Gulf of Mexico/Continental Edge Water alternate semidiurnally in his study. Wennekens (1959) also finds the low salinity influence of Edge Water in the upper 300 m along the western edge of the current and states that its identity is retained north of the Straits.

Niiler and Richardson (1973) show the following seasonal differences in the typical water properties in the Florida Straits: (a) During the summer, water in the upper 100 m warms 3° to 4°C to 27° to 28°C at the surface. In winter there is typically a deeper (~ 100 m) mixed layer near 24°C. (b) At the 200 to 400 m depth range on the eastern edge of the current the wintertime profiles typically show an excess of 18° water which is absent in the summer. Niiler and Richardson suggest that this is advected into the Windward Passage near 20°N and subsequently flows through the Florida Straits. By contrast, Gunn and Watts (1982) have suggested that the flow into the Windward Passage is lower in winter/spring in association with the presence of an Antilles Current than in summer/fall when no Antilles Current was observed. An approximate advective flow-through time scale from the Windward Passage to the Florida Straits is 1 month, hence such a lag would not explain this discrepancy. It seems possible at these depths that the 18° water could join the eastern edge of the Gulf Stream via the Old Bahama Channel along northern Cuba. This channel is 25 km wide with a sill depth of 670 m. Wennekens (1959) observes that waters along the offshore edge of the current north of the Old Bahama Channel have the standard $[\sigma_t O_2]$ characteristics of Western Atlantic Water (Sargasso Sea) rather than the oxygen deficiency associated with the water of Caribbean origin, as would be consistent with a tongue of water joining the flow through that channel. (c) More AAIW participates in the circulation through the Florida Straits in the summer than in the winter. The transport of water colder than 12°C increases from 2.7 Sv in the winter to 5.6 Sv in the summer. Water colder than 7°C is typically present as shallow as 300 m on the Miami Terrace in the summer. Lee and Mooers (1977) show similar seasonal and higher frequency trends from their examination of current and temperature variability near the bottom in the Florida Straits.

Sections off Chesapeake Bay typically show, in addition to patches of SUW and AAIW (Fig. 11), a near-surface entrainment of cold fresh shelf waters from north of Cape Hatteras (Ford et al. 1952, Fisher 1972, Stefanson et al. 1971).

A filament of this cold water is particularly noticeable along the northern edge in springtime, when the temperature contrast is greatest. As it leaves the shelf, it is confined to the upper 100 to 200 m, and can cause temperature inversions near the north wall down to 8 °C (10° to 12 °C, later in the summer) which are characterized by salinities lower than in Slope Waters. This water apparently has excess density, as the inversions can adjust to greater depths (~ 400 m) in ~ 150 km downstream.

Warren and Volkmann (1968) discuss the water mass structure in a section near 69 °W in which two intermediate depth layers of low salinity are observed: (1) one in a temperature range 4.5° to 6 °C with $\sigma_t \sim 27.6$ and S < 34.88‰, low enough to cause temperature inversions associated with Subarctic Intermediate Water, and (2) another in the temperature range 3.6° to 4 °C with $\sigma_t < 27.75$ to 27.80 and S ~ 34.96‰, associated with Labrador Sea Water. Fuglister's (1963) Gulf Stream '60 sections (east of 69 °W) also show these intermediate depth salinity minima of northern origin.

Gatien (1976) has reexamined the Gulf Stream '60 sections and categorized the waters to the north of the Stream by their [TS] distinctions into (1) Warm Slope Water in the upper 300 m just north of the Stream, and (2) Coastal Water in the upper 100 m farther north with (3) Labrador Water beneath it along the continental margin down to ~ 1200 m depth.

McCartney et al. (1980) discuss two sections taken along 55 °W a few months apart which differ vastly in several water mass structures. They show that after a severe winter, patches of 18° water, Subarctic Intermediate Water, and Labrador Sea Water are all present, whereas they were absent in the preceding survey. Both sections illustrate the presence of high salinity Mediterranean Water at ~ 1200 m and Denmark Straits Overflow Water at the bottom, indicated by its negative anomaly in silicate.

Clearly the temporal information available in the Florida Straits gives valuable input to the study of the general circulation. It would be extremely useful to have a high quality comprehensive (T, S, O_2, Si transport, velocity structure, etc.) set of observations systematically repeated along other sections such as Chesapeake-Bermuda, near the transport maximum (~ 68 °W) and south of the Grand Banks (~ 50 °W).

6.5 Dynamics

In this section we examine aspects of the vorticity balance in the Gulf Stream and discuss the implications of the potential vorticity distribution with regard to the general circulation and as a signature of instability. We also discuss measurements pertaining to the energy fluxes between the "mean field" of the Gulf Stream and its fluctuations. These relationships and implications are rather preliminary given the currently available measurements; however, a brief discussion here seems appropriate to focus the questions, since new measurement capabilities now combine to offer the means to address these problems.

Ertel (1942) demonstrated that if dissipative effects and diabatic processes are insignificant, a parcel of water conserves its potential vorticity, $\pi = \omega_a \cdot \nabla \sigma_\theta$, where π is the projection of the absolute vorticity ω_a onto the gradient of the potential density. Thompson and Stewart (1977) discuss the broad range of applicability of this conservation theorem to ocean currents. Since temperature (or more precisely potential temperature) is also conserved, we use the following convenient expression for the potential vorticity in the Gulf Stream:

$$\pi \simeq (f + \zeta) \frac{\partial T}{\partial z} - \frac{\partial v}{\partial z} \frac{\partial T}{\partial n}, \tag{1}$$

where in normal coordinates $\zeta = \kappa v + \partial v / \partial n$. For convenience in the following discussions, the two terms on the right will be referred to as π_1 and π_2, respectively. In the central ocean where the relative vorticity is small compared to f, π is adequately represented by f $\partial T / \partial z$.

Using π as a conservative quantity, it can serve as a tracer in the ocean. McDowell et al. (1982) have presented maps and sections of the North Atlantic potential vorticity distribution as a study of general circulation patterns. They find regions within which π is relatively uniform, with rapid changes concentrated outside them. South of the Gulf Stream in the main thermocline waters of the wind-driven gyre they find a large region of uniform π. Since this gyre provides source waters for the Gulf Stream, viewing π as a tracer, the Stream might be expected to have at this level a uniform π distribution across it, and the temporal variability from one section to the next might be relatively insignificant. Stommel (1965) showed in a two-layer sense that the observed velocity and thermocline depth distribution are consistent with a uniform π in the anticyclonic portion of the Stream. Robinson (1965) and Spiegel and Robinson (1968) examined the structure of an inertial western boundary current in a stratified ocean and found that a uniform π distribution requires an influx of fluid at all levels from offshore.

The cross-stream distribution of potential vorticity is also of particular interest as a test of predictions by quasi-geostrophic theory regarding necessary conditions for unstable growth of fluctuations (meanders). One condition is that the cross-stream derivative π_y must change sign somewhere across the section. Moreover, the product of π_y with the basic current, $u\pi_y$, must be positive somewhere within the section. These conditions on the π distribution for quasi-geostrophic temporal growth of fluctuations are developed in Stern (1961), Charney and Stern (1962), and most comprehensively in Pedlosky (1964 a, b). Killworth (1980) reviews these developments. One must be cautious in attempts to relate these quasi-geostrophic theories to observational data on a finite amplitude meandering current. Formally the theory requires a basic state current which is slowly changing in space and time, so that linearization applies for perturbation quantities. Yet in the ocean the perturbations can alter the basic state, so that the instantaneous current may have a π distribution which may be distorted from the basic state. Thus we present observations below without claim to know how representative the π distribution is of the energetics of an ensemble of meanders.

There is an interesting contrast between model results such as from Rhines and Young (1982) and supporting observations (McDowell et al. 1982) in which π tends to be homogenized along isopycnal surfaces offshore of the Gulf Stream, and the above instability conditions on the cross-stream *derivative* of π. In fact for all isopycnal layers in and above the main thermocline, $\pi \simeq f\partial T/\partial z$ is several times higher in the Slope Water than in the Sargasso Sea, and a strong gradient, $\pi_y > 0$, must exist somewhere across the Stream. It is interesting to speculate whether Gulf Stream processes (meanders) may react to varying gradients in π to exert control toward a uniform π within isopycnal layers which is then manifest in the recirculation gyre, or whether processes within the gyre (eddies) homogenize π and tend to produce a uniform π Gulf Stream. Altogether, it is of interest to examine the π distribution in a region where meanders are growing rapidly.

Gill (1977) shows the π distribution in the upper main thermocline waters (16°–20°C), calculated from current and temperature profiles across three sections of the Agulhas Current. Conservation of potential vorticity requires that it be a function of streamfunction, $\pi = \pi(\psi)$. He illustrates that $\pi(\psi)$ retains the same form (a) in two sections spaced 200 km apart downstream with very different bottom topography, and (b) at two different times on the same section when the current had shifted laterally by 20 km.

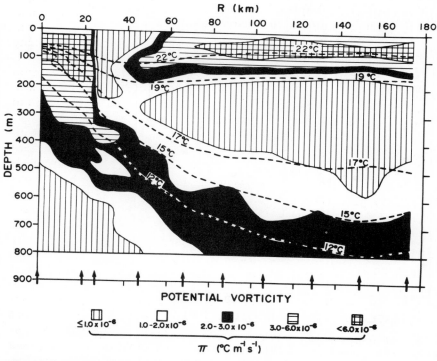

Fig. 12. Potential vorticity π contours on section normal to the Gulf Stream near 73°C. Selected isotherm contours are shown as dashed lines.

The overall π distribution in the Gulf Stream is illustrated by Watts (1983), using density sections for which the path and curvature of the current have also been determined. Conditions are shown under which the constituent terms of π in Eq. (1) may be computed to adequate accuracy by integrating the gradient wind relationship (centripetal term included). Figure 12 shows the resulting π distribution on an XBT section near 73°W normal to the (instantaneous) Stream. Superimposed on the π contours are five selected isotherms. Throughout the anticyclonic portion of the Stream there is clearly a tendency for π to be uniform along isothermal surfaces. It clearly changes vertically in the same manner as the offshore stratification ($\partial T/\partial z$). The 18° water between the 17° to 19°C isotherms is a lens of minimum π, whereas the upper thermocline waters from 12° to 17°C show a local maximum band of π. Near the north wall (15°C/200 m) there is a sharp frontal increase in π for all the waters warmer than 12°C, as already noted from the limiting behavior of $f(\partial T/\partial z)$ in the Slope Water. The π distribution within the 17° to 19°C layer ($\pi_{18} = (\pi_1 + \pi_2)_{18}$) is combined in the left half of Fig. 13 from four such XBT sections at 73°W normal to the Stream, adjusted laterally to make their north wall position (15°C/200 m) their common origin. The right half of Fig. 13 shows the distribution of $\pi_{18} \simeq f(\partial T/\partial z)_{19-17}$ on a section southward through the subtropical gyre. The central gyre has a broad region of uniform π_{18} with a rather symmetric increase both northward across the Stream and southward through the gyre. This is suggestive of an intriguingly simple general circulation pattern in which π is conserved along streamlines around the recirculation gyre.

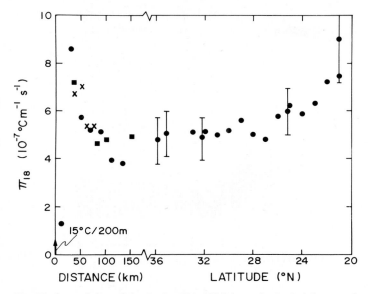

Fig. 13. Potential vorticity in the 17°–19°C layer (π_{18}). *At left* are points from four XBT sections (with different symbols) near 73°W across the Gulf Stream, all adjusted laterally to have a common origin at the "north wall" (15°/200 m). *At right*, with different abscissa scale (but same ordinate for π_{18}) is the distribution of $f(\partial T/\partial z)_{18}$ southward through the subtropical gyre.

Figures 14 and 15 show the constituent terms of π for the two layers, in which in the vertical π has respectively a relative minimum (17°–19°C, π_{18}) and a relative maximum (12°–17°C, π_{15}). The results for the π_{18} calculations are from the same XBT section as in Fig. 12, and for the π_{15} calculations are from a deep hydrographic section made by Warren and Volkmann (1968) in which deep reference velocities were measured by Swallow floats and the Stream path and curvature were determined. Details regarding these sections and calculations are discussed in Watts (1983). In both layers $(f+\zeta)$ decreases by $\sim 50\%$ across the anticyclonic portion of the Stream, but $(\partial T/\partial z)$ increases, with net tendency towards a uniform π_1; hence the gradient tends to be concentrated near the north wall and cyclonic portion of the current. In these examples $\kappa v \sim 10\%$ f and $\pi_2 \sim 25\%$ π in the most baroclinic region. Both layers show that π_y changes sign just offshore of the maximum current (although the case for the 17°–19°C layer is uncertain, see Watts 1983), and $u\pi_y$ is positive in the high velocity core. Hence these distributions would satisfy the necessary conditions for quasi-geostrophic temporal growth of instabilities within the energetic central region of the current.

By contrast, in the Florida Straits the average π distribution determined by Brooks and Niiler (1977) appears to increase monotonically along density surfaces toward the shoreward edge without mid-stream extrema. The current is probably stable along that coast in fact, as the authors discuss on the basis of observed energy fluxes. They state that portions of the variability of the Stream in the Florida Straits may be accounted for by the direct wind variability, or

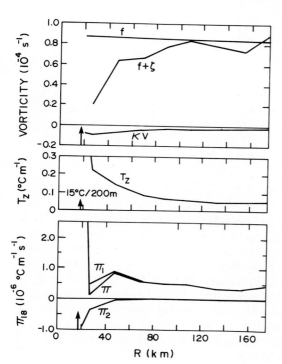

Fig. 14. Potential vorticity and its constituent terms in a 17°–19°C layer on a section normal to the Gulf Stream near 73°W. The "north wall" position is indicated by *arrow*

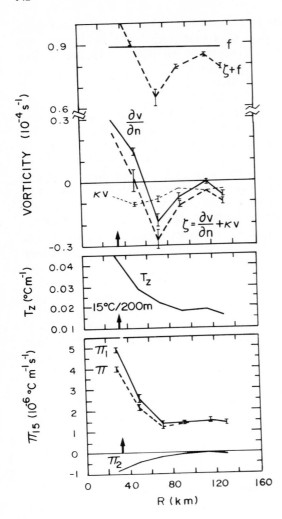

Fig. 15. Potential vorticity and its constituent terms in a 12°–17°C layer on a section normal to the Gulf Stream near 69°W. The "north wall" position is indicated by *arrow*

fluctuations could propagate into that region from other areas. Corroborating evidence off Onslow Bay comes from Webster (1961) and Brooks and Bane (1981), who studied the eddy energy fluxes in the surface currents using GEK, and mid-to-bottom currents using moored current meters, respectively. They assume that the kinetic energy flux $\overline{u'v'}\, \partial \overline{U}/\partial y$ is the most significant term. Both investigations indicate a net eddy conversion of energy from the fluctuations to the mean. This appears to be consistent with the observed decay of meanders downstream of the "Charleston Bump", as was discussed earlier in connection with Fig. 3. Fofonoff (1981) discusses and reviews these studies of eddy-mean flow interactions carefully.

Johns and Watts (1983) analyze the energetics from four current meter records in deep waters under the Gulf Stream off Cape Hatteras. The find a flux of kinetic energy from the mean field to the fluctuations ($\overline{u'v'}\ \partial\overline{U}/\partial y < 0$) associated with periodicities greater than 10 days. Fluctuations in the 4–10 day band return kinetic energy only 5% as quickly to the mean. The potential energy fluxes ($\overline{v'T'}\ \partial\overline{T}/\partial y$) are from the mean to the fluctuations at two moorings along the 3 000 m isobath and roughly equal but opposite further offshore in 3 700 m water for all periodicities (4–90 days). Because these measurements are in deep water where the fluctuations are less energetic, the covariances are typically only 2% of those reported in shallower water off Onslow Bay or in the Florida Straits. The mean field gradients are also low, further reducing the energy transfer rates. Yet the direction of energy transfer within the meander band of periodicities is of interest as a signature that the barotropic instability mechanism may be operative also in the deep currents under a region where the surface meanders are certainly growing rapidly.

The dynamical conclusions from the potential vorticity and the energetics calculations in this section will surely be refined by more appropriate measurements in the near future, specifically designed for these purposes. Nevertheless, the tentative findings in this section suggest an important contrast in the potential vorticity distributions in the upper main thermocline between the Florida Straits, where the current appears to be stable, and the Chesapeake-Bermuda transect, where meanders are rapidly growing. The findings are consistent with theoretical predictions from quasi-geostrophy.

6.6 Summary

The Gulf Stream has been the subject of many studies over the years, due to the intrinsic interest in this powerful current and its relative accessibility to North American oceanographers. It is certainly one of the most completely described currents in the world, yet it is striking how many additional questions remain unanswered. Among topics of current interest are: water transport and water mass variability, heat transport, current structure, near-surface path variability and its relationship to deep current fluctuations, dynamical balances, instabilities, and finite amplitude perturbations. Some of these questions themselves arise due to the level of sophistication of knowledge regarding the Gulf Stream.

To address some of the deficiencies in our present knowledge, such as variability in the water mass structure, will require careful, repeated applications of long-established observational methods. A set of repeated sections on a few key transects of the Stream would add immensely to our knowledge of the natural variability in the Gulf Stream and its relation to the general ocean circulation. The mixing and interleaving of water masses in the Gulf Stream frontal region and the importance of double diffusive processes, while not a subject of this chapter, seems to be a promising subject of inquiry for the general circulation and thermocline studies.

Now there are also available new measurement techniques and technical extensions of old ones, which appear to be particularly suited for answering some of the above questions. Direct observations of the current structure are of great importance, and are now possible with a combination of new current profiling techniques, Lagrangian floats, and current meter mooring designs which permit velocity measurements (almost) within the strong current. Observations of the path and structure of the thermal front are being extended by the use of inverted echo sounders, satellite infrared thermal sensors and altimeters. It seems tractable in the near future to characterize the influence of bottom topography on the Gulf Stream path, to identify the essential instability processes and the associated energetics, and to determine the extent of coupling between the near-surface and deep currents due to stretching imposed by the thermocline slope. Current meter observations near the Stream and numerical studies have indicated some degree of coupling between open ocean eddies and the Gulf Stream mean circulation system as well as meanders, which requires further investigation. Meander dispersion characteristics may be found to vary geographically or depend upon amplitude. Given the very high downstream coherences in meander displacement observed by IES's (thus far in just a 200-km path segment), hope is again raised that the Gulf Stream path may be forecast from upstream conditions.

Recognizing that this discussion of presently tractable questions on Gulf Stream variability is necessarily incomplete and undoubtedly reflects personal biases, the essential point remains that this is an exciting and productive time to study the Gulf Stream.

7. The Northeast Atlantic Ocean

W.J. Gould

7.1 Introduction

In this chapter the northeast Atlantic will be defined as being bounded in the west by the axis of the mid-Atlantic ridge and in the north by a line from the north of Scotland to Iceland. The extreme south of the region will be discussed in the chapter on the tropical oceans.

7.2 General Oceanographic Conditions and Mean Circulation

There are several factors which have made the study of mesoscale activity in the northeast Atlantic more difficult than in the west and it is partly as a result of this that the state of knowledge of such activity is less well advanced. Much of the information concerning the geographic distribution of eddy activity in the oceans of the world has been obtained by the analysis of subsurface temperature data and in particular data from expendable bathythermograph (XBT) probes. Analyses of the type used by Dantzler (1977) have provided indications of areas of high and low mesoscale isotherm displacements and thence of potential energy distributions. Dantzler's analysis extends to much of the northeast Atlantic south of areas in which the isotherm he used (11 °C) outcrops at the sea surface. In the northern parts of the area he studies the analyses must be regarded as subject to possible contamination due to the seasonal erosion of the thermocline. This is known (Robinson, Bauer and Schroeder 1979) to extend to depths in excess of 600 m in some parts of the northeast Atlantic. Thus XBT data may be of limited usefulness in the north.

A second contributory factor to the difficulty of interpretation of temperature-depth information is the presence of the warm, saline Mediterranean outflow originating in the Straits of Gibraltar (Fuglister 1960, Worthington 1976). With its core at around 1 000 m, this water mass introduces enhanced variability over much of the water column. Pingree (1972) showed this influence to extend at least over the depth range 500 to 1 600 m for stations west of Spain and Portugal and northwards into the Bay of Biscay. The high inherent variability in the Mediterranean water complicates the process of determining density structure (and hence the potential energy of mesoscale thermocline displacements) from temperature alone (Stramma 1981).

Eddies in Marine Science
(ed. by A.R. Robinson)
© Springer-Verlag Berlin Heidelberg 1983

For the most part the mean circulation of the basin is weak and is probably best defined in the analyses by Maillard (1981) and Saunders (1982) of averaged and smoothed historical hydrographic station data. Both workers reveal the main part of the upper ocean north Atlantic circulation to cross the mid Atlantic ridge towards the east in two branches, north and south of the Azores (north of 45°N and close to 35°N), which then turn respectively north and south.

7.3 Data Sources

Three types of data will be considered in assessing the mesoscale activity in the area they are:

Hydrographic stations (CTD and water bottle data)
Expendable bathythermograph (XBT) data
Direct current observations

In most cases observational programmes have used these observational techniques in conjunction with one another.

7.3.1 Hydrographic Data

The most extensive coverage of the area is given by the stations worked during the International Geophysical Year and presented in the atlases of Fuglister (1960) and Dietrich (1969) of the southern and northern stations of that survey. Fuglister displays some twelve zonal or near zonal sections between 8° and 58°N. The typical station spacing of 150 to 200 km severely aliasses all but the largest scale features and consequently the only evidence of mesocale (~ 100 km diameter) perturbations of the main thermocline will be apparent station to station noise. The sections, since most of them traverse both the eastern and western basins, may be used for a crude comparison of the relative levels of activity on either side of the mid-Atlantic ridge. On two sections there are features in the eastern basin which could be interpreted as aliassed mesocale displacements of the main thermocline. These are at 28½°W on 16°N and at 30½°W on 36°N. The remaining eastern basin segments are virtually featureless and in no case are the eastern basin sections more energetic than their western basin counterparts.

A further comprehensive survey of a significant portion of the NE Atlantic, but with the added advantage of a smaller station spacing than in the IGY data is reported by the U.S. Navy Oceanographic Office (1973). Four surveys were conducted between October 1970 and October 1971 with STD stations to depths of 2 500 m on zonal sections with a spacing between 225 and 275 km and a station separation on each section of 150 km. The surveys cover the area 30° to 45°N and from 10° to 30°W. Although the station spacing is still larger than one would choose for a mesoscale survey, the data reveal patchiness in salinity and in derived sound velocity with horizontal dimensions of the order of a few hundred kilometers.

The most detailed, basin-wide section is that reported by Saunders (1980, 1982) at approximately 41°N and between 30°W and the coast of Spain. The spacing of CTD stations was 40 km and these were augmented by neutrally buoyant float tracks at approximately 100 km separation. This section does show a well-defined "eddy" between 25° and 30°W, on the flank of the mid Atlantic ridge. It is a warm-cored anticyclonic feature, and produces near-surface geostrophic currents of the order of 15 cm s^{-1} when fitted to the float data at around 1 000 m. The maximum density perturbations occur near 1 000 m (the Mediterranean water core) suggesting that this water mass plays a dominant role in the eddy's dynamics. This eddy is easily the most energetic feature on the section and has a diameter of the order of 400 km. It lies close to an area shown to be one of enhanced potential energy by Dantzler (1977). (this Vol. Chap. 16 Fig. 6).

The northern areas, which probably contain the strongest mean flows, have received considerable attention. In particular the IGY Polar front surveys (Dietrich 1969), and surveys reported by Erofeeva (1972) show the meandering of the polar front to have typical radii of the order of 100 to 150 km and large dynamic height fluctuations.

7.4 XBT Observations

Since the mid 1970's there has been an appreciable effort to investigate the spatial variability of the temperature field in the NE Atlantic by the use of XBT probes. The studies have been made with typical profile spacing of a few tens of kilometers to depths of 500 or 800 m using T4 or T7 probes. The sections

Table 1. Tracks on which closely spaced XBT data have been collected in the NE Atlantic

Location	Probes	Spacing	Published
1. 33°N 33°W to Azores	T7	40 km	Gould; POLYMODE News 16
2. Swath 40°–42°N	T7	29 km	Dugan, Wilson; PMN 16
3. Swath 32°–34°N	T7	29 km	Dugan, Schuetz, Wilson; PMN 30
4. 43°N 10°W to 30°N 17°W	T4	20 km	Huber, Müller; PMN 37
5. 48°N–42°N, 28°W–10°W	T4	40 km	Gould; PMN 43
6. Swath 34°–36°N	T7	30 km	Dugan, Schuetz; PMN 48
7. UK to W. Africa	T4	28–56 km	Henke; PMN 59
8. 52½°–41°N, 10°–30°W	T4 + T7	25 km	Dickson and Gurbutt; PMN 64
9. UK to 50°N 30°W	T4	20–25 km	Wegner; PMN 64
10. UK to 50°N 30°W and Azores to English Channel	T4	35–100 km	Wegner; PMN 70
11. Central NE Atlantic	T4 + T7	25 km	Dickson and Gurbutt; PMN 74
12. UK to West Africa	T4 + T7	40 km	Henke, Zenk; PMN 77

(All references are to articles in the numbered issues of POLYMODE News)

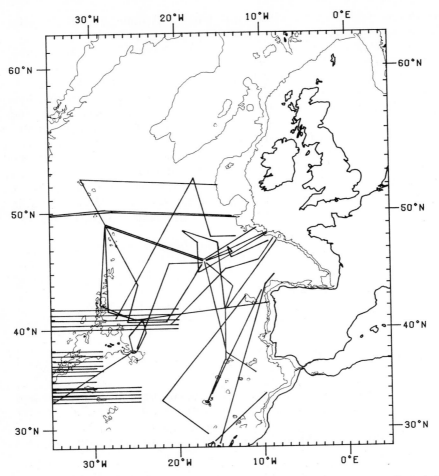

Fig. 1. Tracks along which closely spaced XBT observations have been collected. Most of these are listed in Table 1

which have been collected and published are listed in Table 1 and the positions of the sections are shown in Fig. 1. These reveal that there has been appreciable coverage of the area between 30°N and 50°N. Despite the problems of interpretation mentioned in Section 7.2, it is possible to make some general statements concerning the amplitude of mesoscale temperature perturbations (these defined arbitrarily as having dimensions of the order of 100–200 km and isotherm displacements of order 100 m) on these sections.

The data reveal three distinct areas in which mesoscale features are found. The first, northwest of a line from 40°N 30°W to 50°N 20°W is marked by very large isotherm displacements and is obviously a manifestation of the southern edge of the polar front. The second area is to be found above and to the east of the mid-Atlantic ridge southwest of the Azores (Gould 1976a, 1981). See also Section 7.7. This area is also delineated by Dantzler's (1977) analysis. A third and rather ill-defined area of activity lies in the latitude band 30°–40°N and between 12° and 20°W. In this area mesoscale features have been seen on some of the meridional sections. This is an area close to the Mediterranean outflow and the variability may well reflect the influence of this water mass.

A series of XBT surveys by the MAFF Fisheries Laboratory, Lowestoft UK (Dickson and Gurbutt 1979, 1980) suggest a general decrease of mesoscale activity as one moves from the mid-Atlantic ridge eastward into the eastern basin. Length scales revealed in these surveys appear to remain constant and little or no mesoscale activity is seen east of a line from UK to the Azores and north of 38°N.

7.5 Direct Observation of Currents

The majority of current measurements in the NE Atlantic have been of too short duration to adequately define the mesoscale variability. However since 1978 there has been a major effort to collect longer time series and the success of this operation may be assessed by reference to the data sources cited by Dickson (this Vol. Chap. 15).

The North East Atlantic Dynamics Study (NEADS) has sought to collect 2-year-long records from depths between 600 and 4000 m at widely spaced sites (Fig. 2) in the N.E. Atlantic basin (Gould 1976b) and although many records did not reach the desired 2-year duration, they serve to enable some features of the frequency spectra of mesoscale motions to be determined.

In general Dickson's results show that the kinetic energy levels in the N.E. Atlantic are comparable with those at the least energetic sites in the western basin (close to the centre of the subtropical gyre). Recent work (Gould and Cutler 1980, Dickson, Gould, Gurbutt and Killworth 1982) has shown a significant seasonal modulation of eddy kinetic energy estimates from a wide variety of sites in the NE Atlantic between the latitudes of the Azores and Rockall (35° to 60°N). The effect has been seen at depths from 200 m to greater than 4000 m and at all sites where there is significant bottom slope or roughness. The amplitude of the modulation is in most cases of such magnitude that the KE estimates from long records are dominated by the winter maxima. Dickson (Chap. 15) demonstrates also the marked shortening in time scale of mesoscale motions at abyssal depths as one moves northwards towards the southern end of the Rockall trough.

The frequency content of the eastern basin records can be compared with those in the western basin (Schmitz 1978). Figure 3 shows spectra subjected to the same treatment as used by Schmitz from NEADS sites 3 on the Iberian abyssal plain (42°N 14°W), 4 (north of the Azores at 41°N 25°W) and 5 (46°N 17°W). Each shows a somewhat different energy distribution. Site 3, like many western basin sites, is dominated by the lowest frequencies in the upper layers but at depth shows greatest energy at mesoscale periods (100 day). Site 4,

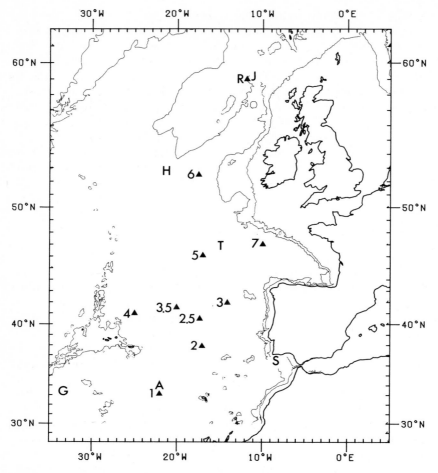

Fig. 2. Positions of NEADS (North East Atlantic Dynamics Study) sites indicated by *numbers*. *Letters* show positions of individual eddy observations according to authors of papers referred to in the text. *A* Armi (1981); *G* Gould (1981); *H* Howe and Tait (1967); *J* JASIN Experiment, Pollard, R. T., T. H. Guymer and P. K. Taylor (1983) Summary of the Jasin 1978 Field Experiment. Phil. Trans. R. Soc. London, A., (in press); *S* Swallow (1969); *T* Tourbillon Experiment, Colin de Verdiere (1980)

which is in generally rough topography and in only 3 500 m of water, shows the deepest records dominated by short (20–50 day) periods and with records above this depth exhibiting dominant 100 day energy. Site 5 is one at which the seasonal modulation of energy has been found and in these records the energy levels in the upper thermocline are greater than at most other eastern basin sites. Once again the records are dominated by the lowest frequencies. At 3 000 m the spectrum is red and not dominated (as in site 3) by the 100 day mesoscale motions.

The statistics of kinetic energy in both the mean flow and in the time varying components of currents at the NEADS sites are shown in Table 2. The data
demonstrate the high, near surface energy at NEADS site 5 and at the Rockall
site; however site 7, near the continental slope shows much lower KE values at
600 m but with quite large (5 cm s^{-1}) mean flows. Some aspect of the mechanism of generation of the seasonal modulation could account for the large difference between the KE values at NEADS 5 and 7 despite their similar latitude.

The current meter records have not been analysed to give estimates of the
vertical modal structure of mesoscale motions, indeed there are in general no
more than four levels observed on any mooring and thus detailed analysis of
only the lowest order modes would be possible. However inspection of the data
time series (Kartavtseff and Billant, 1979, 80, Müller 1981) suggests that much
of the energy is in a sheared barotropic mode, i.e., records have similar appearance at all levels but have reduced amplitudes at depth.

7.6 Satellite Observations

Dickson and Hughes (1981) have studied a cloud free infra-red image of the
N.E. Atlantic from TIROS-N. The area includes much of the Bay of Biscay
southeast of a line from the Ushant peninsular to 46°N 12°W. The enhanced
image shows swirl-like features filling the deep-water part of the area and having typical diameters of 50-100 km. The image is from May, a time of year in
which the seasonal thermocline is well established. The question must be posed
as to whether these surface expressions of eddy-like features are associated
with mesoscale features in the underlying ocean. The horizontal scales are similar to those revealed in the area by satellite-tracked buoys (Madelain and Kerut 1978) and those shown in seasonal surveys of the area using hydrographic
stations (Fruchaud-Laparra, Le Floch, Le Roy, Le Tareau and Madelain
1976).

7.7 Surveys of Individual Eddies

There have been several observations of individual eddies in the NE Atlantic
and in recent years some major experiments have been mounted to study eddy
dynamics in particular areas.

Howe and Tait (1967) report the existence of an eddy near 53°N 19°W and
the data have subsequently been analysed by Brundage and Dahme (1969). For
the most part, their observations were confined to the uppermost 500 m of the
water column, but a few deep stations show major changes in the water column
at depths through the Mediterranean water core. In the upper water column
the eddy is a cold core cyclonic feature elongated in a N-S direction and meas-

Table 2. Statistics of current meter records from long term sites in the NE Atlantic. Nominal depths and positions have been used to avoid complexity due to multiple deployments

Site	Nominal Lat	Long	Nominal depth (m)	Duration (days)	$\bar{u}$	$\bar{v}$	Var u	Var v	KE	$\overline{u'v'}$
1	33°N	22°W	600	323	0.81	−0.38			22.70	−9.95
			1500	176	0.10	−0.86			2.38	−0.35
			3000	323	−0.87	−0.01			3.00	−0.63
			5000	158	−0.43	−0.21			3.20	−0.39
2	38°N	17°W	750	533	−1.20	0.04			28.81	9.80
			1650	432	−0.39	−0.98			4.80	−2.26
			3150	700	0.45	0.02			1.09	0.35
			4250	476	0.79	0.39			0.55	0.13
			5000	152	0.67	1.09			0.44	0.07
2½	40°30'N	17°20'W	500	248	−1.25	−0.01			18.48	11.58
			3000	211	−0.66	−0.03			0.79	−0.07
			4000	248	−0.52	0.08			0.17	−0.07
3	42°N	14°W	600	525	−0.24	0.37	10.83	15.34	13.09	−4.57
			1500	492	−0.71	0.58	4.28	5.26	4.77	−1.26
			3000	469	−0.15	0.30	6.27	5.12	5.70	0.60
			4000	758	−0.94	0.17	6.11	4.62	5.53	0.23
3½	41°30'N	20°W	600	224	0.92	−1.55	17.91	18.86	18.39	7.45
			1000	377	0.56	0.12	4.88	8.90	6.89	−0.24
			1500	238	0.24	−0.22	2.87	5.46	4.17	0.89
			3000	150	−0.26	0.04	1.04	1.43	1.24	0.90
4	41°N	25°W	600	511	0.16	0.76	20.18	13.98	17.08	1.53
			1500	419	−0.07	−0.94	3.50	2.40	2.95	−1.07
			3000	466	−0.79	−0.92	3.07	4.14	3.61	0.94
			3500	688	−0.90	−0.54	4.26	4.36	4.31	−1.24

5	46°N	17°W	600	452	−2.03	1.43	59.47	63.58	61.51	−1.58
			1500	591	−0.56	1.34	8.91	11.05	9.98	4.00
			3000	483	−0.23	0.67	3.04	2.03	2.53	0.24
			4000	362	−0.65	0.77	0.98	1.06	1.02	
			4700	307	−0.64	0.82	1.57	0.87	1.22	
6	52°30′N	17°40′W	3050	830	−0.08	0.28	13.61	10.78	12.19	
			4050	602	−1.70	0.43	18.90	22.59	20.75	
7	47°N	10°W	600	268	5.04	−1.93	19.14	19.58	19.36	−12.34
			1000	268	0.28	−0.73	7.29	6.56	6.93	−1.70
			1500	464	0.08	0.45	5.73	3.30	4.52	−1.27
			4000	732	0.37	−0.33	2.33	1.91	2.12	−0.60
Rockall	59°N	12°W	200	454	−4.89	0.13	209.92	221.22	216.07	6.53
			600	199	−4.18	1.70	105.95	140.68	123.32	−4.46
			1000	423	−3.62	0.70	69.19	89.26	79.23	−3.32
			1500	393	−1.34	−0.64	48.05	44.21	46.13	5.01

$\mathrm{d}k_E / \mathrm{d}(\log_{10} f)\ \ \mathrm{cm}^{-2}\,\mathrm{s}^{-2}$

Fig. 3b ▲

Fig. 3a
▶

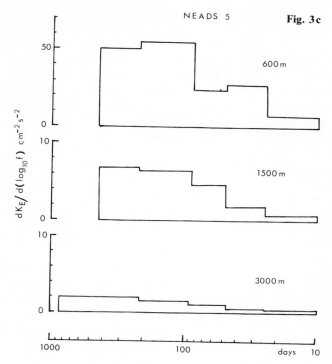

Fig. 3a–c. Frequency spectra of eddy kinetic energy KE. The data have been analysed in a manner described by Schmitz (1978) and display the full band width of the energy estimates. The scaling is such that the area under each portion of the bar chart is proportional to the energy content of that frequency band. **a** NEADS Site 3; **b** NEADS Site 4; **c** NEADS Site 5

uring approximately 200 × 100 km. Howe and Tait speculate that the eddy could have been formed from a pinched-off meander of the N. Atlantic current.

A much smaller eddy-like feature is reported by Swallow (1969) south of Cape St. Vincent (36°20′N 08°40′W). The eddy was revealed in neutrally buoyant float tracks at depths between 1 000 and 1 400 m (close to the maximum salinity layer in the Mediterranean water). The observations were supplemented by a series of hydrographic sections. The small (~ 20 km dia.) anti-clockwise eddy did not have its centre marked with anomonously high salinities, despite being close to the patchy Mediterranean water outflow. It is postulated that local topography may have played a role in the formation of this eddy.

More recently Armi (1981) has reported the presence of intense blobs of Mediterranean water (maximum salinities up to 36.3‰) near 32°N 24°W, well removed from the source in the Straits of Gibraltar, where the typical salinity of the outflowing water is 36.5‰. The blobs are compact, having diameters ~ 50 km. They extend upwards to approx. 700 m and are detectable down to 2 100 m. The dynamics of these eddies is under investigation, as is a mechanism for their generation. Again topographic influence is thought to play a role in their formation.

The previously mentioned cases have been unpremeditated surveys of eddies encountered in the course of other investigations. There have however been at least two relatively recent studies of eddies. The most ambitious is that of the French/British Tourbillon experiment carried out near 47°N 15°W during 1979–80 (Colin de Verdière 1980). The observational programme consisted of a moored array of current meters, repeated CTD surveys on a closely spaced grid and observations using medium range neutrally buoyant floats (Bradley and Tillier 1980) within an area approx. 160 km × 160 km. At the time of writing the major portion of the analysis has been completed but none of the results published. The eddy, however, is of approx. 80 km diameter, anticyclonic, and with a significant structure associated with the entrainment of Mediterranean water around the central core. (J.G. Harvey pers. comm.). It was seen to propagate westwards during the course of the experiment.

A further eddy of similar character was revealed during the JASIN (Joint Air Sea Interaction Experiment) in the summer of 1978 in the north Rockall Trough (near 59°N 12°W). A detailed description of the eddy is given by Pollard (1982).

An extensive investigation of the area of high eddy energy SW of the Azores was carried out in the summer of 1981 (Gould 1981). In this experiment a cyclonic eddy of approximately 100 km diameter was observed to separate from a frontal feature near 33°N 32°W. The eddy was tagged with three satellite-tracked drogued buoys and its westward progress was monitored from its formation in April until the last buoy left the eddy in December near 32°N 37°W (a westward propagation speed of approximately 2 km day^{-1}). Near surface current speeds were at times as high as 40 cm s^{-1} but more typically 20 cm s^{-1} with a rotation period of 20 days. Although not obviously associated with anomalies in the Mediterranean water core, the eddy did penetrate to depths of at least 1 500 m.

7.8 Summary

While eddy activity in the interior of the Northeast Atlantic is less than that over much of the western basin the Northeast Atlantic has been seen not to be devoid of mesoscale eddies. The major branches of the north Atlantic current as they cross the mid Atlantic ridge are areas of relatively high activity (north of 50°N and SW of the Azores).

Very recent work has shown the important role of the Mediterranean water core in the eastern N. Atlantic basin but the relationship of patchiness in the Mediterranean water, to water motions higher in the water column is not yet clear. The process of eddy formation and the areas in which this may occur are not yet fully defined but westward eddy propagation has been observed, suggesting source regions near the eastern boundary. The fate of eddies when they encounter the mid-Atlantic ridge is not yet known. Can the eastern basin be a source region for some types of western basin eddies? The role of Mediterranean water in some western basin eddies (McDowell and Rossby 1978, Kerr 1981) suggests the possibility of their origin east of the mid-Atlantic ridge.

Time scales of eddy motions do not appear to differ greatly from those found in the western N. Atlantic, but the detection of a seasonal modulation of eddy kinetic energy to abyssal depths is at present unique to eastern basin records.

8. Eddy Structure of the North Pacific Ocean

R.L. Bernstein

8.1 Introduction

The North Pacific Ocean is nearly twice as large as the North Atlantic; indeed, in the lower latitudes the Pacific's east-west dimension is about three times that of the Atlantic. Yet both oceans are forced by roughly similar westerly winds in midlatitude, which separate the easterly trade winds to the south from the polar easterlies to the north. Both oceans respond in their wind-driven circulation by forming a set of basin-wide gyres with intensified western boundary currents, the basic character of which has been known for many years. More recently, oceanographers have determined that these large-scale currents frequently develop mesoscale (50 to 500 km) instabilities which are dynamically important.

The relative importance and ubiquitous nature of mesoscale eddy variability first began to be appreciated in the 1960's, with the discoveries of Swallow and Crease in the North Atlantic. As extensive eddy-directed experiments began to be conducted there, the question immediately arose as to the comparative nature of the eddy field in other oceans. In the years since, this question has been approached primarily through the re-analysis of existing historical data, and through additional experimental work. This chapter reviews some of the results obtained through these efforts in the North Pacific.

8.2 Coastal Regions

Mesoscale variability has long been a familiar feature in hydrographic data collected along the coasts of Japan and California. Figure 1, is an excellent example of a mesoscale disturbance in the California Current. Figure 2, a 4-month sequence of dynamic height maps, from data collected in 1952, shows several cyclonic and anticyclonic eddies, with dimensions of 100–200 km. One cyclonic feature intensifies along the Baja California coast, breaks away, and moves slowly west over the 4-month sequence. In the early 1960's a series of direct velocity measurements of the same phenomena were made (Reid et al. 1963, Reid 1965). The eddying nature of the entire California Current system has been summarized by Owen (1980), with particular emphasis on the semi-permanent cyclonic eddy which occurs off southern California between Point Conception and San Diego. Abundant historical hydrographic data shows that

Eddies in Marine Science
(ed. by A.R. Robinson)
© Springer-Verlag Berlin Heidelberg 1983

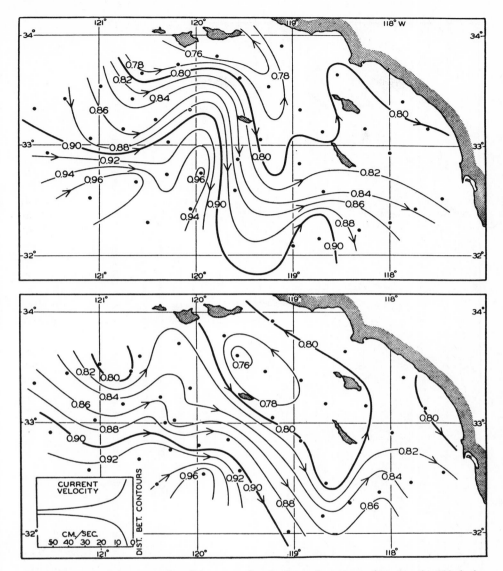

Fig. 1. Geopotential topography of the sea surface in dynamic meters referred to the 500-decibar surface according to observations off southern California on April 4 to 14, 1940 *(upper),* and on April 22 to May 3, 1940 *(lower).* (Sverdrup et al. 1942, p. 454)

this Southern California Eddy seasonally varies in its intensity, reaching its maximum strength in summer, and becoming quite weak in winter. Its location and stability appears to be determined by the coastline geography.

A much stronger counterpart to the Southern California Eddy occurs in the coastal area near Japan. This extremely well-studied feature of the Kuroshio

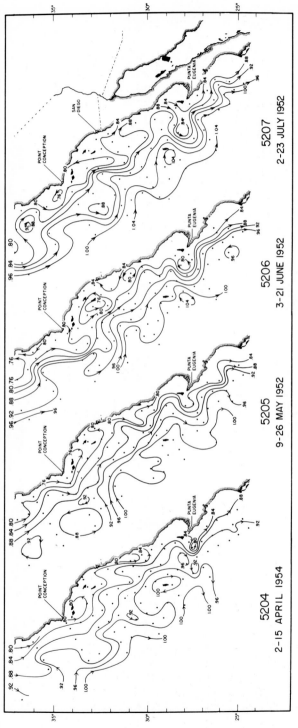

Fig. 2. Sequence of maps of dynamic topography from four CalCOFI surveys of the California Current, April to July, 1952. (Bernstein et al. 1977)

Current is a quasi-stable, 300-km sized cyclonic meander, which develops in the main path of the flow along the south coast of the island of Honshu. The meander and its development has long been studied, and is known to maintain itself for periods of several years or more (Taft 1972). The Kuroshio is now believed to have two stable configurations, with and without the meander, but the governing dynamics which cause the system to change from one state to the other is still not completely understood (Robinson and Taft 1972, White and McCreary 1976).

Slightly further north in the far western Pacific, but still within 500 km of Japan, Kitano (1975) examined the properties of 154 warm-core eddies observed in the Kuroshio and Oyashio Currents, between 1957 and 1973. Their average diameter was 140 km, and they moved in a north or northeasterly direction at speeds of 0.6–4.0 km day^{-1}, following the contours of bathymetry along the continental slope.

8.3 Mid-Latitude Open Ocean

In the rest of the North Pacific, away from the well-sampled regions near Japan and California, little was known about mesoscale structure until recently. In an important sequence of hydrographic cruises, Roden (1970, 1972, 1977, 1979) examined the fronts and eddies over much of the western Pacific, finding frontal transitions generally occurring within a span of 50 km, and eddies typically 150–200 km in diameter. Cheney (1977), and Cheney et al. (1980) observed similar features with aircraft (with air-expendable bathythermographs), ships (hydrographic data), and satellite-tracked drifting buoys.

Dantzler (1976) and Wyrtki et al. (1976) used all historical hydrographic data, and ship-drift data, respectively, to seek out geographic regions of strong or weak mesoscale variability. Perhaps not surprisingly, they found maximum variability concentrated in a relatively narrow zonal band in the western Pacific, where the Kuroshio extends eastward as a jet-like feature along 35°N for several thousand kilometers from Japan. In the same area, Wilson and Dugan (1978) obtained and analyzed data from two long swaths of expendable bathythermograph (XBT) sampling, carried out by 6 or 7 U.S. Naval vessels each sailing along parallel tracks about 45 km apart. Several strong disturbances of approximately 200 km size were encountered. Kenyon (1978) presents results of a single zonal vertical section along 35°N, with XBT's dropped every 20 km between 122°W and 141°E, in which strong eddies dominate the region west of the dateline. Bernstein and White (1977) summarized the results of numerous zonal and meridional vertical sections in the North Pacific, which showed intense eddy or wavelike disturbances in the western half of the Pacific compared with the eastern half, which die out just east of the dateline (Fig. 3).

Recent results by Bernstein and White (1981) from an ongoing ship-of-opportunity XBT program, show that it is possible to map the thermocline expression of these features and follow them from one month to the next, as they

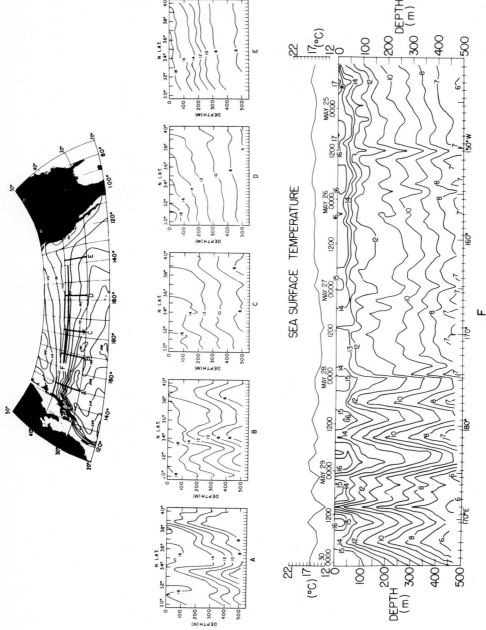

Fig. 3. *Upper* Locations of six vertical sections of temperature, labelled *A–F*, perpendicular and parallel to the axis of the North Pacific Current, superimposed on the long-term mean of dynamic topography of the surface. *Middle, lower* Vertical sections of temperature, from various STD, XBT data sources. (Bernstein and White 1977)

propagate slowly westward at about 4 km day^{-1}. Two years of such maps were time-averaged together, eliminating the traveling eddies, and the resulting long-term mean reveals several stationary meanders in the path of the Kuroshio Extension. The meanders have roughly 500 km wavelength, 100 km amplitude, and conform in some cases to the underlying bathymetry. The Shatsky Rise, a prominent but deep bathymetric feature in the western Pacific, clearly has a permanent influence on the flow above it.

Overall, the eddy characteristics of the Kuroshio appear to be substantially similar to those of the Gulf Stream. Meanders develop and grow in both systems to form cut-off warm and cold eddies north and south, respectively, of the mean path. Once detached, these eddies move toward the west. Both systems exhibit this same type of activity over a region extending for several thousand kilometers zonally, and about 1 000 kilometers meridionally.

In addition to being generally weaker in the eastern Pacific, mesoscale activity also has a tendency to increase in the central Pacific as one proceeds south into the subtropics. This general trend appears most clearly in a long meridional hydrographic section from Alaska to Hawaii (Royer 1978). Eddies are weak or absent between 52°N and 34°N, with typical amplitudes of 5 dyn-cm and lateral dimensions about 40 km. Between 34°N and 22°N the eddies grow in both amplitude and dimension to 10–15 dyn-cm and 150–200 km. Yet even at 15 dyn-cm such eddies are 3 to 6 times weaker than those occurring in the western Pacific near the Kuroshio and its extension.

Observations of eddies north of 35°N in the central Pacific have been obtained by Kirwan et al. (1978), using satellite-tracked drifting buoys. One drifter made four revolutions around an anti-cyclonic eddy at 36°N, 162°W (Fig. 4). Its diameter was about 100 km, not much smaller than those in the western Pacific, but the rotational speed was about three times smaller than typical western Pacific eddies. It propagated west-southwest at 2 km day^{-1}.

Observations of eddies south of 35°N in the eastern mid-latitude region are abundant. Bernstein and White (1974) analyzed data from several sources, including a sequence of 16 monthly hydrographic cruises at 25°N near Hawaii.

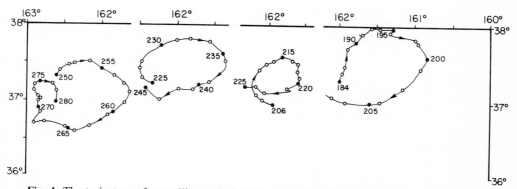

Fig. 4. The trajectory of a satellite-tracked drifter from day 184 through 280 of 1976. During this period the drifter completed four revolutions about an eddy which migrated west-southwest at about 2 cm^{-1}. (Kirwan et al. 1978)

Eddies appear regularly in the data, with lateral dimensions of 200 km, propagating west at nearly 4 km day^{-1}. The disturbances in the main thermocline are typically 40 m, compared with 150–300 m in the western Pacific. Recently, a detailed hydrographic investigation by Roden (1980) showed the existence of cyclonic and anti-cyclonic eddies north and south, respectively, of the subtropical front at 30°N, north of Hawaii.

In addition to tracking drifting buoys, satellites have contributed other information on North Pacific mesoscale phenomena. Along the California coast, satellite infrared scanners delineated the sea surface temperature boundaries and time evolution of an anticyclonic meander developing into a detached eddy, as confirmed by coincident ship hydrographic measurements (Bernstein et al. 1977). The same type of scanner data was used to describe the development of eddies just off Central America, in the Gulf of Techuantepec, by strong coastal winds (Stumpf and Legeckis 1977). In the western Pacific, satellite infrared sea surface temperature, paired with visible channel sunglint information, exhibits strong eddy and frontal behavior of the Kuroshio and Oyashio currents (La Violette et al. 1980). Also with visible channel data, Solomon and Ahlnas (1978) observed eddies in the Kamchatka Current, using pattern of ice floes.

8.4 Tropical Region

In the tropical Pacific, observations of eddy-like phenomena are relatively rare. Most available information comes from a recent two year duration program, in which aircraft and ships executed monthly vertical sections of temperature and density between Hawaii and Tahiti (Wyrtki et al. 1981). In addition numerous satellite-tracked drifting buoys were deployed. In that experiment, and poleward of the westward flowing North Equatorial Current, 300 km diameter cyclonic eddies were found to move west at about 10 km day^{-1}. Only a few such features passed through a given meridian in a one year period. In the South Pacific, poleward of the South Equatorial Current, these eddies appeared less energetic, except in the vicinity of island chains (Patzert and Bernstein 1976). Energetic eddies also did not occur within the strong zonal North Equatorial Current, as demonstrated by Bernstein (1978), using long zonal temperature sections along the high velocity core of the flow.

The North Equatorial Counter Current is a strong but seasonally varying eastward flow running across the entire tropical North Pacific, in the latitude range 5–10°N. At its peak strength (September–January), it develops meanders with an east-west wavelength of 1000 km, and a north-south amplitude of 100 km. The meanders propagate west, upstream to the mean flow, at a speed of 25 km day^{-1}. The meander structure was first discovered by Legeckis (1977) using satellite infrared imagery, in the eastern tropical Pacific. It was subsequently observed as fluctuations in island tide guage records of sea level in the central tropical Pacific by Wyrtki (1978), and in drifting buoy tracks by Patzert and

McNally (1980). Theoretical Studies (Cox 1980) indicate that these pertuba-
tions propagate eastward and downward into the deep ocean, where variability
with similar time and space scales has been observed using deep-moored current meters (Harvey and Patzert 1976).

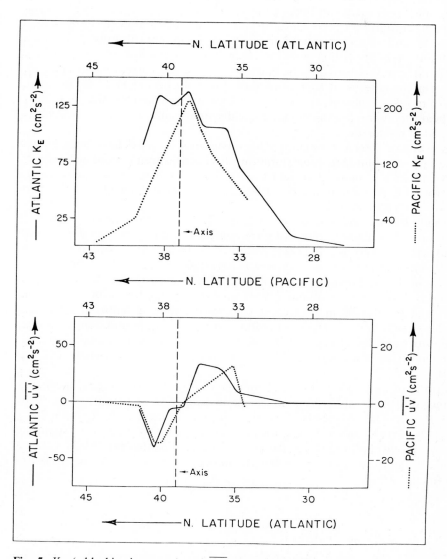

Fig. 5. K_E (eddy kinetic energy) and $\overline{u'v'}$ (off-diagonal Reynolds stress component) as a
function of latitude for sections in the western North Pacific (*dotted line*, ordinate scale on
right) and western North Atlantic (*solid line*, ordinate scale on *left*). The latitude scales are
shifted so as to line up the axis of each mid-latitude jet. (Schmitz 1981b)

8.5 Direct Current Measurements

Only a limited amount of work has been done in the North Pacific, extending beyond basic phenomenological description. Eddy covariance statistics have been determined, however, in a few cases. Szabo and Weatherly (1979), using GEK-estimates of surface current, found significant transfer of kinetic energy both to and from the mean flow in the Kuroshio south of Japan. Nishida and White (1982) found a convergence of cross-stream eddy momentum flux from both north and south into the Kuroshio Extension, contributing to its mean momentum. Schmitz (1981) compared their results, which were based entirely on XBT data, to that found in the North Atlantic using current meter data, and found similar patterns of flux convergence in both oceans (Fig. 5). Bernstein and White (1982), also in the Kuroshio Extension, examined the meridional eddy heat flux there, finding climatologically significant amounts of heat transfer by eddy processes.

The above efforts at estimating eddy covariance statistics have been limited by the lack of abundant, long duration, direct measurements of velocity, by SOFAR floats and moored current meters. These tools have been used extensively for eddy research in the North Atlantic, but are only beginning to be applied to the North Pacific.

Results from a few long duration North Pacific current meter arrays have been reported in the literature. All were confined to the vicinity of Japan, with one exception: Taft et al. (1981) obtained a 19-month record near 30°N, 158°W, north of Hawaii, with current meters 100 and 1 200 m off the bottom. They found weak fluctuations of a few cm s^{-1}, over eddy time scales of 70 to 150 days. Taira and Teramoto (1981) analyzed 9 month records at 33°N, 139°E, and 35°, 139°E, in the Kuroshio south of Japan, at depths of 250 m and 1 670 m. They found velocity fluctuations, with a 33-day period dominant both in the upper and lower layers of the Kuroshio, having nearly equal amplitude (≈ 20 cm^{-1}) in both layers; this is consistent with early results by Taft (1978).

A program of long duration current meter arrays has just begun, specifically directed toward exploratory mesoscale investigation in various parts of the North Pacific. One Woods Hole array is in place along 152°E, between 28°N and 41°N. Several other arrays are planned for various longitudes spanning much of the mid-latitude North Pacific. It thus appears that the 1980's will be a period of significant progress in Pacific mesoscale research, just as the 1970's were for the Atlantic.

9. Subpolar Gyres and the Arctic Ocean

G.T. Needler

9.1 Introduction

The northern parts of the world's oceans, and in particular the Arctic Ocean, remain in many ways unexplored regarding their basic oceanographic properties and the role of quasi-geostrophic eddies in these regions is certainly far from clear. On the other hand, in various local areas information is available about particular oceanographic systems and a variety of eddy types has already been identified. In this chapter some of these are discussed in terms of the areas where they are found. While the presentation is not intended to be exhaustive, it will be seen that enough is known to show that quasi-geostrophic motions play a variety of major roles in the dynamics of the northern oceans.

The small-scale upper ocean solitary eddies discovered in the Beaufort Sea AIDJEX data are discussed and brief mention is made of limited measurements of low-frequency variability in other regions of the Atlantic Ocean. Observations and theories for open-ocean eddies, convective cells, and boundary current instabilities in the northwestern North Atlantic are presented. It is then shown that similar features have been seen in the Norwegian and Greenland seas as have instabilities on the polar front and pulsations of the deep water overflows through Denmark Strait. Lastly, some features of the northeast Pacific are briefly mentioned.

9.2 The Arctic Ocean

According to Hunkins (1974), the existence of "high speed transient currents" in the upper pycnocline of the Arctic Ocean was noted as early as 1937 by P.P. Shirshow using the drift of the first Soviet ice research station, NP-1. Belyakov (1972) has discussed the data from this and other Soviet ice stations as have Galt (1967) and Bernstein (1971) from the U.S. Fletchers Ice Station, T-3. More extensive information on upper ocean currents was obtained during the 1970 and 1072 AIDJEX Pilot Studies, which were carried out north of Alaska and the data from which have been analyzed and interpreted by Newton et al. (1974) and Hunkins (1974). They identified five discrete baroclinic "eddies" in this early AIDJEX data. All the high speed events mentioned above are listed in Table 1 along with their location and maximum speed. The location of the events is shown in Fig. 1 where they are plotted using the identifying code given in the table.

Eddies in Marine Science
(ed. by A.R. Robinson)
© Springer-Verlag Berlin Heidelberg 1983

G.T. Needler

Table 1. High speed transient currents, their location and maximum speed

Code letter	Experiment	Approximate date	Latitude	Longitude	Maximum current speed (cm s^{-1})
j	NP-1	1937	Somewhere along drift track		—
a	NP-2 (Ser 2)	Aug 1950	79°00'N	170°00'W	34
b	NP-2 (Ser 4)	Jan 1951	80°30'	163°00'	20
c	NP-2 (Ser 5)	Mar 1951	81°30'	171°30'	24
d	T-3	Jul 1965	75°30'	142°00'	45
e	AIDJEX 1970	Mar 1970	71°20'	136°30'	22
f	AIDJEX 1972	Mar 1972	76°00'	149°00'	25
g	AIDJEX 1972	Mar 1972	75°00'	148°00'	30
h	AIDJEX 1972	Mar 1972	75°00'	148°00'	31
i	AIDJEX 1972	Mar 1972	75°00'	151°00'	Not available

During the March–April 1972 AIDJEX Pilot Study, oceanographic measurements were made from a triangular array of three ice stations that were separated by about 100 km. During the 5 weeks of the experiment, the stations drifted to the southwest about 100 km at an average speed of 2.4 cm s^{-1}, which is consistent with what is known about the large-scale circulation at their loca-

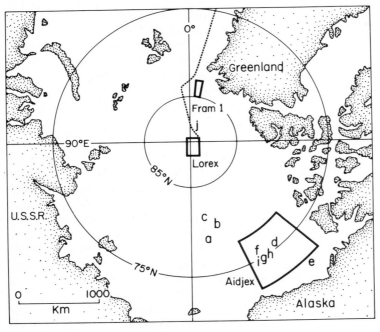

Fig. 1. A map of the Arctic Ocean showing the location of experiments referred to in the text

tion to the south of the Beaufort Sea gyre. Excellent satellite navigation meant that all current measurements taken relative to the motion of the ice could be corrected for the ice motion. The basic data set consisted of hydrographic data at all three stations, and fixed and profiling current measurements in the upper few hundred meters of the water column. The array was however designed for the purpose of examining air/ice interactions and the air and water boundary layers and not primarily for deeper oceanographic features.

The analysis of the 1972 AIDJEX data set is complicated by the ice motion, which itself depends on the surface currents, and, although it was possible to accurately correct water velocity measurements for this ice motion, estimates of the horizontal scale of transient features had to be based on the assumption that oceanographic features were relatively stationary during the time it took the ice station to cross them. The data, however, quite clearly showed four separate high current events or eddies that lasted from one to four days with maximum velocities of about 40 cm s^{-1} and calculated horizontal scales of 10–20 km. There was no coherence to be seen during these periods on the 100 km scale of the ice stations. Two of the events were thought to be cyclonic and two anticyclonic. The horizontal velocity in the eddies showed a relatively sharp maximum between 100 and 150 m with the velocity being small or negligible in the upper mixed layer under the ice and decreasing sharply to the maximum sampling depth of 170 m (Hunkins 1974). The fixed current meters when available at 150 m showed the passage of the eddies. Those at 500 m showed negligible currents. The data set was also used to make comparisons between the observed currents and the geostrophic velocity calculated from the hydrographic data assuming the eddies were stationary. Small station spacing was acceptable for this calculation since positions were accurately determined from the satellite navigation. Corrections were also made in the computed geostrophic profiles by taking account of centrifugal effects and a "gradient" current was obtained. A comparison of the observed, geostrophic, and gradient currents is given in Fig. 2 and one can see that the agreement is in general good. The computed currents decrease rapidly at depths below the maximum of the observed currents and negligible horizontal gradients are to be seen in the density data at depths beyond 250 to 300 m.

Hunkins (1974) examined several possibilities, including baroclinic instability, for the formation of the eddies. He suggested, as was later confirmed by more detailed calculations of Hart and Killworth (1976), that growth rates with the mean shears and depths associated with the area of observation would be too small to indicate local generation. This led to the hypothesis that they must be generated at the shallow shelf regions by the west where shears are higher. This possibility also was supported by the observation that the T-S relationship of the eddies was consistent with that of the shelf waters and different than that in the observation area. Hunkins also addressed the question of low velocities near the surface since most generation mechanisms would give maximum velocities there. He suggested that frictional effects at the ice-water interface had reduced the surface values in the frictional layer on the Ekman spin-down time scale and conjectured (Hunkins 1980) that the time scale for decay of the velocity maximum in the pycnocline would be months-to-years using typical diffu-

G.T. Needler

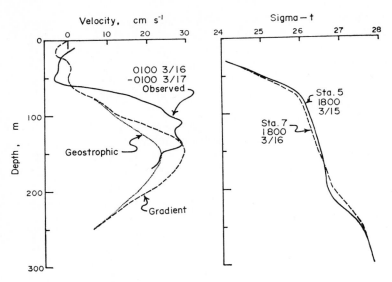

Fig. 2. Comparisons of the observed current with calculated geostrophic and gradient currents and of the density profile at the center of the eddy with that outside. (Reprinted from Hunkins 1974)

sion times. Such a time scale would allow advection of the eddies around the Beaufort Sea gyre from the indicated formation region.

During the main AIDJEX experiment, which was carried out from April 1975 to April 1976 in the area indicated in Fig. 1, a very much more extensive data set was obtained. Daily STD casts to 700 m and twice-daily velocity profiles to 200 m were taken over the year period from each of four camps. The data have been analyzed by Manley (1981) in much the same fashion as were the data from the earlier pilot experiment and the conclusions drawn from the earlier work have been mostly confirmed. A total of 169 eddies were identified. They had velocity maxima laying between 100 and 150 m, were geostrophic within measurement errors, had horizontal scales of about 20 km, and had a maximum observed velocity of 58 cm s^{-1}. The full AIDJEX data set showed however that the eddies are almost exclusively (98%) anticyclonic. During the experiment the T-S relationship of the eddies was again found to be different than that of the local water mass but both warm and cold core eddies were found. Manley (1981) has shown that the eddy T-S characteristics are consistent with summer and wintertime waters of the Chukchi Sea, suggesting their formation along the Alaskan Shelf-Slope front where water can be introduced into the Beaufort Sea at the appropriate density level. The role of the eddies in the transport of heat and salt is being investigated.

The areal extent of the Beaufort Sea type eddies has not been determined. Only two recent data sets can be applied to this question. One comes from the month-long FRAM 1 drift (see Fig. 1) in April–May 1979 when no eddies were observed (Manley and Hunkins 1980). If the eddies were present during the

FRAM drift with the density seen during AIDJEX, one would have expected to observe four during the month period of the experiment. The second data set comes from LOREX 79 (Pounder 1980) where daily velocity, temperature, and conductivity measurements were taken to a depth of 1 400 m or to the bottom during the drift of ice station ICEMAN across the Lomonosov Ridge. Although the data have not been exhaustively analyzed, Beaufort Sea-type eddies are not readily apparent. It remains to be determined, however, whether or not their existence is really limited to the Beaufort Sea and, if it is, whether this reflects their lifetime or simply the pattern of advection away from the source region and around the Beaufort Sea gyre.

The LOREX data (Pounder 1980) do exhibit velocity profiles that increase with depth to a maximum of as large as 10 cm s^{-1} at about 400 m and then decrease gradually with depth to the deepest measurements. Of the total of 40 daily profiles, 9 exhibited weak currents and the remainder velocity profiles of the type just described but with varying directions. To date no analysis has been performed to determine how the current changed direction as ICEMAN drifted across the Lomonosov Ridge into the Fram Basin. The possibility that large-scale baroclinic eddies of low mode number were observed remains. During LOREX near-bottom velocity measurements were also obtained (Aagaard 1980) from two moorings in water depths of 1 440 m and 3 565 m over a period of longer than a month. Instruments were placed 25 m and 200 m off the bottom. The analysis of the data shows the currents of up to 12 cm s^{-1} with apparent bottom intensification and a broad spectral peak at frequencies of about a cycle every 3 days. Aagaard suggests the existence of trapped bottom modes at these frequencies. The vertical scale, however, is difficult to check from the data and the ice field shows ever larger motions at similar frequencies. In this case also, clear interpretation of the observations awaits further analysis and perhaps a more comprehensive data set.

Except for the near-surface Beaufort Sea eddies, the Arctic Ocean remains relatively unexplored regarding the existence of eddies, their nature, and frequency.

9.3 Labrador Sea and Northwestern Atlantic

Perhaps the first search for evidence of mid-ocean eddies in the northwestern Atlantic was conducted by Gill (1975) when he analyzed the BT records of various ocean weather stations. He found that, after removing seasonal effects of heating in the upper 250 m, a signal remained in the temperature field which was essentially independent of depth and showed temporal variations at periods between the period over which the data were smoothed and a year. This he took to be indicative of eddies with depth scales much greater than 250 m. The energy levels of the eddies varies significantly from one station to another and appeared to be connected to their positions relative to the center of oceanic gyres and to major fronts and currents. Although Prangsma (1977) suggested that Gill's results for those stations near major fronts could have been in-

fluenced by the movement of the weather ships, the existence of quasi-geostrophic motions in the records appears clear. Wyrtki et al. (1976) have examined the existence of eddies using a different data source. They have taken the
records of drift currents as observed by merchant ships and constructed maps
of mean and eddy kinetic energy per unit mass for the North Atlantic and the
world's oceans. Because of the limited data base this could only be done by 5°
squares for the northern North Atlantic, but the analysis gives typical values of
eddy kinetic energy of 400 cm^2 s^{-2} as compared to 20 cm^2 s^{-2} for the mean
flow for the whole region north of the Gulf Stream extension. Wyrtki et al.'s
analysis gives results qualitatively consistent with the results given by Gill at
the weather ship locations.

An earlier analysis of the Ocean Weather Station Bravo data was carried
out by Lazier (1973) who was concerned with the mechanisms involved in the
formation of Labrador Sea water. Figure 3 shows the variation of the observed
density and temperature during two two-week periods in 1964 and 1965. In
both cases, deep isopycnal surfaces rose towards the surface as water passed by
the weather ship or it steamed in and out of a region of deep convection. Consideration of the temperature field indicated that motion of the ship relative to
the water column had to be occurring. At the bottom of each section in Fig. 3 is

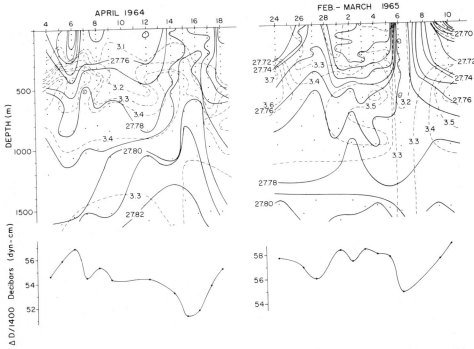

Fig. 3. Distributions of potential density and potential temperature during two observing periods at Weather Ship Bravo. Time variations of dynamic height relative to 1400 dbar are
shown at the bottom of each plot. (Reprinted from Lazier 1973)

plotted the dynamic height of the sea surface relative to 1400 m. Changes of dynamic height of about 3 dyn – cm occurred as the intense eddies passed by. Lazier notes that such changes are typical of observations throughout the year but it is only in the late winter that they are associated with deep isopycnals approaching the surface since in summertime the same changes in dynamic height can arise from small movements of the isopycnals in the much stronger density gradients that exist at that time of year.

It was not until a cruise by the CSS HUDSON in March 1976 that the convective-like events of the type noted by Lazier (1973) from the Ocean Weather Station Bravo were investigated in order to determine their role in the formation of Labrador Sea water. Through analysis of vertical and horizontal current and CTD data obtained during this experiment, Clarke and Gascard (1983) showed that enhanced air/sea exchange in the western side of the Labrador Sea during the period of the experiment resulted in the development of a gyre in the region offshore from the Labrador Current system. The gyre served to retain water in a local region where it could be cooled by the cold continental air mass just after it moved off the ice pack covering the continental shelf. This led to the formation of denser and denser water in the upper part of the gyre that was apparently mixed and stirred by mesoscale processes with a scale of 30 to 50 km. Convection was observed to take place on very short spatial scales, but the densest and most homogeneous water columns were contained within small anticyclonic eddies of about the local Rossby radius of deformation scale, 10 km, and with depth of greater than 2000 m. These changed position within the large-scale features. Peak vertical velocities as large as 9 cm s^{-1} were observed. Clarke and Gascard noted the similarity of their observations to those obtained in the northwestern Mediterranean Sea and interpreted dynamically by Gascard (1978). Gascard attributes the small eddy scale to baroclinic instability of the mesoscale velocity field and the small eddy circulation for preventing the newly formed waters from spreading out on isopycnal surfaces. The small-scale convective eddies observed in the Mediterranean were cyclonic in simple agreement with geostrophy since the water within the eddy is denser at all depths than its surroundings. As mentioned above, however, the Labrador Sea small-scale eddy was anticyclonic and only contained water denser than its surroundings at its upper levels. A mechanism for the anticyclonic behaviour is given by Clarke and Gascard (1982b) in terms of the depression of the deepest isopycnals measured.

The general problem of convective "chimney" formation in the ocean has been investigated by Killworth (1979) who also discussed in some detail the observation of one event seen in the Weddell Sea (Gordon 1978). His work shows the necessity of "preconditioning" a water mass for deep convection in the open ocean by some mechanism which could create a larger scale feature with significant velocity shears. He also examined the likelihood of observing such convective events in a region such as the Greenland Sea through considerations of the amount of bottom water formed, the lifetime of a typical eddy, and the number of observations available. He concluded that by 1978 there was an 82% possibility that no chimney would have been observed there and indeed none had been. The lifetime of convective eddies remains a matter of some doubt. The Weddell Sea event seen by Gordon (1978) had survived at least from one winter season to the following summer. The lifetime of the Labrador Sea convective cells is unknown, although there is some indication that the new Labrador Sea water formed in 1976 was spread out at depth between Cape Farewell and the Tail of the Grand Banks by the following summer.

To the west of the Labrador Sea off the edge of the continental shelf lies the Labrador Current which is composed of fresh cold surface waters from the Canadian Arctic overlying warmer, more saline waters that have entered the Labrador Sea by the East Greenland Current. Satellite photographs of the Labrador Current reveal the typical instabilities and wavy motions now well known as features of many boundary currents. Three subsurface current meter moorings and associated hydrographic data were collected by the Atlantic Oceanographic Laboratory in 1976 at the time of the deep convection experiments discussed above. An analysis of these data by Allen (1979) has provided information on the spatial and temporal variability of the Labrador Current. Typical mean southward currents of 30–50 cm s^{-1} near the surface and 20–40 cm s^{-1} near the bottom were observed across the current system. Low-frequency oscillations with typical amplitudes of 25 cm s^{-1} were apparent in all records. Allen (1979) interprets the data to show bottom-trapped Rossby waves with periods of 4 to 8 days and the appropriate theoretical correlations between the temperature and velocity fields. At larger periods, 8 to 10 days, his analysis shows weakly baroclinic topographic Rossby waves with horizontal scales of about 100 km as determined by satellite data. In the bottom flow, alongslope variations in the velocity field at periods of 2 to 3 days were to be seen. These, Allen suggests, could arise from instabilities of the strong flows generated by the longer period topographic waves. The T-S distributions and dynamic topography obtained from the hydrographic data also showed large-scale on-offshore excursions of at least 10–15 km of the Labrador Current edge over time scales as short as 20 h and along-current scales of 30–50 km.

9.4 The Norwegian and Greenland Seas and Their Overflows

The Norwegian and Greenland seas exhibit many of the eddy features of the northwest North Atlantic discussed above. The only exception is penetrative convection which, as was pointed out, is unlikely to have been seen, considering the number of wintertime observations available. Low-frequency variability in the central Norwegian Sea was observed, for example, by an expedition of the P.P. Shishov Institute of Oceanology of the USSR Academy of Sciences during July to September 1975 (Kort et al. 1977). A seven-mooring array with an outside dimension of 60 km was located in the frontal zone with the Norwegian Current on the east and the East Greenland Current on the west. A complex pattern of currents in space and time was observed and partially described kinematically as a system of two eddies (one cyclonic and one anticyclonic). The eddies were thought to be quasistationary and to have originated in the Norwegian Current.

The Norwegian Current itself has been the subject of many investigations and one of the earliest by Helland-Hansen and Nansen (1909) led to the observation of large-scale variability in the hydrographic station data. A major experiment involving six moorings deployed in a cross at water depths of 500 to 870 m was carried out during 6 weeks in the summer of 1969 (Horn and Schott 1976). Station data were collected repeatedly around the array. The current records typically showed considerable low-frequency variability such as can be seen in the low-passed current components shown in Fig. 4 taken from Schott and Bock (1980). Energy spectra of the currents showed peaks in the two to three day period range. Mysak and Schott (1977) considered a number of simple baroclinic and barotropic instability models for the Norwegian Current in an effort to explain the fluctuations. Although the vertical phase coherence, which can be seen in Fig. 4, suggested that a barotropic instability model might be suitable, it was found that the observed period could only be obtained for unreasonably-high Rossby numbers. Instead a two-layer baroclinic instability model with realistic parameters was found to provide a plausible explanation for the dominant fluctuations. More recently Schott and Bock (1980) have used the data to calculate the energy interaction terms for baroclinic and barotropic instability and to compare them to the transfers suggested by the Mysak and Schott model calculations. They found that, although at the moorings energy is indeed transferred into the perturbations from the mean flow by baroclinic instability, it flows back to the mean flow at a larger rate by barotropic transfer mechanisms. Directional spectra yielded wavelengths in agreement with baroclinic model predictions but questions remained as to the direction of phase propagation and its implications regarding the group velocity of the waves.

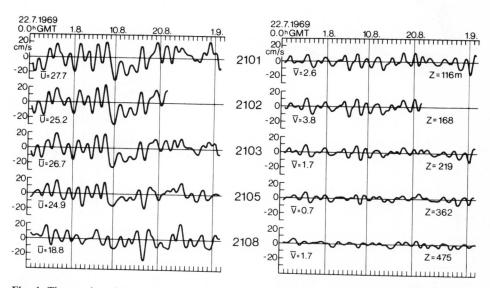

Fig. 4. Time series of low-passed current components, parallel u and normal v to the current axis at the central mooring. (Reprinted from Schott and Bock 1980)

Schott and Bock were unable to investigate the effects of wind forcing over the Norwegian Sea but noted the availability in the wind field of significant energy in the frequency band of the fluctuations.

To the southwest of the Norwegian Sea lies the Iceland-Faroe Ridge and the polar front which separates the cold fresh arctic and subarctic waters from the warmer saltier waters of the North Atlantic. Through the lowest points of the ridge, as well as through Denmark Strait, the arctic waters flow out into the deep North Atlantic. Over the ridge the polar front undergoes large excursions on a variety of time scales. Hansen and Meincke (1979) compiled various positions of the front from the data of hydrographic surveys carried out for fisheries investigations from 1951 and 1957. Their analysis is shown in Fig. 5. From a number of data sources Hansen and Meincke present evidence for the existence of eddies in this region that exist with both cold-fresh and warm-salty cores relative to the surrounding water. The vertical extent of the eddies is usually 100 to 200 m, but they are often visible over the complete depth of 400 to 500 m. Temperature-salinity analysis shows their origin to be at the polar frontal zone. The meanders of the polar front itself has implications regarding

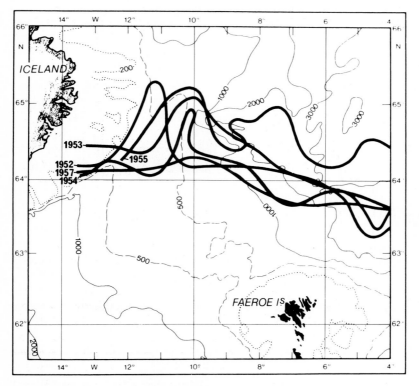

Fig. 5. Locations of the polar front in the surface layer (0 to 25 m) in May and June and of the years 1952 to 1957. The lines correspond to the 35.0‰ isohaline, which represents the zone of maximum horizontal gradients. (Reprinted from Hansen and Meinke 1979)

the flow of arctic and subarctic waters over the ridge and down the slope into the less dense waters on the Atlantic side since in its average position along the ridge crest it serves as a barrier to such flows. It is not known how great this intermittent overflow is in comparison to the more regular flow through the narrow gaps in the ridge. From the data Hansen and Meincke were unable to determine the mechanisms that lead to the formation of the frontal eddies, but suggest both barotropic-baroclinic instability mechanisms and the tendency of the mean flow to follow isobaths. The latter could lead to the emanation of the eddies into the interior from the bottom layer around features of the observed scale.

The overflows of waters through the Denmark Strait between Iceland and Greenland were measured during a 1-month experiment carried out in the summer of 1973 by the Atlantic Oceanographic Laboratory. Analysis of the current meter records and hydrographic date (Smith 1976) shows that although the overflow existed throughout the experiment, it exhibits a strong spectral peak at a period of 1.8 days and that the fluctuations intensify in the downstream direction and are highly correlated over the entire flow at the southern end of the Strait. The temperature, speed, and direction at a mooring downstream from the crest is shown in Fig. 6. In order to explain this phenomenon Smith used a quasi-geostrophic two-layer model for channel flow with a sloping bottom and obtained a most unstable wave with a 80 km wavelength and 2.1 day period, in close agreement with the observed currents. Smith points out that his linear model cannot explain the finite nature of the observed flow but his predicted results are clearly consistent with the data. He also reports that in a separate model it was demonstrated that finite-amplitude baroclinic instabilities in a two-layer fluid may take the form of a train of discrete vortices propagating in the downstram direction. This would appear as an amplitude oscillation to a fixed observer and help explain some of the intermittency found in the measured currents at low frequencies.

9.5 The Northeast North Pacific and Bering Sea

The final observations of eddy activity described in this chapter are taken from the North Pacific and the Bering Sea. In the latter, observations of variability on the Alaskan continental shelf have been made by several authors. Recently Kinder and Coachman (1977) have described observations of an isolated eddy of high-salinity water nestled in the outer reaches of the Pribilof Canyon and partially in water depths of greater than 1 000 m. The T-S characteristics of the eddy were those of the Bering Slope Current and the authors attribute its formation as evolving from a pinching off of a meander of this current in a manner similar to that which occurs when the Gulf Stream forms warm eddies in the Slope Water. Similar eddy events have been observed in the northern Gulf of Alaska and reported by Royer et al. (1979). They describe a persistent anticyclonic 100 km feature lying off the continental shelf and relate its formation to instabilities of the Alaska Current in the region.

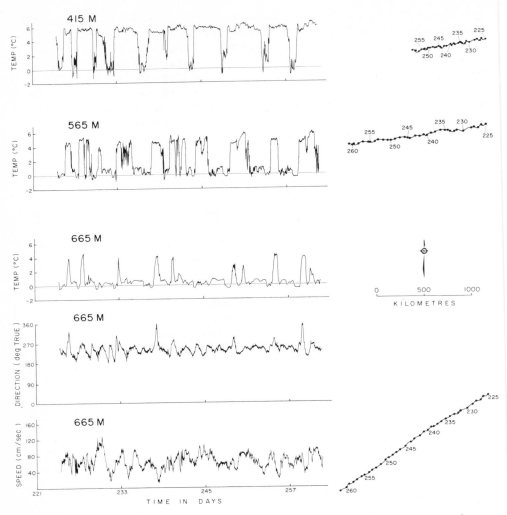

Fig. 6. Temperature, speed, and direction records and progressive vector diagrams at a downstream mooring in Denmark Strait. (Reprinted from Smith 1976)

To the south, oceanographic variability has been measured at Ocean Weather Station P since 1956 and along a section from the continental shelf to the weather station since 1959. This data set is one of the longest available for studies of variability in sub-polar gyres. Variability along the section was first analyzed by Fofonoff and Tabata (1966) who were interested in relating the transport across the section to changes in the large-scale curl of the wind stress. They found, however, that eddy-like features and apparent counter-currents made the analysis difficult. The nature of the variability can be seen in their space-time representation of dynamic height anomaly shown in Fig. 7. It is clear that even with the large data base the smallest individual features in the dynamic height are directly related to the sampling rate in space and time.

The Station "P" data have been used by many investigators as a source of information on mixed-layer variability but its use for the examination of fluctuations on quasi-geostrophic time scales has been limited because of the sampling scheme used for STD casts. This arises because between 1956 and 1969 only one of the two weather ships occupying the station took hydrographic casts and regular six-week gaps thus appear in the data for the periods when

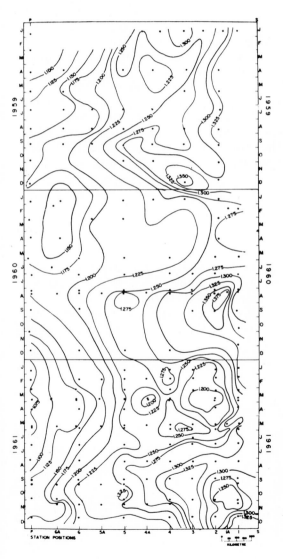

Fig. 7. Dynamic height anomaly at the surface relative to the 1000 m surface (0/1000 decibar) along the Line to Station "P". The differences in the anomaly at a given time are proportional to the geostrophic current at the surface relative to that on the 1000 m surface. However, differences in time have no simple dynamical interpretation. (Reprinted from Fofonoff and Tabata 1966)

the station was occuppied by the second ship. Ross and Needler (1976) analyzed data available up to 1971 and found that, although some evidence existed
in the hydrographic data for fluctuations of the main pycnocline on periods of
two or three months, it was not possible to be sure that this did not arise from
the leakage of low-frequency energy into this frequency band in the spectral
analysis because of the regular gaps of the data base at similar frequencies.
During the past 10 years, data have been collected from both ships so that a record now exists which should overcome this problem, but the author is una-
ware of any recent attempts to use it to examine quasi-geostrophic oscillations.

The large-scale circulation of the northeastern North Pacific has been examined by Tabata (1976) using a number of different data sources. He presents
maps of geostrophic anomolies for a given season which show much of the
same variability apparent in Fig. 7 for the section to Ocean Weather Station P.
Tabata's and other analyses often show a pronounced thermal anomaly in the
northern Gulf of Alaska. Recently, Willmott and Mysak (1980) have used a
continuously stratified inviscid linear model for a quarter-plane region in order
to investigate whether these eddies could be explained by atmospheric forcing.
The first and second mode planetary waves generated by the model are reflected off the boundaries and cause a number of eddies which in general do
not consist of closed streamlines. They also find that the size of the eddies is increased dramatically if the quarter-plane is shifted from the north-south direction.

9.6 Summary

This chapter provides a brief description of a variety of quasigeostrophic oscillations in the northern oceans. Some of these can be associated with formation
mechanisms and a much smaller number are known to play a certain role in the
dynamical balances of the oceanography of their region. In most cases, much
better data sets and modelling efforts will be needed to resolve the role of eddies in the north.

10. Tropical and Equatorial Regions

G. Siedler

10.1 Latitudinal Changes of Mesoscale Processes

The major observational programmes aimed at exploring the three-dimensional mesoscale eddy structure of the oceans were performed outside the equatorial regions. Experimental sites were chosen in the tropics at latitudes higher than 10° (Koshlyakov and Monin 1978) or at mid-latitudes (MODE Group, 1978; Robinson 1980). There exist, however, some XBT (Expendable Bathythermograph) sections through the equatorial belts that were obtained for studying mesoscale structures. In addition experiments with other objectives were performed in the tropics, such as GATE (Siedler and Woods 1980, Philander and Düing 1980) in the Atlantic. We will check here whether some of these experiments can supply useful information on mesoscale structures in the tropical regions. The available data are not sufficiently complete to differentiate between eddy and wave dynamics, but it will be shown that the kinematic properties of fluctuations are consistent with eddies outside the equatorial belt.

It cannot be expected that mesoscale disturbances of the oceanic circulation will have similar properties near the equator and at mid-latitudes. Due to the decrease in baroclinic times scales and the simultaneous increase in horizontal space scales with diminishing latitude (Lighthill 1969), disturbances will obtain higher phase speeds, and they may be trapped in the equatorial belt due to the symmetric change of the planetary vorticity. This can result in a dominance of orderly wave-like motions near the equator, while eddy dynamics will govern the mesoscale processes at higher latitudes (see Weisberg et al. 1980). For further details the reader is referred to the review of tropical ocean modeling that was published by Moore and Philander (1977).

The change in wave scales when approaching the equator is demonstrated in Fig. 1 (Philander 1979). Rossby wave periods decrease towards lower latitudes. The period range with forced waves separating Rossby and inertial waves narrows down, and equatorward of about 3° latitude free waves can exist at all frequencies.

An adequate measure for separating the region of distinct equatorial dynamics from higher-latitude dynamics is the latitude where the internal Rossby radius of deformation λ equals the distance from the equator.

$$\lambda^* = \sqrt{c/\beta} \tag{1}$$

Here c is the phase speed of the baroclinic wave, β is the derivative of the Coriolis parameter f, and the asterisk indicates that λ is taken at this particular la-

Eddies in Marine Science
(ed. by A.R. Robinson)
© Springer-Verlag Berlin Heidelberg 1983

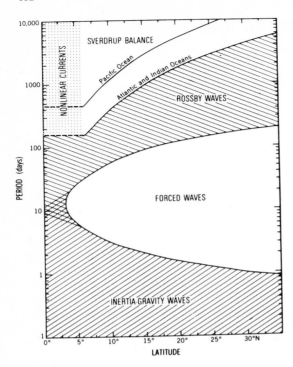

Fig. 1. Periods at which various types of baroclinic waves or large-scale current systems are possible, as a function of latitude. (After Philander 1979)

titude. For the first baroclinic mode we obtain $\lambda^* \approx 350$ km, corresponding to 3° latitude. Philander (1979) concludes that equatorward of approximately 5° latitude one should expect the region of distinct equatorial dynamics.

Outside this equatorial belt an estimate of the horizontal scales of meso-scale motions can be obtained from the Rossby radius of deformation λ in the following form:

$$\lambda = \frac{\bar{N} H_T}{f} \tag{2}$$

$\bar{N}$ is the mean Brunt-Väisälä frequency of the baroclinic layer, and H_T is the thickness of this layer. For typical North Atlantic values at latitude $\theta = 30°$, $H_T \sim 700$ m, $\bar{N} \sim 2\pi \times 3$ cph we find $\lambda_{30} \sim 50$ km. Proceeding towards lower latitudes the thickness of the baroclinic layer decreases and the Brunt-Väisälä frequency increases, approximately compensating in equ. 10.2, and λ is mainly controlled by the change of f. For example: $\theta = 10°$, $H_T \sim 400$ m, $\bar{N} \sim 5$ cph, $\lambda_{10} \sim 140$ km or $\lambda_{30}/\lambda_{10} \sim \sin 10°/\sin 30°$.

The characteristic scale of eddies should increase approximately proportional to $(\sin \theta)^{-1}$ from the subtropics towards the equatorial belt. An increase in eddy scale r (center to maximum velocity) towards lower latitude is found when comparing results from POLYGON-70 and MODE-1 observations in the

Atlantic given by Koshlyakov and Monin (1978) and from hydrographic arrays in the Indian Ocean (Koshlyakov et al. 1970):

$\theta = 28°N$ r = 90– 95 km

$\theta = 16.5°N$ r = 100–120 km

$\theta = 12°N$ r = 100–150 km

While this may be an indication for such a change in scales, the data base is not good enough for obtaining evidence from observations at this time. The existence of mesoscale fluctuations can, however, be demonstrated by the meridional XBT sections in Figs. 2 and 3.

In the following we will inspect data from tropical regions for the occurrence and typical dimensions of mesoscale structures. When analysing meridional variations near the equator, one has to keep in mind that the density field varies strongly with the transition to opposite-direction steady zonal mean currents, and spatial scales related to the mean currents cannot always be separated from eddy-scales.

10.2 Eddy-Resolving Arrays in the Tropics

Arrays of hydrographic stations and of moored current meters were applied by U.S.S.R. groups in the tropics on several occasions for studies of mesoscale variability. Results will be presented here from two hydrographic surveys in 1967 in the Indian Ocean (Koshlyakov et al. 1970) and from current meter arrays supplemented by hydrographic measurements in 1970 in the Atlantic Ocean (Brekhovskikh et al. 1971).

The maps in Fig. 4 originating from the 1967 experiment centered at 12°N, 65°E are based on horizontally smoothed density fields obtained from hydrographic data. Geostrophic currents were calculated for 1500 dbar reference levels. The figures on the left side give stream lines for the surface, 150 m and 800 m levels from the first observational period in January–February, the figures on the right side present the corresponding maps for the second observational period two months later in March–April.

The patterns resemble those that can be expected for mid-ocean eddies, with diameters of 200 to 300 km. The eddy structure during the first period is strongly correlated in the vertical, with the eddy centers at the same position at all three depth levels. In the second set of observations, the cyclonic eddy in the upper layers can hardly be identified at the 800 m level, and its center seems to be displaced to the east with increasing depth. If we assume that the southern eddy in series A propagates to the west and is found again in series B, we obtain a displacement speed of 2–3 cm s^{-1}.

In the later experiment POLYGON-70 in the tropical Atlantic, a cross-shaped array of current meter moorings centered at 16°30'N and 33°30'W was repeatedly deployed for a total observational period of 6 months. Grachev and Koshlyakov (1977) presented results of an objective analysis of the mesoscale

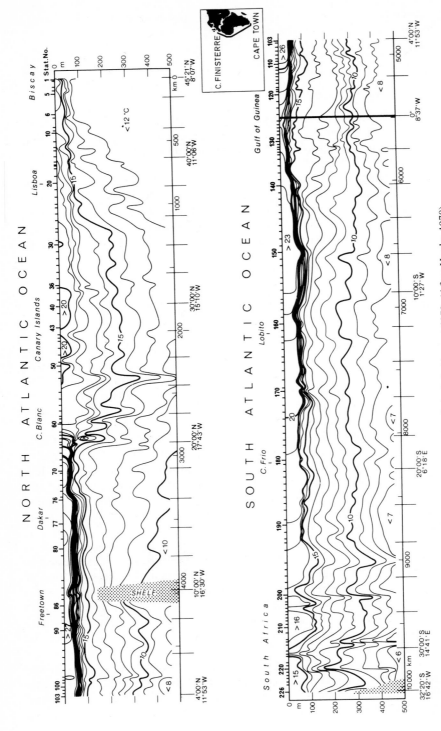

Fig. 2. XBT section obtained in the eastern part of the Atlantic in July/August 1978. (After Henke 1978)

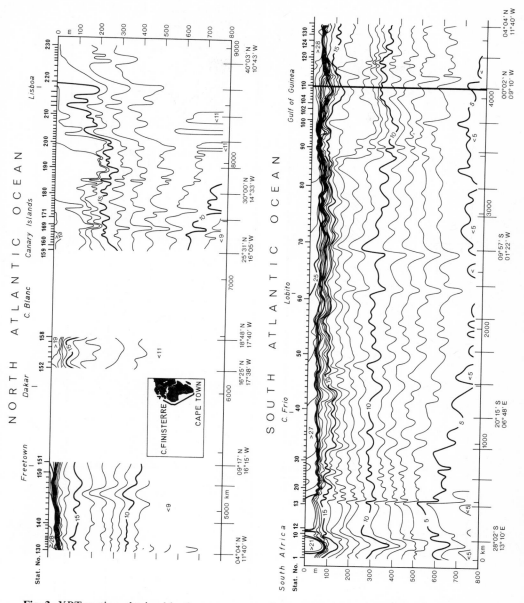

Fig. 3. XBT section obtained in the eastern part of the Atlantic in March 1980. (After Henke and Zenk 1980)

current field (see also Koshlyakov and Monin 1978). Figures 5 to 7 are taken from that analysis, demonstrating the eddy structure at 300 m, 600 m and 1000 m. Features are strongly correlated in the vertical, and an ap-

G. Siedler

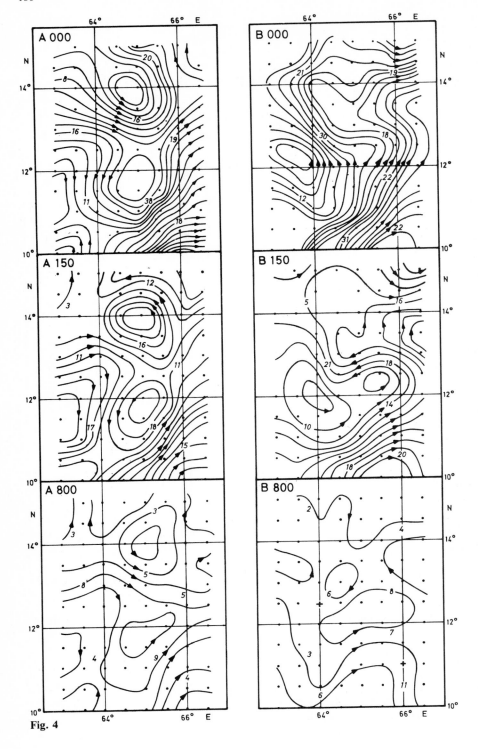

Fig. 4

proximately westward propagation with a speed of 5 cm s^{-1} is found in this time series. Typical eddy diameters are 200 to 240 km.

Wyrtki (1967) used closely spaced hydrographic station arrays in the Pacific near Hawaii to determine structure functions for isotherm depths. They have a typical peak near 200 km and are supposed to be related to geostrophic eddies of similar dimensions.

10.3 Mesoscale Features in Sections

Some information on typical mesoscale variability in the tropics can be extracted from the GATE 1974 data set. During the three phases of that experiment (I: 26 June–16 July; II: 28 July–16 August; III: 30 August–19 September) hydrographic and current meter data were gathered along the 23.5°W meridian (Bubnov et al. 1979, Bubnov et al. 1980). Results from these observations will be presented here. Studies of the equatorial current at several longitudes were performed during Phase II that led to the detection of major wave phenomena in the equatorial belt (see Philander and Düing 1980, Siedler and Philander 1982). A dominating oscillation found was a horizontal meandering of the Equatorial Undercurrent, with a period of 16 days and a zonal wavelength close to 2000 km. The most energetic fluctuation had a period of about one month and a wavelength of approximately 1000 km. Similar results were obtained in the equatorial Pacific where wavelike structures of a zonal temperature front occurred between 1° and 3°N, with a wavelength of about 1000 km and a period of 25 days (Legeckis 1977b). These observations strengthen the confidence in the assumption arrived at in the beginning that orderly wave-like motions and not eddies will be dominant in the equatorial belt. We will therefore exclude the ±5° latitude region when discussing meso-scale structures in meridional sections.

A temperature section obtained during GATE Phase I is presented in Fig. 8. The large-scale slopes in the upper thermocline are related to the large-scale wind-driven circulation with the westerly South Equatorial Current south of approximately 4°N and the westerly North Equatorial Current north of 10°N. The Equatorial Undercurrent can be recognized at the equator, marked by the vertical spreading of isotherms. The latitude belt where wave dominance can be expected is indicated above. For an inspection of mesoscale structures outside this belt, the upper ocean density sections for GATE Phases I–III are shown in Fig. 9. Both in the northern and in the southern parts of these sections isopycnal depths fluctuate by tens of meters vertically and with typical

◀ **Fig. 4.** Streamlines obtained from hydrographic surveys in 1967 in the tropical Indian Ocean. *Numbers* indicate speeds of geostrophic currents relative to 1500 dbar in cm/s. *A000, A150* and *A800* give the geostrophic current fields at the surface, 150 m and 800 m depth for the first observational period, *B000, B150* and *B800* represent the corresponding fields 2 months later. (After Koshlyakov et al. 1970)

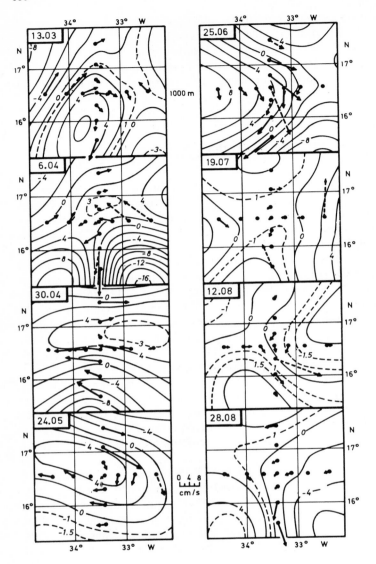

Fig. 5. Objective analysis of the velocity field obtained from POLYGON-70 moored current meters at 1000 m depth. Day and month numbers are given in *upper left* corners, *dots* are at position of moorings, *arrows* indicate low-pass filtered observed currents, *dashed line arrows* were derived from interpolation. (After Grachev and Koslyakov 1977 and Koshlyakov and Monin 1978)

horizontal scales of 200–300 km. Horizontal scales below 200 km can not be resolved due to the large station distances poleward of 5° latitude. The difference in the current field inside and outside the equatorial belt is demonstrated by the velocity stick plots in Fig. 10. They represent temporal variations of daily

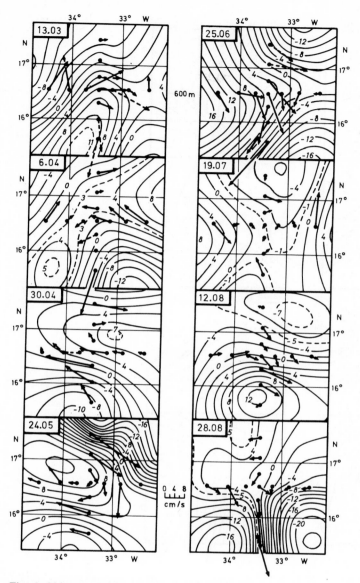

Fig. 6. Objective analysis of the velocity field obtained from POLYGON-70 moored current meters at 600 m depth. Day and month numbers are given in *upper left* corners, *dots* are at position of moorings, *arrows* indicate low-pass filtered observed currents, *dashed line arrows* were derived from interpolation. (After Grachev and Koshlyakov 1977 and Koshlyakov and Monin 1978)

averaged currents at 10 m, 75 m and 150 m depth that were measured on moorings along 23.5°W. Wave-like disturbances with high velocities can be recognized in the equatorial region, while low currents are observed at higher latitudes. Unfortunately, it is hardly possible to identify any typical time scales

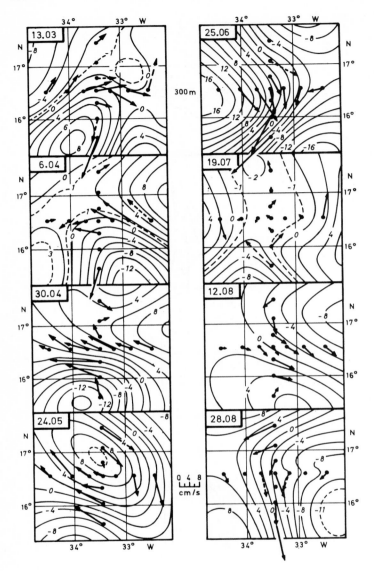

Fig. 7. Objective analysis of the velocity field obtained from POLYGON-70 moored current meters at 300 m depth. Day and month numbers are given in *upper left* corners, *dots* are at position of moorings, *arrows* indicate low-pass filtered observed currents, *dashed line arrows* were derived from interpolation. (After Grachev and Koshlyakov 1977 and Koshlyakov and Monin 1978)

outside the equatorial belt in these data. A better guess of time scales related to eddies in the tropics can be obtained from stick plots gained in the Pacific that are given in Fig. 11 (Halpern 1979). Strong fluctuations with time scales of 1–2 months seem to appear in most of these series.

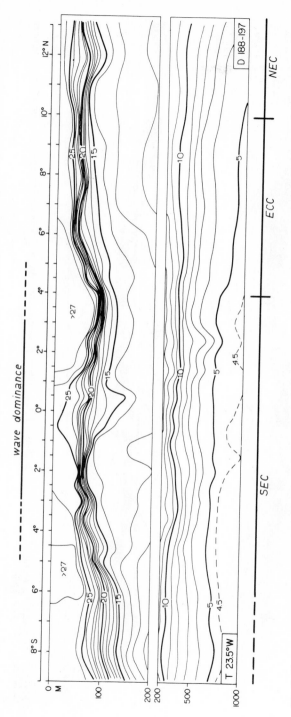

Fig. 8. Temperature section along 23.5 °W for GATE Phase I. Station positions are indicated by *tick marks* along the top. Mean zonal currents are indicated below: *SEC* South Equatorial Current; *ECC* Equatorial Counter Current; *NEC* North Equatorial Current. (After Bubnov et al. 1979)

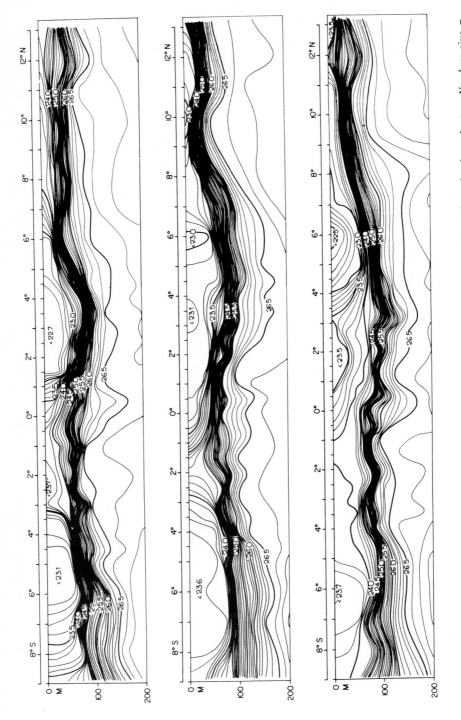

Fig. 9. Density sections along 23.5°W for GATE Phases I–III. Station positions are indicated by *tick marks* along the top. *Numbers give* σ_t. (After Bubnov et al. 1979)

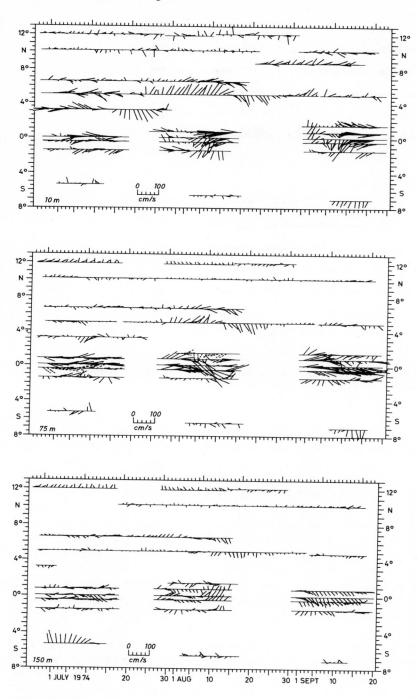

Fig. 10. Velocity stick plots from low-passed moored current meter data along 23.5°W for GATE Phases I–III at three depth levels. (After Bubnov et al. 1979)

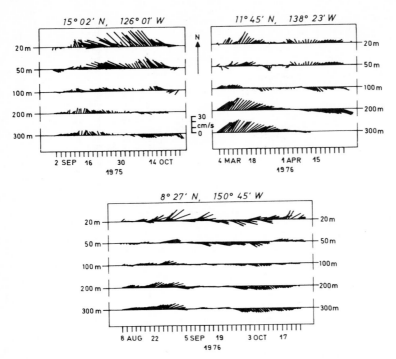

Fig. 11. Velocity stick plots from low-passed moored current meter data at three latitudes and five depth levels in the Pacific. (After Halpern 1979)

A major effort was made within the framework of the U.S. NORPAX program to observe the variability of the equatorial current system in the Pacific. Part of this investigation was the "Tahiti Shuttle Experiment" in 1977–1978 including repeated AXBT (Airborne Expendable Bathythermograph) measurements on flights along 150° and 158°W between Hawaii and Tahiti (Wyrtki et al. 1981). An example of the temperature distribution observed is given in Fig. 12 (Patzert et al. 1978). Barnett and Patzert (1980) performed an empirical orthogonal function (EOF) analysis on these data and for the principal modes arrived at time scales of 2–3 months or longer and at meridional space scales of 1000–2000 km. However, this analysis included the whole region covered by the sections, and the results present the combined effects of equatorial wave and extra-equatorial eddy processes. The scales are larger than expected. Dantzler's (1976) structure function analysis of dynamic height across the thermocline along zonal sections in the North Pacific shows that peaks at approximately 200 and 500 km horizontal length scale occur in the tropical data.

   In order to identify individual mesoscale structures in the region outside the equatorial belt and to visualize their variations in time, depth deviations of two isotherms from the mean depth at each longitude of the "Tahiti Shuttle Experiment" are plotted in Figs. 13 and 14. Isotherm depth variations having a

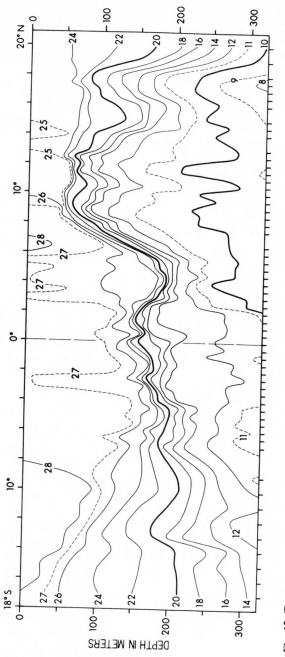

Fig. 12. Temperature section from the Tahiti Shuttle Experiment obtained with AXBTs along 150°W on 3 January 1978. (After Patzert et al. 1978)

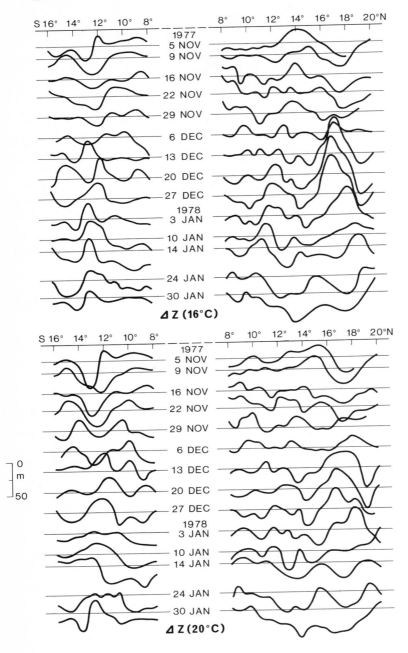

Fig. 13. Deviation outside the equatorial belt of the 20° and 16°C isotherm depths from the mean depth obtained by averaging over all Tahiti Shuttle Experiment data along 150°W.

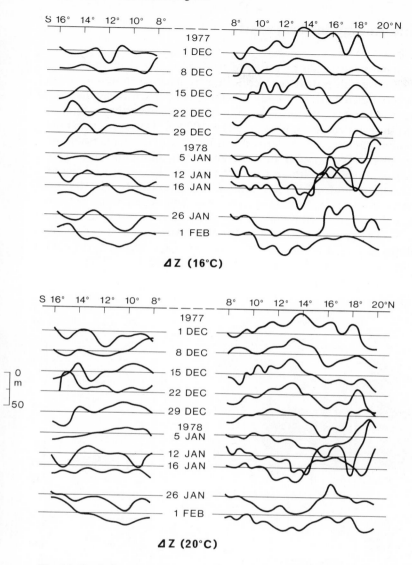

Fig. 14. Deviation outside the equatorial belt of the 20° and 16°C isotherm depths from the mean depth obtained by averaging over all Tahiti Shuttle Experiment data along 158°W.

scale of approximately 200 km and occurring repeatedly in succeeding sections can be found in both figures. See, e.g., the negative deflection of the 20°C-isotherm between 12° and 14°S at 150°W or the positive deviation of the 16°C-isotherm at 16° to 19°N. The signals can be seen for about 1 month each. Such fluctuations are consistent with mesoscale eddies drifting in zonal direction with a few cm s^{-1}.

10.4 Mesoscale Energy in the Tropics

Several attempts have been made to estimate the eddy energy in the ocean. Wyrtky et al. (1976) used ship drift data collected by hydrographic offices to determine the regional distribution of eddy kinetic energy. Their basic data set, however, had the disadvantage that it represented averages over approximately 400 km and could thus not resolve eddies of a few 100 km. Dantzler (1977) determined eddy potential energy from vertical isotherm displacements based on XBT data. The results of Wyrtki et al. indicate an energy minimum in the central sub-tropical oceanic gyres, with an increase towards the equatorial region by a factor of 2–3. Dantzler also finds a minimum in the central North Atlantic, particularly on the western side, but the increase towards the equator is generally less than by a factor of 2, although specific areas may have higher values as shown in Fig. 15. The energy maximum at the West African coast is possibly related to upwelling areas. Eddy formation off an upwelling region in the eastern tropical Pacific was shown already by Stumpf and Legeckis (1977).

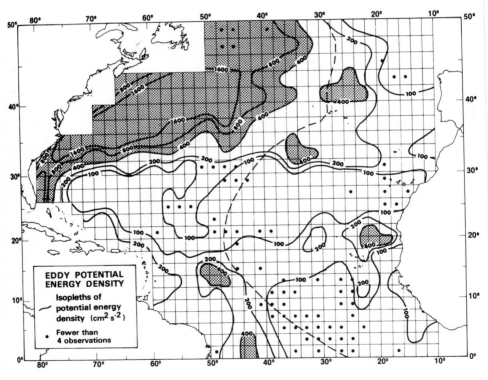

Fig. 15. Eddy potential energy estimated from mean-squared thermocline displacement in the North Atlantic. *Dashed line* is the approximate axis of the Mid-Atlantic Ridge. (After Dantzler 1977)

The term "eddy energy" may however be misleading in this context. Deviations from the mean flow or from the mean thermocline depth include eddy and wave signals as well as long-term adjustments, e.g., in an annual cycle. Emery et al. (1980) have shown that closed cells probably representing mesoscale eddies in XBT data in the sub-tropical North Pacific contribute to only about one half of the isotherm deflections. Furthermore, the strong waves in the equatorial belt apparent in Fig. 10 and the annual variations of the zonal equatorial current system will dominate the mean deviation data in the equatorial belt. It is probable, however, that an increase in eddy energy which was found along the western and northern parts of the main currents of the gyre circulation will also be found in the southern part, the North Equatorial Current region, if it were possible to eliminate non-eddy components from the observed variance.

We conclude that the available observations provide evidence for mesoscale eddy structures in the tropical oceans. Variability in the equatorial belt between approximately 5°N and 5°S, however, is apparently dominated by wave processes. There is indication that the horizontal scales of eddy structures may increase in meridional direction from the subtropics towards the tropics.

11. Eddies in the Indian Ocean

J.C. Swallow

11.1 Introduction

This chapter is concerned with that part of the Indian Ocean north of about 20°S. Most of the area, from about 10°S northwards, comes under the influence of the monsoons and has a strongly variable seasonal circulation. As in most parts of the ocean, the only available overall view of eddy properties in this region is the one provided by Wyrtki, Magaard and Hager (1976). They calculated eddy kinetic energy per unit mass for each 5° square of the world's oceans, using all available ships'drift observations of surface current. In the Indian Ocean, their values of eddy kinetic energy are relatively high – over 1000 $cm^2 s^{-2}$ along most of the equator, and over 2000 off Somalia; nowhere north of 10°S is the eddy kinetic energy less than 600 $cm^2 s^{-2}$. Before attempting to compare eddies in this region with those in others, we need to estimate how much of this energy is related to the seasonal changes in circulation. In doing this, some characteristic subsections of the region will be identified, and in the rest of the chapter the eddy properties of each of these subsections will be briefly described.

11.2 Kinetic Energy of the Seasonal Circulation

This can be estimated roughly from the monthly mean surface current vectors, shown for every 2° square in which observations existed, in the KNMI Atlas of the Indian Ocean (KNMI 1952). The estimate is rough because the distribution of observations is uneven through the year and, off the main shipping routes, data are sparse. In Fig. 1 the annual cycle of surface current is shown derived from these observations in 12 2° squares. Although the squares were chosen to be reasonably well sampled and representative, in some months at some squares there were no more than ten observations. From the atlas, vector mean directions were read off by protractor and combined with the stated magnitudes to give east and north components (u and v) of current. The kinetic energy per unit mass associated with the seasonal circulation was then calculated for each square as $\frac{1}{2}\langle u'^2 + v'^2 \rangle$ where the u' and v' are residuals after subtracting annual means. These kinetic energies ($cm^2 s^{-2}$) are listed in Table 1 below and compared with the kinetic energy values given for the same places by Wrytki et al. (1976).

Eddies in Marine Science
(ed. by A.R. Robinson)
© Springer-Verlag Berlin Heidelberg 1983

Table 1. Comparison of kinetic energies

Area		KNMI		Wyrtki et al.	
		"Seasonal" K.E.	Mean K.E.	Eddy K.E.	Mean K.E.
1.	17°N, 65°E	60	4	600	20
2.	11°N, 61°E	140	2	600	20
3.	11°N, 69°E	40	10	600	20
4.	7°N, 51°E	1 200	900	2 000	500
5.	17°N, 89°E	130	12	700	20
6.	13°N, 81°E	480	34	1 000	30
7.	1°N, 63°E	910	34	1 000	100
8.	1°S, 73°E	560	360	1 000	100
9.	3°N, 81°E	340	4	1 000	50
10.	15°S, 67°E	80	240	600	100
11.	15°S, 85°E	30	110	600	100
12.	11°S, 49°E	400	1 500	1 200	200

One cannot expect to make any very precise comparison of these values. Besides the limitations of the basic data, the values from Wyrtki et al. (1976) have been picked from small parts of two global charts (Fig. 1a, b, Chap. 15) with suitably wide contour intervals. Their data were averaged over 5° squares, versus 2° for the KNMI atlas. The two sources are based on different data sets, though in both cases for this part of the ocean most of the observations are from the period 1918–1939. Even so, the mean K.E. estimates are not very different, allowing that 20 $cm^2 s^{-2}$ was the minimum contour in the global chart of Wyrtki et al. The only serious discrepancy, at position 12 (11°S, 49°E) in the northern branch of the south equatorial current just at the northern tip of Madagascar, is the result of averaging over different horizontal scales in a narrow strong current. Comparing the "seasonal" kinetic energy estimates to the total eddy kinetic energy, in the regions where the latter is weakest [central Arabian Sea (1, 2, 3), central Bay of Bengal (5), mid-ocean part of south equatorial current (10, 11)] the seasonal contribution accounts for less than a quarter, more usually only about one tenth, of the total eddy K.E. Most of the latter must belong to what are normally thought of as "eddies" elsewhere in the ocean, that have been averaged out in forming the monthly vector means, or (less probably) to year-to-year differences in the seasonal circulation. In the regions of relatively strong currents [Somali Current (4), off Madras (6), near the equator (7, 8, 9) and off Cape Amber (12)] a much larger proportion, about half, of the total eddy K.E. is accounted for by the seasonal signal. The remainder left to be explained by "normal" eddies is not very different from what was found in the quiet areas. It is still fairly high, comparable to what is shown by Wyrtki et al. (1976) for much of the Western North Atlantic not too near the Gulf Stream.

For those readers who are familiar only with the usual summer and winter pictures of Indian Ocean surface currents, shown as insets in Fig. 1, it may be instructive to look more closely at some of those seasonal cycles. In the central Arabian Sea (1, 2, 3) and Bay of Bengal (5) there is a simple annual cycle,

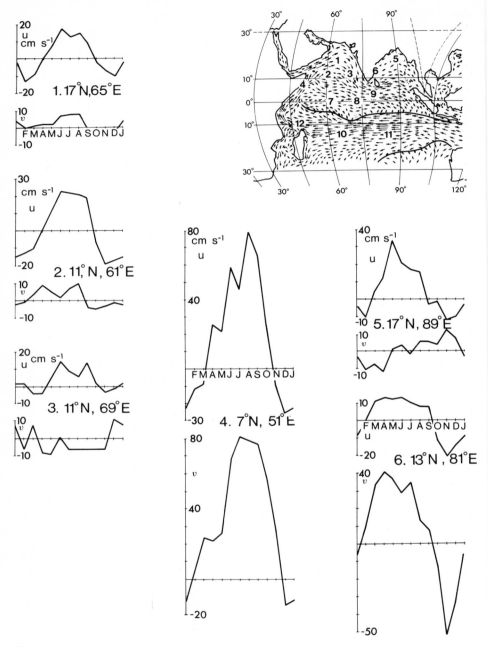

Fig. 1a

Fig. 1. Monthly mean eastward (u) and northward (v) components of surface current, in 12 2° squares. (KNMI Publ. 135). Location chartlets also show mean circulation in **a** July **b** January

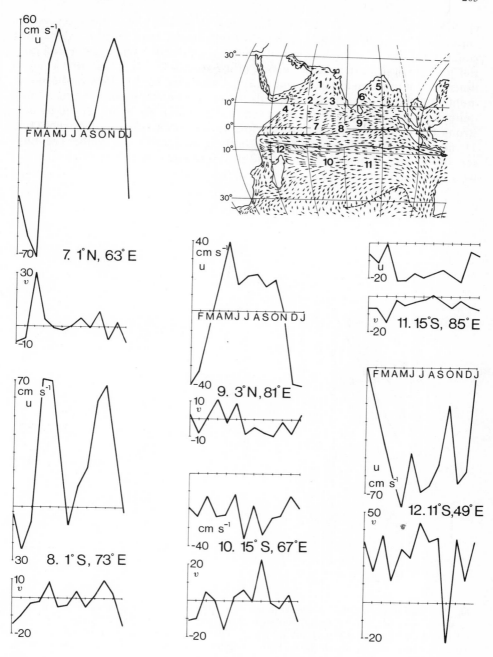

Fig. 1 b

clockwise in northern summer and the reverse in winter. The stronger reversing currents off Somalia (4) and off Madras (6) show the same pattern. Along the equator, though, (7, 8) there are two maxima of eastward velocity, the equatorial jets which appear at each change of the monsoon. Note also that conditions are not uniform along the equator, the westward surface flow in the early months of the year is stronger towards the west. These equatorial features are quite narrow; at 3°N, 81°E (9) there is a simple annual cycle, influenced by the strong reversing current south of Ceylon. In the south equatorial current (10, 11) no clear annual signal can be seen in these data. Off Cape Amber (12) observations are few in some months and therefore noisy, but there does seem to be a decrease in westward speed early in the year.

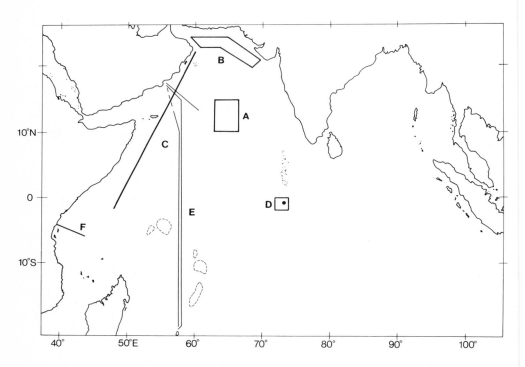

Fig. 2. Location chart for some observations referred to in text. *A* area of survey reported by Koshlyakov et al. (1970); *B* area of survey reported by Das et al. (1980); *C* tanker XBT track referred to by Bruce (1979); *D* site of Knox's (1976) measurements (Gan); *E* sections for which dynamic heights are shown in Fig. 3. Neighbouring banks indicated by dotted 200 m isobath; *F* section shown in Fig. 11

Fig. 3. Dynamic height variations at selected pressures, relative to 2 000 dbar, on two sections ▶ in long. 58°E. **a** 10 March–4 April, 1964, **b** 24 May–15 July 1964. See Fig. 2 for locations

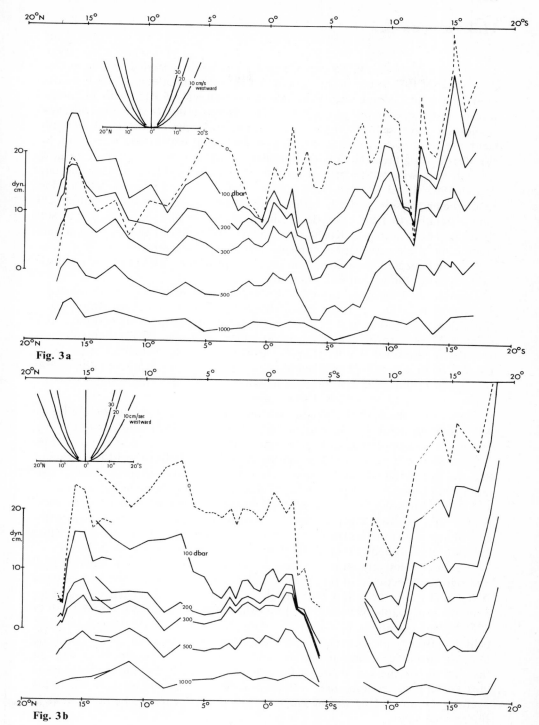

Fig. 3 a

Fig. 3 b

11.3 Arabian Sea

One of the early POLYGON experiments, specifically designed to look at variable currents, took place in this region in 1967 (Stockman et al. 1969, Ozmidov et al. 1970). An L-shaped array of 7 moorings was set out, each with 11 current meters under surface buoyancy, but its dimensions (100 km side) and duration (2 months) were such that a good description of the main features of the eddy field was not expected. At the same time, hydrographic surveys were made of the surrounding area. The region bounded by $10°–15°N$, $63°–66°30'E$ (Fig. 2) was covered twice with a grid of stations 55 km apart, in late January and late March 1967. These showed (Koshlyakov et al. 1970) eddies some 200–300 km in diameter, closely packed, with surface geostrophic velocities of about 20 cm s^{-1} relative to 1 200 dbar, decreasing downwards to about 5 cm s^{-1} at 800 dbar. In another paper, Koshlyakov et al. (1972) found that, after suitable smoothing, these geostrophic currents agreed with the directly observed ones within the expected accuracy (3 cm s^{-1}) except in the top 50 m where significant ageostrophic velocities can occur.

Another hydrographic survey in this region, in which eddies were resolved, was reported by Das et al. (1980). It covered part of the northern Arabian Sea (Fig. 2) with a 55-km grid of stations, some of it occupied in February and some in March–April 1974. They found eddies of about 200 km diameter, and a density distribution that would give surface geostrophic velocities of about 20 cm s^{-1}, relative to 800 dbar. Roughly circular patches, indicative of eddies 100 to 200 km in diameter, can be seen in satellite-derived infrared images of the sea surface in that area (O.B. Brown, pers. comm.).

Other than these, the only data with suitable horizontal resolution are BT or XBT sections, notably those organised by J.G. Bruce (1979, and personal communications) from tankers of opportunity. The tanker track is marked on Fig. 2. That part of it north of $12°N$ can be regarded as belonging to the Arabian Sea, the rest to the Somali Current region. These sections often show large variations in thermocline depth (the $20°C$ isotherm depth changing by as much as 100 m) but on isolated sections with no supporting evidence the interpretation is uncertain. A fairly clear example of part of a warm-core eddy to the northeast of Socotra could be seen when three nearly synoptic XBT sections in June 1978 were combined (Bruce et al. 1980). It was about 500 km in diameter.

For such relatively large features, it is worth while looking at conventional hydrographic sections. In the Arabian Sea, the two surveys done mainly by the ATLANTIS II in 1963 and 1965 are still the best available. Their near-surface dynamic topography was described by Bruce (1968). East of $57°E$ and north of $5°N$, there were no strong features in the surface topography. Maximum geostrophic surface currents relative to 1 000 dbar were about 30 cm s^{-1}, with no marked difference between summer and winter monsoons. On these sections the station spacing was however fairly wide – typically 140 km. Even in more closely sampled sections such as those in long $58°E$ shown in Fig. 3, where the average station spacing was 80 km, eddy activitiy appeared to be fairly weak in the northern part.

Possible evidence of a small, more intense, eddy in the Arabian Sea can be seen in one of the drifter trajectories in Fig. 4. Near 9°–10°N, 60°–61°E there are two elliptical loops suggesting an eddy 180 × 80 km in diameter, with average surface speed 40 cm s^{-1} and period 12 days. This was in early September

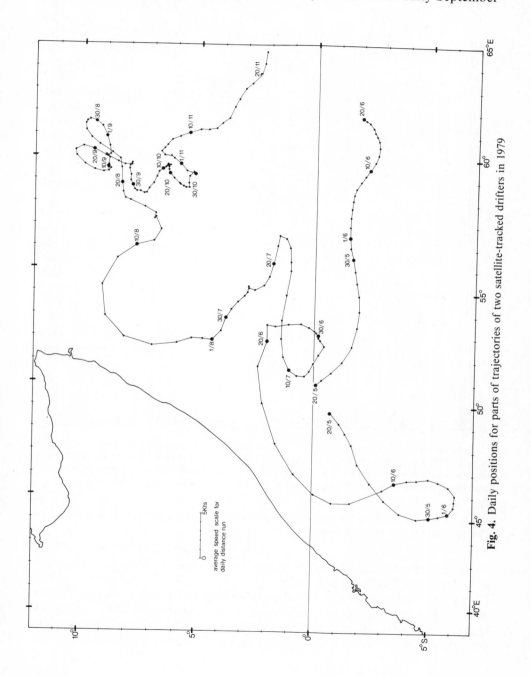

Fig. 4. Daily positions for parts of trajectories of two satellite-tracked drifters in 1979

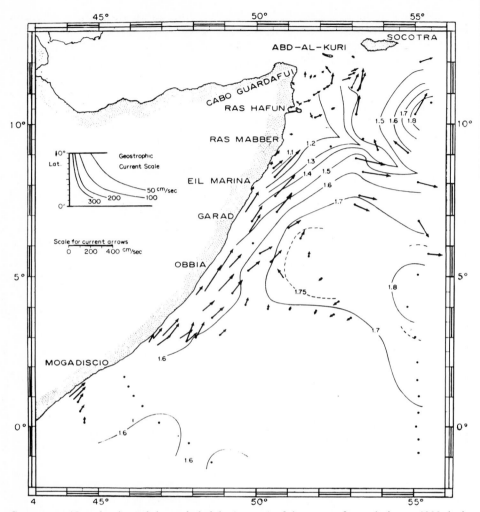

Currents at 10 m depth, and dynamic height (metres) of the sea surface relative to 1000 decibars. Dots indicate stations on which the dynamic height contours are based. To use the scale
of geostrophic currents, pick off the spacing of an adjacent pair of height contours and measure that distance from the left-hand edge of the scale, along the appropriate latitude.

Fig. 5 a

Fig. 5. Surface currents in two states of development of the Somali Current. **a** August 1964.
(Swallow and Bruce 1966); **b** June 1979

1979, towards the end of the SW monsoon, and not far from the strong offshore branch of the Somali Current.

To summarize, it seems that eddies are likely to be found anywhere in the
Arabian Sea, with horizontal dimensions in the range 200–500 km, vertical extent of some hundreds of metres, and typical surface currents of 20–30 cm s^{-1},
consistent with the eddy kinetic energy estimate for that area by Wyrtki et al.

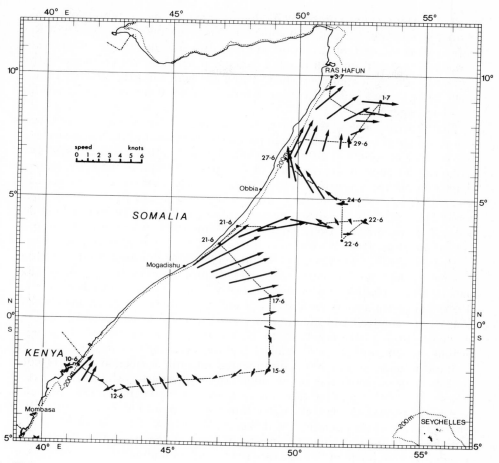

Fig. 5 b

(1976). Away from the region of the Somali Current, eddy energy does not seem to be much greater in the southwest monsoon (relatively strong winds) than in the northeast (relatively weak). However, the whole area is poorly sampled. For example, we have no description of the oceanic eddy field under the intense transient atmospheric "monsoon vortex" that formed in June 1979 in the central Arabian Sea (Krishnamurti et al. 1980).

11.4 Somali Current

The strong seasonally reversing boundary current off Somalia and its associated eddies are by now fairly well known. The main eddy has been known for a long time and its surface expression was clearly described by Findlay (1866). In his words: "To the south of Sokotra, at a distance of 150 miles, is a great whirl of current, caused probably by the interposition of the island; or, it may be, that shoal water exists at that spot; it commences about the parallel of Ras Hafun, when the current strikes off to the eastward to the 55th meridian, then to the southward, to the 5th parallel, whence it again curves up to the northeastward, forming a complete whirl. At the northern limit the velocity is very great, being 4 to 5 miles per hour, while at its southern extreme it is only ¾ to 1 mile per hour."

That describes one state of the eddy field, seen at a late stage of the southwest monsoon (Fig 5a). In it, a large part of the transport of the boundary current is being recirculated after turning offshore at 8°–10°N. Bruce (1979) estimates that the geostrophic transport in the upper 400 dbar of the eddy, relative to 400 dbar, is approximately 40 million $m^3 s^{-1}$. The asymmetry in surface currents, strongly eastward in the north, but weak westward in the south, mentioned by Findlay, is of course due to the strong offshore Ekman transport superimposed by the southwest winds. In 1979 the eddy was seen to develop in situ in late June, (Fig. 5b) within two weeks of the onset of strong southwest winds there (Düing, Molinari and Swallow 1980a). At that time, to the south of its position the boundary current was already turning offshore at 3°–4°N. That state of affairs continued until mid-August when, as described by Brown et al. (1980) mainly from satellite-derived sea surface temperature maps, the circulation pattern changed and the boundary current began to run continuously up the coast, turning offshore at 9°–10°N. Whether the lower-latitude branch of current moved north as a dynamical feature, eventually coalescing with the northern branch, or whether it disappeared and the associated patch of cold upwelled surface water was simply advected up the coast by the boundary current, is not clear. In either case, the near-surface water in the eddy, initially Arabian Sea water of relatively high salinity, was replaced by lower salinity water from south of the equator. The tanker XBT sections of Bruce (1979) suggest that the eddy formed similarly in situ in each of the preceding four years, within the same range of latitude (4°–12°N) and with about the same diameter (400–600 km). It decayed slowly, with the relaxation of the SW monsoon, being just detectable in the temperature field in December, or January of the next year.

When the boundary current turns offshore at 3°–5°N, as is often observed early in the SW monsoon and sometimes later, part of it is recirculated in another clockwise eddy. See Fig. 5b and the drifter trajectory in Fig. 4, and Regier and Stommel (1976). In the latter reference, drifters revealed an elliptical eddy roughly 700 km × 300 km, centred at 1°N, 50°E, with its major axis parallel to the coast and average surface speed of about 70 cm s^{-1}. Being so close to the equator, this eddy is not as clearly apparent in the temperature structure (Bruce 1979) as in surface currents.

In late April and early May 1979, before the Somali Current had crossed the equator, the eddy could be seen as a dome in the thermocline south of the equator (Fig. 6) and presumably migrated northwards with the northward progression of boundary current. Farther east in the same figure, there appear to be waves or unstable meanders in the offshore part of the boundary current, before it merges with the eastward equatorial surface jet. To the northeast of

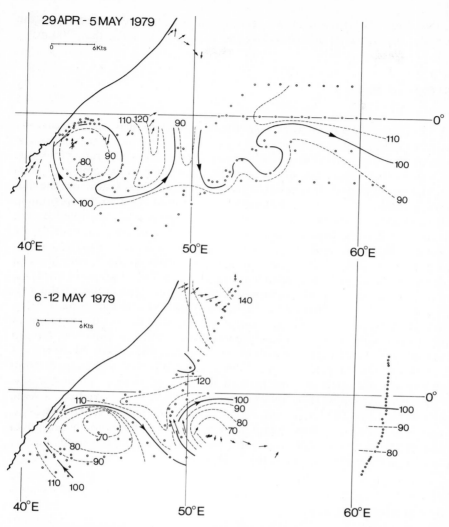

Fig. 6. Surface current vectors and contours of the depth (m) of the 20 °C isotherm, when the Somali Current turned offshore near the equator

the main eddy, east of Socotra, Bruce (1979) points out another smaller clock-wise eddy between 12 °N and 15 °N. In general, the observed pattern of eddies recurring annually in close relationship to the development of the local winds and the boundary current itself, seems to be in good qualitative agreement with the models described by Anderson (1980) and, for the early stages of the SW monsoon, Delecluse and Philander (1981). In the main eddy, typical surface speeds are about 1 m s^{-1} (Fig. 5), but can exceed 3 m s^{-1} in the parts that coin-cide with the boundary current, leading to correspondingly high values of eddy kinetic energy. Velocities decrease rapidly downwards through the thermocline but, as Bruce's (1979) temperature sections show, when fully developed the main eddy penetrates to at least 400 m and the others to more than 200 m.

11.5 Bay of Bengal

The writer does not know of any observations in the Bay of Bengal specifically intended to reveal eddies. Lafond (1957), describing the general features of the circulation there, remarked that "the surface circulation is generally clockwise from January to July and counterclockwise from August to December ... During the transition periods it is believed that large eddies break off from the major flow patterns." Lafond and Lafond (1968) pointed out two depressions in the thermocline, interpreted as clockwise eddies, in sections occupied by the ANTON BRUUN in March–May 1963. One of these, centred near 18°N, 89°E, was 200–300 km in diameter, with a depression of the 20°C isotherm of about 80 m. The other, similar in size, was in deep water offshore from Madras. D.P. Rao (1977) mapped the dynamic topography at several pressure levels relative to 1000 dbar using hydrographic data from the 33rd cruise of the "Vityaz" (January–March 1961). Stations were spaced 150–200 km along sections 400 km apart, with some infilling – too widely spaced for good resolution of eddies. The contouring shows closely packed high and low cells 300–400 km across (only two station spaces) with average geostrophic speeds 20–30 cm s^{-1} at the surface, decreasing to about 5 cm s^{-1} at 500 dbar. The bathythermograph sections occupied in July 1979 in the Bay of Bengal, plotted in the MONEX BT data report (Das 1980), suggest that there may have been a weak clockwise eddy 400–500 km across, centred near 14°N, 89°E, where the 20°C isotherm was depressed 30 m. Again, as in the case of the Arabian Sea, there seem to be no observations of the oceanic response in deep water to the intense transient atmospheric cyclones of this region.

11.6 Equatorial Zone

The equatorial regions of all three oceans are dealt with in Chapter 10, but it seems appropriate to include some observations pertaining to the equatorial Indian Ocean here. In a series of weekly PCM profiles made near Gan (1°S, 73°E; position 8 in Fig. 1) lasting more than 2 years, Knox (1976) found that the dominant feature of the near-surface currents was the twice-yearly eastward jet. Monthly mean components of surface currents from his data (Knox and McPhaden 1980) are compared with those from the KNMI atlas in Fig. 7. The differences in the later part of the year are probably due to abnormally strong zonal winds in late 1973; the agreement in the earlier part of the year lends confidence in the use of historical ship-drift data (KNMI atlas) and suggests that excess speed in the PCM data, due to being alongside islands, is small. Knox found that the eastward jet started with very little delay after the

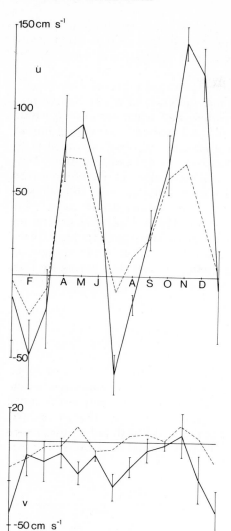

Fig. 7. Monthly mean eastward (u) and northward (v) components of surface current, from Knox and McPhaden (1980) at Gan *(solid lines)* and climatological mean data *(dashed lines)* for the 2° square centred at 1°S, 73°E, from KNMI (1952). *Error bars* are standard deviations of the monthly means.

onset of eastward winds along the equator. Luyten et al. (1980) noticed similar behaviour in the trajectories of satellite-tracked drifters along the equator. With the relaxation of eastward winds, the drifters moved slowly westward and then became stagnant. It was remarkable that these regions of stagnation were over high rough topography, either the Carlsberg Ridge (65° to 70°E) or the 90°E Ridge. Below the thermocline, when the eastward surface jet was present, current profiling has revealed several bands of westward flow, some 20 to 40 cm s^{-1} in speed and 100 to 200 m thick, in the upper 2000 m (Luyten et al. 1980, Luyten 1981). From a year-long array of moored current meter records,

Luyten and Roemmich (1982) find that these subsurface jets, like the surface current, have a predominantly semi-annual period and a large zonal scale. They are associated with a downward energy flux suggestive of forcing from the surface by the zonal winds.

The equatorial undercurrent, a sub-surface eastward maximum of velocity in the upper thermocline, is present only intermittently in the Indian Ocean, typically in February and March in the western half when meridional winds are weak. Knox (1976) saw it then, in one of the 2 years of current profiling at Gan. From a combination of current measurements and observations of periodic patchiness in the salinity maximum associated with the undercurrent, Nelepo et al. (1979) inferred the presence of lateral oscillations in the undercurrent with wavelength 800–1000 km, period 16–20 days and amplitude 200–300 km. Wave-like disturbances can be seen in some of the surface drifter trajectories near the equator (e.g., Fig. 4). Gonella, Fieux and Philander (1981), describing the trajectories of a group of buoys in the eastern equatorial Indian Ocean, suggest that the deceleration of the eastward equatorial jet in the autumn of 1979 was associated with a westward-propagating Rossby wave having a phase speed of 50–60 cm s^{-1}.

11.7 South Equatorial Current

This broad westward surface current occurs between latitudes 7°S and 20°S, approximately, originating to the northwest of Australia, and dividing into two branches at the coast of Madagascar near 16°S. Eddies in the source region off Australia, and in the southern branch when it passes south of Madagascar, are described in Chapter 12. The northern branch is well illustrated by the tracks of some of Cresswell and Golding's (1979) satellite tracked drifters (Fig. 8).

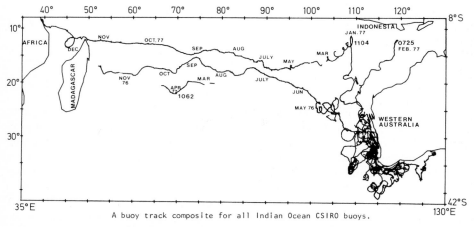

A buoy track composite for all Indian Ocean CSIRO buoys.

Fig. 8. Trajectories of drifters in the South Equatorial Current. (Cresswell and Golding 1979)

The average westward speed was approximately 20 cm s^{-1}, consistent with the mean K.E. at positions 10 and 11 (Fig. 1). Various kinds of fluctuations can be seen. The clockwise eddies in trajectory 1 104 in Jan.–Mar. 1977 have a diameter of approximately 100 km and 20 days period. The deviations to the south in the two southernmost trajectories in Sept.–Oct. 1976 and Mar.–Apr. 1977, seem to have occurred over featureless deep water, though the smaller deviation between Oct. and Nov. 1976 coincides with passing over the Mascarene

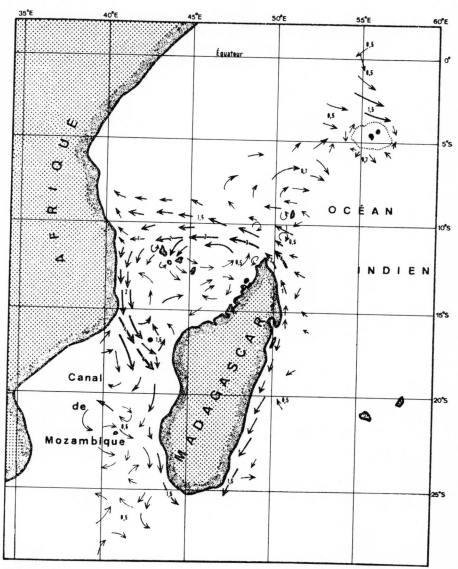

Fig. 9. Summary of GEK current observations. (Piton and Poulain 1974)

ridge. In the northernmost trajectory, after passing north of Madagascar the buoy made a complete anticlockwise loop round Mayotte Is before running aground in Tanzania (Cresswell 1979).

Hydrographic sections across the south equatorial current (several in Wyrtki 1971, see also Fig. 3) often show irregular variations of surface dynamic height equivalent to about 20 cm s^{-1} relative to 1 000 dbar, suggestive of eddies with diameters 200–400 km. The most marked of those features in Fig. 3, near 11°S, was downstream of a gap in the Mascarene ridge. It evidently penetrated down to at least 500 dbar.

Several eddies have been seen in the surveys made in the Mozambique channel and between Madagascar and the Seychelles, by members of the OR-STOM laboratory at Nosy-Be, Madagascar. Some of their GEK observations are synthesized in Fig. 9, from Piton and Poulain (1974). They detected several

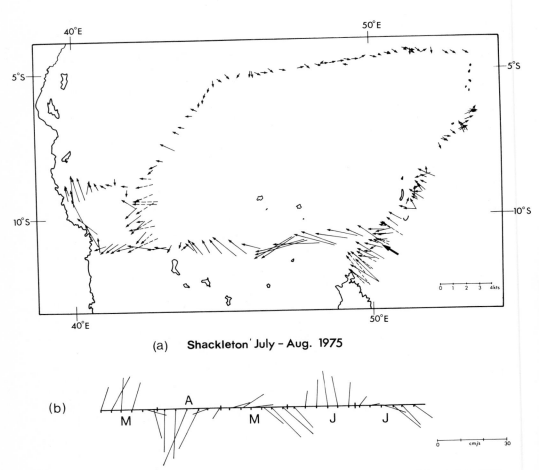

(a)   **Shackleton' July – Aug. 1975**

Fig. 10. a Surface current vectors, July–August 1975. **b** 5-day mean vectors, March–July 1975, from a current meter at 513 m depth. Position of mooring is marked by the *heavy arrow* in **a**

times the anticyclonic eddy near the Comoro Is., later encountered by one of Cresswell's drifters mentioned above, and a cyclonic eddy between it and the coast of Madagascar. In this region, as in the Somali Current, it seems that eddies tend to recur in nearly the same positions, with the same sense of rotation.

The wave-like nature of fluctuations in that branch of the current crossing the northern end of the Mozambique Channel can be seen in the surface currents of Fig. 10, with an apparent wavelength of 500 km. Records of currents from a nearby mooring (one shown as a stick diagram in Fig. 10) were dominated by an oscillation of approximately 50 days period and almost the same amplitude as the wave in the surface current.

11.8 Deep Currents in the Indian Ocean

Not much has been said above about variability of currents at depth, except for some remarks about geostrophic velocities tapering off downwards in the upper few hundred metres. At least two cases have been noted, though, of discrete subsurface eddies or rotating blobs of relatively deep water. One of these, reported by Bruce and Volkmann (1969), was a clockwise eddy at 9°N, 53½°E off the Somali coast, with a velocity maximum exceeding 50 cm s^{-1} at 600–700 m depth and a diameter of approximately 250 km. This was seen in the

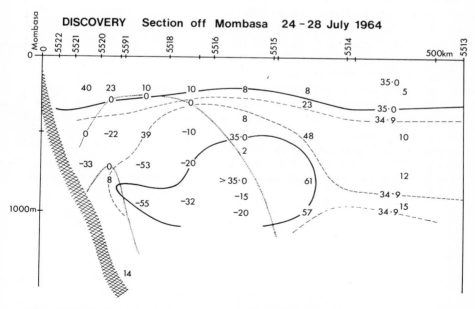

Fig. 11. Geostrophic currents (cm s^{-1}) through section marked F in Fig. 2, relative to directly observed components at 200 m depth, and selected salinity contours

northeast monsoon and was tentatively identified as a relic of the main eddy of the preceding southwest monsoon. What may have been a similar case, though rotating anticlockwise, was seen on only a single section and is shown in Fig. 11. Velocities exceeded 50 cm s^{-1} at 800–900 m, and the apparent diameter was 250 km. The region of high velocities was characterized by higher salinities, of Red Sea origin.

Several short samples of deep currents were collected in the neighbourhood of the Somali Current in the summer of 1979. From these, Leetmaa et al. (1980) obtained transverse correlation length scales of 90 ± 20 km at 700 m depth and 50 ± 15 km at 2000 m. At the same depths, the rms velocity magnitudes were 18 cm s^{-1} and 10 cm s^{-1}. At the surface, the corresponding values were 160 ± 40 km and 85 cm s^{-1}.

11.9 Significance of Indian Ocean Eddies

To a great extent (80%?) the Somali Current and the inshore part of the main eddy off Somalia are one and the same thing. The transport increases dramatically northwards from the equator (Swallow and Bruce 1966) because so much water is being recirculated in the eddy. Like Findlay, we should think of it as the Great Whirl, with a small boundary current, rather than a major current which happens to have an eddy on its offshore side. When there is a branch of the current turning offshore at 3°–4°N, the recirculation eddy for that branch is centred close to the equator. In June 1979 (Fig. 5b) there was a centre near 1°S, 47°E and correspondingly shallow thermocline, in a clockwise eddy south of the equator. Although no biological observations were being made there, it was obvious from the colour and smell of the surface water that the region near that centre was quite productive. Though not as dramatic as the wedges of cold upwelling water near the coast, such cyclonic centres may perhaps contribute significantly to the biological resources of the region. Similarly, cyclonic eddies may play a significant role in the interaction with the atmosphere in the Arabian Sea, by raising the thermocline locally and enabling surface temperature anomalies to be generated more readily by downward mixing. Of course, the eddies themselves are generated by the action of the winds, and the whole process must be very complex, but even such simple qualitative speculation points to the need to take account of the presence of eddies in attempting to work on such problems.

12. The South Pacific Including the East Australian Current

A.F. Bennett

12.1 Introduction: The Subtropical South Pacific

The oceanic region considered in this chapter is the subtropical South Pacific, bounded by latitudes 23°S and 45°S and by longitudes 150°E and 290°E. It borders the tropical South Pacific and the Southern Ocean, which are considered in Chapters 10 and 14 respectively.

The large scale hydrographic features of the region have, to the meager extent that they are known, been described in a stimulating review by Warren (1970). There appeared to be no information about mesoscale circulation available at the time that review was prepared. The global analysis of oceanographic data by Levitus and Oort (1977) also demonstrated the extreme paucity of data in the region. Indeed, the subregion from New Zealand to Chile is perhaps the worst-surveyed part of the world ocean. An appreciable amount of mechanical bathythermograph data (MBT) has been collected along shipping routes, but there is far less expendable bathythermograph data (XBT) and, were it not for the trans-Pacific sections at latitudes 28°S and 43°S obtained during the "Scorpio" Expedition (Stommel et al. 1973), there would be even less salinity and density data (SD); see respectively Figs. 3, 4 and 2 of Levitus and Oort (1977). It is unfortunate that only the latter two types of measurements (XBT and SD) can detect variability below the upper mixed layer. It should also be noted that, east of New Zealand, the hydrographic station spacing of the SCORPIO expedition did not resolve mesoscale eddies.

Some indirect evidence for eddy variability in the subtropical South Pacific has been obtained by Wyrtki et al. (1976), who inferred the values of surface currents from log-book records of ships' drifts off set courses. The records represented velocities averaged over 24 h and hence over a ship track of about 400 km (assuming a typical ship speed of 10 kt or 400 km/day). Wyrtki et al. (1976) sorted the records into $5° \times 5°$ squares, and calculated the sample mean and the corresponding "eddy" kinetic energies per unit mass in each square. They displayed contours of the two fields in their Figs. 3 and 4 respectively, which are reproduced here as Figs. 1 and 2. The eddy kinetic energy in the mid-ocean subtropical South Pacific was about 0.04 $m^2 \ s^{-2}$. Similar values were found in other sub-tropical mid-ocean regions. The contours of eddy kinetic energy were far more involved than those of mean kinetic energy, but the apparent length scales of variability of the former were of necessity in excess of 5°–10°. Wyrtki et al. (1976) based their study on no less than 4 million records, but it must be assumed that relatively few records were made in the subtropical

Eddies in Marine Science
(ed. by A.R. Robinson)
© Springer-Verlag Berlin Heidelberg 1983

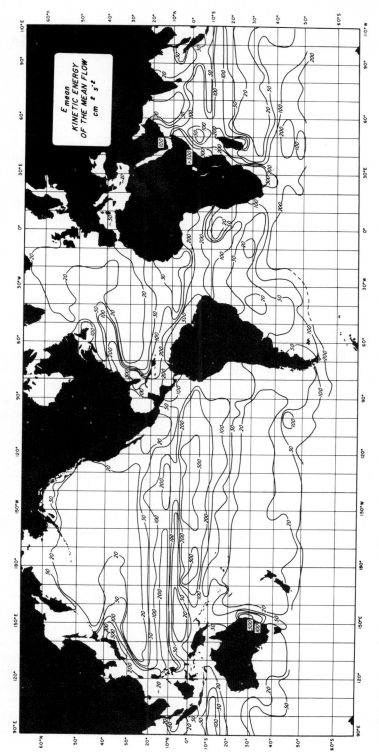

Fig. 1. Kinetic energy per unit mass of the mean flow for the world oceans based on 5° square averages. (After Wyrtki et al. 1976). See Fig. 1a, Chap. 15 for an enlargement

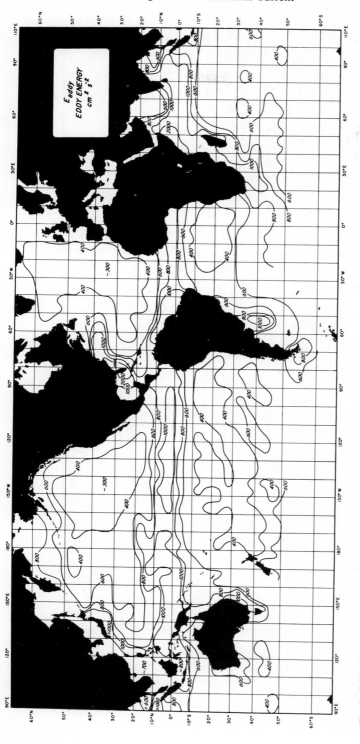

Fig. 2. Eddy kinetic energy per unit mass for the world oceans based on 5° square averages. (After Wyrtki et al. 1976). See Fig. 1b, Chap. 15 for an enlargement

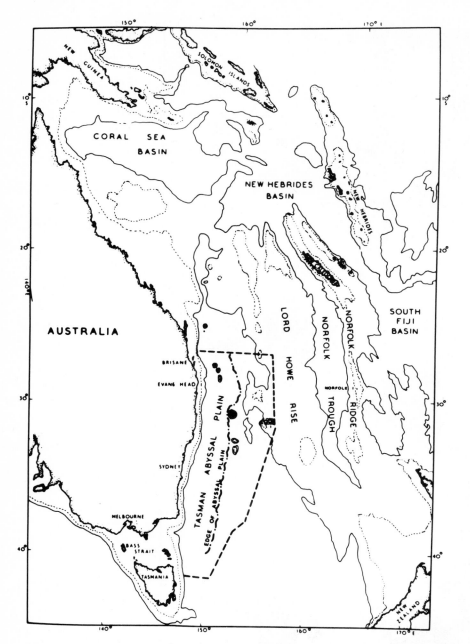

Fig. 3. The Tasman and Coral Seas, with simplified bathymetry. The *thin dotted* and *continuous lines* are, respectively, the 1000-m and 3000-m contours. The area in which most cruises to study the East Australian Current have taken place (to 1970) is enclosed by the *dashed line*. (After Hamon 1970). • Lord Howe Island

South Pacific. Thus the details of Fig. 2 in that region should be regarded with caution. There is also a difficulty in inferring the variability of deeper currents from ship drift velocities, which are an unknown combination of geostrophic (deep) current, nongeostrophic (surface) current and windage.

There is far more eddy-resolving data in the subregion west of New Zealand, which is known as the Tasman Sea (see Fig. 3). First, the stations in the "Scorpio" sections were spaced in some places only 10–15 km apart. Second, there has been an extensive series of observation programs designed to resolve eddies in the Tasman Sea, since at least as early as 1960. More recently mesoscale features have been detected in the sub-surface circulation northwest of New Zealand.

This chapter is essentially a review of the Tasman Sea observations, the resulting theories of eddy formulation, and the wider significance of eddies in the region. The observations are reviewed in Section 12.2. The picture which emerges is one of warm, shallow, anticyclonic eddies shed regularly by the East Australian Current (EAC) after the current separates from the Australian coast around 33°S. Thus the eddies are similar to Gulf Stream rings in formation mechanism if not entirely so in structure. The separated EAC most probably meanders eastward as far as New Zealand, in the role of the Tasman Front and associated jet postulated by Warren (1970). Theories of eddy formation are reviewed in Section 12.3. They include: the meandering of a free inertial jet due to steering by topography and the β-effect; and the baroclinic instability of a western boundary current. Possible influences of mid-ocean and EAC eddies on subtropical South Pacific mean circulation, on Southern Hemisphere climate and on Tasman Sea biology are discussed in Section 12.4. Future prospects for eddy research are surveyed in Section 12.5.

12.2 Observations: The Tasman Sea

12.2.1 The East Australian Current

12.2.1.a The Dynamic Height Field

Evidence for the existence of anticyclonic eddies in the EAC appeared in the dynamic topographies obtained by Hamon (1961) and Wyrtki (1962b). The rapid variability of the eddy field was demonstrated by Hamon (1965a), in particular by his Figs. 5 and 6, reproduced here as Figs. 4 and 5. These are two charts of surface dynamic topography relative to 1300 decibars, (hereafter denoted h_{1300}) obtained by cruises only 4 weeks apart. The intense anticyclonic eddy centered near 152°E, 35°S (Fig. 4) moved to 153°E, 37°S (Fig. 5), allowing the very tight poleward meander in the EAC to expand into a broad high. The eddies typically have diameters of about 250 km. Figure 6 shows a meander which was apparently in the process of shedding an anticyclonic eddy.

Subsurface currents, calculated geostrophically from station pairs, decrease monotonically with increasing depth. At 250 m they reduce to half their surface

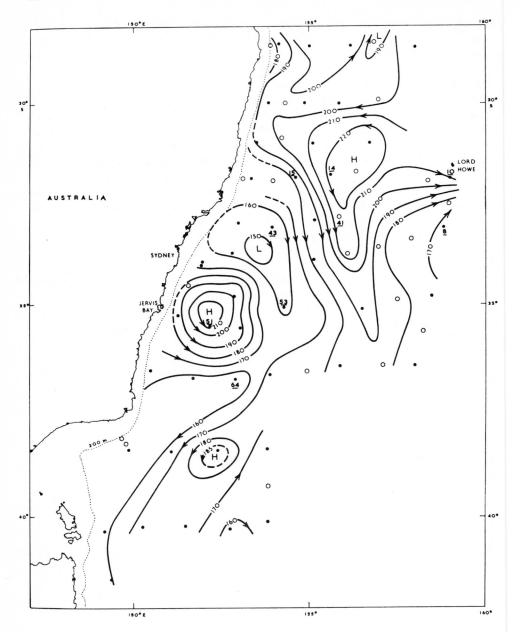

Fig. 4. Surface dynamic topography, relative to 1300 dbar, for HMAS GASCOYNE Cruise 1/
64 (January 13–February 6, 1964). • Stations with sampling to at least 1300 m. ○ Bathythermograph stations to 285 m. (After Hamon 1965a)

values. Hamon's choice of a reference level at 1300 dbar led to counter currents at 2000 m, so he suggested that the reference level should be closer to
Wyrtki's choice of 1750 dbar, which is the level of minimum oxygen concentration. There is substantial volume transport in the eddies. For example, between
stations 51 and 64 in Fig. 4, and between the surface and 1300 dbar, the transport was 35×10^6 m^3 s^{-1}. Further evidence of EAC eddies was obtained by Boland and Hamon (1970). Again they regularly found anticyclonic eddies about
250 km wide in the surface dynamic topography, near the Australian coast at
around 35°S. They reported one of the few winter cruises in the region, up to
that time. Although no closed eddy was mapped, the variation of dynamic topography was as strong as has been observed in summer.

Cyclonic eddies are observed only very rarely. Hamon (1965a; his Fig. 1)
located one at 29°S, close to the Australian coast. Andrews et al. (1980; their

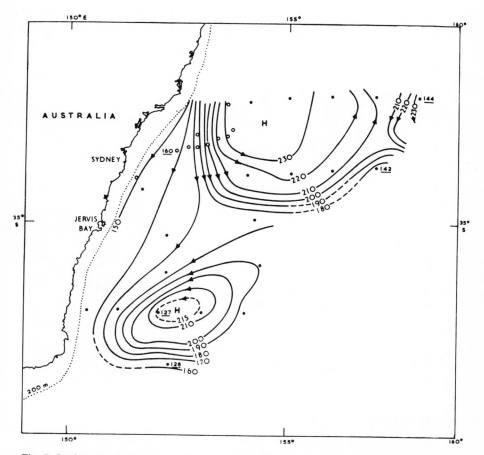

Fig. 5. Surface dynamic topography, relative to 1300 dbar, for HMAS GASCOYNE Cruise 3/
64 (March 18–25, 1964). ● Stations with sampling to at least 1300 m. ○ Bathythermograph stations to 285 m. (After Hamon 1965a)

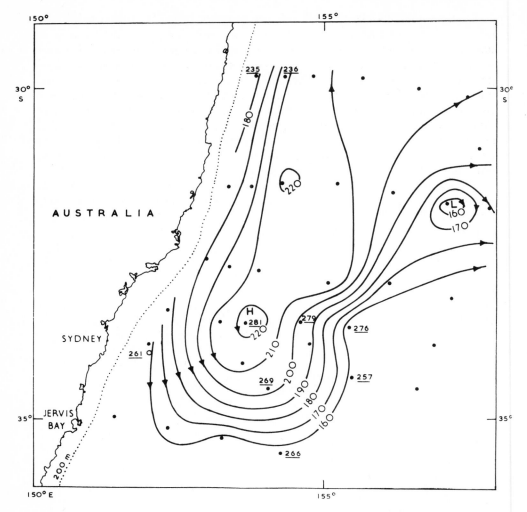

Fig. 6. Surface dynamic topography, relative to 1300 dbar, for HMAS GASCOYNE Cruise 6/
64 (September 16–27, 1964). (After Hamon 1965a)

Fig. 1a) gave evidence of a cyclonic eddy further offshore, possibly formed by
a northward meander of the EAC after the latter had separated from the coast
near 33°S.

12.2.1.b Temperature Structure

Vertical profiles of temperature are shown in Fig. 7, after Hamon (1965a). In
contrast with Gulf Stream eddies, the temperature differences between EAC
eddies and their environment are very much smaller at the surface than at 500
m. Hamon noted that occasionally they even have opposite signs.

The depth of the mixed layer covers a wide range in the EAC area. Hamon (1968a) found that at many winter stations the depth exceeded 200 m, while at many summer stations it was less than 20 m. On the other hand, both extremes were observed in a relatively small area on one late winter cruise. Hamon suggested that "... the very deep and very shallow mixed layers are convergence and divergence effects, respectively, associated with local dynamic features of the strong circulation pattern. Alternative mechanisms (Ekman convergence or divergence, convective overturn in later winter in the case of deep layers, or surface heating in summer in the case of shallow layers) would be expected to produce more uniform results in an area so small in relation to mean weather patterns."

Very deep surface mixed layers are usually observed in the centres of EAC eddies. Contours of mixed layer depth in an anticyclonic eddy observed by Andrews and Scully-Power (1976) are shown in Fig. 8. Surface convergence during mixed layer deepening, as suggested by Hamon (1968a), would entrain water from the environment into the eddy, thereby explaining the small surface temperature differences observed between EAC eddies and their environment. Subsurface mixed layers appear in EAC eddies in summer. An XBT section through such an eddy is shown in Fig. 9, after Nilsson and Cresswell (1980). When this same eddy was observed in the previous winter, it had only a surface mixed layer at 18 °C. In Fig. 9, the 18 °C layer is seen to have been "capped" by a new surface layer at 20°–21°C, created by summer heating.

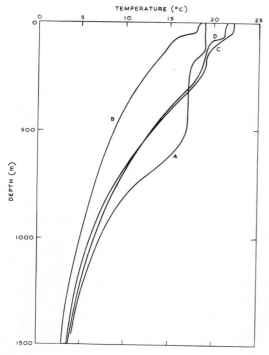

Fig. 7. Temperature as a function of depth at three stations in eddies (curves A, C, D) and one station outside the eddies (curve B). Curves C and D are from stations 51 and 127 Figs. 4 and 5, respectively. (After Hamon 1965a)

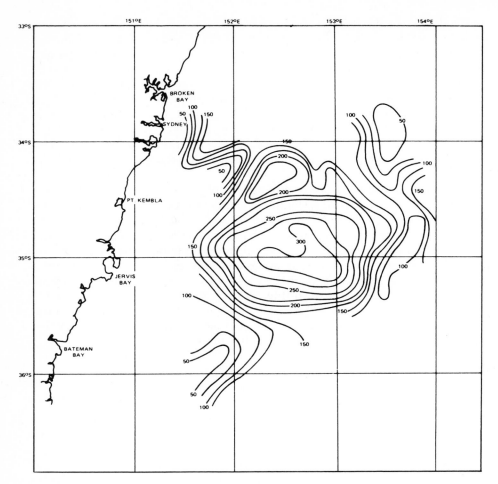

Fig. 8. Contours of mixed layer depth (m). Contour interval is 25 m. (After Andrews and Scully-Power 1976)

12.2.1.c Direct Current Measurements

Surface currents have been measured using geomagnetic electro kinetograph (GEK). A typical result is shown in Fig. 10, after Hamon (1965a). It was obtained on the cruise which produced the surface dynamic topography shown in Fig. 6. Current speeds are as high as 1.5 ms^{-1}, and current directions are in general along contours of h_{1300} except near the continental slope. Let v be the downstream surface current velocity, and let x be a cross-stream coordinate increasing to the right, looking downstream. Then the cross-stream shear ($\partial v/\partial x$) is typically 10^{-4} s^{-1} on the right hand side of the current maximum, and 2×10^{-5} s^{-1} on the left hand side.

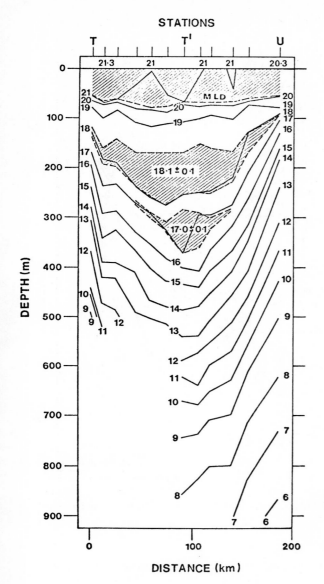

Fig. 9. An XBT temperature section diagonally across eddy A (see Fig. 11), from HMAS IMBLA Cruise 3/ 77. (March 14–15, 1977). *Hatched areas* show surface and subsurface mixed layers. (After Nilsson and Cresswell 1980)

Some measurements of deep velocities down to 3 300 m were obtained by Boland and Hamon (1970) using acoustically tracked subsurface floats. Quoting from Boland and Hamon: "It is clear...that the greatest measured velocities (>0.1 ms^{-1}) were obtained under strong surface currents and were approximately in the same direction as the surface current. Below 3 000 m, or in regions of small surface currents there is little coherence between the surface

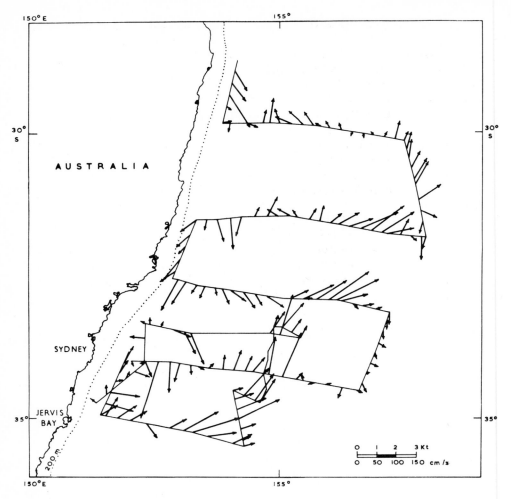

Fig. 10. GEK surface currents, for HMAS GASCOYNE Cruise 6/64. To be compared with Fig. 6. (After Hamon 1965a)

and deep currents". The floats confirmed that 1300 dbar is too shallow a reference level in the area.

Satellite-tracked surface drifting buoys have now become an important tool for locating EAC eddies. Cresswell (1976) reported some preliminary observations. These were difficult to interpret due to drogue loss somewhere along the buoy path. The mean kinetic energy densities of the buoy motion, during the early months of its drift, ranged from 10 to 70 Jm^{-3}. Since 1977, spar and torpedo-shaped buoys have been tracked in the Tasman Sea for as long as a year. (Nilsson and Cresswell 1980). Both types have 4.5 m parachute drogues set 20 m below the surface. The close relationship between buoy path, GEK measurements, and surface dynamic topography is shown in Fig. 11.

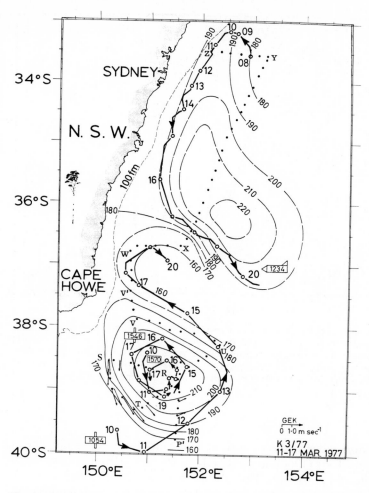

Fig. 11. XBT track, and surface dynamic topography relative to 1300 dbar, from HMAS KIMBLA Cruise 3/77. XBT stations P'→Z: March 13–17, 1977. Buoy tacks and GEK vectors are superimposed. Days of the month are shown with the buoy tracks. The *numbered symbols* represent spar or torpedo buoys. (After Nilsson and Cresswell 1980)

12.2.1.d Mean Sea Level Fluctuations

The sea level at Lord Howe Island (see Fig. 3) has variations, in its monthly mean, of up to 0.6 m. These may be due to moving current patterns. The low frequency spectrum in the range 0.0026–0.5 cycles per day has been computed by Hamon (1968b) and is shown in Fig. 12. The plot is energy-preserving. There appears to be a broad maximum around 0.006 cycles per day (170-day period) which, surprisingly, is not present in the sea level spectrum at Sydney. Spectra at Honolulu and Canton are also shown, for comparison.

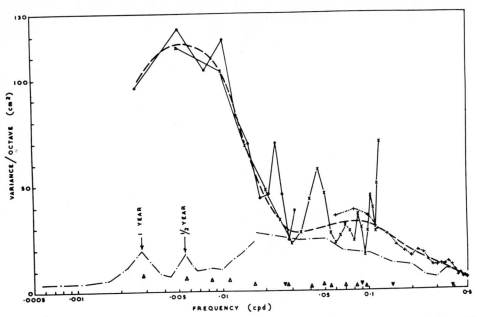

Fig. 12. *Solid lines* spectrum of sea level at Lord Howe Island (32°S, 159°E approx.) from 12-h means (*plus signs*), 4-day means (*crosses*), and 16-day means (*solid dots*). *Dashed line* visually estimated mean spectrum. *Dot-dash line* spectrum of sea level at Sydney, from monthly means, July 1957 to December 1958 (0.017–0.5 cpd). *Open triangles* spectrum of sea level at Honolulu. *Solid inverted triangles* spectrum of sea level at Canton Island. (After Hamon 1968b)

12.2.1.e Scales

Estimates of time and space scales of current variability in the EAC have been made by Hamon and Kerr (1968), using merchant ship data from cruises north and south along the 200 m isobath. They found that the spatially lagged correlation coefficient decreases monotonically with increasing lag, with a minimum value of about 0.2 for lags in excess of 100 km. The temporally lagged correlation coefficient appears to be significantly negative for lags between 30 and 40 days, suggesting a weak periodicity in the system with periods of about 70 days. No evidence of longshore movement of current patterns was found in the correlation coefficient lagged both in space and time. More recent estimates made by Hamon et al. (1975) using larger sets of ship data confirm the earlier length scale estimate of 100 km, but increase the time scale estimate to 100 days. The lagged correlation coefficients are shown Figs. 13 und 14. With the larger data set, a mean southward "speed" of 0.1 ms^{-1} for current patterns was evident at the edge of the continental shelf.

Much larger length scale estimates were obtained by Hamon and Cresswell (1972). They calculated the structure function S(l) of surface dynamic topography relative to 1 300 m, using the formula

$$S(l) = \overline{\{h_{1\,300}(\underline{r}) - h_{1\,300}(\underline{r}+\underline{l})\}^2},$$

where $l = |\underline{l}|$ and the overbar denotes an average over station pairs in the same cruise, and over 14 cruises. Results from the EAC are shown in Fig. 15, together with results off the West Australian coast, and off Hawaii. Curve B in Fig. 15 was derived from XBT data at 240 m using an $h_{1\,300}/T_{240}$ linear regression formula. Note that for large l, S should approach $2\sigma^2$ where σ^2 is the variance of $h_{1\,300}$. The broad maxima in curves A and C for l around 250 m indicate dominant wavelengths around 500 km. Hamon and Cresswell attributed the difference between curve A (east and north lags) and B (east lags only) to anisotropy in the EAC. They showed that the shapes of the curves are roughly consistent with theoretical structure functions for the sinusoidal wave

$$h_{1\,300} = \sin(k\,x) \,,$$

where x increases eastward.

A long series of XBT sections along 34°S, from 151°E to 156°E, has been described by Boland (1979). The sections were made every two weeks from

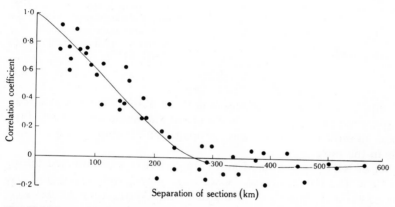

Fig. 13. Correlation coefficient as a function of distance along coast, averaged over 2 years. (After Hamon, et al. 1975)

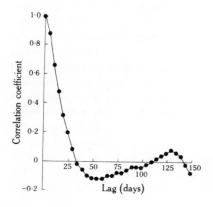

Fig. 14. Correlation coefficient as a function of time, averaged over all coast. (After Hamon et al. 1975)

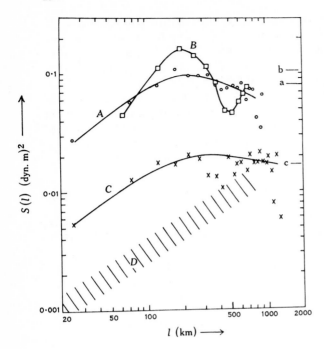

Fig. 15. Structure function $S(l)$ as a function of distance l between station pairs. *A* off east Australian coast, 30°–40°S., 145°–160°E., from hydrographic data. *B* off east Australian coast from XBT data along 34°S, approx. *C* off west Australian coast, from hydrographic data. *D* off Hawaii, for comparison. *a, b, c:* $2\sigma^2$ for the same data used for *A, B, C* respectively. (After Hamon and Cresswell 1972)

July 1969 to July 1975. An east-west length scale of about 500 km was apparent, and the spectrum of T_{240} at 154°E had a broad maximum around 140 days. Also, there was an indication of westward propagation of temperature anomalies, at speeds in the range 1–3 ms^{-1}.

Another set of XBT data (Boland and Church 1981) was collected during 14 cruises in 1978. The set confirmed the scale estimates given above, and also indicated the evolution of the EAC during that year. Anticyclonic eddies were present in the southern part of the system, but having no systematic pattern in their movements. One asymmetric eddy appeared to develop into an interconnected pair, which rotated around each other in the anticlockwise sense.

In summary, the EAC shows great variability, with length and time scales estimated to lie in the ranges 100–500 km and 70–170 days, respectively. It should be noted that the wide range of length scale estimates was obtained using a variety of estimators. These are: the first minimum of the correlation function (100 km), the eddy diameter (250 km), and twice the structure function maximum or wavelength (500 km). Some rationalization of these estimates may be made, if it is accepted that the correlation function minimum should be about half the eddy diameter, which in turn should be about half a wavelength. Nevertheless considerable scatter should be expected even only if due to the limited size of the data set. Very long time series (decades) would be required for statistically reliable estimates even of mean fields in the area. The variability of the EAC was emphasized by Thompson and Veronis (1980) in their analysis of hydrographic data from a single cruise. Using the inverse method of

Wunsch (1977), they concluded that there was *no* EAC at the time of the cruise. Clearly, a mean EAC is yet to be defined.

12.2.2 The Tasman Front

Denham and Crook (1976) constructed sections from XBT observations collected on cruises which crossed either the Lord Howe Rise or the Norfolk Ridge (see Fig. 3). They found subsurface thermal fronts in the vicinity of the two bathymetric features, but concluded that what had been observed was in fact a single front which meanders between the two features.

Stanton (1976) reported by hydrographic survey of the Norfolk Ridge area. The surface temperature field is shown in Fig. 16. A temperature front centered on 17°C was crossed six times; the greatest gradient detected crossing the front was 2.2°C in 11 km. The surface dynamic topography relative to 1000 dbar is shown in Fig. 17. A strong zonal flow may be seen, closely following the path of the 17°C surface isotherm (which Stanton calls the Mid-Tasman Convergence). GEK measurements indicated that the flow along the current axis was far more jet-like than could be inferred from the spacing of the hydrographic stations. Nevertheless the maximum GEK velocity and surface geostrophic velocity were both about 0.38 ms^{-1}. Stanton computed that geostrophic transport

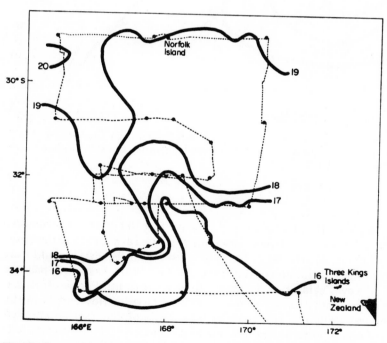

Fig. 16. Sea surface isotherm (°C) northwest of New Zealand in August to September 1973 (*dashed line* is ship's track). (After Stanton 1976)

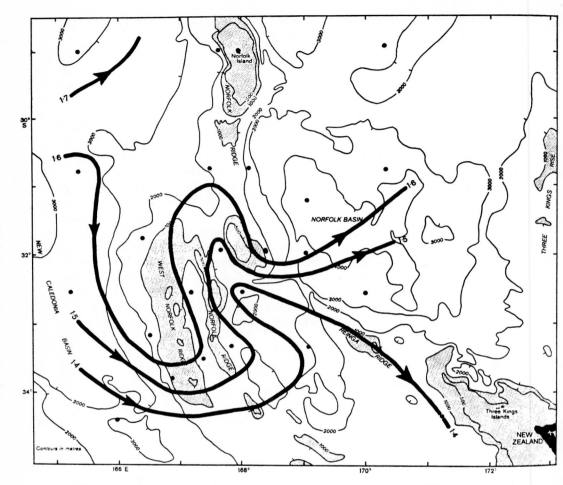

Fig. 17. Geopotential topography (dyn m) of the sea surface relative to 1000 dbar northwest of New Zealand in August to September 1973. (Sketch bathymetry and station positions are also shown). (After Stanton 1976)

of the flow to be about $10\text{--}14 \times 10^6$ m^3 s^{-1}, relative to 1300 dbar. He concluded that the observed jet was the eastward end of the trans-Tasman zonal flow postulated by Warren (1970). The latter has pointed out that the EAC would have to separate from the Australian coast around 33 °S and flow eastward, past the North Island of New Zealand, in order to return to the South Pacific. If it continued any further south then it would eventually have to flow either northward or southward along the west coast of New Zealand. Such *eastern* boundary currents are unsupportable, on dynamical ground.

Andrews et al. (1980) interpreted their trans-Tasman XBT data as evidence for the zonal jet theory. However, the sub-surface thermal structure near the

Norfolk Ridge revealed by their survey was only tenuously connected to the detailed structure which they had mapped near the Australian coast. Andrews et al. (1980), using an h_{1300}/T_{450} regression formula, estimated the net eastward transport in the general direction along the Tasman Front to be about 15×10^6 m^3 s^{-1}. This is consistent with Stanton's estimate. An estimated further 20×10^6 m^3 s^{-1} flowing southward along the Australian coast as the EAC was presumably recirculated within an off-shore eddy. Stanton (1981) reported data from XBT stations evenly spaced across the Tasman Sea, supporting the tentative conclusions of Andrews et al. (1980).

Heath (1980) argued that the zonal jet observed by Stanton was not fed by the EAC, but was instead a local intensification of the circulation north west of New Zealand, caused by wind driving or topographic effects. He also pointed out that Warren's arguments against the existence of eastern boundary currents would have to be qualified if topography were taken into account. The actual effect of topography on a baroclinic flow is not readily estimated. The finding of Thompson and Veronis (1980) that there was no EAC during one cruise and hence no transport source for a zonal jet at that time further indicates that a satisfactory dynamical model for such a jet will have to be highly sophisticated.

The only substantial theories of mesoscale circulation in the Tasman Sea concern the EAC. These will be reviewed in the next section.

12.3 Theories

12.3.1 A Line Vortex and a Wall

A kinematic explanation was offered by Hamon (1965a) for the southward drift of anticyclonic eddies along the East Australian coast. He pointed out that the sense of motion is that of a positive line vortex, moving in a semiinfinite region bounded by a rigid wall. However, more recent observations by Nilsson, Andrews and Scully-Power (1976) showed that eddies have remained at about the same latitude for months, indicating that a more complex, dynamical explanation is required.

12.3.2 A Free Inertial Jet

Godfrey and Robinson (1971) adapted to the EAC a theoretical model of Gulf Stream and Kuroshio meanders. The model, which had been developed by Robinson and Niiler (1967), represents a meandering current by a free inertial jet. The path of the jet is determined by an approximate, time-dependent, cross-stream averaged vorticity equation expressed in path coordinates. The controlling influences are: path curvature at an upstream latitude, the variation of the Coriolis parameter with latitude, and the bottom topography. A cross-stream profile of downstream velocity must be assumed. Godfrey and Robinson chose a profile based on data of Hamon (1965a).

By numerical integration of the path equation, it was demonstrated that a nondivergent jet could travel about 1 000 km southward along the East Australian coast, beginning at 25°S and leaving the coast around 33°S to form large

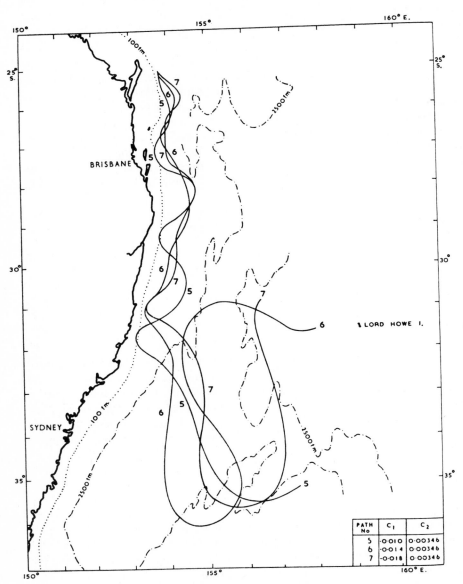

PATH No	c_1	c_2
5	-0·010	0·00346
6	-0·014	0·00346
7	-0·018	0·00346

Fig. 18. A series of nondivergent paths. All three paths start at the same point and travel in the same direction, but with different initial curvatures. Bottom velocities range from 0.091 ms^{-1} to 0.235 ms^{-1}. (After Godfrey and Robinson 1971)

amplitude meanders or to cross over itself (Fig. 18). Such a current would tend to follow the Lord Howe Rise northward into shallow water or be reflected back by it. Although the model was instructive, it was regarded by Godfrey and Robinson as less useful for the EAC than for the Gulf Stream and Kuroshio. A number of approximations are made in the derivation of the path equation. These include the assumption that the path radii of curvature were much larger than current widths (about 150 km), and that certain advection and curvature terms were negligible in comparison with others. Both of these assumptions could be tested using the results of the numerical integrations and shore-based tide gauge data, and both were found to be false.

12.3.3 Baroclinic Instability of a Western Boundary Flow

Godfrey (1973) compared the EAC with the western boundary flow in Bryan and Cox's (1968a,b) multi-level, primitive equation numerical ocean model. Agreement could not be expected to be close. The model basin has neither bottom topography nor continental shelf, there was no seasonal variation in either the thermal driving or the wind driving, and the model diffusion processes are probably not realistic for the EAC area. Nevertheless, there are some points of qualitative agreement.

First, the western boundary flow in the model sheds baroclinic eddies every 50 model days, spaced about 900 km apart. The short time scale is just consistent with EAC observations (see Fig. 10), but the length scale is too large. However, the model day is about $\sqrt{10}$ earth days. Godfrey argued that, if the eddy spacing were the interior deformation radius (which is inversely proportional to the Coriolis parameter), then a length scale of about 300 km would be more appropriate for comparison with observations. Such a figure is well within the observed range of 100–500 km. It may be noted that, if the horizontal length scale of the vertical shear in dynamic height is also assumed to be the interior deformation radius, then the baroclinic advective time scale (which is the interior deformation radius divided by the thermal wind scale) is proportional to the rotation period. Thus, given the above assumption, it was consistent of Godfrey not to adjust the eddy shedding interval, expressed in model days, prior to comparison with observations. A difficulty with this argument is that the baroclinic advective time scale is the time scale for the growth only of infinitesimal, inviscid, baroclinic waves on a current without horizontal shear, on an f-plane.

A second point of comparison between the model and the EAC is the presence of nearshore upwelling. This occurs in the model in a band about 90 km wide, and fluctuates in space and time with the development of eddies. Godfrey interpreted Rochford's (1972) observations of enrichment events as indicating similar upwelling in the EAC, although in a band only half as wide as in the model.

12.4 The Significance of Eddies

12.4.1 Physical Significance

A basic motivation for the study of mesoscale eddies has been the estimation of their role in large-scale mixing in the ocean, especially the mixing of vorticity, momentum and heat. It is appropriate to consider first the role of the eddies in mixing vorticity.

12.4.1.a The Sverdrup Vorticity Balance

The magnitudes of the terms due to the β-effect and the terms due to the eddy stresses, in the dynamical equation for the vertical component of vorticity, may be tentatively estimated using the velocity and length scales obtained by Wyrtki et al. (1976). The result is that the terms due to the eddy stresses are over an order of magnitude smaller than the β-term. The estimate does not assume near-isotropy of the normal stresses ($\overline{u'u'} \approx \overline{v'v'}$, in the usual notation) nor does it assume relatively small shear stresses ($\overline{u'v'} \ll (\overline{u'u'})^{1/2} (\overline{v'v'})^{1/2}$). Such assumptions would lead to even smaller estimates of the effectiveness of eddies in mixing vorticity. This subject has been discussed briefly by Bennett (1978), and at length by Harrison (1980).

The conclusion from this result is that the classical Sverdrup balance, between the meridional transport and the curl of the surface wind stress, should be reliable. Meyers (1980) has calculated the Sverdrup transport using the wind stress field derived by Wyrtki and Meyers (1975). At 28 °S, he found a transport between Chile and 160 °E of $26 \times 10^6 \, \text{m}^3 \, \text{s}^{-1}$ northwards. This is generally consistent with estimates of southward geostrophic transport in the EAC, relative to 1 300 dbar, of $12–43 \times 10^6 \, \text{m}^3 \, \text{s}^{-1}$ (Hamon 1965a). Meyers also stated that the northward transport of the subtropical gyre at 28 °S between Chile and Australia, as observed during the "Scorpio" Expedition, was about $26 \times 10^6 \, \text{m}^3 \, \text{s}^{-1}$, and inferred that the "Sverdrup transports are a good representation of the observed subtropical gyres in the Pacific.". This inference is confusing: first, no choice of reference level was indicated, and second, the geostrophic transport between Chile and Australia should, like the Ekman transport at 28 °S, be very small. If it is the case that the geostrophic transport between Chile and 160 °E is $26 \times 10^6 \, \text{m}^3 \, \text{s}^{-1}$, relative to some sensibly and independently chosen reference level, then Meyers' inference would be reasonable and the significance of eddies in the section-averaged vorticity balance would have to be regarded as being small.

12.4.1.b The Geostrophic Momentum Balance

The relative significance of the eddy stresses in the horizontal momentum equations is smaller than their relative significance in the vertical vorticity equations, by a factor which is the ratio of the length scale of variablity of the eddy stresses to the planetary radius. Hence the role of eddies in the geostrophic momentum balance is even less important than their role in the Sverdrup vorticity balance.

12.4.1.c The Heat Balance

Estimates of poleward heat flux in the combined Southern Hemisphere oceans at 30°S range from 1.6×10^{15} W *away* from the South Pole (Bennett 1978) to 2.6×10^{15} W *towards* the South Pole (Trenberth 1979). The former estimate was based on oceanic data; the latter on atmospheric data. Two estimates of poleward heat flux in the South Pacific Ocean at 30°S are 1×10^{15} W (Bennett 1978) and 2×10^{15} W (Hastenrath 1980, atmospheric data), both *towards* the South Pole.

It is appropriate here to examine the role of eddies in the actual heat flux, and their role in estimates of the heat flux based on oceanic data. The two roles may not be the same, since the oceanic data may not resolve the eddies. Arguments based on the geostrophic nature of the circulation within eddies, and also based on their vertical structure, indicate that the contribution by eddies to the actual net poleward heat flux is insignificant. The arguments assume a mid-ocean eddy field with scales observed in the North Atlantic, for want of any suitable data from the South Pacific. On the other hand it may be shown that such an eddy field introduces considerable uncertainty into heat flux estimates made using the relatively widely spaced hydrographic stations in the SCORPIO sections (Bennett 1978).

It might be expected that the more intense EAC could contribute significantly to the poleward heat flux. In fact the close correlation between h_{1300} and T_{450} in the EAC area indicates that, with sufficient accuracy, the isopleths of dynamic height anomaly at each depth have the same shape as the isotherms. Hence the essentially baroclinic, geostrophic flow in an eddy does not transport any heat. The only way in which an eddy can transport heat is by net displacement. The rate of displacement southward has been estimated to be at most 0.1 ms^{-1}. The vertical temperature profiles (Fig. 7) show that the heat in an eddy is contained in the upper kilometre, and is associated with a temperature constrast of about 5° between the eddy and the environment. Thus an eddy 250 km wide would contribute a southward heat flux of 0.5×10^{15} W. The eddy would take about 30 days to cross a latitude, and the next eddy would arrive about 100 days later, so the time-averaged heat flux is about 10^{14} W. It would appear that EAC eddies do not contribute significantly to the oceanic poleward heat flux. They may, however, have a significant impact on local climate.

12.4.2 Biological Significance

The first information on biological structure in an EAC eddy has been deduced by Scott (1978) from nitrate, silicate and available oxygen utilization data. These indicate that the biomass and productivity in the eddy must be much smaller than in the environment. The animal population in the eddy should also be different to that in the environment; irrespective of the productivity in an eddy, the relatively stable physical and chemical structure in an eddy should favour animals such as coelenterates which have prolonged life cycles. The eventual decay of the eddy would release these animals into the environment. This may explain the irregular arrival of swarms of Portuguese Men-of-War (*Physalia physalis*) on New South Wales beaches.

Further insight on the productivity of EAC eddies was offered by Tranter et al. (1980), who reported a five-fold increase in productivity in an eddy during late winter (September–November). They speculated that the isothermal core of the eddy was subjected to convective overturning when the surface waters in the eddy were exposed to lower air temperatures after the eddy drifted south-ward. The overturning, so Tranter et al. (1980) argued, brought nutrients from deep in the core up to the depleted shallow layer (75 m) where the light level was high enough to allow photosynthesis to take place. For a comprehensive discussion of biological aspects of eddies, the reader is referred to Chapter 22.

12.5 Future Research

12.5.1 The Mid-Ocean

There appears to be no hydrographic data describing the eddy field in the mid-ocean subtropical South Pacific. It seems unlikely that a complete description of the eddy field in such an enormous ocean region will ever be obtained. Carefully designed surveys of selected areas would be invaluable for estimating the significance of eddies in the dynamics of the region, and for the construction and qualitative validation of numerical models.

A considerable number of free, surface buoys drifted through the region during the First Global Weather Experiment (Garrett 1980). Virtually none of them were drogued, so it is particularly difficult to estimate the influence, if any, of geostrophic currents on their paths. It should also be noted that the buoys were distributed to provide a meteorological observation system, and no two were launched within 500 km of each other. Assuming that the most ener-getic eddies have horizontal length scales of about 100 km, it may be shown that the buoy paths can be used to estimate *absolute* diffusion statistics, but not *relative* diffusion statistics (Middleton 1979). Had two buoys been launched to-gether, they would have an average taken 6 months to drift apart to the 500 km spacing required for the FGWE. In summary, if the full potential of drifting buoys in oceanography is to be realised, then the design of the buoys them-selves and also the strategy of their deployment will have to be decided by oceanographers.

12.5.2 The East Australian Current

The scales and gross structure of EAC eddies have been roughly defined. Given that eddies appear at the rate of about two per year, it will be some time before a statistically reliable description of the eddy field will be available. Bottom topography and especially baroclinic instability appear to be fundamental in the dynamics of eddy formation and propagation, so truly representative numerical models will have to be most complex.

In the short term, the most fruitful area for research may be the exploration of the influence of EAC eddies upon the meteorology, biology, chemistry and geology of the Tasman Sea. Although eddies may not play an important role in the climate, it would seem a priori that they most probably do have an important role in short range and medium range weather development. The influence of eddies upon biological productivity has already been described.

The influence of eddies upon marine sediments is only now becoming apparent. Godfrey et al. (1980) noted a striking similarity between, on the one hand, the northernmost boundary of the area of major deposition of fine silt and clay on the New South Wales (NSW) continental shelf, and on the other hand the mean axis of the EAC after separation from the NSW coast around 33°S (see Fig. 19). The coincidence between the two lines of demarcation is all the more remarkable, considering that the rivers of the *north* coast of NSW dischare greater quantities of fine sediment than do the rivers of the *south* coast.

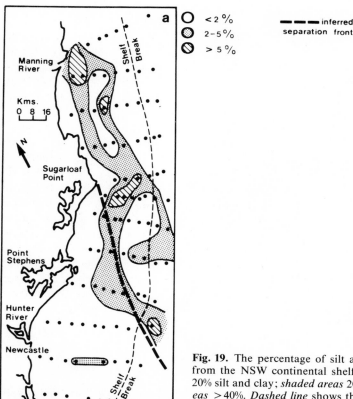

Fig. 19. The percentage of silt and clay in sediments from the NSW continental shelf, *stippled areas* 10%–20% silt and clay; *shaded areas* 20%–40%; *darkened areas* >40%. *Dashed line* shows the inferred line of the East Australian Current. (After Godfrey et al. 1980)

The inference is that the higher kinetic energies within the EAC meanders and eddies enable them to hold fine sediments in suspension. If this inference is correct then the sediment patterns, which "may be expected to have been maintained since sea level reached its present position some 5000 years ago ... indicate that the EAC has had this flow pattern for some considerable time ..." (Godfrey et al. 1980).

It is to be hoped that the study of eddies will, in the near future, yield other equally fascinating insights about their qualitative significance in general oceanography.

13. Eddies in the Southern Indian Ocean and Agulhas Current

M.L. Gründlingh

13.1 Introduction

Present-day knowledge of the large-scale circulation in the Indian Ocean orginates mainly from the cruises executed during the International Indian Ocean Expedition (IIOE), the first (and only) multinational effort aimed at increasing the sparse data coverage of this ocean. Because of the expanse of the area to be covered, the investigation was directed at the large-scale oceanographic characteristics. However, very few of the cruises ventured into the "mid-ocean" area (south of 20°S and between 50° and 100°E) so that the atlas compiled of the IIOE data (Wyrtki 1971) reveals a much lower station density in the central part of the Indian Ocean than elsewhere.

This preponderance of stations along the African, Asian and Australian coasts has been enhanced during the past decade through regional surveys guided by national or local interest. These coastal areas have become the scene of regular, conventional hydrographic surveys as well as of satellite-orientated investigations. It is mainly along the coasts that sufficiently strong surface thermal fronts occur that can be recorded by satellite infra-red equipment, and it is also along the continental boundaries that satellite-tracked buoys have been deployed. The object of this chapter, namely to discuss vortices in the southern Indian Ocean, is therefore inevitably reduced by the availability of data to a review of mesoscale motions in the smaller areas along the eastern and western perimeters of the ocean.

The chapter is in two sections, one on the eastern part of the Southern Indian Ocean, mainly the West Australian Current area, and one on the western part, the Agulhas Current and its associated circulation. Meridionally the area is bounded roughly by 20°S and 40°S although there are many areas within these limits about which very little is known as far as vortices and other mesoscale features are concerned. The chapter includes a discussion of already-published results (especially as far as the South-East Indian Ocean is concerned) and a presentation of some original, unpublished data on the South-West Indian Ocean. The review is descriptive since a comprehensive theoretical treatment of vortices in the South Indian Ocean has not yet been undertaken. The discussion should enable an assessment to be made of the relative significance of eddies in this part of the world, and possibly provide a guide for future research in the area.

Eddies in Marine Science
(ed. by A.R. Robinson)
© Springer-Verlag Berlin Heidelberg 1983

13.2 South Eastern Indian Ocean

Information on eddies in the vicinity of the West Australian Current has been
extracted from published results and an attempt has been made to make this
synthesis as comprehensive as possible. For proper orientation of those unfamiliar with the circulation in the south-east Indian Ocean, a brief summary of the
large-scale features of this area may be appropriate.

The area to the northwest of Australia (see Fig. 1), can be considered the
"origin" of the South Equatorial Current. It is in this region that a confluence
occurs of water masses from the Banda and Arafura Seas and from the subtropical regions to the south (Rochford 1961, 1962, 1969, Gentili 1972). Although
typically located at 11°S. The South Equatorial Current is displaced meridionally in sympathy with the monsoon season, moving northward closer to the
coast of Java during the south-east monsoon, and away to the south during the
north-west monsoon. During the latter period (November to March) the area
between the South Equatorial Current and the coast of Java is taken up by the
Java Current, and eastward-setting current of low velocity (3–25 cm s^{-1}). In
contrast to this weak flow, the South Equatorial Current has velocities as high
as 96 cm s^{-1} (Hamon 1965b) at 110°E and its width is 200–300 km.

To the south of the South Equatorial Current, and centred at about 32°S.
119°E there is a quasi-permanent, cyclonic gyre separating the synotic-scale,
equatorial flow that forms part of the anti-cyclonic circulation in the south Indian Ocean from the mesoscale, coastal circulation off the West Australian
coast (Wyrtki 1962a). Along its western flank the gyre transports water equatorward (the West Australian Current) while at about 30°S there is a flow first toward the Australian coast, then southwards along the coast. Cresswell and
Golding (1980) differentiated between the high-salinity water flowing toward
the coast at 30°S in summer and turning southward along the coast (Andrews
1977, Hamon 1965b), the low-salinity Leeuwin Current which flows southward
in autumn and winter above the continental slope, and the shelf currents that
are seasonal and possibly wind-generated. In the light of the large degree of
season and spatial variability, the concept of a consistent, northward-flowing
West Australian Current seems to fade and is replaced by circulation features
of much smaller dimension. Hamon and Cresswell (1972) reported a characteristic horizontal scale of about 500 km for the area off Western Australia, and
length scales of this magnitude are also visible in the "planetary wave fields"
of Rochford (1961, 1962) and Andrews (1977). It is mainly these variable, looping, currents that seem to constitute the major eddy-generating mechanism in
the area.

To the south, the south-east Indian Ocean is "delineated" by the Subtropical Convergence, a region of strong temperature and salinity gradients located
between 36°S and 40°S (Rochford 1962). A consistent eastward-flowing current, the Antarctic Circumpolar Current, is encountered only further southward (Wyrtki 1971).

13.2.1 Vicinity of the South Equatorial Current (A. Fig. 1)

During the second half of 1976, a free-drifting buoy tracked by the NIMBUS VI satellite was transported from its deployment area, namely, the West Coast of Australia, towards the South Equatorial Current (see Chap. 11 for a representation of the drift track). In December 1976 and January 1977 (at 12°S, 108°E) the track of this buoy described a series of clockwise loops that were evidently the result of the buoy having become trapped in an eddy which was advecting toward the southwest. From the data of Cresswell and Golding (1979) it is estimated that the eddy was moving at about 10 cm s^{-1} while the buoy was rotating at about 20 cm s^{-1}. The buoy continued its oscillating motion for about 2 months.

The impression is gained from the subsequent drift track of this buoy, namely, a consistent westerly drift right up to the African continent, as well as a similar transoceanic drift of another buoy the year before (Cresswell and Golding 1979), that the eddy under consideration was embedded in the South Equatorial Current.

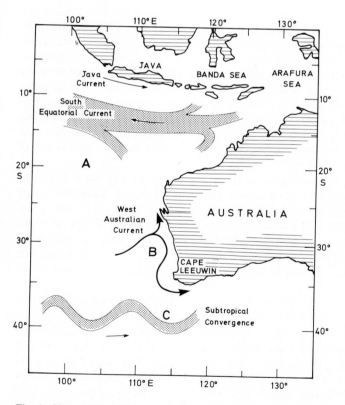

Fig. 1. Chart of the Southeast Indian Ocean, with a conceptual impression of the circulation. A, B and C are referred to in the text and indicate regions where eddies have been reported

Slightly westwards and to the south of this eddy (14°S, 100°E) a vortex also appeared in data collected in 1959 and 1960 (Wyrtki 1962a). The 1959 vortex had an anticyclonic rotation and, from the dynamic topography, seems to have been much larger than the cyclonic eddy observed in 1977. In terms of dynamic height anomaly of the surface above the 1750 m level, this eddy was the most intense of all eddies observed in the Southeast Indian Ocean up to 1963 (Hamon 1965b).

13.2.2 West of Australia (B. Fig. 1)

This area seems to contain an abundance of eddies, and most reports of vortices originate from observations in this region. In a review of cruises off the West Australian coast between 1965 and 1969, Hamon (1972) confirmed the large degree of variability that existed in this area. His data revealed several weakly defined mesoscale eddies, most of which seem to have been confined to the upper few hundred meters of the ocean. The results of four cruises in 1972 and 1973 off the west coast of Australia were treated by Andrews (1977) who found evidence of various cyclonic and anticyclonic eddy-like patterns. These eddies were embedded in a larger-scale flow (the vortex of Wyrtki 1962a, and Hamon 1965b) directed towards the Australian coast at 29°–31°S, southwards down the coast and around Cape Leeuwin. The eddies were weakly developed, but the data revealed some consistency in their positions. This agreed more or less with the reports of mesoscale eddies by Hamon (1972) although positive correlation is difficult in weakly defined fields of this kind. Although Andrews (1977) did not provide any estimates of the (geostrophic) current velocity or of the volume transport of the eddies, the flow on which the eddies were superimposed had an average transport of 8.4×10^6 m^3 s^{-1} relative to 1300 m. We estimate that the volume transport associated with eddies was of the order of 1×10^6 m^3 s^{-1}.

Another example of the type of eddy encountered off the West Coast of Australia can be seen in the data of Golding, Cresswell and Boland (1977) (see also Webster, Golding and Dyson 1979). In December, 1974 the R.V. SPRIGHTLY traversed a cyclonic eddy situated about 170 km WNW of Perth (Fig. 2). Although the deep hydrographic stations were too sparse to provide an estimate of the geostrophic velocities or volume transport, the GEK surface currents revealed speeds of up to 90 cm s^{-1} on the perimeter of the eddy (Golding and Symonds 1978).

During 1976 several free-drifting buoys were deployed west of Australia and tracked by the NIMBUS VI satellite. Two of these buoys became trapped by an anticyclonic eddy (see Fig. 3) about 200 km west of Perth (Cresswell, 1977) and remained inside the eddy for many loops, rotating with a period that varied from 4 to 7.5 days. Surface drift velocities at this time ranged from 20 to 100 cm s^{-1}. After a while the buoys were ejected from the eddy. These eddies seem to be more conspicuous than the eddies discussed above, but this could be attributed to the dramatisation brought about by buoy tracks. The following characteristics of the eddy are noteworthy:

a) The eddy seemd to undergo geometric pulsations or beats, i.e., the eddy seemed to contract and expand with a period of about 1 month. One of the buoys (Number 0415) remained inside the eddy for more than 2 months, without any consistent sign of weakening of the eddy.

b) There is evidence from the temperature sensors on board the buoys that the surface temperature of the eddy decreased steadily by 2.5°C in about 2 months. This could, of course, have been a seasonal change during the approach of the southern winter. Also interesting is an isolated report that the eddy was 2°C warmer than the environment.

c) The eddy seemed to remain more or less stationary for about one month, then started advecting in the NNE direction at 5 cm s^{-1}. The movement of the eddy (NNE) did not coincide with any of the directions of the buoys leaving the eddy (S and NW).

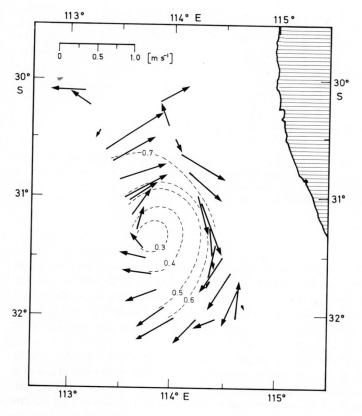

Fig. 2. Current velocity vectors and isotachs from the data of the R.V. SPRIGHTLY between 17 and 20 December, 1974, indicating a cyclonic, elliptic eddy of about 200 km diameter. (After Golding and Symonds 1978)

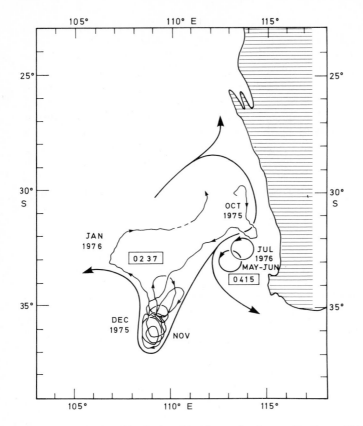

Fig. 3. A cyclonic eddy depicted by the track of a satellite buoy 0237 between October 1975 and January 1976 (Cresswell and Vaudrey 1977) and an indication of the position of an anticyclonic eddy, also encountered by a satellite buoy 0415 (Cresswell 1977). A conceptual representation of the composite circulation has been added by the author

Further to the south, another satellite-tracked buoy became entrained in an anticyclonic eddy (Cresswell and Vaudrey 1977) situated at 36°S, 109°E (see Fig. 3). Since the buoy remained trapped by the eddy for only a few revolutions, little detail can be derived from its drift track. Drift velocities calculated over 24-h periods attained a maximum of about 90 cm s^{-1} while the average drift speed inside the eddy was approximately 20–25 cm s^{-1}.

An area that still falls within the geographic frame of this chapter, but of which very little is known (judging from published results), is the Great Australian Bight. Cresswell and Golding (1980) drew attention to three cyclonic eddies that were described by satellite-buoy tracks in the western part of the Bight. These eddies were associated with, and on the seaward side of, the eastward-flowing Leeuwin Current, but most of the eddies propagated westwards at 6–8 km d^{-1}.

13.2.3 Vicinity of the Subtropical Convergence (C, Fig. 1)

The drift track of a NIMBUS VI satellite buoy at 39°S, 115°E revealed some cyclonic and anticyclonic loops between December 1975 and March 1976. A hydrographic survey of the area was executed while the buoy was still in the vicinity, and the result of this investigation confirmed that the buoy had entered an eddy. It further provided insight into the subsurface circulation and the position of the Subtropical Convergence (Cresswell et al. 1978).

The average drift velocity of the buoy in the cyclonic eddy was approximately 40 cm s^{-1} and a comparison with the hydrographic results revealed that the buoy crossed the Subtropical Convergence on occasion but returned to the subtropical side before moving northwards to the Australian coast. From the few loops of the buoy track inside the cyclonic eddy, a westward progression of the eddy at 6 cm s^{-1} was derived.

In a discussion of eddies, three questions immediately come to mind: Where do they come from? What are their characteristics? Where do they go? Observational evidence usually succeeds in partly answering the second question but the origin of eddies and their eventual fate are problems of which the solutions require retrocasting events, theoretical goundwork and a large amount of painstaking survey. Since the fortuitous location of an eddy (this is what normally happens during the initial stages of an investigation of vortices) provides a mere glimpse of the total life cycle of an eddy, tentative conclusions only can be reached about other elements of the eddy's existence.

13.2.3.a Origin

Very little, if any, evidence exists that can throw light on the mechanism by which the eddies in the southeast Indian Ocean are generated. Where the existence of eddies was revealed by satellite-tracked buoys, little or no subsurface hydrographic data were obtained to augment the surface drift pattern. The impression is gained from the abundance of eddies that the West Australian Current System at times seems to *consist* of a collection of weak eddies which seem to be easier to identify than the Current itself. The weak circulation constituting the West Australian Current (i.e., the equatorward return of the South Indian Ocean gyre) is further fragmented by the bottom topography (Andrews 1977). However, more observations augmented by theoretical considerations will obviously aid in determining the origin of the eddies.

13.2.3.b Characteristics

In contrast to the general circulation off the West Coast of Australia, which seems to resemble a weakly developed eastern boundary current, the eddies observed in this area are relatively intense. Surface velocities of up to 100 cm s^{-1} have been reported for both cyclonic and anticyclonic eddies, and typical lifetimes of some eddies are estimated to be several months or longer. Hydrographic data suggest that the depth to which the eddies extend could be a few hundred meters only, and that the volume transport of a typical eddy is of the order of 1×10^6 m^3 s^{-1}. There further seems to be a high concentration of eddies within 500 km of the coast (see Cresswell and Golding, 1979 and 1980) but this could be due to the preponderance of buoy tracks in the area.

13.2.3.c Propagation and Decay

The available evidence does not uniquely identify the general propagation direction of the eddies. Since the background flow in which the eddies are embedded is slow, it can very well be that the eddies will follow rather erratic courses without consistent overall displacement. Two observational "time series" of eddy propagation are available, both originating from the entrapment of drifting buoys inside an eddy: Cresswell (1977) did not find any weakening or meaningful propagation of an eddy observed of the south-west coast of Australia, while Cresswell and Golding (1979) reported the south-westerly propagation of an eddy in the South Equatorial Current.

13.3 South Western Indian Ocean

Guided by the availability of data, the "Southwestern Indian Ocean" will be accepted as indicating that area between 20° and 50°E and between 20° and 40°S. All major currents in this region will be assumed to belong to the "Agulhas Current System", a label that includes that part of the East Madagascar Current flowing around the southern tip of Madagascar, the Mocambique Current ejecting southwards from the Mocambique Channel, the Agulhas Current, the Agulhas Return Current and all other currents forming part of the circulation in this area (see Fig. 4). During the past decade, the Agulhas Current System has been investigated quite intensely and it is therefore appropriate that a present-day concept of the circulation in this area be given (see Gründlingh and Lutjeharms, 1979).

Very little is known about the dynamics of the two northerly members of the System, namely, the East Madagascar Current and the Mocambique Current. Harris (1972) estimated that they contribute about 14% and 49%, respectively, to the volume of the Agulhas Current, while Duncan (1970) put their relative contributions at more or less equal levels. The least that one can conclude from these estimates is that there exists a high degree of variability in the southwestward flux of these two currents.

The Agulhas Current starts, per definition, south of Madagascar at the confluence of the East Madagascar and Mocambique Currents. The Agulhas is approximately 100 km wide (Pearce 1977) and 1 000 to 2 500 m deep (Duncan 1970). Pearce also indicated the similarity between the Agulhas Current and the other large western boundary currents of the world. From the point of its origin, the Agulhas intensifies in the southwesterly direction down the southeast coast of southern Africa. Maximum current speeds (directly measured) are about 200–300 cm s^{-1} (see Duncan's comment in Schumann 1976) while evidence has been found of an increase in volume transport from 70×10^6 m^3 s^{-1}

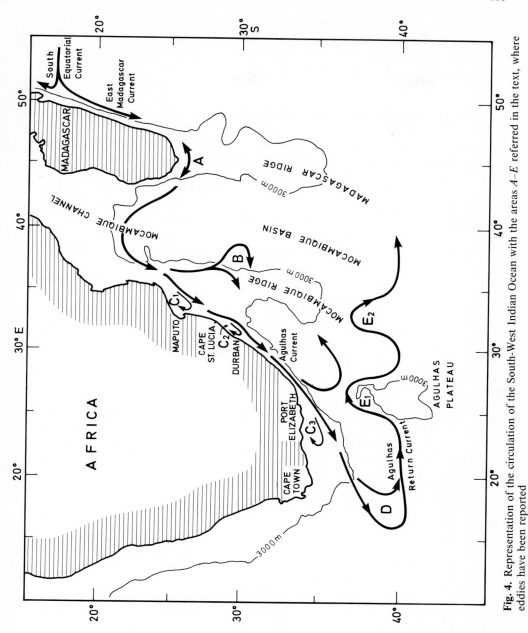

Fig. 4. Representation of the circulation of the South-West Indian Ocean with the areas *A–E* referred in the text, where eddies have been reported

to 136×10^6 m^3 s^{-1} south of the continent (Gründlingh 1980). As it separates from the continental shelf at about 36°S it starts to fragment and at 40°S impinges on the Subtropical Convergence. At this point it returns sharply eastwards to form the Agulhas Return Current.

13.3.1 Eddies South of Madagascar (A, Fig. 4)

Because of the absence of any coastal bights and promontories, the main area of vortex formation seems to be situated south of Madagascar where the East Madagascar Current separates from the coast. An indication of the shedding of eddies can be seen from the tracks of free-drifting buoys in the area (Lutjeharms, Fromme and Duncan, 1980); an especially good example is found in the track of TIROS satellitetracked buoy that crossed the Madagascar Ridge and rotated inside a small cyclonic eddy of about 60 km diameter (Fig. 5). The average drift speed of the buoy during this period was approximately 45 cm s^{-1}.

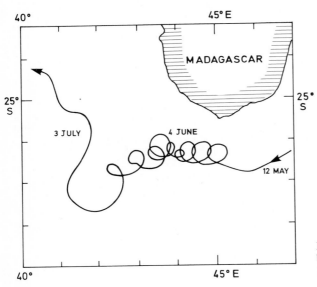

Fig. 5. Drift of a satellite-tracked buoy in a cyclonic eddy South of Madagascar in 1979

13.3.2 Larger Southwestern Indian Ocean (B, Fig. 4)

The first feature resembling a cyclonic eddy in this area appeared in the data of Harris (1970b, H in Fig. 6). Present-day knowledge allows this upheaval of the isotherms (Fig. 7) to be positively identified as a cyclonic eddy. Due to the large spacing of the oceanographic stations (~ 150–200 km), the size of the eddy can be estimated only at approximately twice the station interval, namely

300–400 km, while the eddy extended vertically to about 2350 m (estimated depth of the level of no motion). The volume transport of the eddy relative to this depth is 45×10^6 m³ s⁻¹.

The second observation of a similar eddy occurred again quite accidentally, in 1975. On the return leg after deploying a satellitetracked buoy, the R.V. MEIRING NAUDÉ executed a line of stations that traversed a cyclonic eddy (C Fig. 6) approximately 200 km in diameter (estimated from the vertical displacement of the 10°C isotherm, Fig. 8). Geostrophic velocities relative to 900 m peaked at 30 cm s⁻¹ in 300 m depth (Gründlingh 1977), and the volume transport was 11×10^6 m³ s⁻¹ (relative to 900 m) within 110 km from the eddy centre. This expedition was quite fortunate as far as locating eddies is concerned, since the track of the buoy during the first 10 days of drift, combined with the hydrographic results in the immediate vicinity of the deployment, indicated that the buoy had been deployed on the perimeter of another cyclonic eddy of approximately the same size (eddy I, Fig. 6). The maximum drift velocity of the buoy during the first ten days was 68 cm s⁻¹, and the buoy was subsequently ejected from the eddy.

It was on the evidence of these observations in 1962 and 1975 that an investigation was initiated to determine the origin and characteristics of the eddies

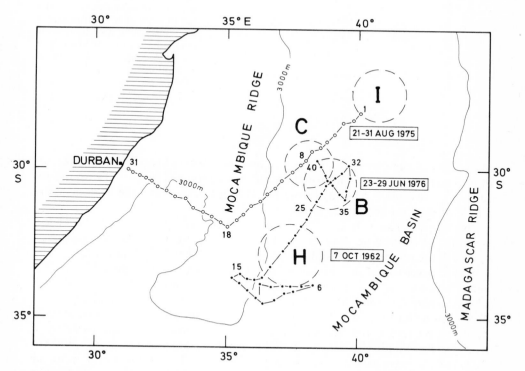

Fig. 6. Cruise tracks with eddies located up to 1976; *H* after Harris (1970B), see Fig. 7, *C* and *I* after Grundlingh (1977) see Fig. 8; *B* see Fig. 9

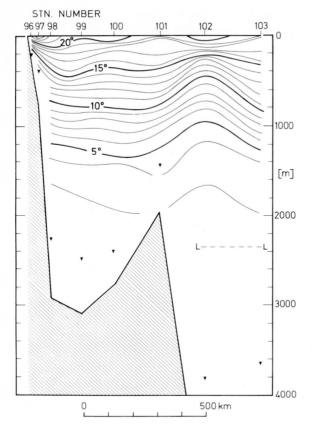

Fig. 7. Vertical temperature section on a line South-east of Durban in October 1962 (after Harris 1970b), showing a cyclonic eddy centered at station 102 (see *H* in Fig. 6). *L–L* indicates the level of no motion used to calculate the volume transport between stations 102 and 103

in this part of the ocean. Many eddies have since been observed in this area, although very little of this information has been reported to date. From the results of these cruises, an eddy located in 1976 is briefly presented here as an example of the vortices in this region.

During a cruise in June 1976, the R.V. MEIRING NAUDÉ traversed a cyclonic eddy centred at about 31°S. 39°E (see Fig. 6). The eddy was approximately 200 km in diameter and extended beyond 1 000 m in depth (Fig. 9). The T/S characteristics of the surface waters (Fig. 10) showed that the eddy was situated between water of a subtropical origin (cooler, more saline) and water of a more tropical nature (warmer, less saline). Reference to Darbyshire's (1966) distribution chart led to the tentative conclusion that the vortex was alien in its environment and had originated to the west or northwest of the position in which it was identified. The geostrophic volume transport relative to 1 000 m (Fig. 11) was estimated at 22×10^6 m^3 s^{-1} through the northwesterly sector of the eddy, and only 9×10^6 m^3 s^{-1} in the southwest. The large amounts of water exchanged between the eddy and its surroundings was taken to indicate that

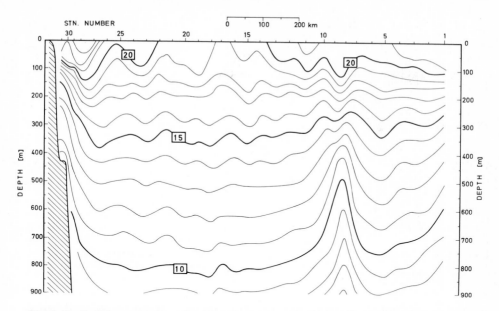

Fig. 8. Vertical temperature section on a line running Southeast from Durban and then North-east, 27–31 August 1975 (see Fig. 6 for location). A cyclonic eddy (*C*) is centered at station 8

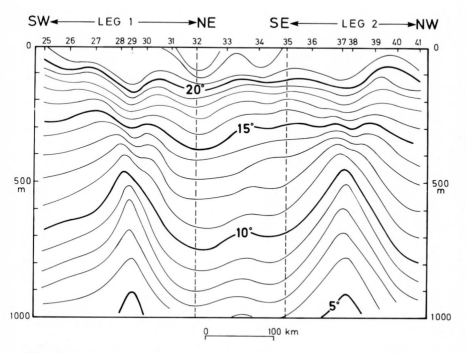

Fig. 9. Two temperature sections across eddy *B* (see Fig. 8 for station disposition)

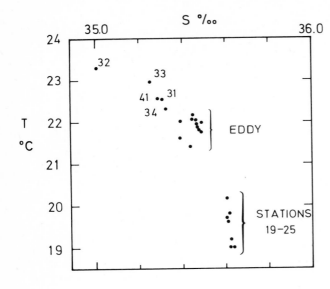

Fig. 10. Surface TS characteristics in the vicinity of eddy *B*, showing that the eddy was situated between the tropical water of station 32 and the subtropical water of station 19–25

the vortex was not free-drifting but was still connected to an eastward-flowing "host" current.

From other eddies observed subsequently in more or less the same position in the Mocambique Basin, volume transports of up to 48×10^6 m^3 s^{-1} and kinetic energies of 3.3×10^{15} J have been calculated. Evidence was found that the

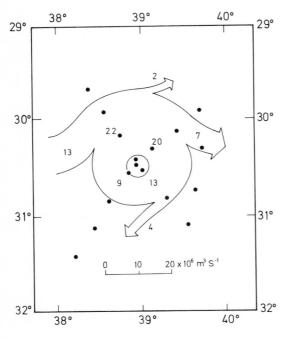

Fig. 11. Volume transport (in units of 10^6 m^3 s^{-1}) in eddy *B*. *Dots* are station positions (see Fig. 6 for locations)

eddies advect, although the speed at which this occurred could be estimated in one case only at about 6 cm s^{-1}. As far as the origin of the eddies is concerned, it was possible to establish that the escarpment on the eastern side of the Mocambique Ridge induced cyclonic eddies in the southeastward-flowing Mocambique Ridge Current (B, Fig. 4).

Although all these eddies were observed offshore (i.e., to the east) of the Mocambique Ridge, an eddy of very similar nature has been observed well *inshore* of the Ridge. A free-drifting buoy tracked by the EOLE satellite in July 1973 circulated in a cyclonic eddy of at least 50 km in diameter at about 31°S, 33°E. The buoy completed four revolutions during which its average velocity (~ 50 cm s^{-1}) compared with the currents measured directly from the deployment vessel (Stavropoulos and Duncan 1974).

13.3.3 Adjacent to the Agulhas Current (C₁ to C₃, Fig. 4)

The eddies classified in this group are almost all shear eddies and are by and large confined to the continental shelf area. Many of the eddies have been observed on the shelf between Cape St. Lucia (28°30'S) and Port Edward (31°S). In this region, the shelf widens from a few kilometers to about 40–50 km, and the concomitant separation of the Agulhas Current from the coast causes vortices to be shed onto the shelf (see Malan and Schumann, 1979). There are indications that these eddies are created just south of Cape St. Lucia (Gründlingh 1974) and then propagate to the south. Other observations (e.g., Pearce 1975, 1977) have been unable to resolve the direction of propagation. These vortices are approximately 20–30 km in diameter and display surface velocities of up to 1 m s^{-1}. A similar cyclonic eddy with a diameter of about 50 km was detected by radiation thermometer in June 1968 (Fig. 12). It is interesting to note that the two radiation thermometer surveys were done about one week before the loss of the oil tanker WORLD GLORY on the 13th June 1968, when subsequent oil slicks were carried ashore, probably by a similar vortex.

Until recently, it has always been considered that the eddies inshore of the Agulhas Current can be generated and sustained only by a current-shelf interaction. It has now been "discovered" that the Agulhas exhibits large meanders, with 200 to 300 km amplitudes, and that these meanders involve eddy-like features between the Current and the coast (Gründlingh, 1979). Infrared images seem to indicate that the meanders, and therefore the eddies too, advect southward at about 20 cm s^{-1} (Harris et al. 1978).

In the area south of Port Elizabeth (C₃, Fig. 4) the shelf widens onto the Agulhas Bank, while at the same time the Agulhas Current starts separating from the coast. Multiple instabilities occur and shear edge eddies are formed between the rapidly-moving Agulhas Current and the more or less stagnant water on the Bank. These eddies have been shown to be a characteristic of the area (see, e.g., Bang 1970a), but very little is known about the dynamics of the eddies. Lutjeharms (1980a, b) indicated the role played by these eddies after the collision between two fully laden oil tankers VENOIL and VENPET in 1977.

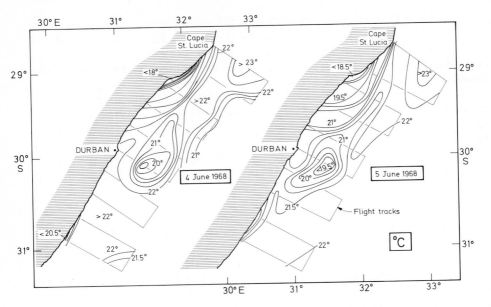

Fig. 12. Small cyclonic eddies inside the Agulhas Current (C$_2$, Fig. 4) as illustrated by airborne radiation thermometer data (Snyman pers. comm.)

13.3.4 The Agulhas Retroflection Area (Fig. 4 D)

As the Agulhas Current undergoes an intense eastward deflection through approximately 135°, an anticlockwise vortex is created that seems to be a semi-permanent phenomenon in the Cape waters, (see Dietrich 1935). The fact that variable currents could be expected in this area was realised during the previous century (Rennel 1832, Mühry 1864), when unwary seafarers were often surprised by the rather unpredictable behaviour of the currents, sometimes with disastrous consequences.

In 1964, an anticlockwise eddy observed in this area was regarded as the westward extremity of the Agulhas Current (Duncan 1968). The eddy was centered at 40°S, 15°E and was about 200 km wide and more than 2000 m deep. The maximum volume transport and geostrophic velocity of the eddy relative to 1 000 m was 29×10^6 m^3 s^{-1} and 38 cm s^{-1}, respectively, (Duncan 1970). There were clear indications that the vortex was an occlusion of the Agulhas Current loop and that the eddy was embedded in a general eastward flow. This eddy can be considered the Agulhas Current counterpart of the Gulf Stream "ring" (see Chap. 3).

Satellite-tracked buoys that traversed this area (Gründlingh 1978, Harris and Stavropoulos 1978) also exhibited anticlockwise eddies at the westward termination of the Agulhas Current. These eddies were approximately 150 km in diameter and the surface velocities obtained from the satellite buoy tracks exceeded 1 m s^{-1}.

An eddy was also observed here during the first large-scale, multiship operation to be conducted in South African waters, the "Agulhas Current Project" in 1969 (Band 1970a, b). The results of this project indicated that sporadic occlusions of the Agulhas loop could occur, severing an anticyclonic vortex from the main axis of the Current. The eventual fate of an eddy created in this way is highly speculative, but it is not unlikely that the eddy could become entrained in the flow up the South African west coast and thus introduce warm, subtropical water in the colder, upwelling area west of the Cape (see Lutjeharms 1981).

13.3.5 The Vicinity of the Agulhas Plateau (E_1, E_2. Fig. 4)

After the retroflection at 40°S, 15°E the Agulhas Return Current flows quasizonally eastwards until it reaches the Agulhas Plateau. This plateau, situated at 27°E, presents a barrier to the zonal flow between 37° and 42°S, as it rises by about 2500 m above the surrounding plain of 4500 m.

As the Agulhas Return Current meets the obstruction, it becomes deflected clockwise around the perimeter (E_1, Fig. 4). Bang (1970a) reported ship sets of up to 3 cm s^{-1} in the northern part of this loop. Because of the proximity of the northward-and southward-flowing currents inside the loop, the lateral shear creates clockwise eddies of various sizes, some only a few tens of kilometers across (Fig. 13). It has also been suggested that the Agulhas Plateau can generate planetary waves (Harris 1970a). The northward intrusion of cold water also features quite prominently on satellite infrared images of this area (Harris et al. 1978, Lutjeharms 1978, 1981). It is of course possible that a southward displacement of the Subtropical Convergence at about 42°S could redirect the Agulhas Return Current past the southern limit of the Agulhas Plateau thereby obviating the creation of eddies on the Plateau, although this is probably the exception.

A satellite-tracked buoy (Gründlingh 1978) also described another clockwise eddy (E_2, Fig. 4) to the east of the Agulhas Plateau. The cause of this eddy has not been unambiguously explained yet but it is likely to be related to the perturbation by the Agulhas Plateau. Due to the nutrient wealth of the eddies, they present fertile production areas for plankton (Zoutendyk 1970).

13.4 Discussion

One of the conclusions that can be drawn from this brief summary of vortices in the southern Indian Ocean concerns the abundance of eddies. Although the review is incomplete because of the numerous gaps that exist in our data base, almost every sector of this region contained evidence of mesoscale eddies (in contrast to the limited number observed up to 1976 – see Swallow 1976). Both cyclonic and anticyclonic eddies have been observed with depths ranging from a few hundred meters to more than 2000 m. It was not expected that the eddies

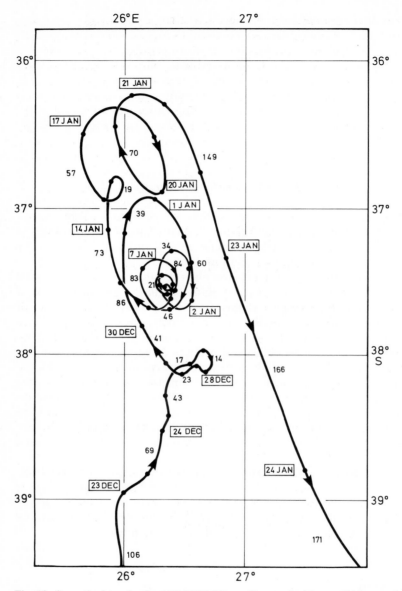

Fig. 13. Smoothed track of a NIMBUS VI satellite-tracked buoy 1210 over the Agulhas Plateau (E₁ Fig. 4) in 1975 and 1976. Figures accompanying the track are drift velocities (in cm s^{-1}) between position fixes (*dots*)

in this part of the world would be different from vortices observed anywhere else, and the data failed to reveal the existence of what could be called a Southern Indian Ocean eddy.

The majority of eddies discussed in this chapter seem to fall into one of two groups: First, there are those in which the bottom topography or the shape of the coastline plays a fundamental role. As examples of these can be considered most of the eddies in the southeast as well as southwest Indian Ocean. Second, there is that special type of eddy that is created when a large meander occludes, and which is commonly referred to as a *ring*. Although the bottom topography certainly also plays a role in the generation of rings, the characteristics of rings are so fundamentally different from those of "normal" eddies, that they have attracted special attention throughout the world (see Chap. 3). As far as the present chapter is concerned, existing evidence seems to indicate that rings are formed only south of the African continent.

In contrast to the relative abundance of information on the circulation in the southeast and southwest Indian Ocean, there still exists only a small amount of data originating from the central parts of the southern Indian Ocean. Isolated sections have been executed across this region but, understandably, the station spacing on these expeditions has not been aimed at resolving mesoscale features. This gap in the data must be considered the prime reason behind our lack of knowledge on vortices in this area. Considering the wealth of eddies at each end of the 8 000-km oceanic expanse between Africa and Australia, it seems unlikely that there are no vortices in between.

Although, as stated above, there is no real distinction between the eddy types in the Indian Ocean and, say, the North Atlantic, it may be of interest to compare some of the major characteristics. It must, however, be stated that oceanographic research in the south Indian Ocean has literally and figuratively skimmed only the surface as far as eddies are concerned. From this point of view it may be unfair to compare the observations in the Indian Ocean with the much-studied North Atlantic, but this process could assist in identifying the future course of eddy research to be adopted in the southern Indian Ocean.

Before starting with the comparison, an important point to make is that the study of vortices such as the Gulf Stream rings (see Chap. 3) and the multinational programmes like MODE and POLYMODE were all initiated after the Gulf Stream itself had been intensely investigated. Although some of these mesoscale studies required and involved a real-time cognizance of the position and behaviour of the Gulf Stream, the steady-state characteristics of this current were already well documented. In addition, a continuous influx of mostly temperature data from the scientific, naval and commercial sectors enabled the U.S. Naval Oceanographic Office to compile a regular review of the Gulf Stream *(The Gulf Stream Monthly Summary)*. This provided a close-to-ideal platform of synoptic-scale information from which to launch the various mesoscale investigations.

A condition such as this is difficult to foresee for the southern Indian Ocean, mainly because of the small number of ships making quasi-zonal, open-ocean passages across this expanse between southern Africa and Australia. Detailed information on the mid-ocean currents of the area, for both economic and military use, is therefore wanting.

A comparison between the rings formed in the southern Agulhas Current area and the rings of the North Atlantic shows the significant difference: Whereas the generation of rings in the North Atlantic is spread almost uniformly over quite a large region, the rings of the Agulhas Current are located in more or less fixed positions. So, e.g., the southwestward extremity of the Agulhas, i.e., the retroflection area, consistently sheds anticyclonic eddies such as the one reported by Duncan (1968). East of the Agulhas Plateau the rings seem to be most cyclonic, although the possibility of a warm-core, anticyclonic eddy being created there and subsequently entering the relatively cold environment of the Southern Ocean cannot be excluded. Data from satellite-tracked buoys that have become temporarily trapped inside one of these rings indicate that there is no real difference in size between the rings from the two hemispheres, and virtually no deep hydrographic data exist for the Agulhas rings to permit an even superficial comparison. Since surface speeds are compatible, there is possibly also geostrophic agreement.

As far as comparing the southeastern with the southwestern Indian Ocean, eddies in the east seem to be shallower although not weaker in terms of surface speed than those in the west. The volume transports definitely seem to be higher off the African continent than off Australia, while the number of eddies seem to be on par.

The data available for the central part of the southern Indian Ocean are still insufficient in quantity to reveal the characteristics of any mesoscale motions in that area or to make any meaningful comparisons.

Suggestions for future research in the field will obviously involve two parallel approaches: First, our knowledge about the synoptic-scale, quasi-stationary currents in the area will have to be increased quantitatively and qualitatively. This will provide the indispensable basis from which to proceed onto the smaller-scale motions, since it will delineate areas of greater or lesser variability and provide temporal and spatial estimates of the intensity of the steady-state flow. Second, the subject of acquiring more information on eddies can be initiated simultaneously, with the following separate goals:

a) To determine the mode, frequency and area of eddy generation.

b) To implement the technique of tracking free-drifting eddies, probably through satellite technology, thereby determining where the eddies go, how they interact with the general circulation, and how and where they dissipate their energy or become entrained into other currents.

c) To determine the dynamics (horizontal and vertical motion transport) and structure (density, chemistry, biology) of the eddies and their time rate of change. It is especially in this field that profitable use can be made of methods and models developed for the North Atlantic.

It must, however, be considered that, after the multinational research effort of the IIOE, future programmes will probably be designed by and within the overall capabilities of the countries bordering the southern Indian Ocean. It will therefore not be possible to conduct expeditions and investigations on the same scale as those in the Atlantic and Pacific.

14. The Southern Ocean

H.L. Bryden

14.1 Introduction

During the 50 years between 1925 and 1975, much effort was put into obtaining a large-scale, long-term average description of the Southern Ocean. Summaries of this description are generally presented as maps of dynamic height contours around the Southern Ocean to show the predominantly zonal circulation (Fig. 1) and as a schematic diagram of the circulation for a typical vertical-meridional section to show the meridional flux of water mass properties (Fig. 2). While such summaries are useful, they effectively suppress any temporal or small spatial scale variability and the suggested smoothness can be misleading. For example, the lack of small scale variability in these summaries perhaps led to two independent attempts to measure the transport of the Antarctic Circumpolar Current through Drake Passage with a few current meters deployed for a few days, which not surprisingly in view of later measurements of variability in the region produced transports differing by more than 250×10^6 m^3 s^{-1} (Reid and Nowlin 1971, Foster 1972, Mann 1977). A useful summary of the large-scale descriptive effort of the last half-century in the Southern Ocean is provided by Gordon, Molinelli and Baker (1982).

Since 1975, there has been a concerted effort to measure, describe and understand the variability on temporal scales longer than a few days and on spatial scales longer than 10 km. Such variability will be referred to as eddies in this work. This effort was spurred in large part by a determination to measure the transport of the Antarctic Current through Drake Passage (ISOS Executive Committee 1977). Such a transport measurement requires first that the eddy scales be described and then that sufficient measurements be made so that eddy fluctuations are properly averaged out in space and time. In the process of describing the observed variability, it was found that the eddies provide an efficient mechanism for meridional transport of various properties and they have a potentially important role in the dynamics of the circumpolar circulation. Baker, Nowlin, Pillsbury and Bryden (1977) gave an initial description of the variability observed in Drake Passage and Baker (1979) listed many of the recent observations in the Southern Ocean. This chapter is intended to summarize the observations of eddies in the Southern Ocean, to examine the energy sources and sinks for observed eddies, to review the effects of eddies including their transport of various properties and their role in the energetics of the circumpolar region, and to assess the implications of the observed eddies for models of circulation in the Southern Ocean.

Eddies in Marine Science
(ed. by A.R. Robinson)
© Springer-Verlag Berlin Heidelberg 1983

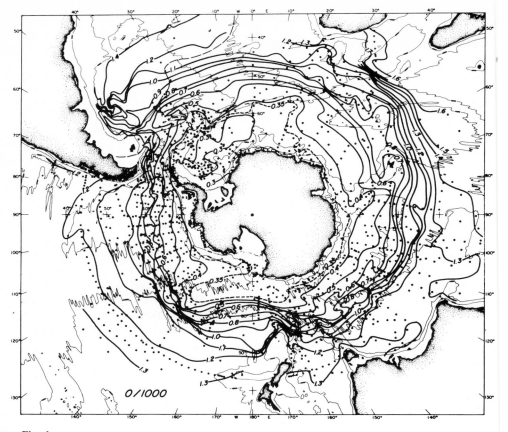

0/1000

Fig. 1

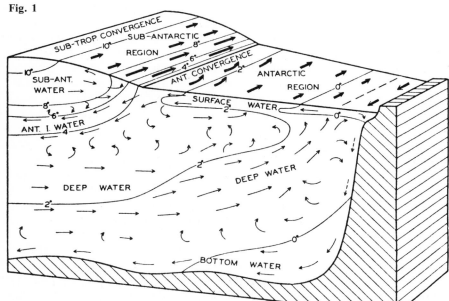

Fig. 2

14.2 Observations of Eddies in the Southern Ocean

While temporal and spatial variability has often been suggested as a potential cause of anomalous measurements in the Southern Ocean, interpretations of observations before 1975 were generally made with at least an underlying assumption of smooth and steady flow. [An exception is Mackintosh's (1946) description of the polar front as an "unstable boundary" with "twists and loops".] The first modern evidence of eddy variability was the discovery of multiple cores of high geostrophic velocity by Nowlin, Whitworth and Pillsbury (1977) in their closely spaced hydrographic sections across Drake Passage. These cores are about 50 km wide, are separated by 150 km regions of nearly no geostrophic currents, appear to extend throughout the water column, and mark the transition between water masses. After Bryden and Pillsbury (1977) showed that the observed variability in year-long current records made in conjunction with the closely spaced hydrographic sections could account for the large difference in previous estimates of transport through Drake Passage, a systematic study of the scales of eddy variability in Drake Passage was begun.

From the initial current measurements, Bryden and Pillsbury (1977) showed that the eddies have a typical time-scale of 14 days, defined by the zero-crossing of the autocorrelation function, and are not correlated over distances of 80 km, which was the shortest horizontal separation. In a remarkable study, Sciremammano (1979) demonstrated the vertical coherence of the eddies by showing that meanders in the polar front observed from satellite photographs (Legeckis 1977) could be clearly seen in velocity and temperature measurements at 2700 m depth. Finally, using 3 years of measurements including a small-scale cluster array in the central Passage, Sciremammano, Pillsbury, Nowlin and Whitworth (1980) estimated the horizontal scale of the current fluctuations to be 30 to 40 km, defined by the zero crossing of the transverse velocity correlation function, and the horizontal scale of temperature fluctuations to be 80 km. Sciremammano et al. (1980) also showed that the variability in zonal velocity is approximately equal to that in meridional velocity at all depths and that eddy kinetic energy in the central Passage decreases from 200 $cm^2 s^{-2}$ at 300 m depth to 20 $cm^2 s^{-2}$ at 2700 m depth with bottom intensification below as it increases to 35 $cm^2 s^{-2}$ at 3500 m depth. Above 1500 m depth, the kinetic energy of the mean flow is generally larger than the eddy kinetic energy and below 1500 m mean kinetic energy and eddy kinetic energy are comparable.

▲ **Fig. 1.** Anomaly of geopotential height of the sea surface relative to the 1000-dbar level, expressed in dynamic meters. Stations used in preparing this chart are shown (Gordon, Molinelli and Baker, Journal of Geophysical Research, 83, p. 3025, 1978, copyrighted by the American Geophysical Union)

◀ **Fig. 2.** Schematic representation of the currents and water masses of the Antarctic regions and of the distribution of temperature (Sverdrup, Johnson, and Fleming, The Oceans: Their Physics, Chemistry, and General Biology, © 1942 renewed 1970, p. 620. Reprinted by permission of Prentice-Hall Inc., Englewood Cliffs, N.J.)

The eddy variability in Drake Passage has been characterized in three ways. Nowlin, Whitworth and Pillsbury (1977) characterized the variability as due to cores of high zonal velocity with relatively narrow meridional scale which migrate north and south past moored current meters. The second characterization is due to Joyce and Patterson (1977), who observed the formation of a cyclonic ring at the polar front which then drifted northeast away from the front. The third characterization of the eddy variability is one of wave-like meanders of the polar front which propagate zonally at approximately 10 cm s^{-1} as observed in satellite photographs by Legeckis (1977). Sciremammano et al. (1980) attempted to distinguish between these characterizations. They suggested that the meridional migration of high velocity cores occurred on seasonal time scales and therefore was less important than the variability associated with isolated rings and meanders. Because objective maps of streamfunction for the 1977 cluster array measurements show many meanders and only a few isolated rings (Wright pers. comm.), it is suggested that the eddy variability in central Drake Passage is due principally to wave-like meanders of the polar front.

While Drake Passage has been the site of intense observations during the past 5 years, eddy variability has been observed in many other regions of the Southern Ocean. Gordon, Georgi and Taylor (1977) found a highly meandered polar front in their synoptic survey of the western Scotia Sea. Gordon (1978) observed a cold, deep, cyclonic eddy in the Weddell Sea which he cited as evidence of deep convection and hence formation of Antarctic Bottom Water. Duncan (1968) and Gründlingh (1978) described eddies south of South Africa. Savchenko, Emery and Vladimirov (1978) in a survey of the polar front south of Australia found a cyclonic ring similar to that described by Joyce and Patterson (1977) in Drake Passage. In an early study, Gordon (1972) described intense spatial variability in the region where the Antarctic Circumpolar Current interacts with Macquarie Ridge south of New Zealand. From laboratory and numerical models of this interaction, Boyer and Guala (1972) attributed this variability to the shedding of eddies as the current flows around and over Macquarie Ridge. From recent measurements southeast of New Zealand, Heath, Bryden and Hayes (1978) observed large meanders and an anticyclonic eddy with a 35 cm s^{-1} velocity at 2 000 m depth (Fig. 3). In an analysis of historical hydrographic data from the Southern Ocean, Lutjeharms and Baker (1980) found evidence of eddy variability with horizontal scales of 100 km or less throughout the Southern Ocean. They found the strongest variability near the polar front in regions where the Antarctic Circumpolar Current crosses prominent bottom topography and near the Falkland and Agulhas Currents.

The horizontal scale and surface velocity for each observation of a ring are presented in Table 1. While rings may not be the dominant form of eddy variability in the Southern Ocean, they are dramatic and hence attract enough observations so their characteristics can be described. Because rings and meanders are observed to have similar scales and velocities in Drake Passage, it is likely that these characteristics are typical of the scales and velocities of eddy variability in the various regions. The spatial scale, taken to be the radius of the ring, varies only from 30 km in Drake Passage to 100 km south of South Africa. Surface velocity varies from 30 cm s^{-1} in Drake Passage to 90 cm s^{-1} south of

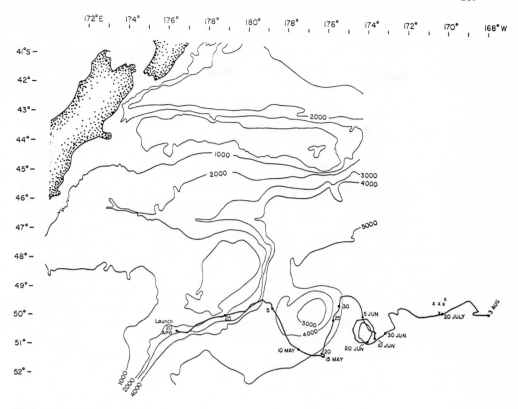

Fig. 3. Path of satellite-tracked surface buoy southeast of New Zealand. Currents measured in the array marked by crosses in late July were 35 cm s^{-1} at 2 000 m depth

Table 1. Characteristics of Eddies Observed in the Southern Ocean

Region	Horizontal Scale (km)	Surface Velocity (cm s^{-1})	Investigator
Drake Passage	30–40	30–40	Joyce and Patterson (1977)
South of South Africa	100	40–90	Duncan (1968) Gründlingh (1978)
South of Australia	35	40	Savchenko, Emery and Vladimirov (1978)
Southeast of New Zealand	50	40	Bryden, Cresswell, Hayes and Heath

South Africa. Because every description of a ring indicates that evidence for a ring extends to the greatest measurement depths, it is likely that the eddy variability extends throughout the water column in a coherent manner. In summary, the characteristics of eddies observed in the Southern Ocean are:

1. Eddies are found everywhere;
2. Eddy spatial scales vary from 30 to 100 km;
3. Eddy surface velocities are typically 30 cm s^{-1} or greater;
4. Eddies are vertically coherent from surface to bottom.

14.3 Eddy Generation and Decay

In contrast to other regions of the world ocean, there has been substantial progress in identifying the sources and sinks of eddy energy in the Southern Ocean. Since Bryden (1979) noted that eddies in Drake Passage transport heat poleward or down-gradient and thereby convert available potential energy contained in the mean density distribution into eddy kinetic plus potential energies, the generation mechanism for Southern Ocean eddies has been identified to be the baroclinic instability process. Early models suggested that the instability criterion is satisfied by the mean density distribution in Drake Passage (de Szoeke 1977) and that the observed propagation of eddies could be predicted by applying standard instability models to Drake Passage conditions (Fandry 1979). McWilliams, Holland and Chow (1978) identified some eddies in their numerical model of the Southern Ocean with baroclinically unstable modes but these modes were not the dominant form of eddy variability in their numerical ocean. In a careful comparison between predictions of a linearized baroclinic instability model and observations in Drake Passage, Wright (1981) found the predicted temporal and spatial scales and vertical structure for the most unstable wave to be in remarkable agreement with the observed scales and vertical structure of the dominant eddy variability.

The signature of the baroclinic instability process in Drake Passage observations consists of the following properties:

1. Poleward, or down-gradient, eddy heat flux. Sciremammano (1980) confirms that every current meter record in Drake Passage exhibits poleward eddy heat flux (Fig. 4). Measurements southeast of New Zealand also exhibit poleward eddy heat flux of similar magnitude above 2000 m depth. Patterson (1977) has suggested that the heat flux may be equatorward downstream of Drake Passage if rings eventually rejoin the Antarctic Circumpolar Current. Sciremammano (1980) uses a similar argument for meanders to caution that the eddy heat flux may be of different sign in different regions. After the generation and decay of eddies is discussed later in this section, these objections to determining the sign of eddy heat flux can be moderated.

2. Eastward phase propagation at approximately 10 cm s^{-1}. This propagation, which is in the direction of the mean flow, has been noted for Drake Passage observations by Fandry (1979) and Wright (1981) and is also evident in the measurements southeast of New Zealand. In Drake Passage this propagation implies the existence of a critical level at about 1500 m depth where the mean velocity equals the propagation velocity.

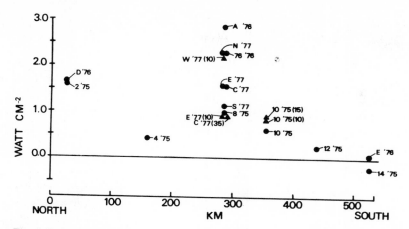

Fig. 4. Poleward heat flux for each current meter record in Drake Passage at 1000 m depth or below: solid circles denote current meters between 2000 and 2800 m depths; solid triangles denote current meters at other depths with depths indicated in parentheses in hundreds of meters (Sciremammano, Journal of Physical Oceanography, 10, p. 843, 1980)

3. A vertical phase difference such that deeper velocities occur before shallower velocities at the same mooring (Fig. 5). This upward phase propagation has been noted by Bryden (1979), Pillsbury, Whitworth, Nowlin and Sciremammano (1979) and Wright (1981). There appears to be essentially no vertical phase difference in temperature.

4. Satisfaction of the instability criterion due to the presence of eastward velocity shear and accompanying poleward density gradient near the bottom counterbalancing the generally positive meridional gradient of potential vorticity, $Q_y > 0$, throughout the water column. This was first suggested by de Szoeke (1977), contradicted by Fandry (1979), who suggested that $Q_y < 0$ throughout the water column was balanced by eastward shear at the surface, and confirmed by Wright (1981) who carefully reconsidered the distribution of Q_y and determined that Fandry was misled by his relatively inaccurate exponential fits of velocity and Brunt-Väisälä frequency. The instability is thus due to the interaction of the flow with the bottom boundary.

5. The nearly in-phase relationship between poleward velocity and temperature, especially at depth (Fig. 5, Bryden 1979, Wright 1981). Near the surface, temperature and poleward velocity may be close to 90° out of phase so that the surface flow is approximately along isotherms. At depth, however, the in-phase relationship of poleward velocity and temperature implies that the horizontal flow crosses isotherms.

This difference between streamlines and isotherms is crucial to the baroclinic instability process and associated poleward eddy heat flux. Failure to recognize this difference has led Sciremammano (1980) to suggest that there may be no net heat flux associated with meanders as seen in satellite photographs since he implicity assumes these meanders are simultaneously streamlines and

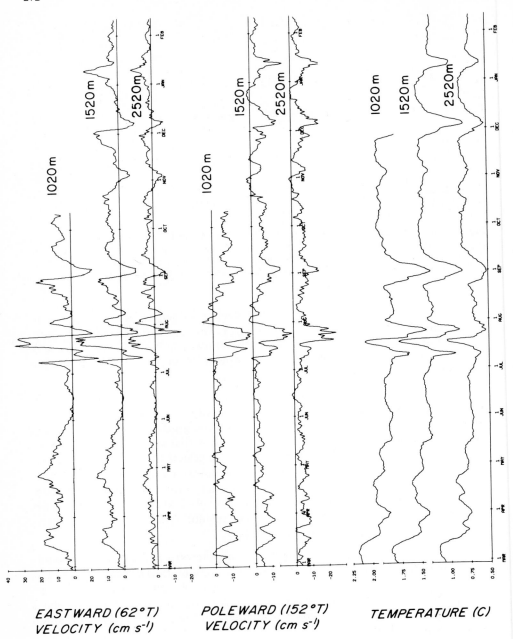

EASTWARD (62°T)
VELOCITY (cm s⁻¹)

POLEWARD (152°T)
VELOCITY (cm s⁻¹)

TEMPERATURE (C)

Fig. 5. Eastward velocity, poleward velocity and temperature plotted versus time for current meter records at 1 020, 1 500 and 2 520 m depths in central Drake Passage. Note vertical coherence, vertical phase difference such that deeper velocities lead shallower velocities, and the nearly in-phase fluctuation of temperature and poleward velocity (Bryden, Journal of Marine Research, 37, p. 16, 1979)

isotherms. From Drake Passage measurements (Bryden 1979), a schematic picture can be constructed which illustrates the difference between streamlines and isotherm (Fig. 6). The eddy variability is taken to be a wave-like meander with meridional displacement about a mean zonal line. Since eddies are observed to propagate eastward, the positive x-direction can be considered either as time or eastward distance. Isotherms have similar meridional displacements but are 90° different in phase from the meander-streamline as is typical for Drake Passage observations at depth. The horizontal flow clearly crosses isotherms, also assumed to be isopycnals for simplicity. In the instability process, there is also a vertical component of the flow such that equatorward flow is downward and poleward flow is upward. The horizontal flow across isotherms is partially balanced by this vertical flow but the remainder causes the meander to grow in the downstream direction or in time. Near the sea surface and bottom, isotherms and streamlines can become more nearly in phase since there can be no vertical motion at the surface or at a flat bottom so that all cross isotherm flow must cause growth in the meander. Presumably, when the meander displacement from its mean zonal position divided by its zonal wavelength ex-

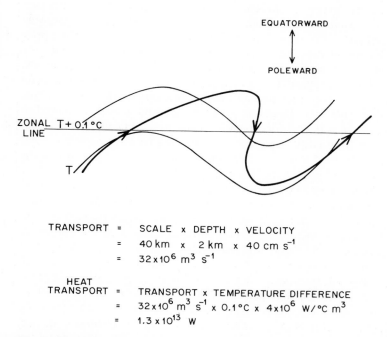

Fig. 6. Schematic relationship between streamlines and isotherms observed at depth in Drake Passage. Eddy variability is taken to be a wave-like meander with meridional displacement about a mean zonal line. Note that poleward flow is warmer than equatorial flow. For typical velocity and temperature fluctuations and scales in Drake Passage, this meander would transport 32×10^6 m³ s⁻¹ and have a temperature difference between poleward and equatorward flow of 0.1 °C. The resulting poleward heat transport would be 1.3×10^{13} W and 30 such meanders in the circumpolar region at any time would account for the required poleward heat transport of 4×10^{14} W across the polar front

ceeds some critical value, the meander pinches off to form an isolated ring as described by Joyce, Patterson and Millard (1981).

In this instability process, energy is taken from the large-scale potential energy distribution and put into eddy-scale kinetic and potential energies and eddy-scale temperature anomalies are displaced into an ambient environment of different temperature. The kinetic and potential energies and anomalously warmer or colder water exist now on an eddy scale of order 100 km but are not well-mixed into the environment. The mixing must take place on a smaller scale. In a remarkable set of experiments, Joyce, Zenk and Toole (1978) and Toole (1981) showed that this mixing to the next smaller scale in the upper 750 m of the water column is due to cross-frontal intrusions of horizontal scale of order 1 km and of vertical scale of order 50 m. These intrusions appear to gain their energy by transporting heat down the eddy temperature gradient and thereby converting eddy potential energy into intrusion-scale potential and kinetic energy. That such intrusions are observed both in the frontal region associated with a meander of the polar front and on the edge of an isolated ring suggests that this mixing of eddy-scale energy and temperature anomalies occurs continually. While the statistics of their estimates were not satisfactory with only two synoptic surveys of these intrusions, Joyce et al. (1978) estimated a down-gradient intrusive heat flux which is approximately half as large as Bryden's (1979) estimate of down-gradient eddy heat flux. At least in the upper part of the water column then, these intrusions appear to be responsible for mixing eddy-scale energy and temperature anomalies down to smaller scales. Analysis by Toole (1981) suggested further that the mixing from intrusion-scale to the molecular scale is accomplished by the mechanism of salt fingers. Thus, the elements of a cascade from largest to smallest scale have been observed in the Southern Ocean: in a baroclinic instability process, eddies transfer large-scale potential energy into eddy energy; intrusions then transfer eddy potential energy into scales of order 1 km, and salt fingers eventually mix the intrusions into the ambient environment.

Since eddies gain energy throughout the water column as indicated by down-gradient eddy heat flux at all depths while intrusions are observed only in the upper 750 m of the water column, there must be another sink for eddy energy below 750 m depth. McWilliams, Holland and Chow's (1978) numerical model of the Southern Ocean suggested that eddies transfer energy downward so that energy put in at the sea surface by wind can eventually be dissipated in the deep ocean by bottom friction or by topographic drag. Actually, by choosing their model parameters so that bottom friction rather than lateral friction was the dominant dissipation mechanism, McWilliams et al. (1978) virtually imposed this vertical energy flux. If lateral friction had been dominant, presumably the eddies would have transported energy meridionally away from the Antarctic Circumpolar Current by means of down-gradient momentum fluxes as Gill's (1968) model required to dissipate the energy input. By whatever mechanism, the energy input by wind must be dissipated and the near equivalence of energy input by wind and of the conversion of available potential energy into eddy energy (Bryden 1979) suggests that the eddies are important in the dissipation process. Because there is no observational evidence to date of a significant eddy momentum flux which would transport energy meridionally out of the circumpolar region and because poleward eddy heat flux can be interpreted as downward eddy energy flux under certain conditions (McWilliams et al. 1978), it appears likely that eddies do transport energy downward so that bottom friction or topographic drag can help to dissipate the energy put into the water column by wind.

14.4 Effects Eddies and Implications for Models the Southern Ocean

The poleward eddy heat flux, which is part of the signature of the baroclinic instability process, is a potentially important factor in the heat budget for the Southern Ocean. From calculations of air-sea energy exchange, Gordon (1975) has estimated that 4×10^{14} W of heat is lost to the atmosphere by waters south of the polar front. Extrapolating measured eddy heat fluxes in Drake Passage to the entire circumpolar region, Bryden (1979) showed that eddy heat flux could accomplish all of the poleward heat transport required to balance this heat loss by Antarctic waters. Recent measurements southeast of New Zealand (Heath, Bryden and Hayes 1978) exhibit a poleward eddy heat flux above 2000 m depth similar to that in Drake Passage. Using observed temperature-salinity and temperature-silica relationships, de Szoeke (1978) and Jennings (1980) in similar calculations suggested that the implied eddy salinity and silica fluxes are important in the salt budget and silica enrichment of Antarctic waters. Thus, eddies are observed to transport significant amounts of heat, salt and silica across the polar front with no net meridional mass or volume transport.

This importance of eddies in transporting temperature, salinity and silica across the polar front suggests a reconsideration of the classical vertical-meridional circulation (Fig. 2) proposed by Sverdrup (1933) and Deacon (1937). Because the meridionally-directed arrows on this circulation are based on property distributions, these arrows should be taken to indicate property transport rather than mass transport. For example, an equatorward arrow indicates equatorward transport of properties typical of the Antarctic region across the polar front into the subantarctic region. This property transport has traditionally been assumed to be due to a net mass or volume transport across the polar front, but in view of the observed eddy fluxes it could be due to eddy processes with no net mass transport. In view of the large number of direct measurements necessary to estimate accurate zonal averages, it is unlikely that a fully satisfactory test between these two mechanisms can be achieved. However, present indications suggest that the eddy transport mechanism is more important.

The primary argument that the property transport cannot be achieved by meridional mass transport is that there can be no zonally-averaged geostrophic meridional velocities above about 2000 m depth. Because the Southern Ocean is uninterrupted by land barriers in the latitude band of Drake Passage between about 55° and 60°S, there can be no zonally averaged zonal pressure gradient and hence no zonally averaged, geostrophic meridional velocity above the depth of major bottom topographic features in this latitude band, that is above about 2 000 m. If the arrows in Fig. 2 represent zonally averaged meridional flow, this flow would have to be ageostrophic above 2 000 m depth which seems unlikely. It is possible that meridional property transport could be accomplished by the large-scale geostrophic flow associated with the Antarctic Circumpolar Current which has regions of significant deviation from zonal flow, most notably its equatorward excursion into the southwest Atlantic (Fig. 1). If the property transport were due to such large-scale circulation, one might consider the arrows in Fig. 2 to be associated with actual mass transport. de Szoeke and Levine (1981), however, have shown that the large-scale geostrophic flow as determined from Gordon, Molinelli, and Baker's (1982) collection of hydrographic stations accomplishes no net heat transport across the polar front within an error of $\pm 2.3 \times 10^{14}$ W. They also have estimated that the wind-driven Ekman heat transport is 1.5×10^{14} W equatorward and suggested that eddy processes are therefore likely to be responsible for the required poleward heat transport.

The primary arguments that the meridional property transport is accomplished by eddy processes are that eddy heat transport and salinity and silica transports in the Southern Ocean are observed to be of the right sign and magnitude for the global budgets to be balanced and that such transports across the polar front are expected to be part of the baroclinic instability process by which eddies are generated. Further support for the importance of eddy transport can be derived from Stommel's (1980) consideration of global sources and sinks of heat and freshwater for the ocean, in which he suggested that because of vertical stability requirements the vertical-meridional circulation could not accomplish the necessary heat and freshwater transports across the region of the polar front and that mixing or eddy processes must be important mechanisms of transport across the polar front.

Thus, all indications are that the meridional property transport across the polar front above about 2 000 m depth is accomplished by eddy processes. Conceptually, the meridionally directed arrows above 2 000 m in Fig. 2 should be considered as eddy property fluxes and not as actual mass transports. Below 2 000 m where there are major topographic barriers, there can be net zonal pressure gradients and hence net meridional mass transports so that the deep arrows in Fig. 2 may represent actual meridional mass transport. In particular, it would seem likely that the equatorward wind-driven Ekman transport of about 28×10^6 m^3 s^{-1} (de Szoeke and Levine 1981) is compensated by a net poleward transport below 2 000 m depth of 28×10^6 m^3 s^{-1}.

Eddies also appear to be important in the dynamics of the circulation in the Southern Ocean. Models of the Antarctic Circumpolar Current have traditionally had difficulty in producing a circulation which resembles that in Fig. 1 without imposing very large horizontal or vertical mixing coefficients (Stommel 1957). In particular, models have had difficulty in balancing the energy and eastward momentum put into the water column by the large eastward wind stress in the latitude band associated with the Antarctic Circumpolar Current. The observed conversion of available potential energy into eddy energy at approximately the same rate as energy is put into the water column by wind (Bryden 1979) suggests that the dissipation of eddies, provided energy is transferred to smaller scales and not back into the large-scale circulation, can account for the necessary energy dissipation in the circumpolar region. As discussed above, eddy energy does seem to be transferred to smaller intrusions in the upper part of the water column. Thus, eddies appear to effect the necessary energy dissipation in the Southern Ocean.

Previous hypotheses for balancing the eastward momentum input by wind have included eddy transport of eastward momentum meridionally away from the Antarctic Circumpolar Current (which would also involve radiation of energy away from the current and hence help with energy dissipation), bottom friction and topographic pressure drag (Munk and Palmén 1951). While no significant eddy momentum flux has been observed, such momentum flux might occur downstream of major topographic obstacles such as Drake Passage or Macquarie Ridge as suggested by laboratory and analytical models (Boyer and Guala 1972, McCartney 1976). For bottom friction to balance momentum input by wind would require large bottom velocities which have not been observed. Topographic pressure drag seems the most likely mechanism for balancing the momentum input by wind. Indeed, if the wind-driven Ekman-layer equatorward transport were balanced by geostrophic poleward flow below 2 000 m depth, the pressure force on the topography associated with this poleward flow would exactly balance the wind input. This deep circulation could be maintained against frictional dissipation by the downward eddy energy flux, which McWilliams, Holland and Chow (1978) showed to be an important aspect of their numerical models of the Antarctic Circumpolar Current. Since under certain conditions poleward eddy heat flux can be related to downward eddy energy flux, it appears that eddies do act to maintain the deep circulation and hence play a role in balancing the eastward momentum input by wind.

Thus, eddies have profound effects on the circulation in the Southern Ocean which have important implications for models of this circulation. Eddies are observed to effect poleward heat, salinity and silica fluxes across the polar front and the logical implication is that eddies effect all meridional property transports across the polar front above about 2 000 m depth with no net mass transport. This suggests that meridionally directed arrows on the traditional diagram of vertical-meridional circulation should be interpreted as eddy fluxes and not as mass transports. Eddies also appear to effect the energy dissipation required to balance the energy input by wind and by downward energy flux to maintain a deep circulation which acts to balance the momentum input by wind through topographic pressure drag. This suggests that realistic models of circulation in the Southern Ocean must include eddy processes to effect the required energy and momentum balances.

15. Global Summaries and Intercomparisons: Flow Statistics from Long-Term Current Meter Moorings

R.R. Dickson

15.1 Introduction

Despite its title, this chapter does not aim to repeat the detailed local accounts of eddy distributions and dynamics already provided by the various regional authors. These authors know their own areas and data sources intimately, will have used a variety of direct and indirect evidence to assess eddy activity in their regions and in any event the regional chapters were not available at the time of writing. By contrast, this chapter examines only one particular type of data – long-term current meter records – in order to provide a summary of kinetic energy statistics for as much of the world ocean as possible in the hope that meaningful regional characteristics will emerge.

The statistics used are those of eddy kinetic energy per unit mass (K_E) and, for comparison, the kinetic energy per unit mass of the mean flow (K_M) derived from long-term, low-passed records from the open ocean, continental rise or slope (but excluding shelf). Definitions are conventional:

$$K_M = \frac{1}{2}((\bar{u})^2 + (\bar{v})^2) \tag{1}$$

$$K_E = \frac{1}{2n}(u'^2 + v'^2) \tag{2}$$

Where $u'^2 = \Sigma(u - \bar{u})^2$ and $v'^2 = \Sigma(v - \bar{v})^2$.

In this context then, the term "eddy" does not necessarily imply a closed circulation cell, but covers all types of low-frequency variability about the mean from whatever cause, with periods longer than the filter cut-off ($\sim 2\,\text{d}$) and shorter than the general circulation time-scales.

It is obvious that any intercomparison of records makes certain demands on the data. In theory at least, records should be of adequate length to ensure reasonable stationarity of statistics; the records should also be of standard length, and processing techniques (e.g., filters) should be comparable. In practice it is doubtful whether these requirements are met for the bulk of the data and in some instances it is clear they are not. For example, the long period (4–5 months) fluctuations in current directions at the NEADS 1 site on the Madeira Abyssal Plain (North East Atlantic Dynamics Study; T.J. Müller, pers. comm.) mean that even with available records of 345–510 d, the statistics may be non-stationary. We cannot overcome this problem with present-day lengths of record. The best we can do is to counter the problem by using only records of a certain duration. The choice is subjective. Initially this study aimed at a mini-

Eddies in Marine Science
(ed. by A.R. Robinson)
© Springer-Verlag Berlin Heidelberg 1983

mum record length of 9 months but this was later reduced to 7 months as a better compromise between statistical stability and geographical coverage. The
few records of extreme duration are equally problematical, and have been
shortened where possible to preserve comparability with the bulk of the data
set. The MODE records of up to 801 d, for example, are represented here by
selected sets of around 430 d duration. Finally, since the statistics used stem
from a wide range of sources the processing and especially the low-pass filtering techniques employed will differ. Most, however, have a high frequency cut-
off at around 2 d and in view of the well-known spectral energy gap at this frequency, observed at a number of sites in the ocean (Fofonoff and Webster
1971, p. 433, Thompson 1977), it was assumed that the bulk of records would
remain comparable. (Where known, filters with markedly different characteristics are mentioned in the text tables against the records to which they apply.)
There is, however, no such well-defined energy gap in records from the equatorial zone and even in mid- and high-latitudes one can identify records in which
a 2 d cut-off is inappropriate. The most obvious examples are the MONA 5
and 6 records from the Denmark Strait (Aagaard and Malmberg 1978, Aagaard, pers. comm.) where the primary spectral peak lies at 1.5–2.5 d.

   Though these questions of comparability are real enough, it goes without
saying that the main drawback to any "global summary" of flow statistics is the
sheer absence of *any* sort of long-term record over vast areas of the world
ocean. The available records are overwhelmingly those of the North Atlantic,
the Drake Passage and the equatorial zone with other ocean areas contributing
very few observations for comparison. Even this coverage would have been impossible but for the generous provision of unpublished statistics from very recent recoveries at a wide range of sites. Many of the equatorial observations
are in this category for example and without the preliminary statistics of the
NEADS Group the North Atlantic coverage would be almost completely restricted to its western basin. The author's debt to the originators of these data
will be obvious from the number of "in press" or "pers. comm." references
listed as data sources in the text.

   The preliminary nature of many records, the partial geographical coverage
and the problems of comparability have together determined both the form of
this chapter and the degree of detail which can be expected in its conclusions.

15.2 Global Kinetic Energy Estimates from Ship-Drift Analysis

Wyrtki et al. (1976) have provided the only available representation of the
ocean's kinetic energy field on a global scale. Using 4 million individual estimates of ship-drift collected world-wide between 1900 and 1972, they compute
the kinetic energy of the mean flow and its fluctuations as 5° square averages
for the world ocean (Fig. 1a and b) and as 1° square averages for the North At-

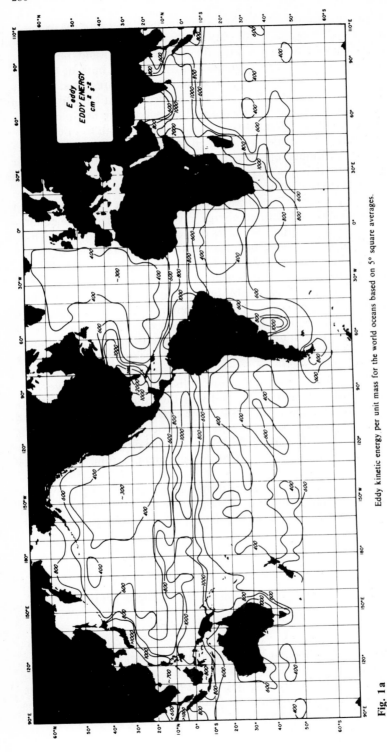

Fig. 1 a

Eddy kinetic energy per unit mass for the world oceans based on 5° square averages.

Fig. 1. Kinetic energy of the mean flow and eddy kinetic energy (both per unit mass) for the world ocean based on 5°-square averages. (Wyrtki, Magaard and Hager 1976)

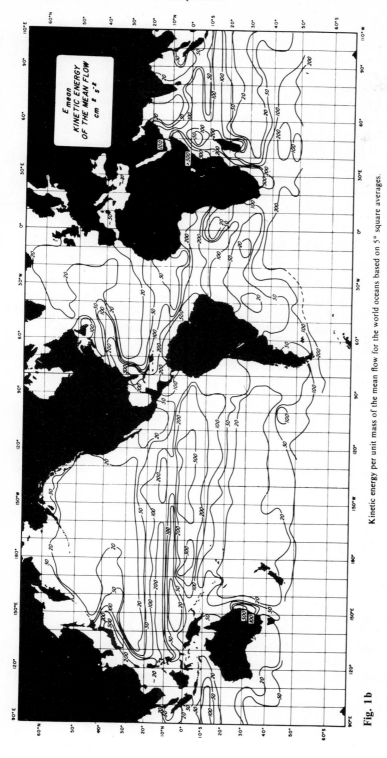

E mean
KINETIC ENERGY
OF THE MEAN FLOW
cm² s⁻²

Fig. 1b   Kinetic energy per unit mass of the mean flow for the world oceans based on 5° square averages.

lantic (see also Hager 1977). As Wunsch (1981) points out, there are obvious pitfalls in using this type of data base as well as problems in comparing these estimates with flow statistics derived from long-term current measurements. The ship-drift estimates are uncorrected for windage, apply only to the surface layer, are to some extent smoothed in forming area averages, and include some contribution from high frequency motions. Estimates of K_E from direct measurements, on the other hand, are fixed-point subsurface values from time series which are normally low-passed to remove periods of less than 1 or 2 d.

Despite these differences, the global patterns of eddy- and mean-flow kinetic energies revealed in the ship-drift analysis provide a convenient framework for interrelating our sparse and widely spaced set of direct observations, even though comparisons *between* the different types of kinetic energy estimates must be confined to broad-scale generalities. The generalities shown in the ship-drift analyses can be summarized briefly as follows:

1. there is a striking general similarity between the distribution patterns of K_E and K_M world-wide;

2. K_E maxima of over $1\,000$ cm^2 s^{-2} (5° square averages) are found along the western boundaries of each ocean in either hemisphere and in an almost continuous belt along the equator. With the exception of the Brazil Current, these are also sites of K_M maxima. The few observations from south of 50°S indicate that both K_E and K_M increase in the Circumpolar Current;

3. K_E and K_M minima are located in the central and eastern subtropical gyres, with a slight increase along the eastern and poleward edges of these gyres;

4. K_E/K_M ratios are normally greater than unity, reaching maxima of 20–40 in the central and eastern subtropical gyres. However, values as low as ~ 0.5 are found close to the Gulf Stream where both K_E and K_M are high. (Not evident in the 5° square averages of Fig. 1 but see Wyrtki et al. 1976, p. 2645.)

These, then, are the global tendencies against which the direct observations are compared in the sections which follow.

15.3 The North Atlantic

15.3.1 Introduction

A total of 269 long current meter records has been identified for the North Atlantic between 10°N and 65°N. Most (89%; Fig. 2a) have durations of between 220 and 520 d, though a few shorter and longer records are also included, either to cover data-poor areas or because no finer breakdown of ultra-long records was available. The flow statistics for all 269 records are listed in Tables 1–16.

From the histogram of sampling depths (Fig. 2b), five depth layers were selected for plotting purposes which reflect peaks in the depth distribution of the

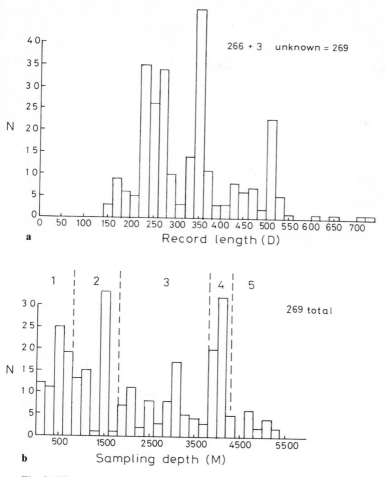

Fig. 2. Histograms of **a** record length and **b** sampling depth for the 269 long-term current meter records from the North Atlantic

data and which make reasonable sense from an oceanographic standpoint. For example, layer 1 (from 0–800 m) was chosen to reflect conditions in and above the thermocline while layer 4 (3800–4300 m) is intentionally narrow to preserve the comparability of the many records from ~ 4000 m depth. For each of the five depth layers, Figs. 3a–o illustrate pan-Atlantic distributions of K_E, K_M and K_E/K_M ratio, though readers should be aware of the following comments:

1. The ratio K_E/K_M is regarded as a much more variable and less meaningful quantity than either K_E or K_M alone. As one example, the Atlantic data set contains a number of records for which K_M is vanishingly small (say, 0.2 cm^2 s^{-2} or less) which are all treated identically in the K_M plots (as zeros) but which make major and largely meaningless differences in K_E/K_M ratio. For

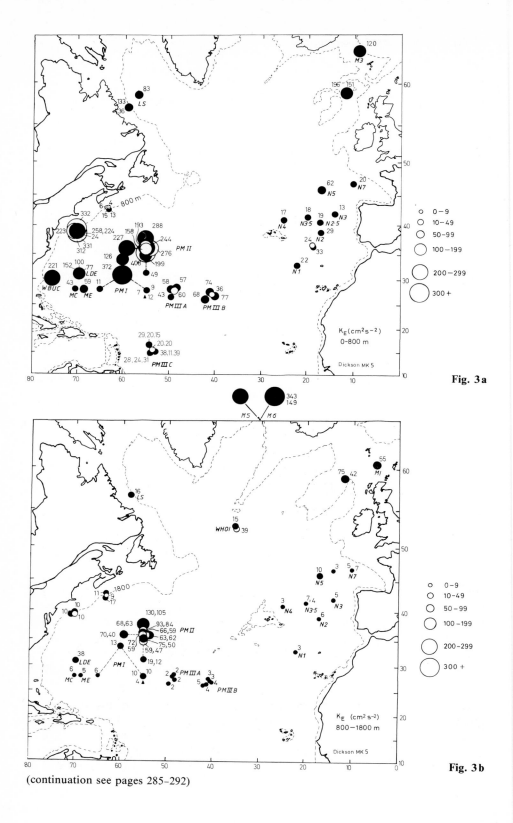

Fig. 3a

Fig. 3b

(continuation see pages 285–292)

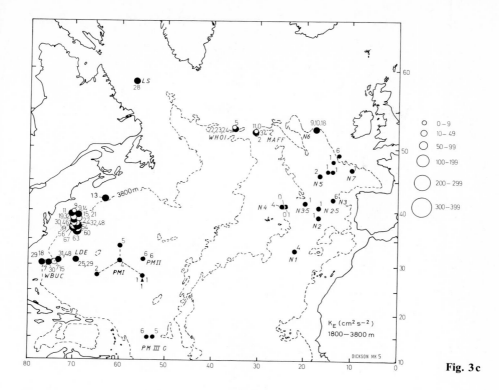

Fig. 3c

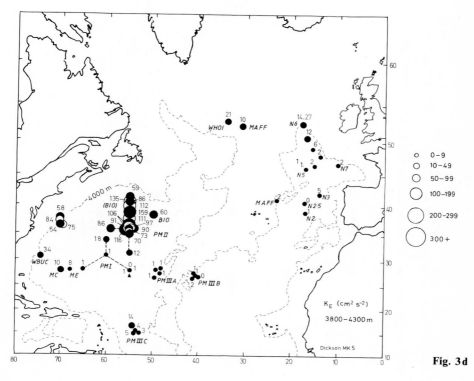

Fig. 3d

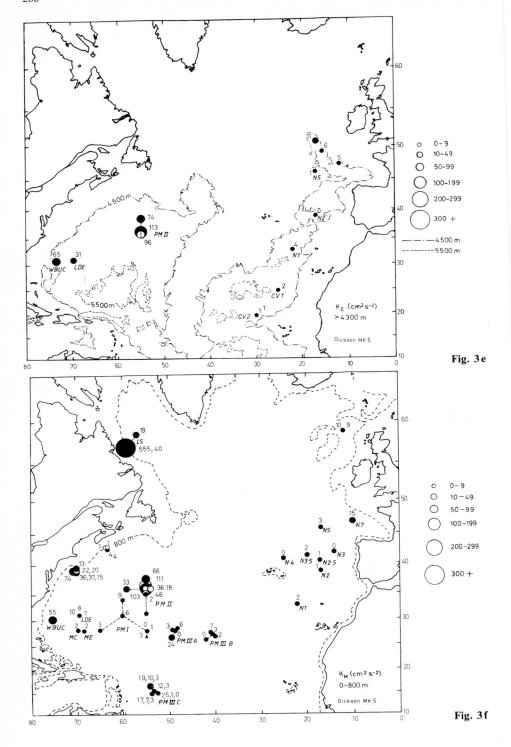

Fig. 3e

Fig. 3f

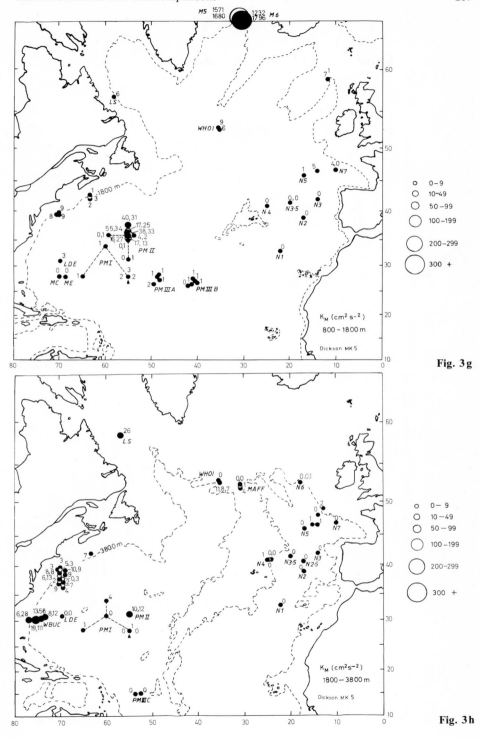

Fig. 3 g

Fig. 3 h

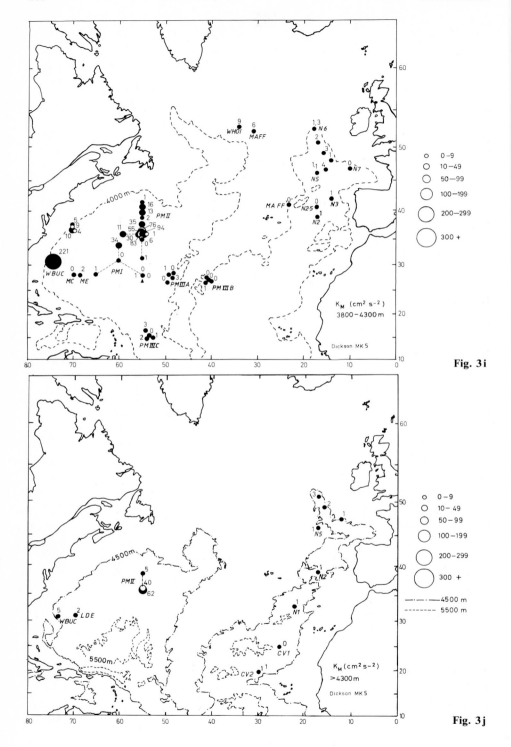

Fig. 3i

Fig. 3j

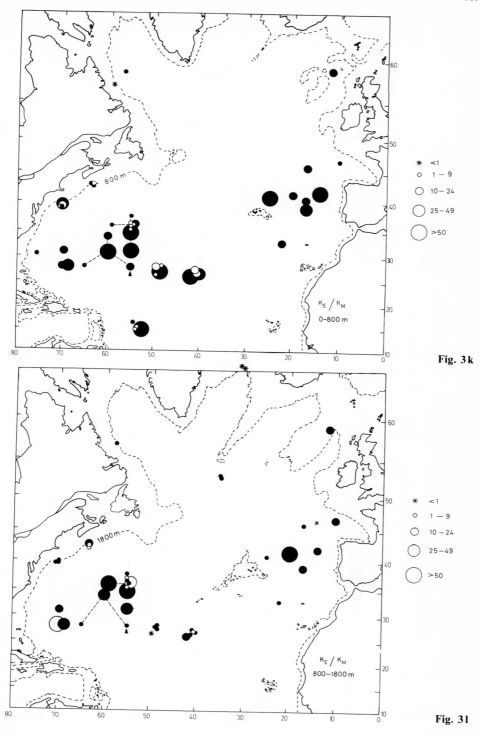

Fig. 3 k

Fig. 3 l

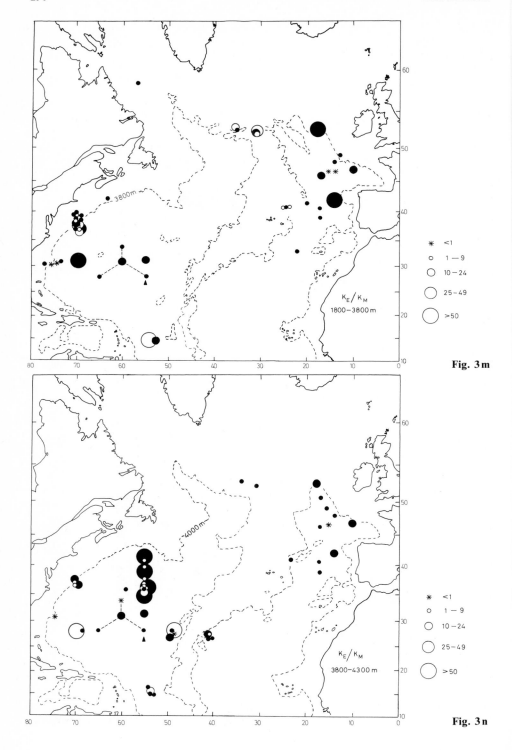

Fig. 3 m

Fig. 3 n

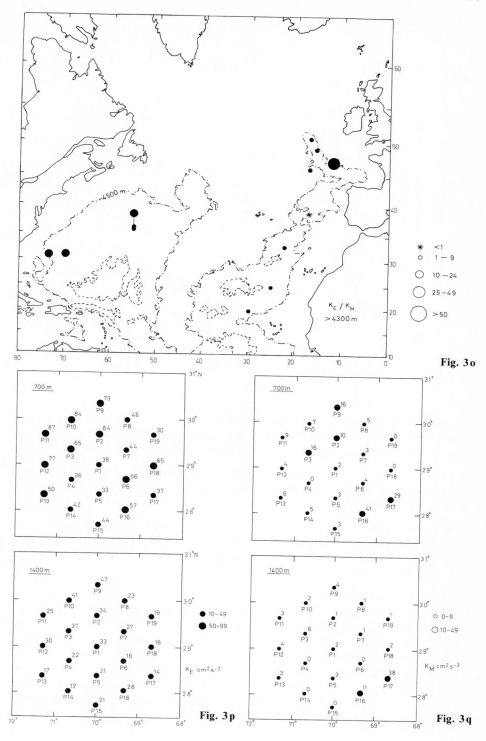

Fig. 3o

Fig. 3p

Fig. 3q

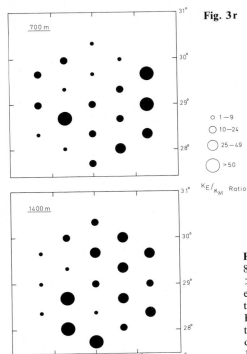

Fig. 3 r

○ 1−9
○ 10−24
○ 25−49
◯ >50

K_E/K_M Ratio

Fig. 3. North Atlantic flow statistics in the 0–800, 800–1800, 1800–3800, 3800–4300 and >4300 m depth layers. **a–e** Eddy kinetic energy (K_E) per unit mass, **f–j** Kinetic energy of the mean flow (K_M) per unit mass, **k–o** K_E/K_M ratio. **p–r** K_E, K_M and K_E/K_M ratio for the Soviet Polymode Array at 700 and 1400 m depth. (Corrected: see footnote to Table 14b.)

this and other reasons, the distributions of K_E/K_M ratio are intended merely to give a general *regional* impression of the values prevailing in a given depth layer. It would be misleading to compare values on a record-by-record basis;

2. Tables 1–16 include a number of short records with durations of less than 7 months. If these are also included in the plots they are identified by a bracket in the "duration" column;

3. where more than one value of K_E or K_M is shown in the plots for a given mooring and depth layer, it indicates either that records were obtained from a range of depths on that mooring or from the same depth on successive moorings. A search of the tables will indicate which.

4. *in addition to* the 269 records described above, a further 38 records are now available from the Soviet POLYMODE Array (19 moorings with records from 700 and 1400 m depth, set in a 300 km square centred on 29°N 70°W, and occupied from August 1977–August 1978). Because of the known problems in making accurate measurements of flow using Savonius-rotor current meters on a surface-buoyed mooring, these records have different characteristics to the other records discussed in this chapter; although a correction factor is available (Polloni, Mariano and Rossby 1981) and has been used in the calculation of flow statistics (Table 14b; T. Rossby pers. comm.), these statistics are not in-

cluded in the Atlantic plots (Fig. 3a–o), but are shown separately in Fig. 3p–r.

5. The means ($\bar{u},\bar{v}$; hence K_M) are known with less reliability than the variances. Errors in both depend on the spectral content of the record in relation to record length, but for variance the error fraction converges much more rapidly than that for the mean. Since many existing records have dominant periods $\geqslant 100$ d, durations $\leqslant 300$ d, and means which are small compared with the fluctuating component, the difference between the reliability of the two estimates may be significant (Gould pers. comm.).

6. The energetics of the eddy field are not wholly defined by its kinetic energy. Though it is not discussed here, the reader should be aware that eddy potential energy (P_E) may be as fundamental a quantity as eddy kinetic energy (K_E) from the standpoint of establishing an eddy "climatology".

15.3.2 Ocean-Scale Patterns of Kinetic Energy

Figure 3a–o is itself intended to form the main summary of flow statistics here since for reasons of space it would be impracticable to describe more than their most general characteristics. However, the bald statistics themselves conceal the fact that two records with the same K_E value may be dominated by very different time and space scales of variation. In Sections 15.3.3 and 15.3.4, an attempt is made to describe these differences of scale along meridional transects of a few stations through the western and eastern basins. These transects are intended to be illustrative, however, not representative.

The broadscale characteristics of Fig. 3a–o are as follows:

1. A first crucial point is that, for a given depth layer, the K_E and K_M statistics from different sources appear to be regionally self-consistent, despite the many possible problems of comparability discussed earlier. Thus, although "misfit" stations can certainly be found (e.g., the anomalously large thermocline K_E at POLYMODE I 548; Fig 3a), the distributions shown appear to be systematic at regional, basin and gyre scales.

2. The overall K_E distribution, so far as the coverage extends, is qualitatively consistent with the pattern derived from ship-drift analysis (cf. Figs 1 and 3a) and, with perhaps one point of difference, suggest much the same conclusions as those reached by Wunsch (1981, p. 343) in his global overview of "low frequency variability of the sea": "The mesoscale eddy field is highly inhomogeneous spatially – both in the horizontal and the vertical. Direct wind forcing and instability of the interior ocean are unlikely to provide much energy to the mesoscale. Much of the total mesoscale energy is found in the immediate proximity to the western boundary currents in the region where they are going seaward. It seems, therefore, that these regions are the generators of much of the eddy field energy." Though quite possibly linked dynamically to the eastward-going jet, the interior is comparatively quiescent.

The relative "quiet" of much of the eastern basin, at all depths, is one of the principal features of Fig. 3a–e. Absolute K_E minima at thermocline and abyssal depths in the western basin are found at the POLYMODE I "Triangle" site (7–12 and 0–1 $cm^2 s^{-2}$ respectively). In the eastern basin the present restricted coverage places these minima in or around the Iberia Abyssal Plain – at the NEADS-3 site in the thermocline (13 $cm^2 s^{-2}$) and at the NEADS-$2\frac{1}{2}$ site (0 $cm^2 s^{-2}$) at abyssal depths. The changes in K_E structure from the interior of the western basin across the Gulf Stream system and through the eastern basin are described in more detail below (Sects. 15.3.3 and 15.3.4).

One feature which the ship-drift analysis fails to pick up even qualitatively is the general increase in K_E north of 40°N in the eastern basin. The increase is most clearly seen in the 0–800 m layer (Fig. 3a) but is also detectable at abyssal depths (Fig. 3d and e). This is also perhaps the "one point of difference" to Wunsch's conclusions since a significant part of this increase does appear to be contributed by direct atmospheric forcing, even at depth (see Sect. 15.3.4).

3. As predicted from ship-drift, values of K_M are almost everywhere lower than K_E (Fig. 3f–j) so that K_E/K_M ratios are overwhelmingly greater than unity (Fig. 3k–o). However, there is no consistent evidence for high ratios in the interior and low ratios close to the Gulf Stream as the ship-drift evidence had seemed to suggest. Some slight indication of this tendency could perhaps be argued for the 0–800 m layer (Fig. 3k) but one might equally well stress the greater overall uniformity of values in that layer compared with at greater depths. There does appear to be some consistent evidence however that values of K_E/K_M ratio are less depth-dependent immediately south of the Gulf Stream than in either the interior of the western basin, or the eastern basin as a whole.

15.3.3 A Transect of the Western Basin

Analysing energy statistics from the western basin, Richman et al. (1977) report "an energetic space-time continuum that shows a remarkable spatial inhomogeneity. Some features ... change over distances comparable to the eddy scale itself; other order-of-magnitude changes in energy level occur vertically from thermocline to deep water; and others occur over the subtropical gyre scale." What this and other studies have successively shown, however (notably the detailed analyses by Schmitz 1974, 1976, 1977, 1978, 1980 and Luyten 1977) is that this " intricate inhomogeneity" also has geographic "system". Whether this apparent system is representative or not awaits future observations; as Schmitz (1980) points out "the North Atlantic circulation is inadequately described at the present time, still less well understood and possibly very complex in spatial and temporal scales ... We are still in the process of exploration."

At present the generalities of eddy kinetic energy distribution in the western basin are summarized as follows. They are described below for a transect of the principal flow regimes of the basin: first the rough topographic region at the deep interior of the subtropical gyre [represented by the POLYMODE I "triangle" site (PM I △)], then running north up 55°W across the region of "weakly depth-dependent flow" (Schmitz 1980) which includes both the westward recirculation and a zone of eastward flow south of the Gulf Stream, then through the Gulf Stream axis itself up the lower and upper Continental Rise at 70°W (Luyten 1977), and finally to the Continental Slope (Schmitz 1974). This transect is characterized by the following changes:

1. K_E *Amplitude.* From PM I $\triangle$ at the interior of the subtropical gyre, which shows the lowest values of eddy kinetic energy yet measured in the western basin, K_E increases by over two orders of magnitude at 4000 m and by a factor of $\times 30$ in the thermocline as the Gulf Stream is approached from the south along 55°–60°W (Schmitz 1978). The increase is progressive but not uniform. Starting with values less than 1 cm^2 s^{-2} at PM I$\triangle$, K_E at ~ 4000 m depth shows a first order of magnitude increase abruptly on crossing the southern boundary of the deep recirculation at around 32°N (Worthington 1976), accompanied by a sharp increase in K_M (Schmitz 1976, 1977). From this point northward along the POLYMODE II Array, abyssal K_E increases by a second order of magnitude to a maximum of ~ 150 cm^2 s^{-2} at around 38°N, some 100–200 km south of the Gulf Stream axis (Schmitz 1978). At thermocline depths, the equivalent overall increase along this transect is from 9 cm^2 s^{-2} at PM I $\triangle$ to 290 cm^2 s^{-2} at PM II 08 (37.5°N 55°W), the most energetic site at which sampling covered a full range of depths (Schmitz 1978). Further north still, instruments confined to ~ 4000 m depth indicate a steady decrease in K_E under the Gulf Stream at 55°W and up the Continental Rise to PM II 12 (41.5°N 55°W). Switching to the Rise Array at 69°–70°W, Luyten describes a similar almost-linear decrease of K_E with latitude below the thermocline, but with the meridional eddy component $\overline{(u'^2)}$ showing a much sharper attenuation than the zonal $\overline{(v'^2)}$, in response to the increasing bottom slope.

2. *Dominant Frequency.* The dominant time-scale of low frequency motions appears to decrease progressively northward along the transect under discussion. PM I $\triangle$ in the deep interior of the gyre is dominated by the longest eddy time-scales in both the thermocline and the deep water (Schmitz 1980), sufficiently long that "little more than one realisation was obtained" in the 9-month records (Richman et al. 1977). At the MODE sites further west (~ 28°N 70°W), closer to the edge of the interior region, Schmitz (1978) shows that these "secular" time-scales of > 100 d period are again dominant in the thermocline but that there is a shift towards the temporal mesoscale (~ 50–100 d) at depth. Crossing into the recirculation zone (Owens, Luyten and Bryden, 1982) and at the southern margin of the Gulf Stream (PM II 08; Schmitz 1978), the temporal mesoscale now dominates spectra from both the thermocline and the deep water, and in the comparison of spectra from 4000 m at PM II 08 and PM II 12 it is clear that this shift to higher dominant frequencies continues up the Continental Rise at 55°W (Schmitz 1978, his Fig. 7; see also the results for PM II 11 contained in Schmitz and Holland, in press, their Table 2). Luyten's (1977) analysis of data from the Rise Array at 70°W confirms this shift in energy towards higher frequencies at depths below the thermocline on moving up the Continental Rise from the Gulf Stream (and also up the Continental Slope, Schmitz 1974, 1978). The lower Rise (south of the 4000 m isobath) is dominated by meridional bursts of flow of some 30 d duration with essentially the same

spatial scales as the mean flow ($\sim$ 150 km meridional; $<$ 50 km zonal). On the upper Rise (north of the 4000 m isobath) fluctuations with a period of $\sim$ 30 d are still present but the frequency of their occurrence decreases with increasing distance from the Stream, suggesting their origin to be a meandering or pulsing of the mean flow (Luyten 1977). The increased bottom slope ($\geqq 5 \times 10^{-3}$) now restricts cross-slope motions at these longer periods but permits topographically controlled oscillations with periods of 5–15 d which Thompson (1971, 1977) has associated with topographic Rossby waves.

3. *Vertical Structure*. Schmitz (1978) compares normalized vertical distributions of K_E for three of the above sites which span the observed range of variation in K_E. The interior site (PM I $\triangle$) with the lowest absolute values of K_E also showed the most depth-dependent vertical K_E structure while the most energetic site (PM II 08) showed the least depth-dependent relative vertical distribution of K_E. The MODE Centre site was an intermediate case. These results are in qualitative agreement with theoretical arguments by Rhines (1977) which suggest that as energy levels increase (approaching the Gulf Stream system in this case) there is a tendency for motions to become more barotropic. However, Rhines' theory also predicts this same tendency on moving from rough to smooth topography; PM I $\triangle$ is over rough topography while PM II 08 is comparatively smooth. Schmitz (1978) and Wunsch (1981) discuss the spectral content of these changes in vertical K_E structure.

While, even in this well-studied ocean area, future observations along different transects may well suggest other or more complex horizontal and vertical structures, it is heartening to note that eddy-resolving gyre-scale numerical models now reproduce several important observed characteristics of the eddy field (Schmitz and Holland 1982): they show good quantitative agreement between observed and modelled K_E distributions in relation to the Gulf Stream system (especially at depth) and even reproduce the characteristic frequency distributions of K_E at the MODE site (shifting from secular time-scales in the thermocline to mesoscale dominance at depth) and in the recirculation (mesoscale dominance at both thermocline and abyssal depths).

Comparisons with the ship-drift estimates of Wyrtki et al. (1976) however show only a general similarity with many important differences in detail. For example, though Wyrtki et al. report a "plateau" in K_E of around 400 cm^2 s^{-2} in the region 28°–34°N 55°–65°W Schmitz (1976) points out that 500 m values in fact range from 10 cm^2 s^{-2} near 28°N, 55 and 65°W to 100–400 cm^2 s^{-2} near 31°–34°N, 60°W. Other points of dissimilarity are noted by Schmitz (1978, p. 301) and Schmitz and Holland (1982).

15.3.4 A Transect of the Eastern Basin

Though the long-term mooring network in the eastern basin is now quite extensive (Fig. 3a), many of the results are very recent and under analysis. However, a sufficient number of kinetic energy spectra have now been published to construct a simple meridional transect of eddy characteristics through the eastern basin. Apart from data collected by the present author the only unpublished spectra used are those for NEADS-2 (38°00'N 17°00'W; North East Atlantic Dynamics Study) which forms the southernmost station in the transect. Data for this site were kindly supplied by Dr Tom Müller, Institut für Meereskunde, University of Kiel.

The transect to be described runs north from the NEADS-2 site on the flanks of the Azores-Portugal Ridge to the NEADS-3 site (42N 14W) on the Iberian Abyssal Plain (Gould this Vol.) and thence to NEADS-5 (46°N 17°W) in the foothills of the mid-Atlantic Ridge, close to their junction with the Porcupine Abyssal Plain (Gould, this volume). From there a line of deep moorings "P", "O" and "N" along the centre of the Porcupine Abyssal Plain continues the transect northwards to the NEADS-6 site (52°30'N 17°45'W) on the Continental Rise south-east of Rockall Bank (all Dickson, unpublished), and the transect ends with mooring I4 in the northern Rockall Trough (Gould and Cutler 1980) and MONA 1 and 3 on the flanks of the Iceland-Faroe Ridge (Willebrand and Meinke 1980).

The NEADS-2 site in the eastern subtropical gyre shows characteristics similar to the PM I △ site in the interior of the western basin. The (energy-preserving) kinetic energy frequency spectra for four depths between 780 and 4220 m (Fig. 4a; T.J. Müller, pers. comm.) are dominated by long time-scales at all depths while K_E values are small (0–6 cm^2 s^{-2} below 1500 m) and are markedly depth-dependent. (Though not differentiated here, peak eddy energies in the longest records (at 780 and 3170 m) are found at periods well in excess of 100 d.) Further north at NEADS-3, Gould (this Vol. his Fig. 6) shows that long period motions ($\tau \geqq 100$ d) continue to dominate at all depths, but with a relative shift of eddy energy to the longest accessible periods (180–770 d) with decreasing depth.

By the latitude of NEADS-5, K_E at thermocline depths has increased by a factor of 2–4 over equivalent values at NEADS-2 and -3 (Fig. 3a) and is beginning to show signs of increase at intermediate depths also (Fig. 3b). As Gould shows (this Vol. his Fig. 5) this northward increase in K_E is partly contributed by a seasonal input of energy at relatively high frequencies (> 0.04 cpd) arising through winter forcing of some (unspecified) type. These two tendencies – the increase in K_E, eventually over the full depth range, and the increasing influence of seasonal forcing – become major features of the transect from this point northwards.

Figure 5 (Dickson unpublished) gives a preliminary impression of time-scales and eddy energies at abyssal depths as our transect continues northward to the head of the Porcupine Abyssal Plain (see Fig. 6 for locations). Only the meridional component of the low-passed ($\tau > 2$ d) current is shown, though the essential characteristics of the flow are also reflected in the zonal component. K_E values for the *complete* record are shown against each time series indicating a steady northward increase from 1–2 cm^2 s^{-2} at P, 5–6 at O, 11–12 at N and 18–27 at NEADS-6. The time series plots themselves show that over smooth topography, in a reasonably uniform range of abyssal depths and over a common 1-year (nominal) time-base, fluctuations about the mean speed not only increase in amplitude but alter (shorten?) in period northwards from "P"

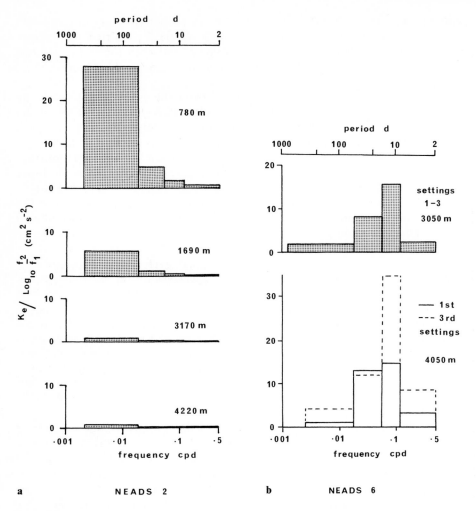

Fig. 4. Plots of decadal spectral estimates, energy preserving by area, for **a** NEADS-2 (38°N 17°W; T. J. Müller, pers. comm.) and **b** NEADS-6 (52°30′N 17°45′W). The observation depths are indicated

to show a fairly well-defined ∼ 45 d period at "N". Between "N" and NEADS-6, a distance of 200 km, there is an abrupt change to an energetic 10 d dominant time-scale of variation as the bottom slope increases up the Continental Rise. This 10 d variation has been present throughout the 32-month occupation of this site. In general the flow characteristics at the NEADS-6 site are closely similar to the results from the American Continental Rise reported by Luyten (1977). Each has an intermittent ∼ 30 d peak in eddy energy in the along-isobath component but not in the cross-isobath component (see earlier discussion, p. 296) and each shows an additional peak at 5–15 d periods asso-

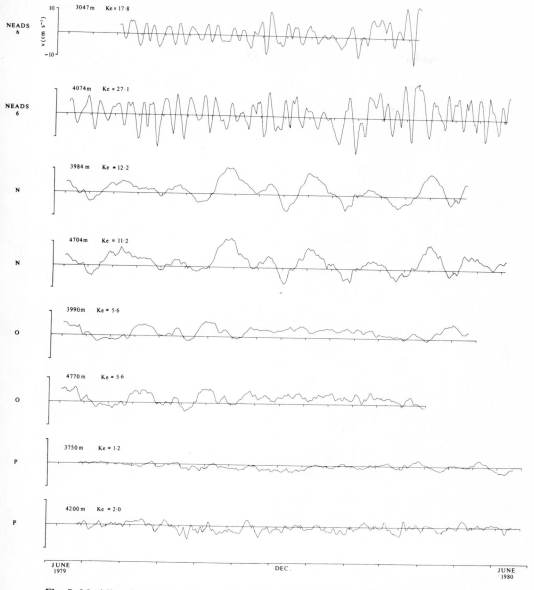

Fig. 5. Meridional current component (cm s^{-1}) from low-passed records at four sites (P, O, N and NEADS-6) on the Porcupine Abyssal Plain. The sampling depth and eddy kinetic energy values are shown for each record.

ciated, in the case of the US data, with topographic Rossby waves. This shift to higher frequencies on the Rise is well illustrated by the decadal kinetic energy frequency spectra for NEADS-6 (Fig. 4b; Dickson unpublished); the bands of periods shown are identical to those used for NEADS-2 (Fig. 4a). Despite their

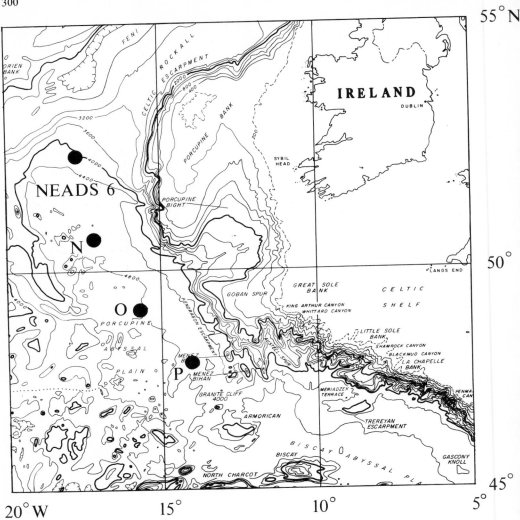

Fig. 6. Location of moorings P, O, N and NEADS-6 on the Porcupine Abyssal Plain.

great depth, the NEADS-6 records also appear to exhibit a seasonal variation in eddy kinetic energy, in keeping with the results described above for NEADS-5 and below for I4 (Dickson et al., 1982). These results are still under investigation.

No data from the Porcupine Abyssal Plain are yet available at intermediate or thermocline depths, but on reaching I4 in the Rockall trough it is clear that K_E at these depths has increased markedly over the equivalent values at NEADS-5 (e.g., $10 \to 42$ cm^2 s^{-2} at 1500 m, $61.5 \to 151$ cm^2 s^{-2} at 600 m; Fig. 3a,b, Tables 1 and 16). The most remarkable characteristic of the I4 records is the dominance of the seasonal signal in K_E and its great depth of penetration.

Gould and Cutler (1980) show a winter/spring increase in eddy energy at all levels to 1500 m at the dominant period of 15 d.

For their respective sampling depths eddy kinetic energies at MONA 1 and 3 on the flank of the Iceland-Faroe Ridge are comparable to those of the Rockall Channel Figs (3a, b). At this northern end of our transect the dominant periods of eddy motions (~ 10 d) are tentatively ascribed to baroclinic instabilities at the polar front (Willebrand and Meinke 1980). Only occasionally, during the passage of storms (Meinke 1975), is wind generation thought to play the dominant role.

Thus in qualitative agreement with the western basin our transect of the eastern basin indicates two general – and at present rather poorly based – tendencies in the data. First, as Gould (this Vol.) points out, eddy energies tend to increase northward; second, there is some indication of a decrease in dominant eddy time-scales northward, though this change is discontinuous rather than clear-cut. Dominant periods > 100 d are still found over the full depth range at the latitude of NEADS-5. (Gould this Vol.) From NEADS-5 northward the indications are that the decrease in eddy time-scales is at least partly contributed through seasonal forcing of some type and that this seasonal influence penetrates to considerable depths. These observations have not yet been interpreted in terms of a forcing mechanism. However, Müller and Frankignoul (1981) have investigated the role of stochastic wind forcing in generating oceanic geostrophic eddies via a simple ocean model with a realistic model wind-stress spectrum. They claim that their model predicts qualitatively the amplitudes and space- and time-scales of the observed eddy field in mid-ocean regions far removed from strong currents, and suggest that direct wind forcing may well be the dominant forcing mechanism in these regions.

15.4 The North Pacific

Figure 7 gives the location of all existing current meter records of any duration in the North Pacific, while Table 18 lists their flow statistics. The disparity in coverage between the Pacific and Atlantic is obvious. Of the 25 Pacific records listed (cf. 269 for the North Atlantic) 11 are of less than 7 months' duration and 20 are from depths greater than 4000 m. Thus the degree of intercomparison that can be achieved is slight.

15.4.1 The Kuroshio

Eight near-bottom current meters were placed on the continental slope under the Kuroshio by Taft in 1971 (Taft 1978). All records are short (23–103 d) but the four longest are used here to provide at least some comparison with the equivalent layers of the Gulf Stream. The prime difference, as Taft points out, is the fact that eddy kinetic energies appear to be weaker under the Kuroshio than along the Atlantic continental slope, especially in the deeper layers. Ex-

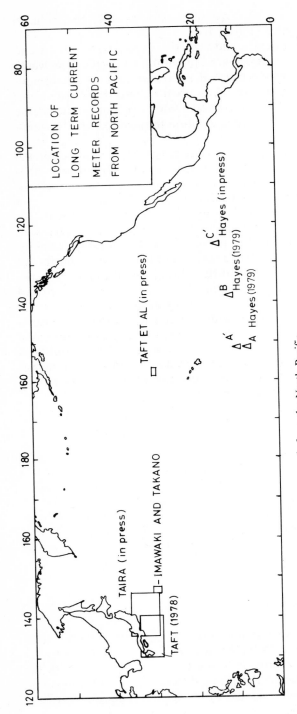

Fig. 7. Location of long-term current meter records from the North Pacific

tending the comparison made by Taft, Table 17 compares the Kuroshio data with statistics from both the Rise Array (Luyten 1977, Thompson 1977) and the Western Boundary Undercurrent Array (Rhines pers. comm.).

In the shallowest of the three layers considered ($\sim 1\,800\text{--}2\,000$ m) eddy kinetic energies for the Kuroshio are roughly equivalent to those of the west Atlantic arrays, but in the two deeper layers, the eddy field near the Kuroshio appears to be much less energetic than at the two Atlantic sites. Moreover, Taft's statistics refer to periods > 1 d while those from the west Atlantic are for periods > 2 d. Quite clearly, these differences may merely reflect the short, sparse, and scattered nature of the Kuroshio data, but as Taft points out it may also reflect the influence of the meridional Izu-Ogasawara Ridge ($140°$E) whose sill depth of around $2\,000$ m may obstruct the westward propagation of deep eddies from the west Pacific to the Shikoku Basin.

All three long records (1, 4, 7) from the slope show a dominant fluctuation with a period of 20 d, based on the first zero crossing of the autocorrelation function. Fluctuations in the cross-isobath flow at two of the short records (6, 8) support this result. This 20 d period is slightly longer than the 5–15 d range of energetic periods reported by Luyten (1977) for the fluctuations on the upper continental rise north of the Gulf Stream (Taft 1978).

Taira (in press) describes two records of ~ 9 months' duration from the relatively shallow crest of the Izu Ridge (Fig. 8; Table 18b). The record "Hachijojima West" (HW) is from the western face of the Izu Ridge, below the main thermocline (1 670 m), where the Kuroshio runs in from the south after its southward meander off Enshunada. (The Kuroshio has been in this meander mode since 1975; Taft's earlier measurements refer to the alternate mode when the Kuroshio flowed parallel to the continental slope.) The "Oshima West" (OW) measurements, from the main thermocline at 250 m depth, are representative of the Kuroshio branch which crosses the Izu Ridge via the near-shore Oshima West Channel.

The strong topographic influence of the Izu Ridge on these records rules out any meaningful comparison with records from the west Atlantic, where no

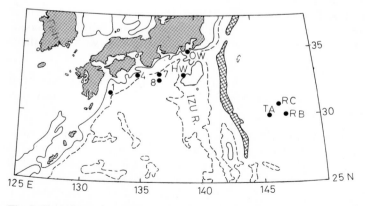

Fig. 8. Location of long-term mooring sites in the western North Pacific (Taft 1978, Taira, pers. comm., Imawaki and Takano, pers. comm.)

such feature exists. At both sites, the topographic control is shown by the anisotropic nature of the long-period velocity fluctuations ($\tau > 10$ d) with dominant fluctuations parallel to the bottom contours. The records from the two sites are very different in character however. At OW, in the constrained flow of Oshima Channel, the record is completely dominated by the mean flow, while at HW the kinetic energy is predominantly in the long-period fluctuations.

There are similarities and differences also in the time scales of low-frequency variability. Both records show a spectral peak at ~ 30 d in general agreement with Taft's records from the continental slope, but the OW record alone shows an additional peak at around 100 d (Taira, in press). Taira suggests that this difference in time scales may reflect the position of OW in the main thermocline since, from the analysis of isotherm displacements, *space* scales are also known to be larger above the main thermocline than below (i.e., about 320 km above 350 m cf. 240 km below 700 m).

15.4.2 Western Pacific

Three long-term moorings have so far been recovered from the deep layer of the West Pacific. They lie more than 500 km east of the Izu Ridge and a similar distance south of the Kuroshio extension (Fig. 8). Record TA, briefly described by Taira (in press) is a 167 d near-bottom series located at the foot of a gentle topographic slope in 5795 m depth. 120 km to the east, sites RB and RC have provided 5 records of 215–691 d duration, also from the deep layer (4000–5180 m depth) but in this case between 1 and 2 km from the bottom. Though unpublished, some preliminary statistics from these moorings have kindly been provided by Drs Imawaki and Takano.

Flow statistics from all three sites are listed in Table 18 and show the following characteristics:

1. Eddy kinetic energies from five of the six records are in good agreement, ranging from 5.5–8.3 cm^2 s^{-2} ($\tau > 1$ d). These values are comparable with eddy statistics from the MODE sites but higher than the 0–1 cm^2 s^{-2} observed at depth in the deep-interior region of the West Atlantic (e.g. the southern sites of the Polymode I array; see Figure 3d). Other points of similarity with these deep West Atlantic records are,

2. the weak kinetic energies of the mean flow. With the conspicuous exception of the single record at TA, K_M estimates at RB and RC are consistently less than 1 cm^2 s^{-2} (cf Fig. 3i);

3. the tendency of records to be dominated by longer period motions on moving from the continental slope towards the interior. Using the maximum entropy spectral method, Imawaki and Takano show that the five records at RB and RC have dominant periods of between 60 and 130 d. Schmitz (1978) identified a well-defined peak in the temporal mesoscale at 4000 m near MODE Centre while Gould, Schmitz and Wunsch (1974) also showed a dominant period of 50–100 d in the deep water data from the pre-MODE arrays.

15.4.3 East-Central Pacific

Taft, Ramp, Dworski and Holloway 1981 report flow statistics from a cluster of moorings set beneath the subtropical gyre of the east-central Pacific (Fig. 7, 9). Two successive arrays located on slightly hilly topography (100 m above the mean level of 5900 m) provided seven long records of ~9 mo duration, all from 100 m above the bottom (Table 18d). Two of the moorings (M11, M21) are sufficiently close to provide a 19-month merged record. Statistics of the low-passed data ($\tau > 1$ d) show the following characteristics:

1. The statistics are not stationary over the maximum record length of 19 mo.

2. Eddy kinetic energies and (with one exception) kinetic energies of the mean flow are extremely weak, with time-space averages of 1.6 cm^2 s^{-2} and 0.5 cm^2 s^{-2} respectively (excluding K_M from record M16). These values are comparable to the lowest values from the deep interior of the West Atlantic (Fig. 3d and 3i), and with a high proportion of zero speeds in the raw data (21%–59%) are very similar in character to MAFF moorings J, K and L from the eastern Atlantic, north-east of the Azores (Table 4). At the latter site the equivalent statistics were $\langle K_E \rangle = 0.8$ cm^2 s^{-2} $\langle K_M \rangle = 0.3$ cm^2 s^{-2} with 29%–88% of zero speeds recorded in the raw hourly values.

3. The low frequency fluctuations show two dominant time scales at about 70 and 150 d. The 150 d time scale was coherent throughout the array and present in both u and v spectra; the ~70 d period had spatial scales of < 100 km and

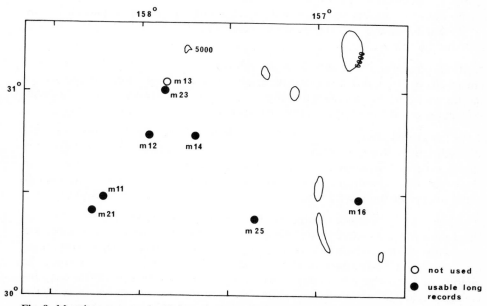

Fig. 9. Mooring array set by Taft et al. (1981) beneath the subtropical gyre of the east-central Pacific

was evident only in the v spectra. (Estimated from Array 1 data only, using the maximum entropy spectral method (Taft et al. 1981).)

4. In contrast to the MODE measurements and those of Hayes (1979) from the tropical Pacific, variances of meridional and zonal velocity components are approximately equal.

15.4.4 Central and Eastern Tropical Pacific

The final group of moorings to be considered are those reported by Hayes (1979; 1982) for the central and eastern tropical Pacific. The records are all from the near-bottom layer and derive from two deployments at three general sites, referred to here as A, B and C (first deployment, Hayes 1979) and A' and C' (second deployment, Hayes, 1982). From these the three long-term moorings discussed here are A, B and C' (Fig. 7).

Flow statistics from a selection of depths are listed in Table 18e. Conveniently, Taft et al. (1981) have already provided a comparison with their arrays further north.

1. Under the North Equatorial Current (sites B and C) K_E's of 8 and 11 cm^2 s^{-2} at 30 and 200 m above the bed were five to seven times larger than the space-time average K_E from beneath the subtropical gyre. Under the North Equatorial Countercurrent however (site A) K_E's were almost as weak as those reported by Taft et al. (1981).

2. Hayes' measurements show an intermittent bottom-intensification of K_E which does not appear to be present in the results of Taft et al. (1981). Hayes (1982) attributes this to small-scale bottom-trapped motions generated by the hilly topography.

3. In keeping with the results from the subtropical gyre, Hayes' (1979) measurements from 30 m above the bottom at all three sites show a dominant period of about 60 d in the meridional components. At two of Hayes' sites the u components showed longer dominant time scales of 100 and 180 d.

4. In his global comparison of spectra from long-term current meter records, Wunsch shows that the spectra from Hayes' site C contain the usual (global) characteristics: an isotropic band at high frequencies ($\tau < 50$ d) with a -2 slope, an eddy-containing band between 50 and 100 d, and a predominance of zonal energy at the lowest frequencies (Wunsch 1981, his Fig. 11.15A).

Thus, although the available Pacific records are sparse and of relatively short duration, they cover a sufficient range of flow regimes to identify certain similarities with the North Atlantic records, if only in very general terms (e.g., the decrease in eddy energies and increase in time scales on moving from the vicinity of the western boundary current and continental slope to the deep interior). Certain physical dissimilarities between the two ocean basins (e.g., the presence of the Izu-Ogasawara Ridge in the west Pacific) are perhaps also reflected in these data, but with present coverage and lengths of record, the evidence for this remains extremely tentative. One major deficiency at present remains the lack of any long time series from the most energetic part of the west Pacific (the Kuroshio Extension; Nishida and White, 1982) though plans to meet this deficiency are underway (Schmitz and Niiler pers. comm.).

15.5 The Equatorial Zone

The equatorial zone for present purposes is defined as the area ~ 200 km north and south of the equator where, as the Coriolis parameter vanishes, eddy motions are dominated by a spectrum of equatorially-trapped waves (internal inertial-gravity, mixed Rossby-gravity and Kelvin; see Eriksen 1980). Although long-term equatorial records have now been recovered from all three oceans, many of the recoveries are very recent and still under analysis, and the published data are largely confined to the western Indian Ocean. Thus although very real differences exist in the forcing and geometry of the three equatorial oceans (Philander 1979) the effects of these differences on flow statistics have not yet been determined. Nevertheless, some first-order characteristics of the equatorial zone as such have been identified.

As Eriksen (1980) and Knox (1981) point out, there is no energy gap at subinertial frequencies in equatorial or near-equatorial spectra, in contrast to midlatitudes. Nevertheless, the records to be described have been filtered to remove frequencies higher than ~ 2 d so flow statistics remain comparable to those from higher latitudes (Table 19).

Figure 10 illustrates the location of available records from the western Indian Ocean, while Fig. 11 shows the depth distribution of K_E, K_M and K_E/K_M

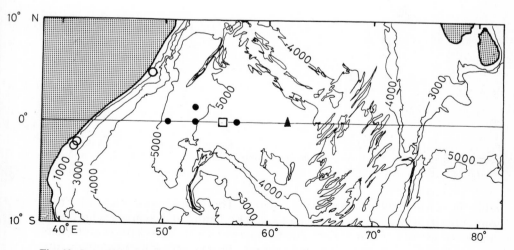

Fig. 10. Location of long-term moorings in the western equatorial Indian Ocean. *Open circles* Schott (pers comm); *closed circles* Luyten 1982; *square* Knox (1981 and pers. comm.); *triangle* Fieux (pers. comm.)

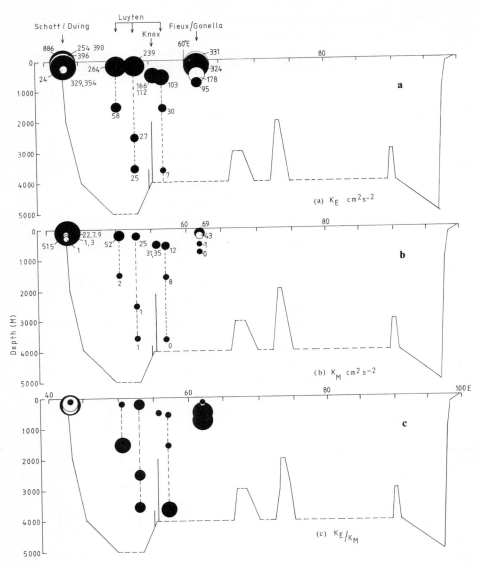

Fig. 11. Vertical equatorial section of **a** $K_E cm^2 s^{-2}$, **b** $K_M cm^2 s^{-2}$, and **c** K_E/K_M ratio for the western Indian Ocean. Data from Schott (pers. comm.), Luyten (1982), Knox (1981 and pers. comm.), and Fieux (pers. comm.)

ratio for stations lying close to the equator. The westernmost station is from long-term deployments at the Somali coast by Düing and Schott (Schott pers. comm.) while the remainder are due to Luyten (1982), Knox (1981) and Fieux and Gonella (pers. comm.). The only other published records known to the author are those from the Gulf of Guinea (Weisberg et al. 1979) and the short ($\sim$80 d) series included in the GATE Atlas (Düing, Ostapoff and Merle

1980b), though further unpublished statistics have kindly been provided by Knox and Halpern (pers. comm.) for the central equatorial Pacific at 152–153°W.

The equatorial section for the Indian Ocean (Fig. 11) shows the following characteristics:

1. As in the shipdrift analysis of Wyrtki, Magaard and Hager (1976), near-surface values of K_E are generally as large or larger than those observed near the Gulf Stream. Though short, the GATE records from the Atlantic sector at 10°W (Düing, Ostapoff and Merle 1980) appear to confirm that high near-surface K_E values are also characteristic of the equatorial Atlantic ranging from 298–713 $cm^2 s^{-2}$ at depths less than 100 m.

2. In more detail, values of K_E for the western Indian Ocean appear to increase towards the western boundary (Fig. 11; see also Luyten 1982). The value of 886 $cm^2 s^{-2}$ observed at 85 m depth on the Somali coast (mooring M1, Table 19, Schott, pers. comm.) is the second largest K_E value yet observed in the world ocean. The largest value lies nearby at 4°55′N (Table 19). This is qualitatively (but not quantitatively) consistent with the suggestion of Wyrtki et al. that the area of the Somali Current should form both a local and a global maximum in eddy kinetic energy, partly as a result of the major seasonal change in the strength of the mean flow. Another possible reason for the observed westward intensification of K_E, suggested by Luyten (1982) is the reflection of Rossby waves at the western boundary, giving rise to Rossby waves of smaller scale but larger amplitude; this trend in K_E may also be reinforced by attenuation of the deep eddy field at the easternmost full-depth mooring (596; Table 19) through its proximity to the Carlsberg Ridge ($\sim 65°E$). As Luyten points out, it is the deeper two instruments of 596 that are anomalously low in K_E.

3. Dominant frequency. Weisberg et al. (1979) appear to have been the first to show a characteristic difference in spectral shape between zonal and meridional current components in their records from the equatorial Atlantic at 3–4°W. They show a pronounced peak in v with an approximate 1-month periodicity, followed by a cut-off at longer time scales, while the u spectra are essentially red at low frequency.

A similar behaviour is shown in the energy-preserving kinetic energy frequency spectra from the western Indian Ocean. Luyten (1982, his Fig. 10) shows that for ten records of 201–231 d duration, the zonal spectra are red with maximum energy at periods of 50 d or longer and with zonal energy predominating over meridional at these long periods. Meridional spectra tend to show peak energy at periods of ~ 10–50 d (pronounced in the near-surface records) and in this band meridional variability normally predominates over zonal for records on the equator (but not off it; see Eriksen 1980, his Fig. 7). Similar characteristics are shown in two 6-month records described by Knox (1981) from 0°N 56°E, in one-year records recently recovered by Fieux and Gonella from 0°N 62°E (Fieux, pers. comm.), and in the 13½month series provided by Knox and Halpern (pers. comm.) from the central equatorial Pacific (0°N 152–153°W). This (apparently) typical spectral shape for equatorial records is illustrated in Fig. 12a, b, c, d.

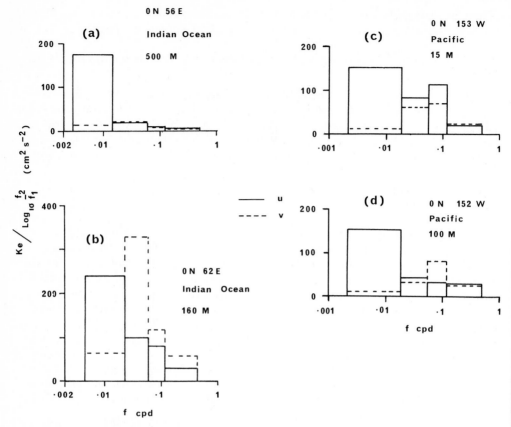

Fig. 12. Plots of decadal spectral estimates, energy preserving by area, for **a** 0°N 56°E, 500 m (Knox 1981), **b** 0°N 62°E, 160 m (Fieux pers. comm.), and **c,d** 0°N 152°–153°W, 15 and 100 m (Knox and Halpern pers. comm.)

In the case of the two records from 0°N 56°E, the ratios of zonal to meridional eddy kinetic energy are close to unity for periods $< {\sim}20$ d, but increase to 6.3 and 8.4 respectively at longer periods. Knox, following Eriksen's (1980) model, suggests that this characteristic is due to the fact that at these low frequencies, gravity waves are practically non-existent and Kelvin waves (for which $v \equiv 0$) contribute an increasing share of the variance (see also Luyten 1982).

Published equatorial records are insufficiently long to determine whether an annual signal provides a further point of comparison with records from mid-latitudes.[1]

---

[1] Since this was written, Eriksen has described strong quasi-annual variability in deep currents near the equator from two-year current measurements near the Gilbert Islands, 0°17′N 173°55′E. (Eriksen, C.C., 1981. J. Phys. Oceanogr. *11* (1): 48–70)

15.6 South Atlantic

The only available long-term data set from the South Atlantic stems from two successive deployments by WHOI in the vicinity of the Rio Grande Rise, which acts as a zonal barrier between the Argentinian and Brazilian Basins, to the west of the mid-Atlantic Ridge. On each deployment (of ~4 mo and ~11 mo duration respectively), four of five moorings were set as a transect to cover the cross-Rise flow through the Vema Channel while the fifth was located within the Channel a few miles downstream (i.e. to the north). Though unpublished, preliminary statistics from this array have kindly been provided by Nelson Hogg, WHOI; the nine longest combined records are listed in Table 20 while Fig. 13 illustrates the distribution of K_E, K_M and K_E/K_M ratio in the Vema Channel for the eight records below 3500 m depth.

On the transect the deepest layer of the Channel is occupied by a constrained and intense mean flow with the two deepest records showing mean northward flows of 21–22 cm s^{-1} over record lengths of 356 and 468 d. Not surprisingly K_M values in the deepest layer exceed K_E by almost an order of magnitude both on the transect and at the "downstream" site (Fig. 13a, b). Moving upward from the deep "notch" of the Channel to depths (<4100 m) where the flow is less intense and presumably less constrained, K_M itself decreases by an order of magnitude, becoming more nearly equivalent to the prevailing values of K_E, but is exceeded by K_E at only one of the eight sampling depths shown (Fig. 13c). The K_E range of 7–37 cm^2 s^{-2} (with a maximum in the centre of the Channel) is similar to that found at equivalent depths in the Drake Passage (next section).

15.7 High Latitudes

15.7.1 The Antarctic Circumpolar Current

Long-term data sets are now available from two key regions of the circumpolar current (Fig. 14). First, 17 moorings deployed in the Drake Passage between 1975 and 1978 have provided 31 records of ~1 year duration (Pillsbury et al. 1977, Sciremammano et al. 1978, Nowlin et al. 1981) and ten 7-month records have also been obtained from a cluster array south-east of New Zealand in the lee of Macquarie Ridge (Bryden pers. comm.). Seventeen-month records have now been recovered from the latter array but were not available at the time of writing.

Flow statistics from these moorings are listed in Table 21 and are compared in Figs. 15–17, a and b. These figures share a common depth scale but the horizontal scales differ by a factor of about 10.

The lower panels form a composite section across the Drake Passage using data from moorings 2, 4, 8, 10, 12 and 14 (1975), D, A and E (1976), N, C, E and S (1977), and Y (1978). Justification for combining multi-year statistics into

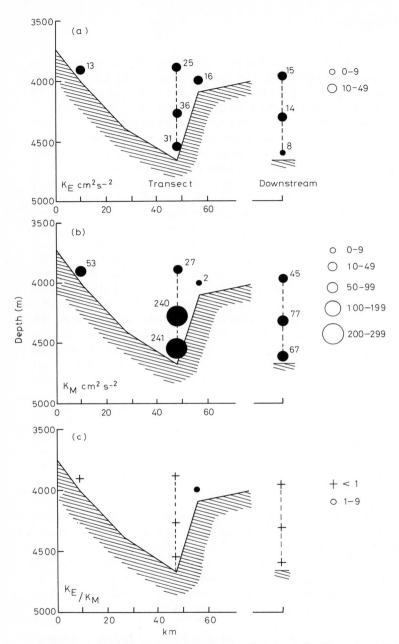

Fig. 13. Vertical distributions of **a** K_E cm^{2}s^{-2}, **b** K_M cm^{2}s^{-2} and **c** K_E/K_M ratio along a west-east transect of the Vema Channel (Rio Grande Rise, South Atlantic) and at a point a few miles downstream (N. G. Hogg, pers. comm.)

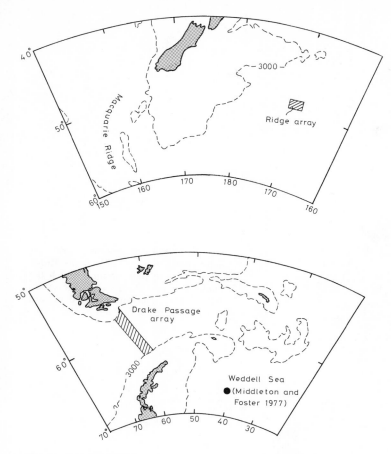

Fig. 14. Location of long-term mooring arrays in the Antarctic Circumpolar Current region

one plot lies in the fact that spectra from deep records at the central passage location do not significantly differ from one another, even over several years (Nowlin et al. 1981). Together with results from detailed analyses by Nowlin et al. and Wunsch (1981) these plots show the following characteristics:

1. Where full-depth coverage is most complete – in the centre of the passage close to the Polar Front – K_E increases sharply above 1 500 m so that "near surface" values are about an order of magnitude greater than at depth. With the shallowest observation at 282 m and with restricted spatial coverage, it is difficult to extrapolate a meaningful surface value from these profiles, but as Nowlin et al. point out, the maximum of 205 cm^2 s^{-2} at 282 m is "not inconsistent" with the surface ship-drift estimates of K_E (about 800 cm^2 s^{-2}) derived by Wyrtki, Magaard and Hager (1976). At the southern end of the section remote

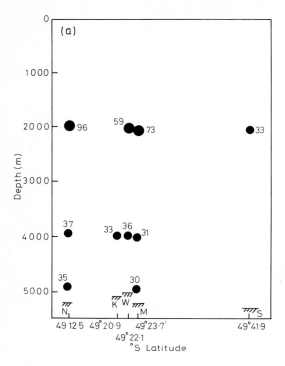

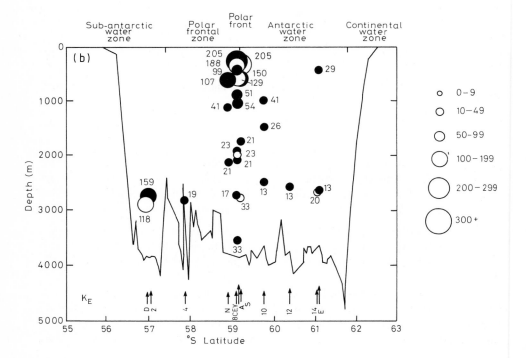

Fig. 15. Depth distribution of eddy kinetic energy (cm^{2}s^{-2}) for **a** the Ridge Array (Bryden pers. comm.) and **b** the Drake Passage Array (Nowlin et al. 1981). The mooring identifiers are shown at the foot of each panel

from the core of the current, the change in K_E with depth is much less pronounced, merely doubling between 2 600 and 460 m depth to a value of 49 cm^2 s^{-2}.

2. In the deep water (> 2500 m), low frequency variability is a maximum in the northern passage, south of Cape Horn. There, K_E values of up to 159 cm^2 s^{-2} are almost an order of magnitude greater than at comparable depths further south, and equal the deep water maximum observed by Schmitz beneath the Gulf Stream. This local increase in K_E is due primarily to greater activity in the meso-scale band at periods of 2–50 d, and is thought to reflect eddy generation through interaction of the circumpolar current system with both local bathymetry and with its northern boundary (Pillsbury et al. 1979, Nowlin et al. 1981).

3. As with K_E, the K_M values at the polar front increase sharply from 2 500 m towards the surface with values in the upper water column (< 750 m depth) exceeding those of the deep water by well over an order of magnitude. The observed maxima of 250–365 cm^2 s^{-2} are very much larger than the area-averaged surface estimates of K_M derived from ship drift (100–200 cm^2 s^{-2}; Wyrtki et al. 1976). As with K_E, the change of K_M with depth is much less pronounced in the southern passage remote from the current core.

4. In the deep water, changes in K_M are slight but there is some tendency for values to increase southward.

5. Over most of the section and in all depth layers K_E and K_M values are roughly equivalent (within a factor of 2–3). Large values of K_E/K_M ratio are only observed at depth in the northern passage where the deep K_E maximum and K_M minimum coincide. In the central part of the section there is a more subtle tendency for the K_E/K_M ratio to increase with depth from values < 1 in the upper ~ 2000 m to values greater than unity in the deepest layers. Significantly, the kinetic energy spectrum for the deep layer suggests the presence of energetic bottom-trapped modes at mesoscale frequencies (Nowlin et al. 1981).

6. Zonal and meridional velocity spectra from the central Drake Passage at 1 519 m are included in Wunsch's (1981) overview of spectra from ten sites in the world ocean. They share the usual "global characteristics": a high frequency band with a -2 slope, energy enhancement in the eddy-containing band and some tendency (though slight in this case) towards zonal dominance at low frequencies.

7. Overall, Nowlin et al. conclude that the large values of both K_E and K_M encountered on the section are consistent with the idea that, in this region, baroclinic instability is converting potential energy of the strong mean flow into fluctuation energy; the conversion rate has been estimated by Bryden (1979).

Though some comparisons of Drake Passage statistics with other regions have been mentioned, the question remains as to whether they are representative of the circumpolar current itself. As Nowlin et al. point out, the Drake Passage presents both a lateral barrier and a topographic obstacle to zonal flow so

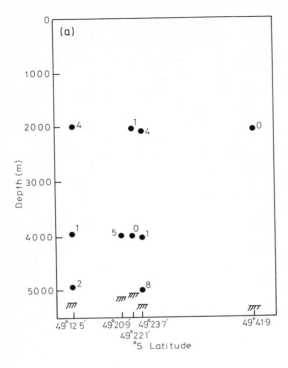

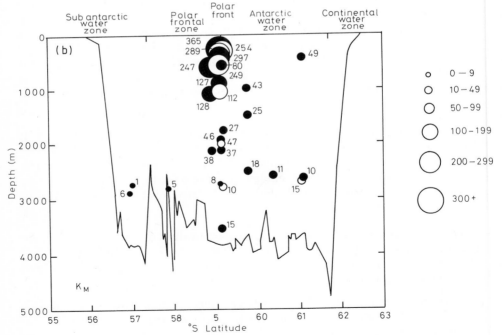

Fig. 16. Depth distribution of kinetic energy of the mean flow (cm^{2}s^{-2}) for **a** the Ridge Array (Bryden pers. comm.) and **b** the Drake Passage Array (Nowlin et al. 1981)

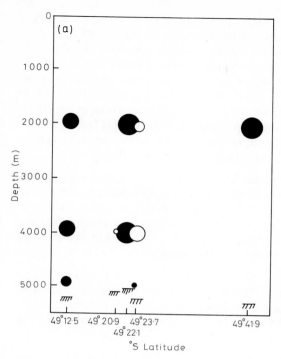

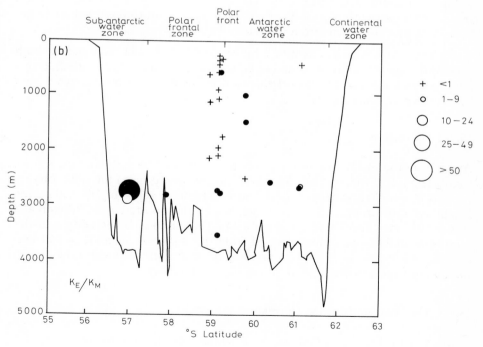

Fig. 17. Depth distribution of K_E/K_M ratio for **a** the Ridge Array (Bryden pers. comm.) and **b** the Drake Passage Array (Nowlin et al. 1981)

that flow statistics and their spatial distributions are likely to be atypical. Bryden's "Ridge" cluster at an open-ocean deep-water site south-east of New Zealand provides a data set for comparison though, in this case also, the flow statistics are unlikely to be "typical". In fact the site was selected because observational and modelling work had suggested that large-scale eddies might be generated through the interaction of the circumpolar current with the Macquarie Ridge upstream. Statistics from the first 7-month deployment are listed in Table 21 and are illustrated in the upper panels of Figs. 15–17.

From his preliminary analysis of these data Bryden (pers. comm.) suggests the following similarities and differences between the two sites:

1. As theoretically expected, Ridge eddies have longer temporal and spatial scales than Drake Passage eddies. At the Ridge site low-passed currents at 2 000 m display an energy peak at ~60 d and a wavelength of about 500 km in contrast to Drake Passage eddies which, in the same depth layer, show an energy peak at around 25 d and a wavelength of roughly 200 km. (Drake Passage observations around 2 000 m depth are largely confined to the 1977 deployment.)

2. Though the depths of observation at the two sites are not strictly comparable there is enough overlap to suggest that Ridge eddies are also the more energetic. Eddy kinetic energies at 2 000 m depth are a factor of ~3 larger than at comparable depths in the Drake Passage (Fig. 15).

3. Since, in addition, the kinetic energies of the mean flow at the Ridge site are very much weaker (of order 0–5 cm^2 s^{-2}) the K_E/K_M ratios observed are larger than in almost all records from the Drake Passage (Fig. 17).

4. Elements of similarity identified by Bryden include the facts that at both sites eddies have similar propagation velocities (8–12 cm s^{-1}, eastward), transport heat poleward and transport eastward momentum northward, away from the circumpolar current.

15.7.2 Weddell Sea

Apart from the ~12 long-term records now obtained by Foldvik and Kvinge from the edge of the Weddell Sea shelf (not discussed here) the only long record from the deep water is a near-bottom (4 490 m) series of 349 d duration from the flat central part of the Weddell Sea basin (reported by Middleton and Foster 1977; see also Table 21c, Figure 14). Though low-passed eddy statistics are not available, even those derived from the *unfiltered* data ($K_E = 3.61$ cm^2 s^{-2}; $K_M = 0.85$ cm^2 s^{-2}) show clearly enough that both the mean flow and its fluctuations are extremely weak in the near-bottom layer.

15.7.3 The West Spitsbergen Current

In the 4 years 1976–79 a total of 12 year-long records were recovered by Aagaard from an array of four moorings deployed across the West Spitsbergen

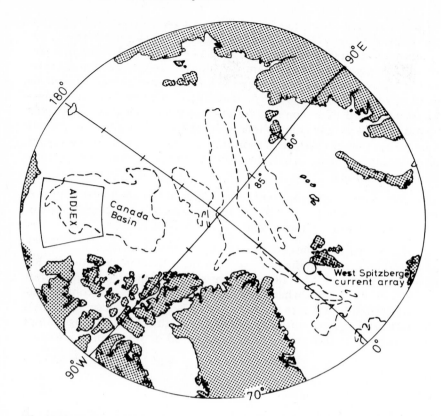

Fig. 18. Location of long-term current meter observations in the Arctic (Aagaard and Hanzlick pers. comm.; Manley and Hunkins pers. comm.)

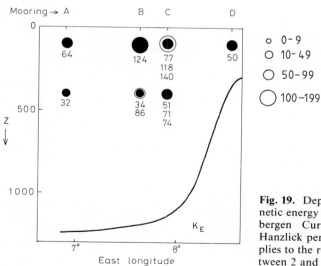

Fig. 19. Depth distribution of eddy kinetic energy (cm²s⁻²) for the West Spitsbergen Current Array (Aagaard and Hanzlick pers. comm.). Note that K_E applies to the restricted band of periods between 2 and 52 d

current (Figs. 18, 19). All four sites (A, B, C, D) were occupied in 1976–77, two (B, C) were reoccupied in 1977–78 and one (C) was continued into 1978–79. All records are from either 100 or 400 m depth, and as with MONA 5 and 6, filtering was done by spectral decomposition, deleting all components outside the desired bands and reconstituting the record. Eddy kinetic energies for a restricted band of periods between 2 and 52 d are listed in Table 22 and are plotted in Fig. 19 (Aagaard pers. comm.). Because of their restricted bandwidth these statistics are not strictly comparable with the bulk of statistics described in this chapter, but they do provide important new information on the spatial and temporal variability of K_E across the section and on the dominant periods of low-frequency motions:

1. Spatially, eddy kinetic energies tend to be larger at the shallower depth of measurement (by a factor of ~2) and are maximal at the two closely spaced stations in the central zone of the current.

2. These spatial changes are obscured by the large annual variations observed at the two central moorings. In the three years of measurement at Station C, K_E at 100 m rose by 82% from 77 to 118 to 140 cm^2 s^{-2} accompanied by an overall increase of 22% at 400 m. Elsewhere, repeated measurements are confined to the 400 m level at Station B where K_E increased by > 150% in successive years. These bald statistics fail to capture the remarkable nature of the change, however, since a progressive shift in the dominant period of variation is also evi-

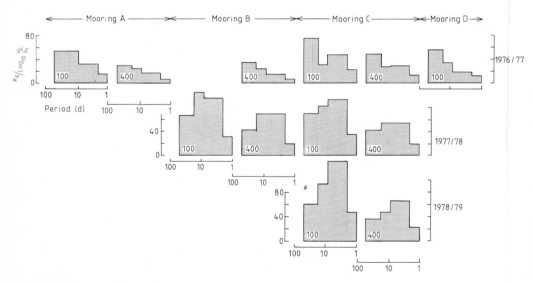

Fig. 20. Plots of decadal spectral estimates, energy preserving by area, for each record of the West Spitsbergen Current Array (data from Aagaard and Hanzlick pers. comm.). *Upper row* refers to the 1976–77 deployment, *middle row* to the 1977–78 deployment and *bottom row* to the 1978–79 deployment; the approximate depth of observation (100 or 400 m) is inserted in each case

dent. Eddy kinetic energies in the period bands 1–2 d, 2–8.3 d, 8.3–17 d and 17–52 d were kindly provided by Aagaard and Hanzlick for each of the 12 records. In Fig. 20 these estimates have been divided by $\log_{10} f2/f1$ where f1 and f2 represent the low and high frequency limits of each band, respectively, so that when plotted against $\log_{10} f$ we obtain spectral plots which are energy preserving by area. The 6 spectra from 1976/77 records are shown in the upper row of Fig. 20. In each case, maximum energy is in the 17–52 d period band, or possibly at longer periods unresolved by the present data. In the 4 spectra from 1977/78 (middle row, Fig. 20) the peak is shared almost equally between the 8.3–17 d and 2–8.3 d bands while in the two records from 1978/79 (bottom row, Fig. 20) the shift of energy to higher frequencies has continued, with energy maxima in the 2–8.3 d band.

This variation on variability – with a progressive interannual change in both the magnitude and dominant frequency of eddy kinetic energy – has been described in some detail since it is the clearest example of such changes in the present data set. With so few ultralong-term records from the world ocean, especially in the near-surface layer, any comment on the likely extent of such changes must await Aagaard's dynamical interpretation of these results.

15.7.4 Arctic Ocean

A unique data set on mesoscale eddy activity beneath the ice of the Arctic Ocean is provided in results from the AIDJEX pilot programme (1970–72) and the full AIDJEX programme of 1975–76. The earlier results were reported by Newton et al. (1974) and Hunkins (1974); later results are reviewed or listed in Hunkins (1980), Manley and Hunkins (pers. comm.) and Manley, Hunkins and Tiemann (1980).

During the main AIDJEX programme, current profiles to 200 m were made twice daily at four camps established on drifting ice near the southern edge of the clockwise gyre of the Canada Basin. Though close to the Alaskan continental slope, the stations overlie deep water in a region unaffected by upstream seamounts (Fig. 18).

After correcting for the drift of the ice station itself (often greater than the translation speed of the eddies), the 2 149 current profiles show the frequent presence of baroclinic eddies (~ 20 km in diameter; swirl speeds up to 58 cm s^{-1}) concentrated in a depth layer which extends from the base of the mixed layer (50 m) to near the foot of the pycnocline (halocline) at around 200 m depth. Leaving aside the other characteristics of these eddies, more fully described by Needler (this Vol.), Fig. 21 provides depth averages of the kinetic energy of the mean flow and its fluctuations for the complete AIDJEX data set (a) and for the two individual station-months which showed the largest and smallest subsurface energy peaks (b and c). Almost all of the energy is contained in the fluctuations with a negligibly small amount in the mean flow. At the subsurface maximum, monthly mean K_E's range over an order of magnitude from a minimum of 20 to a maximum of 228 cm^2 s^{-2} ($\equiv$ ergs cm^{-3}), while

AIDJEX DATA *-*-*-*- CAMP * ALL *
ALL STN KE TOTAL
APR 1, 1975 (91) *VALID* APR 30, 1976 (486)

AIDJEX DATA *-*-*-*- CAMP BIG BEAR
ALL STN KE TOTAL
JUN 1, 1975 (152) *VALID* JUN 30, 1975 (181)

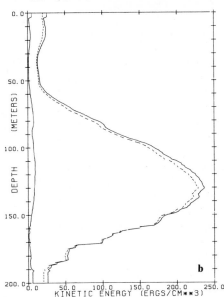

AIDJEX DATA *-*-*-*- CAMP SNOWBIRD
ALL STN KE TOTAL
FEB 1, 1976 (397) *VALID* FEB 29, 1976 (425)

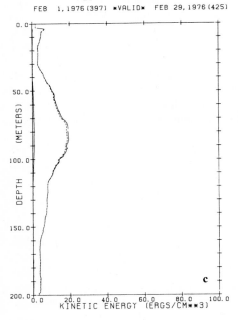

Fig. 21. Depth-averaged distributions of K_M (*solid line* close to the y-axis), K_E (*dashed line*) and total kinetic energy (*solid line*) for **a** the complete AIDJEX data set and **b c** for the two individual station-months which showed the largest and smallest subsurface energy peaks. (Manley and Hunkins pers. comm.)

the "all station" mean shows a broader maximum of 55 cm^2 s^{-2}. Beneath the energetic Pacific water layer (50–200 m) the Atlantic water (200–900 m) is relatively quiescent.

With insufficient vertical shear in the open ocean, baroclinic instability at the shelf-slope front is the favoured candidate mechanism for these eddies (Manley and Hunkins pers. comm.). Since the AIDJEX eddies carry a distinctive and anomalous lens of water at their core which appears to originate from the Chukchi Sea in summer (warm-core eddies) and winter (cold-core) the formation zone is thought to lie along the Alaskan shelf break near Point Barrow where the intrusion of these shelf waters takes place. Dissipation would occur as the eddies are advected north across the Arctic Ocean; hence the "not unexpected" result that no eddies were observed during the month-long FRAM-1 drift (1979) north-east of Greenland, at the opposite end of the transpolar drift stream (Manley and Hunkins pers. comm.).

In his brief comparison of baroclinic eddy characteristics in the Arctic Ocean and West Atlantic, Hunkins (1980) not only highlights the differences in horizontal and vertical space scales (20 km and 200 m respectively for the Arctic compared with ~ 100 km and up to 4000 m for the West Atlantic), but also the fact that Arctic eddies have a definite subsurface kinetic energy maximum. This Arctic characteristic appears related to the presence of an ice cover against which eddies are frictionally dissipated by an Ekman spin-down process.

15.8 Summary

Though intended in itself as a summary of global flow statistics, this chapter has now developed sufficient poundage that a brief additional summary of conclusions seems appropriate. To merely repeat the geography seems a sterile way of doing this, however. More rewarding, but also much more difficult and more tentative, would be to use this reasonably complete collection of data to assess whether the observed kinetic energy "structure" agrees with our preconceptions and, where theory exists, with theory. Naturally such an assessment will rely heavily on the north Atlantic data set and even with that relatively dense coverage, the reader should recall Schmitz's (1980) remark that "we are still in the process of exploration."

15.8.1 Horizontal Distribution of Eddy Kinetic Energy at Midlatitudes

To a large extent, these data confirm the general conclusion of many authors (expressed here by those of Wunsch) that "much of the total mesoscale energy is found in the immediate proximity to the Western Boundary Currents in the region where they are going seaward. It seems therefore that these regions are the generators of much of the eddy field energy" (Wunsch 1981, p. 373). One might broaden this conclusion to include strong currents in general since the

Circumpolar Current is also associated with a local maximum in eddy kinetic energy.

However, these data also show the less-expected result that regions remote from strong currents are not necessarily quiescent. Direct wind forcing seems to play a significant role in generating eddy energy in some regions, such as the Rockall storm belt of the north-east Atlantic. While wind generation of "eddies" is a subject very much under discussion at present (e.g., Muller and Frankignoul 1981, Koblinsky and Niiler 1982, Dickson et al. 1982), the theoretical basis has yet to be fully developed.

Absolute K_E *minima* at thermocline and abyssal depths in the western basin of the Atlantic are found at the POLYMODE I $\triangle$ site (7–12 and 0–1 cm^2 s^{-2} respectively). In the eastern basin the present restricted coverage places these minima in or around the Iberia Abyssal Plain, at the NEADS-3 site in the thermocline (13 cm^2 s^{-2}) and at the NEADS-2½ site (0 cm^2 s^{-2}) at abyssal depths.

15.8.2 Vertical Structure of the Eddy Field at Midlatitudes (with Peter Rhines and Ellen Brown, WHOI)

A series of numerical/theoretical models by Rhines (1977 and pers. comm.; also Salmon 1978) has provided something of the theoretical framework that we need to explain the observed vertical structure of the ocean's eddy field. Essentially these models suggest a monotonic degradation of U_{500}/U_{4000} on ε/δ where $U_{(z)}$ is the r.m.s. velocity at depth z, ε is the Rossby number and δ is the topographic roughness height divided by total depth. In the present context these results suggest that eddies should be more barotropic at high energy levels and over smooth topography yet more baroclinic at low energy levels and over rough topography. Compared with observations from the West Atlantic, these conclusions seemed to hold true both in Schmitz's 1978 study and in his later description of a zone of "weakly depth-dependent flow" along the southern flank of the Gulf Stream (Schmitz 1980). They are also in qualitative agreement with the conclusion of Wunsch (1981, p. 373) that "the major effect of topography seems to be the spin-down of the deep ocean layers relative to the thermocline."

We can now test these conclusions using the expanded set of flow statistics assembled in Tables 1–22. Defining "thermocline" depth as 500 ± 200 m and "abyssal" depth as 4000 ± 200 m we find a global total of 27 moorings with long-term records from *each* of these depths. All are from the North Atlantic. There are, in addition, a further 7 "near misses" from the North Atlantic where records lie slightly outside these depth limits or are of shorter than average duration (i.e., NEADS 1, 2, 3½, 4, PM II 13, Western Boundary Undercurrent 617 and Local Dynamics Experiment 640), and 1 further "near miss" from the remainder of the world ocean (Drake Passage mooring C). Data from these 35 moorings are displayed in Fig. 22.

In the calculations which follow, the eddy kinetic energy ratio $K_{E\,500}/K_{E\,4000}$ is substituted for U_{500}/U_{4000}; this is plotted against the "eddy velocity" at 500 m ($U_{E\,500} = \sqrt{2\,K_{E\,500}}$) which is a measure of ε. The topographic roughness parameter (δ = r.m.s. height variation $\div$ total depth) was obtained for each of the mooring sites by crudely digitising the bathymetry at 10' intervals within a circle of ~ 100 km radius around each mooring. NAVOCEANO topographic charts were used throughout the North Atlantic to ensure self-consistency of depth information; when checked against δ from more detailed charts (NEADS sites) most values agreed within about 10%. The only topography available for the Drake Passage is the transect illustrated in Fig. 15–17. The value of δ was obtained in this case by digitising the bathymetry at 10' intervals, 100 km either side of mooring C.

Figure 22a plots the vertical eddy structure ($K_{E\,500}/K_{E\,4000}$) against eddy velocity at 500 m ($U_{E\,500}$) for the 35 moorings, partitioned into groups from the western and eastern Atlantic; the Drake Passage point (DP) is included in the former. In the western basin plot (left-hand panel) the results already seem to accord well with theory. With the conspicuous exception of PM I 548 – already described as anomalous (p 293) – there does appear to be a significant tendency for the vertical eddy structure to become less depth-dependent (ratios tending towards unity) as the "eddy velocity" increases. The data from the eastern basin, however, show a completely different pattern (right-hand panel, Fig. 22a). There, a narrow range of eddy velocities at thermocline depth is associated with a wide range of vertical eddy structure.

Evidence that this different behaviour between basins might be due to the greater range of topographic roughness at the eastern basin sites is obtained in Fig. 22b, where estimates of vertical eddy structure from all 35 sites are plotted against $U_{E\,500}/\delta$ (i.e., closely similar to the ε/δ parameter considered by Rhines 1977). The results are, encouragingly, in good agreement with the theoretical findings described above, showing the expected tendency towards a monotonic degradation of $K_{E\,500}/K_{E\,4000}$ on $U_{E\,500}/\delta$.

Though some scatter about this tendency certainly remains it should be remembered that:

1. the method of calculating δ is crude at present. For example, it fails to distinguish mean bottom slope from roughness (MODE Centre and MODE East have nearly the same value of δ) and is subject to aliasing where the roughness is at small scales (e.g., the ≤ 20 km topographic scales at PM II 06);

2. the plots include some phenomena that are not geostrophic turbulence and thus should not obey our conjectural rules (e.g., lone Gulf Stream rings, or meanders in the early stages of the Gulf Stream near Cape Hatteras or, perhaps, the Western Boundary Undercurrent);

3. there may be other determinants of vertical eddy structure, not considered here. For example, a baroclinic structure may reflect input of eddy energy at the surface through wind and buoyancy effects, rather than the decay of eddy energy at depth over rough topography.

Under these circumstances Fig. 22b appears to provide an encouragingly good fit with current theory.

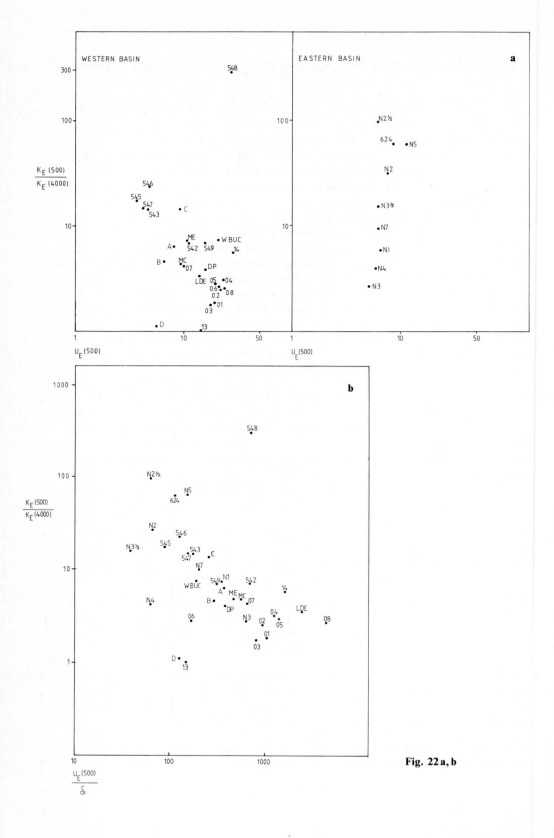

Fig. 22 a, b

15.8.3 Vertical Eddy Structure in Ice-Covered Seas

The AIDJEX results from the Canadian Arctic, while of limited geographical extent, suggest a further characteristic of vertical eddy structure in the case of ice-covered seas; i.e., that where eddy motions exist, the kinetic energy maximum will lie at subsurface depths. This characteristic appears to be related to the presence of the ice cover, against which eddies are frictionally dissipated by an Ekman spin-down process.

15.8.4 Eddy Time Scales at Midlatitudes

With insufficient data to generalise, our information on the temporal structure of eddy motions at midlatitudes is largely restricted to the two meridional transects through the western and eastern basins of the North Atlantic, discussed earlier (pp. 294–301). These and other midlatitude records perhaps allow the one generalisation that the dominant eddy time scale increases to a maximum ($\tau > 100$ d) on moving from the basin margins toward the deep interior of the subtropical gyre. Aside from that, any more complex "patterns" of dominant time scale remain illustrative of a particular place and time but can hardly be described as representative.

The western basin transect towards the Western Boundary Current and Continental Slope beyond was characterised by a shift from 'secular' time scales (> 100 d) at all depths in the deep interior, through a mix of secular (thermocline) and mesoscale (deep layer) dominance at the MODE sites, to mesoscale dominance (50–100 d) at all depths in the recirculation zone and at the southern margin of the Gulf Stream. The shift to higher dominant frequencies continued up the Continental Rise reflecting the successive influences of Gulf Stream meandering ($\tau \sim 30$ d) and increasing bottom slope.

The eastern basin transect was also dominated by "secular" eddy time scales at all depths at southern interior sites. Northward to the latitude of NEADS-5 (47°N) dominant periods > 100 d were still found over the full depth range but from at least 40°N northwards we find evidence of a seasonal (winter-spring) input of eddy energy at temporal-mesoscale (or higher) frequencies which grows to dominate the spectra at all depths as the transect continues north towards and through the Rockall storm belt (Dickson et al. 1982). Superimposed on these basin-wide tendencies are individual records whose

◀ **Fig. 22.** Ratio of eddy kinetic energies at thermocline and abyssal depths ($K_{E\ 500}/K_{E\ 4000}$) versus **a** the "eddy velocity" at 500 m ($U_{E\ 500} = \sqrt{2 K_{E\ 500}}$) for the eastern and western basins of the North Atlantic and **b** $U_{E\ 500}/\delta$ where δ is the topographic roughness parameter for each site. (Preliminary data: Rhines, Brown and Dickson in prep.). Mooring identifiers are as follows: POLYMODE I = 542–549, POLYMODE II = 01–14, POLYMODE IIIB = 624, POLYMODE IIIC = A–D, MODE Center = MC, MODE East = ME, Drake Passage mooring C (1977) = DP, Western Boundary Under Current No. 617 = WBUC, Local Dynamic Expt. No. 640 = LDE, North East Atlantic Dynamics Study = N1–N7 eddy time scales are dominated by local factors, e.g. by baroclinic instabilities at the Polar Front (MONA 1 and 3) or by topographically controlled oscillations (NEADS 6 on the Continental Rise).

Many of the characteristic features of these transects are now explicable by existing theory. Notably, eddy resolving gyre-scale numerical models (Schmitz and Holland 1982) now reproduce the characteristic frequency distributions of K_E at the western basin MODE site (shifting from secular time scales in the thermocline to mesoscale dominance at depth) and in the Gulf Stream recirculation zone (mesoscale dominance at both thermocline and abyssal depths), while the short-period oscillations on the Continental Rise of either basin are now also explicable as bottom-trapped topographic Rossby waves (Thompson 1971, 1977). The seasonal variation in K_E to abyssal depths, observed in the eastern basin, is a relatively recent finding for which an adequate theoretical basis has yet to be developed (see Dickson et al. 1982).

15.8.5 Eddy Time Scales in the Equatorial Zone

Recoveries from the broad equatorial zone are not yet adequate to support generalisations on either the vertical or horizontal distribution of eddy energy, though some geographical "system" (at regional scale) appears to be emerging for the western Indian Ocean (Fig. 11). The one systematic element of eddy climatology for the equatorial belt as a whole appears to concern the time scale of eddy motions.

The INDEX records of Luyten (1982), Knox (1981) and Fieux and Gonella (pers. comm.), the central equatorial Pacific records of Knox and Halpern (pers. comm.), and the results of Weisberg et al. (1979) from the Gulf of Guinea all appear to share elements of a common spectral shape. Typically (see Fig. 12), zonal spectra are red with maximum energy at periods ≥ 50 d and with zonal variability predominating over meridional at these long periods. Meridional spectra tend to peak at shorter periods (~ 10–50 d) and in this band, meridional variability normally predominates over zonal for records on the equator. Eriksen's (1980) model reflects this behaviour, suggesting that this characteristic is due to the fact that at low frequencies, gravity waves are practically non-existent and Kelvin waves (for which $v \equiv 0$) contribute an increasing share of the variance.

15.8.6 Relationship of K_E to K_M Worldwide

To the extent permitted by their restricted coverage, the Eulerian measurements described here do appear to confirm certain of the *general* conclusions reported by Wyrtki et al. (1976) on the basis of ship-drift analysis.

1. The "global" distribution patterns of K_E and K_M are similar; both, for example, show near-surface maxima in the vicinity of the Gulf Stream and its recirculation, at the Somali coast and in the central Drake Passage, with minima in the central and eastern subtropical gyre (North Atlantic).

2. As predicted from ship drift, K_E/K_M ratios are overwhelmingly greater than unity, not merely in the surface layer considered by Wyrtki et al., but at depth also. The principal exceptions tend to be found (though not invariably) at sites occupied by intense and/or constrained flows, e.g., the Western Boundary Undercurrent, Denmark Strait, Vema Channel, Oshima West Channel and Central Drake Passage (Figs. 3 l, m, n, 13c and 17; p. 304).

3. In the only ocean area with an adequate data density (North Atlantic) there is no consistent evidence for high K_E/K_M ratios in the interior and low ratios close to the Gulf Stream as the ship-drift evidence had seemed to suggest. On the other hand, there does appear to be some indication that values of K_E/K_M ratio are less depth-dependent immediately south of the Gulf Stream than in either the interior of the western basin or the eastern basin as a whole.

Table 1. North-east Atlantic Dynamics Study (NEADS)

Identifier, sounding (m)	Originator	Latitude/ longitude	Sampling depth (m)	Duration (days)	K_E (cm² s⁻²)	K_M (cm² s⁻²)	K_E/K_M	Comments
1　5300	IFM (Kiel)	33°00'N 22°00'W	673	510	21.6	2.1	10.3	
			1585	(170)	2.5	0.4	6.3	
			3089	510	3.5	0.4	8.8	
			4770	345	3.7	0.7	5.3	
2　5550	IFM (Kiel)	38°00'N 17°00'W	788	547	28.7	0.8	35.9	Records from first two deployments are separated by several months
			1668	381	5.5	0.40	13.8	
			3168	657	0.8	0.1	8.0	
			4181	433	0.9	0.9	1.0	
			5079	(146)	0.4	0.8	0.5	
2.5　5300	IFM (Kiel)	40°31'N 17°19'W	485	242	19.4	0.8	24.3	
			2945	(202)	0.8	0.2	4.0	
			4050	242	0.2	0.1	2.0	
3　5330	IOS	42°00'N 14°00'W	600	500	13.1	0.10	131.0	
			1500	500	4.8	0.42	11.4	
			3000	470	5.7	0.06	95.0	
			4000	700	4.9	0.46	10.7	
3.5　3700	IOS	41°30'N 20°00'W	600	230	18.4	1.62	11.4	
			950	380	6.9	0.16	43.1	
			1500	225	4.2	0.05	84.0	
			3100	(150)	1.2	0.03	40.0	
4　3634	IOS	41°00'N 25°00'W	600	470	17.1	0.30	57.0	
			1500	430	3.0	0.44	6.8	
			3500	?	4.3	0.55	7.8	
5　4756	IOS	46°00'N 17°00'W	600	450	61.5	3.08	20.0	
			1500	600	10.0	1.05	9.5	
			3000	485	2.5	0.25	10.0	
5　4760	MAFF	46°06'N 17°09'W	4050	362	1.0	0.50	2.0	
			4200	362	0.6	0.33	1.9	
			4710	307	1.2	0.55	2.2	
6(1)　4100	MAFF	52°28.7'N 17°43.9'W	3050	239	9.0	0.19	46.3	Successive deployments
			4050	239	14.4	0.88	16.4	
6(2)　4121	MAFF	52°27.8'N 17°42.0'W	3071	343	9.7	0.01	970.0	Same site
6(3)　4124	MAFF	52°27.8'N 17°42.6'W	3047	248	17.8	0.79	22.6	
			4074	363	27.1	2.79	9.7	
7　4995	COB (France)	47°00'N 10°00'W	600	268	20.1	15.0	1.3	
			1000	268	7.1	4.3	1.7	
			1500	464	4.6	0.10	46.0	
			3000	336	1.3	0.11	11.8	
			4000	732	2.1	0.12	17.5	

Sources NEADS 1, 2, 2½ – T. J. Muller (IFM, Kiel) pers. comm.; NEADS 3, 3½, 4, 5 – W. J. Gould (IOS, Wormley) pers. comm.; NEADS 5 (depths >4000 m), 6 – R. R. Dickson (MAFF, Lowestoft); NEADS 7 – A. Colin de Verdiere (COB, Brest) pers. comm.

Table 2. MONA Records

Identifier, sounding (m)	Originator	Latitude/ longitude	Sampling depth (m)	Duration (days)	K_E $(\text{cm}^2\text{s}^{-2})$	K_M $(\text{cm}^2\text{s}^{-2})$	K_E/K_M	Comments
1 947	IFM (Kiel)	60°35′N 05°09′W	856	253	55.1	?		
3 534	IFM (Kiel)	63°11′N 09°02′W	517	370	120.5	?		Filter has half power period of 30 h
5 1500	IMR (Reykjavik) Univ. Washington (Seattle)	65°17.3′N 30°42.5′W	1400	360	282.1	1571	0.2	
			1475	360	135.7	1680	0.1	
6 1102	IMR (Reykjavik) Univ. Washington (Seattle)	65°10.4′N 30°39.5′W	1002	360	342.9	1232	0.3	K_E estimates obtained by summation of K_E in bands 52–17, 17–8.3, 8.3–2 days[a]
			1077	360	148.8	1796	0.1	

Sources MONA 1 and 3 – J. Meinke (IFM Kiel) pers. comm.; MONA 5 and 6 – K. Aagaard and D. Hanzlick (U. Washington) pers. comm.

[a] Filtering carried out by spectral decomposition, deleting all components outside the desired frequency bands and reconstituting the record.

Table 3. Charlie-Gibbs Fracture Zone (originator: MAFF, U.K.; WHOI, USA)

Identifier, sounding (m)	Originator	Latitude/ longitude	Sampling depth (m)	Duration (days)	K_E $(\text{cm}^2\text{s}^{-2})$	K_M $(\text{cm}^2\text{s}^{-2})$	K_E/K_M	Comments
A 3050	MAFF	52°11.8′N 30°58.2′W	2500	247	10.7	0.16	65.8	
			3000	247	0.4	0.34	1.2	
B 4027	MAFF	52°09.3′N 31°00.2′W	2977	250	33.7	1.37	24.6	
			3977	250	9.6	5.58	1.7	
C 3577	MAFF	52°05.6′N 30°57.5′W	3027	39	6.4	0.2	32.0	Short record – data not plotted
			3527	247	2.2	3.69	0.6	
570 4311	WHOI (USA)	52°42.7′N 33°59.2′W	4227	269	21.4	8.6	2.5	
571 2895	WHOI (USA)	52°53.7′N 35°31.0′W	1007	270	14.6	8.5	1.7	
			2537	35				Short record – data not plotted
			2835	270	5.1	0.3	17.0	
572 3398	WHOI (USA)	52°46.1′N 35°30.0′W	988	270	39.1	6.4	6.1	
			2528	270	21.6	11.0	2.0	
			3060	270	24.2	7.7	3.1	
			3360	270	22.8	7.3	3.1	

Sources Moorings A, B, C – Dickson, Gurbutt and Medler (1980); Moorings 570–572 – Schmitz and Hogg (1978).

Table 4. North-east of Azores (originator: MAFF U.K.)

Identifier, sounding (m)	Latitude/ longitude	Sampling depth (m)	Duration (days)	K_E (cm²s⁻²)	K_M (cm²s⁻²)	K_E/K_M	Comments
J 3753	41°00.3'N 24°27.1'W	2703 3703	(174) 78	0.3 1.8	0.22 0.76	1.4 2.3	Short record – data not plotted
K 3395	41°01.5'N 24°04.5'W	2345 3345	257 339	0.2 0.5	0.06 0.24	3.3 2.1	
L 4096	41°00.4'N 23°17.8'W	4046	340	1.8	0.25	7.2	

Source R. R. Dickson (MAFF, Lowestoft).

Table 5a. MODE (originator: WHOI, USA) MODE Centre

Latitude range (N)	Longitude range (W)	Sampling depth range (m)	409	474	481	503	522	538	542	Duration (days)	K_E (cm²s⁻²)	K_M (cm²s⁻²)	K_E/K_M
28°00'–28°03'	69°39'–69°45'	392– 583				–		–	–	696	51.7	0.8	64.6
28°01'–28°03'	69°39'–69°45'	492– 583				–		–	–	654	51.4	1.6	32.1
28°00'–28°03'	69°39'–69°45'	392– 583				–		–	–	428•	43.3	2.1	20.6
28°01'–28°03'	69°39'–70°07'	1496–1595	–		–		–	–	–	801	7.8	0.3	26.0
28°01'–28°02'	69°39'–70°07'	1496–1595	–		–		–	–	–	704	8.1	0.6	13.5
28°01'–28°03'	69°39'–70°07'	1497–1595	–		–		–	–	–	434•	6.3	0.1	63.0
28°01'–28°03'	69°39'–70°07'	3998–4105	–				–	–	–	432•	10.0	0.2	50.0
28°00'–28°03'	69°39'–70°07'	3960–4105	–				–	–	–	755	8.7	0.0	
28°00'–28°03'	69°39'–69°45'	3960–4105	–		–		–	–	–	657	7.7	0.1	77.0
28°01'–28°03'	69°39'–70°07'	3991–4105	–				–	–	–	643	9.7	0.0	

Mooring 542 is from Polymode Array I. Plotted values are those identified by a dot in the "duration" column.

Source W. J. Schmitz Jr. (WHOI) pers. comm.

Table 5b. MODE (originator: WHOI, USA) MODE East

Latitude range (N)	Longitude range (W)	Sampling depth range (m)	Mooring identifier 430	473	482	502	521	540	Duration (days)	K_E (cm²s⁻²)	K_M (cm²s⁻²)	K_E/K_M
28°09'–28°11'	68°36'–68°42'	370– 509		—	—		—	—	328•	59.2	1.8	32.9
28°09'–28°11'	68°36'–68°40'	370– 509		—	—		—	—	225	72.8	2.7	27.0
28°09'–28°11'	68°36'–68°42'	1385–1524		—	—	—	—	—	578	4.5	0.1	45.0
28°09'–28°11'	68°36'–68°42'	1385–1524		—	—	—	—	—	417•	5.3	0.2	26.5
28°09'–28°11'	68°36'–68°42'	1385–1524		—	—	—	—	—	475	0.8	0.1	8.0
28°09'–28°11'	68°36'–68°42'	3900–4015		—	—	—	—	—	481	5.8	1.7	3.4
28°09'–28°11'	68°35'–68°42'	3875–4015	—	—	—	—	—	—	679	6.7	2.1	3.2
28°09'–28°11'	68°35'–68°42'	3875–4015	—	—	—	—	—	—	562	5.3	2.3	2.3
28°09'–28°11'	68°35'–68°42'	3875–4011	—	—	—	—	—	—	512•	8.2	2.1	3.9

Plotted values are those identified by a dot in the "duration" column.

Source W. J. Schmitz Jr. (WHOI) pers. comm.

R.R. Dickson

Table 6. POLYMODE Array I (originator: WHOI, USA)

Identifier, sounding (m)	Latitude/ longitude	Sampling depth (m)	Duration (days)	K_E (cm² s⁻²)	K_M (cm² s⁻²)	K_E/K_M	Comments
542 5462	28°01'N 69°39'W	492	268	62.2	1.6	38.9	This mooring assumed to be the one referred to as Mode Central in Table 5
		1496	(205)	8.6	1.2	7.2	
		3988	211	9.0	0.2	45.0	
543 5363	27°58'N 64°58'W	498	267	11.3	3.1	3.6	
		998	267	5.9	0.6	9.8	
		1998	267	2.0	0.5	4.0	
		3999	234	0.8	0.6	1.3	
545 6015	27°50'N 55°35'W	492	280	6.8	0.2	34.0	
		990	279	9.5	2.1	4.5	
		1986	279	1.0	0.6	1.7	
		3987	235	0.4	0.3	1.3	
546 5773	27°54'N 54°55'W	526	278	11.6	0.8	14.5	These three moorings together are often referred to as station Δ
		1024	278	9.6	2.4	4.0	
		2021	278	1.2	0.4	3.0	
		4031	233	0.5	0.9	0.6	
547 5785	28°13'N 54°57'W	496	278	8.9	2.9	3.1	
		996	279	3.9	3.1	1.3	
		1996	277	1.3	0.1	13.0	
		4000	236	0.6	0.2	3.0	
548 5550	31°02'N 60°04'W	517	275	372.4	6.1	61.0	
		2018	274	3.8	0.2	19.0	
		4018	211	1.3	0.1	13.0	
549 4687	33°59'N 60°01'W	502	265	126.1	8.7	14.5	
		1012	263	12.7	0.5	25.4	
		2012	232	5.4	4.3	1.3	
		4012	264	18.3	34.2	0.5	

Source W. J. Schmitz Jr (WHOI) pers. comm.

Table 7. POLYMODE Array II (originator: WHOI, USA)

Identifier, sounding (m)	Latitude/ longitude	Sampling depth (m)	Duration (days)	K_E (cm²s⁻²)	K_M (cm²s⁻²)	K_E/K_M	Comments
01 5062	35°54'N 55°05'W	608	298	193.3	110.5	1.7	2nd record of 74 days duration
		1011	404	68.4	55.0	1.2	2nd record of 180 days duration
		1502	524	63.2	33.9	1.9	
		4001	522	106.3	55.1	1.9	
		5000	259	112.7	40.4	2.8	3rd deployment only
02 5345	35°56'N 54°42'W	563	292	243.7	102.5	2.4	2nd deployment only
		974	509	65.6	37.6	1.7	
		1475	509	58.5	33.0	1.8	
		3966	512	97.3	94.1	1.0	
03 5128	35°36'N 55°05'W	669	519	157.6	35.6	4.4	1st deployment only
		1040	219	71.9	15.5	4.6	
		1573	503	58.6	27.3	2.1	
		4068	519	90.9	83.1	1.1	
		5006	255	95.9	62.1	1.5	3rd deployment only
04 5469	35°58'N 53°46'W	595	463	276.1	17.7	15.6	3rd deployment only
		991	246	63.3	4.0	15.8	2nd deployment only
		1498	463	62.1	1.7	36.5	
		3999	423	89.5	0.9	99.4	1st record of 177 days duration
05 5508	34°55'N 55°05'W	620	520	198.7	1.6	124.2	
		1021	520	58.6	0.2	293.0	
		1523	290	47.4	0.9	52.7	2nd record of 71 days duration
		4024	520	70.3	0.1	703.0	
06 5204	35°56'N 59°02'W	612	215	226.7	32.8	6.9	1st deployment only
		1014	501	69.9	0.2	349.5	
		1515	501	40.1	1.0	40.1	
		4013	369	85.5	10.6	8.1	
07 5493	31°35'N 54°59'W	609	380	49.4	0.8	61.8	1st record of 83 days duration
		1011	519	19.3	0.4	48.3	2nd record of 167 days duration
		1513	461	12.2	0.6	20.3	1st deployment in two parts, one of 37 days duration
		1976	243	6.0	0.6	10.0	3rd deployment only

Table 7. (continued)

Identifier, sounding (m)		Latitude/longitude	Sampling depth (m)	Duration (days)	K_E (cm² s⁻²)	K_M (cm² s⁻²)	K_E/K_M	Comments
08	5330	37°29'N 55°00'W	2483	243	6.0	0.5	12.0	3rd deployment only
			4013	519	11.8	0.9	13.1	
			605	516	288.2	66.1	4.4	
			1006	516	130.3	40.3	3.2	
			1508	516	104.8	30.9	3.4	
			4002	516	111.1	34.8	3.2	
09	5344	38°30'N 54°56'W	3991	516	158.6	2.0	79.3	
			5246	265	73.9	5.0	14.8	3rd deployment only
10	5270	39°29'N 55°00'W	3982	516	112.0	12.8	8.8	
11	5174	40°27'N 55°02'W	3978	514	86.2	16.1	5.4	
12	4768	41°29'N 54°59'W	3972	513	58.6	1.0	58.6	
13	5445	36°30'N 55°00'W	654	123	106.1	18.7	5.7	Short record – data not plotted
			1054	261	92.6	17.1	5.4	3rd deployment only
			1553	261	84.0	25.3	3.3	
			4049	261	106.1	75.7	1.4	
14	5487	35°15'N 55°00'W	596	257	408.0	45.7	8.9	
			999	257	74.8	16.8	4.5	3rd deployment only
			1498	257	49.7	13.3	3.7	
			3998	257	72.6	5.5	13.2	
15	5457	35°57'N 55°28'W	3997	235	116.4	30.1	3.9	3rd deployment only

NB: To obtain record lengths comparable with other plotted statistics, data from the third deployment of Polymode Array II have been ignored, except where it is the only data available.

Source W. J. Schmitz Jr (WHOI) pers. comm.; Tarbell, S., A. Spencer and R. E. Payne (1978).

Table 8. POLYMODE Array III Cluster A (originator: WHOI, USA)

Identifier, sounding (m)	Latitude/ longitude	Sampling depth (m)	Duration (days)	K_E $(cm^2 s^{-2})$	K_M $(cm^2 s^{-2})$	K_E/K_M	Comments
628	27°25.6'N	505	31	2.5	0.65	3.8	Short record – data not plotted
4961	47°50.0'W	1489	299	1.7	1.29	1.3	
		3994	(184)	1.0	2.68	0.4	
629	28°01.0'N	203	336	57.3	6.28	9.1	
4954	48°03.3'W	1500	337	1.9	0.52	3.6	
		4006	336	1.5	0.03	50.0	
630	27°51.7'N	200	335	60.4	0.08	755.0	
4895	48°39.4'W	1498	335	1.6	0.60	2.7	
631	27°55.8'N	212	334	58.4	2.54	23.0	
5106	48°52.1'W	4016	335	1.1	0.61	1.8	
632	26°51.8'N	190	333	43.4	23.60	1.8	
4881	49°13.5'W	1488	(164)	2.1	2.30	0.9	
		3993	333	0.7	0.07	10.0	

Source Fu and Wunsch (1979)

Table 9. POLYMODE Array III Cluster B (originator: WHOI, USA)

Identifier, sounding (m)	Latitude/ longitude	Sampling depth (m)	Duration (days)	K_E (cm^{2}s^{-2})	K_M (cm^{2}s^{-2})	K_E/K_M	Comments
623 4251	27°24.8'N 41°07.7'W	128 1426 3927	345 345 345	74.3 3.3 1.2	6.74 0.83 0.07	11.0 4.0 17.1	
624 4372	27°17.5'N 40°45.5'W	529 1528 4028	345 345 (173)	36.1 3.4 0.6	2.54 0.80 0.09	14.2 4.3 6.7	
625 4723	27°14.5'N 40°21.1'W	189 1488 3990	344 343 343	77.4 4.2 0.4	2.13 0.63 0.28	36.3 6.7 1.4	
626 4315	26°52.7'N 41°12.8'W	215 1514 4014	100 342 342	38.5 4.4 1.6	7.90 0.80 0.66	4.9 5.5 2.4	Short record – data not plotted
627 3857	26°09.8'N 41°40.7'W	206 1505	341 341	68.0 4.9	0.35 0.23	194.3 21.3	

Source Fu and Wunsch (1979).

Table 10. POLYMODE Array III Cluster C (originator: WHOI, USA)

Identifier, sounding (m)[a]	Latitude/ longitude	Sampling depth (m)	Duration (days)	K_E (cm²s⁻²)	K_M (cm²s⁻²)	K_E/K_M	Comments
A ~5400	15°02.1'N 54°12.9'W	194	353	28.7	17.1	1.7	
		338	353	23.8	7.3	3.3	
		538	353	31.4	2.5	12.6	
		2538	353	5.8	0.1	58.0	
		4038	353	4.9	1.7	2.9	
B ~5300	15°23.4'N 53°55.2'W	319	355	20.1	11.5	1.7	
		520	355	20.2	2.7	7.5	
		2520	141	–	–	–	short record – data not plotted
		4020	355	4.4	0.3	14.7	
C ~5200	15°11.5'N 53°12.3'W	160	353	38.0	25.3	1.5	
		309	222	10.9	0.6	18.2	
		510	352	39.4	0.3	131.3	
		2508	353	4.7	0.2	23.5	
		4008	353	2.8	0.5	5.6	
D ~5400	16°41.3'N 54°20.4'W	172	354	28.5	18.8	1.5	
		322	354	19.9	9.7	2.1	
		522	354	15.0	2.8	5.4	
		2446	18	–	–	–	short record – data not plotted
		3946	354	14.3	2.7	5.3	

[a] Soundings read from chart, therefore only approximate values. *Source* Koblinsky, Keffer and Niiler (1979)

Table 11. Gulf Stream (originator: BIO, Canada)

Identifier, sounding (m)	Latitude/ longitude	Sampling depth (m)	Duration (days)	K_E ($cm^2 s^{-2}$)	K_M ($cm^2 s^{-2}$)	K_E/K_M	Comments
A ?	40°30'N 55°30'W	4000	?	135	?	?	Duration ~18 months
B ?	38°00'N 50°00'W	4000	?	60	?	?	Duration ~12 months

NB: Preliminary data only. *Source* G. T. Needler (B.I.O.) pers. comm.

Table 12. Western Boundary Undercurrent (originator: WHOI, USA)

Identifier sounding (m)	Latitude/ longitude	Sampling depth (m)	Duration (days)	K_E ($cm^2 s^{-2}$)	K_M ($cm^2 s^{-2}$)	K_E/K_M
616 2993	30°59.9'N 76°39.0'W	1995	356	29.3	6.0	4.9
		2769	356	18.3	28.0	0.7
617 3801	30°31.0'N 75°05.5'W	601	356	221.1	55.1	4.0
		2002	356	6.7	18.7	0.4
		3602	356	30.4	111.2	0.3
618 4002	30°43.2'N 74°10.4'W	2002	353	7.1	12.5	0.6
		3003	353	14.8	57.7	0.3
		3802	353	33.6	221.1	0.2
620 5187	31°03.5'N 73°23.5'W	1958	351	30.7	7.7	4.0
		2958	351	47.9	12.2	3.9
		4967	351	64.5	5.4	11.9

Source P. Rhines (WHOI), pers. comm.

Table 13. Rise Array (originator: WHOI, USA)

Identifier, sounding (m)	Latitude/ longitude	Sampling depth (m)	Duration (days)	K_E (cm²s⁻²)	K_M (cm²s⁻²)	K_E/K_M	Comments K_E	K_M
517 2647	39°11.8'N 70°00'W	193	359	330.8	21.5	15.4	—	—
		197	359	311.7	19.7	15.8	—	—
523 2504	39°25.6'N 69°59.6'W	181	236	332.3	12.9	25.8	—	—
		983	237	10.8	8.4	1.3	9.7	8.8
		1991	237	10.1	3.4	3.0	8.5	3.4
524 2664	39°07.5'N 69°59.9'W	197	236	258.0	35.7	7.2	—	—
		202	236	223.6	37.0	6.0	—	—
		496	236	23.7	14.6	1.6	—	—
		1005	236	11.5	9.3	1.2	10.2	9.3
		2013	235	11.0	4.6	2.4	8.9	4.6
		2512	235	18.8	3.3	5.7	14.4	3.4
525 2759	39°07.1'N 70°32.6'W	195	236	223.1	74.4	3.0	—	—
		997	236	10.8	7.5	1.4	9.9	7.8
		2005	235	13.2	2.7	4.9	11.3	3.0
526 3007	38°47.0'N 70°00.5'W	2006	234	21.4	8.0	2.7	18.5	7.9
		2810	235	40.3	7.3	5.5	32.5	7.5
527 2978	39°09.8'N 68°59.8'W	1977	235	16.8	9.7	1.7	14.9	9.8
		2781	232	25.4	9.3	2.7	21.0	9.2
528 3326	38°35.2'N 69°10.1'W	2329	235	25.9	9.6	2.7	22.4	9.3
529 3480	38°21.4'N 69°59.6'W	2483	234	32.6	6.1	5.3	30.0	5.9
		3283	234	51.8	13.1	4.0	46.2	12.6
530 3815	38°00.5'N 70°00.6'W	2818	241	41.4	1.9	21.8	39.3	1.6
531 3921	38°00.2'N 69°18.5'W	2925	240	35.3	0.1	353.0	31.5	0.1
		3724	239	55.1	3.5	15.7	48.4	3.2
532 4210	37°29.8'N 69°19.9'W	3213	240	58.2	1.5	38.8	53.5	1.8

See note at foot of table

Table 13. (continued)

Identifier, sounding (m)	Latitude/ longitude	Sampling depth (m)	Duration (days)	K_E (cm²s⁻²)	K_M (cm²s⁻²)	K_E/K_M	Comments K_E	K_M
533	37°30.3'N	3182	240	58.5	3.1	18.9	56.1	3.6
4182	70°00.4'W	3981	240	61.9	4.9	12.6	57.8	4.8
534	37°00.4'N	3337	241	69.3	8.7	8.0	67.1	8.7
4339	69°59.8'W	4138	241	87.7	9.3	9.4	83.6	8.9
535	36°59.3'N	3453	239	66.2	7.5	8.8	59.7	7.3
4450	69°19.7'E							
536	36°30.1'N	3466	238	70.0	3.7	18.9	63.3	3.8
4468	69°19.9'W	4267	238	83.7	3.5	23.9	74.8	3.7
537	36°29.8'N	4262	239	57.0	9.8	10.3	53.9	10.3
4463	70°00.0'W							

Slightly different K_E estimates for the Rise Array are listed by Thompson, 1977 (Prog. Oceanog 7, (4): Table 2) and Luyten 1977 (J. Mar. Res. 35, (1): p 58). Thompson's estimates are shown above under "comments" while Luyten's more comprehensive set is listed in columns 5 and 6. The plotted values are a mixture of the two. As in earlier versions of these charts, Thompson's estimates are used below 500 m but are updated with Luyten's values at shallower depths.

Sources Mooring 517 – W. J. Schmitz Jr (WHOI) pers. comm.
All other moorings – Luyten (1977); Thompson (1977).

Table 14a. Local Dynamics Experiment (originator: WHOI, USA)

Identifier, sounding (m)	Latitude/ longitude	Sampling depth (m)	Duration (days)	K_E (cm²s⁻²)	K_M (cm²s⁻²)	K_E/K_M	Comments
640	31°01'N	269	444	152	9.9	15.4	This mooring is the only one from the LDE array for which data are available at present. The K_E values were read off a graph and are therefore approximate. [Owens W.B., Luyten J.R. and Bryden H. 1982, "Moored velocity measurements during the POLYMODE Local Dynamics Experiment".]
5355	69°30'W	516	444	100	7.8	12.8	
		616	444	77	6.5	11.8	
		839	444	38	2.9	13.1	
		2008	307	25	0.1	250.0	
		3004	295	29	0.2	145.0	
		5250	444	31	2.3	13.5	

Source Owens, Luyten and Bryden (1982).

Table 14b. Soviet POLYMODE Array (originator: PP Shirshov Inst. of Oceanol., Moscow)

Identifier, sounding (m)	Latitude/ Longitude	Sampling depth (m)	Duration (days)	K_E $(cm^2 s^{-2})$	K_M $(cm^2 s^{-2})$	K_E/K_M	Comments
1	29°00′N 70°00′W	700 1400	344 330	38 33	1.79 2.23	21.1 14.7	See Footnotes
2	29°39′N 70°00′W	700 1400	374 347	64 34	9.63 0.92	6.6 37.4	
3	29°19.5′N 70°39′W	700 1400	362 334	65 37	15.85 6.24	4.1 6.0	
4	28°40.5′N 70°39′W	700 1400	343 346	36 22	0.13 0.33	274.6 65.4	
5	28°21′N 70°00′W	700 1400	376 374	33 21	2.82 2.53	11.7 8.1	
6	28°40.5′N 69°21′W	700 1400	396 349	53 16	3.80 0.34	14.0 46.5	
7	29°19.5′N 69°21′W	700 1400	373 357	44 27	3.27 0.79	13.2 34.4	
8	29°58.5′N 69°21′W	700 1400	340 401	46 23	4.92 0.52	9.2 45.0	
9	30°18′N 70°00′W	700 1400	344 255	73 47	16.01 4.12	4.5 11.2	
10	29°58.5′N 70°39′W	700 1400	316 314	84 41	6.63 1.93	12.7 20.8	
11	29°38′N 71°18′W	700 1400	331 266	87 25	8.66 2.76	10.0 9.1	
12	29°00′N 71°18′W	700 1400	364 317	77 30	4.29 4.44	17.6 6.8	
13	28°21′N 71°18′W	700 1400	361 322	50 17	5.59 2.81	8.8 6.0	
14	28°01.5′N 70°39′W	700 1400	398 405	42 12	4.71 0.17	8.9 70.2	
15	27°42′N 70°00′W	700 1400	353 281	44 21	3.42 0.34	12.8 59.6	
16	28°01.5′N 69°21′W	700 1400	378 363	57 28	1.40 2.54	40.5 11.1	
17	28°21′N 68°42′W	700 1400	357 363	37 14	1.26 0.37	29.2 37.7	
18	29°00′N 68°42′W	700 1400	308 268	65 16	0.35 1.50	186.4 10.6	
19	29°39′N 68°42′W	700 1400	314 309	30 19	0.08 0.60	372.3 31.2	

Sources Polloni, C., A. Mariano, T. Rossby (1981); T. Rossby (U R I) pers. comm.

Footnotes:

1 These moorings employed surface flotation with a scope (cable length/depth) of 1.05, and were recovered and reset on a monthly basis. Because of known inaccuracies in flow measurements using Savonius rotor current meters on a surface-buoyed mooring, an intercomparison was made between these data and contemporaneous SOFAR float data, using floats passing within 20 km of a mooring (Polloni, Mariano and Rossby, 1981). This showed that the current meter speeds were high by a factor of 1.75 ± 0.4, though there was no bias in direction. For this reason, *current speeds have been divided by 1.75 before computing the K_E estimates listed above and in Fig. 3p* (T Rossby, pers. comm.). A similar procedure was carried out by the present author in the calculation of the accompanying K_M estimates (Data from N.O.D.C.).

2 The period of the Soviet Polymode Array (August 1977–August 1978) coincided partly with the US Polymode Local Dynamics Experiment immediately to the north (31°N 69°30′W; Table 14a).

Table 15. Canadian Continental Slope (originator: BIO, Canada)

Identifier, sounding (m)	Latitude/ longitude	Sampling depth (m)	Duration (days)	K_E (cm²s⁻²)	K_M (cm²s⁻²)	K_E/K_M	Comments
a) Labrador Slope							
245	58°27'N	500	(166)	83.4	18.94	4.4	Short record – data not plotted
3002	56°49'W	2900	(156)	28.5	26.10	1.1	
247	57°12'N	1200	(167)	16.0	6.4	2.5	
1306	58°45'N						
250	57°19'W	100	(170)	133.1	554.5	0.2	
600	59°10'W	500	(170)	36.3	40.5	0.9	
b) Nova Scotia Slope							
S3	42°45'N	500	92	18.4	4.39	4.2	
700	63°30'W	690	257	3.6	0.51	7.1	
S7	42°43'N	690	301	5.6	1.01	5.5	
700	64°00'W						
S4	42°40'N	500	270	14.9	3.69	4.0	
1000	63°30'W	690	(200)	12.7	3.56	3.6	
		990	406	11.3	0.97	11.6	
S5	42°30'N	1534	365	9.1	3.02	3.0	
1554	63°30'W						
S8	42°00'N	1500	299	17.4	1.81	9.6	
2550	63°30'W	2530	(196)	13.4	7.03	1.9	

Source a) R. A. Clarke (BIO) pers. comm.
b) P. C. Smith (BIO) pers. comm.

Table 16. North-east Atlantic, miscellaneous

Identifier, sounding (m)	Originator	Latitude/longitude	Sampling depth (m)	Duration (days)	K_E (cm²s⁻²)	K_M (cm²s⁻²)	K_E/K_M	Comments
N 4754	MAFF	50°42.7'N 17°00.8'W	3984 / 4704	330 / 363	12.2 / 11.2	1.71 / 3.21	7.1 / 3.5	
O 4820	MAFF	49°10.4'N 15°44.6'W	3990 / 4770	333 / 298	5.6 / 5.6	2.61 / 2.41	2.1 / 2.3	
P 4250	MAFF	47°59.5'N 14°06.4'W	3750 / 4200	357 / 357	1.2 / 2.0	0.26 / 0.55	4.7 / 3.6	
Q 4751	MAFF	47°13.6'N 12°12.8'W	4701	349	4.7	0.13	35.8	
R 2099	MAFF	48°59.2'N 12°52.6'W	2049	354	6.0	2.49	2.4	
TOURBILLON 8 4690	MAFF	46°37.3'N 15°24.4'W	3000 / 4000	268 / 268	0.8 / 1.6	0.90 / 3.55	0.8 / 0.4	Statistics ignore first 19 d of data (suspected wave contamination)
TOURBILLON 9 4780	MAFF	46°35.2'N 14°14.5'W	1500 / 3000	(199) / (199)	3.2 / 0.9	5.22 / 1.36	0.6 / 0.6	Statistics ignore first 87 d of data (suspected wave contamination)
I4 1800	IOS	58°50'N 11°40'W	200 / 600 / 1000 / 1500	494 / 468 / 423 / 433	196 / 151 / 75 / 42	10.1 / 9.4 / 6.8 / 1.2	19.4 / 16.1 / 11.0 / 35.0	Uses data interpolated from nearby moorings
5 5521	NAVOCEANO	36°39.0'N 18°40.8'W (mean of two deployments)	102 / 188 / 203 / 289 / 456 / 541 / 1045	168 / 103 / 168 / 103 / (168) / (97) / 79	75.1 / 44.4 / 57.1 / 40.6 / 27.4 / 20.0 / 3.2	28.69 / 10.49 / 25.66 / 9.59 / 20.66 / 9.20 / 3.78	2.6 / 4.2 / 2.2 / 4.2 / 1.3 / 2.2 / 0.8	Data not plotted[a] / combined before plotting[a] / data not plotted[a]
6, 7 5543	NAVOCEANO	36°28.7'N 18°31.2'W	92 / 179	162 / 112	61.8 / 55.4	27.72 / 2.68	2.2 / 20.7	data not plotted[a]

Table 16. (continued)

Identifier, sounding (m)	Originator	Latitude/ longitude	Sampling depth (m)	Duration (days)	K_E (cm²s⁻²)	K_M (cm²s⁻²)	K_E/K_M	Comments
		(mean of two deployments)	193	162	50.0	31.08	1.6	combined before plotting[a]
			446	(162)	24.0	26.46	0.9	
			532	(112)	41.8	8.38	5.0	
			1036	106	8.7	4.07	2.1	data not plotted[a]
CV1 5200	COB/CNEXO	24°49.6'N 25°02.9'W	5190	(178)	1.8	0.46	3.83	
CV2 4885	COB/CNEXO	19°14.0'N 29°47.7'W	4785	(185)	1.0	0.84	1.18	
			4875	(185)	1.2	1.12	1.04	

Sources Moorings N, O, P, Q, R – R.R. Dickson (MAFF, Lowestoft); Tourbillon 8 and 9 – J.W. Ramster (MAFF, Lowestoft); Mooring 14 – W.J. Gould (IOS, Wormley) pers. comm.; Moorings 5 and 6/7 – U.S. Navy NSTL Station, Bay St. Louis, Miss.; CV1 and CV2 – A Vangriesheim, (COB, Brest) pers. comm.

[a] Data from two of the listed moorings (Table 16, Identifiers 5 and 6/7) were omitted from K_M plots through a suspected wave-contamination of the records. Flotation for the first setting of these moorings was close to the surface (90–100 m depth) and gave K_M values an order of magnitude larger than for the second setting, where subsurface flotation was at 170–180 m depth. As a result, the K_M for the full record is itself over an order of magnitude larger than surrounding values in this region and depth layer. K_E statistics may also be affected.

Table 17. Comparison of eddy kinetic energy estimates close to the Kuroshio and Gulf Stream

Record	Depth range (m) (number of observations)	K_E range (cm²s⁻²)
Kuroshio	1803–2116 (2)	9–18
Rise Array	1977–2013 (5)	10–21
WBUC	1958–2002 (4)	7–31
Kuroshio	3690 (1)	4
Rise Array	3337–3981 (5)	55–70
WBUC	3602–3802 (2)	30–34
Kuroshio	4273 (1)	17
Rise Array	4138–4267 (3)	57–88
WBUC	4967 (1)	65

Table 18. North Pacific

Identifier sounding (m)	Latitude/ longitude	Sampling depth (m)	Duration (days)	K_E $(\text{cm}^2\,\text{s}^{-2})$	K_M $(\text{cm}^2\,\text{s}^{-2})$	K_E/K_M	Comments
a) Kuroshio							
1 2266	31°32.2′N 132°28.2°E	2116	74	8.6	0.4	21.50	
4 1953	32°49.0′N 134°43.8′E	1803	103	17.8	2.0	8.90	Energy statistics refer to periods > 1d
7 3840	33°03.0′N 136°34.5′E	3690	64	4.3	4.2	1.02	
8 4423	32°38.6′N 136°32.4′E	4273	30	17.3	11.3	1.53	
b) Kuroshio (Izu ridge)							
3 (HW) 1795	33°02′N 139°00′E	1670	271	34.7	2.9	11.97	Energy values calculated for periods > 30 hours
5 (OW) 466	34°44′N 139°14′E	250	286	78.5	1457.1	0.05	
c) Western Pacific							
TA 5845	30°00′N 145°45′E	5795	167	5.5	34.0	0.16	Energy values calculated for periods > 30 hours
RB 6240	30°00.1′N 147°08.6′E	4000 4500 5000 5180	691 215 415 246	7.6 7.5 8.3 7.9	0.10 0.05 0.02 0.04	76.20 150.00 415.00 197.50	Energy values calculated for periods > 1 day
RC 6180	30°49.8′N 146°40′E	4000	413	15.4	0.83	18.55	
d) East-central Pacific							
M11 5940	30°28.8′N 158°12.3′W	5840	267	3.7	0.5	7.40	
M12 5882	30°47.1′W 157°56.8′W	5782	270	2.0	0.3	6.67	High frequency cut off at 0.04 cph
M14 5867	30°46.5′N 157°41.3′W	5767	267	1.8	0.5	3.60	
M16 5838	30°28.8′N 156°45.6′W	5738	262	2.1	6.4	0.33	
M21 5871	30°25.5′N 158°16.0′W	5771	285	0.6	0.1	6.00	
M23 5927	31°00.1′N 157°52.3′W	5827	283	0.7	0.5	1.40	High frequency cut off at 0.04 cph
M25 5820	30°23.2′N 157°21.4′W	5720	283	0.4	0.6	0.67	
e) Central and Eastern Tropical Pacific							
A 4740	8°28′N 150°49′W	4710	143	2	0.1	20	High frequencies removed by a Gaussian low pass filter with a half-power point at 0.5 f, where f = local inertial frequency
B 4755	11°42′N 138°24′W	4725	197	11	2	5.5	
C′ 4508	14°38′N 125°29′W	4308 4458 4478 4502	156 156 156 156	8.0 10.0 10.0 8.0	0.2 1.4 1.8 1.9	40.0 7.14 5.56 4.21	

Source a) Taft (1978).
    b) K. Taira (O.R.I., U. Tokyo) pers. comm.
      Mooring TA – K. Taira (O.R.I., U. Tokyo) pers. comm.
    c) Moorings RB, RC – S. Imawaki and K. Takano (Geophys. I., U. Kyoto) pers. comm.
    d) Taft, Ramp, Dworski and Holloway (1981).
    e) Hayes (1982).

Table 19. Equatorial Zone

Identifier, sounding (m)	Latitude longitude	Sampling depth (m)	Duration (days)	K_E ($cm^2 s^{-2}$)	K_M ($cm^2 s^{-2}$)	K_E/K_M	Comments
a) Somali current							
M1	02°22'S	85	269	885.8	515.2	1.7	
	41°16'E	138	269	253.6	21.7	11.7	
		196	269	353.5	1.1	321.4	
M2	01°52'S	139	268	389.9	6.8	57.3	
	41°34'E	140	268	396.2	8.5	46.6	
		194	268	329.0	3.0	109.7	
		245	268	23.8	0.5	47.6	
175	04°55'N	137	228	1029.3	240.2	4.3	
	48°47'E	196	110	743.9	33.3	22.3	
		246	228	366.7	93.9	3.9	
b) Index (WHOI)							
593	00°03'N	203	201	263.9	51.09	5.2	
5082	50°28'E	1500	231	58.2	2.21	26.4	
594	00°01'N	201	227	239.1	24.75	9.7	
5074	52°59'E	2508	227	26.9	1.18	22.8	
		3544	227	25.4	1.89	13.5	
595	01°30'N	1500	229	53.7	5.53	9.7	
5117	53°00'E	3542	197	16.2	2.00	8.1	
596	00°00'N	550	223	102.5	12.02	8.5	
4711	57°00'E	1550	223	30.4	8.45	3.6	
		3595	223	7.3	0.21	35.6	
c) Index (SIO)							
1	00°00'N	500	171	165.7	30.90	5.4	
	55°41'E						
2	00°00'N	516	189	111.6	34.52	3.2	
	55°41'E						
d) Index (MNHN)							
A1	00°00'S	160	364	331.3	68.64	4.8	
~4800	61°56.0'E	200	361	324.2	42.67	7.6	
		500	362	177.9	0.51	348.9	
		750	350	95.1	0.42	226.5	Combination of three records with short break between
e) Central Pacific							
1	00°40'N	15	405	303.0	116.9	2.6	
4500	153°00'W						
2	00°00'N	100	405	261.1	4393.6	0.1	
4465	152°00'W	250	405	106.8	119.2	0.9	

Sources a) F. Schott (U. Miami) pers. comm.
        b) Luyten (1982).
        c) Knox (1981).
        d) M. Fieux (Mus. Nat. d'Hist. Natur., Paris) pers. comm.
        e) R. Knox (SIO) and D. Halpern (PMEL, Seattle) pers. comm.

Table 20. Vema Channel (originator: WHOI, USA)

Identifier sounding (m)	Latitude/ longitude	Sampling depth (m)	Duration (days)	K_E ($cm^2 s^{-2}$)	K_M ($cm^2 s^{-2}$)	K_E/K_M	Comments
684/692	30°56.7'S	3490	465	7.2	9.92	0.73	Compass fault. Data not plotted
4009	39°53.9'W	3911	465	12.9	52.53	0.24	
681/689	30°47.2'S	3941	468	25.4	26.92	0.94	
4670	39°32.1'W	4275	468	36.4	239.56	0.15	
		4575	356	31.0	240.73	0.13	
682/690	30°45.2'S	4001	356	15.8	2.35	7.00	
4101	39°27.4'W						
685/688	30°39.9'S	3971	467	15.3	44.57	0.34	
4675	39°49.0'W	4326	467	14.3	77.23	0.19	
		4627	451	7.8	67.35	0.12	

Source: N G Hogg (WHOI) pers. comm.

Table 21. Antarctic Circumpolar Current

	Identifier sounding (m)	Latitude/ longitude	Sampling depth (m)	Duration (days)	K_E $(cm^2 s^{-2})$	K_M $(cm^2 s^{-2})$	K_E/K_M
a) i) Drake Passage 1975	2 3871	57°03.9'S 66°05.7'W	2771	291	159.2	0.6	265.3
	4 3137	57°46.8'S 64°54.0'W	2837	342	19.1	5.0	3.8
	8 3841	59°09.3'S 64°00.0'W	2741	351	17.3	7.7	2.2
	10 3569	59°46.8'S 63°19.0'W	1019 1519 2519	228 352 352	41.3 25.9 13.3	42.8 24.9 17.6	1.0 1.0 0.8
	12 3729	60°23.5'S 63°36.5'W	2604	254	13.4	10.6	1.3
	14 3617	61°03.1'S 61°52.5'W	2667	332	19.9	15.3	1.3
a) ii) Drake Passage 1976	D 3813	57°01.0'S 66°12.0'W	2913	310	118.1	5.7	20.7
	76 3940	59°03.7'S 63°29.4'W	2791	293	34.4	13.0	2.6
	A 3765	59°08.8'S 63°55.6'W	581 2781	253 241	149.7 32.6	80.0 10.3	1.9 3.2
	E 3683	61°02.8'S 61°52.8'W	461 2661	341 308	28.8 12.5	48.6 9.8	0.6 1.3
a) iii) Drake Passage 1977	N	58°55'S 64°06'W	635 1145 2159	311 (184) 300	107.1 40.8 20.8	247.2 127.9 37.7	0.4 0.3 0.6
	E	59°03'S 63°38'W	365 1079 2093	240 258 296	187.5 54.0 21.4	288.9 112.8 37.0	0.6 0.5 0.6
	C	59°06'S 63°50'W	282 586 2010 3521	273 280 (187) 259	205.4 128.5 23.4 33.3	365.4 249.1 46.5 14.5	0.6 0.5 0.5 2.3
	W	59°16'S 64°31'W	550 1060	(171) 266	120.0 43.4	236.0 96.5	0.5 0.4
	S	59°13'S 63°45'W	357 2086	287 253	204.9 20.8	254.4 27.3	0.8 0.8
a) iv) Drake Passage 1978	Y	59°04'S 63°40'W	416 927 1947	(187) (169) (182)	98.9 51.0 22.7	297.0 127.1 46.4	0.3 0.4 0.5
b) Ridge Array	M 5265	49°23.7'S 169°57.0'W	2074 4038 4986	(207) (207) (207)	73.3 30.6 30.1	4.29 0.85 8.24	17.1 36.0 3.7
	N 5240	49°12.5'S 169°53.6'W	2013 3972 4951	(190) (190) (190)	95.8 37.1 35.2	3.51 1.02 1.83	27.3 36.4 19.2
	W 5070	49°22.1'S 170°30.7'W	2040 4012	216 216	59.4 35.7	0.76 0.20	78.2 178.5
	S 5310	49°41.9'S 170°09.6'W	2065	215	32.6	0.03	1086.7
	K 5125	49°20.9'S 170°19.0'W	4000	(152)	32.7	4.76	6.87
c) Weddell Sea	A 4540	66°29.3'S 41°02.6'W	4490	349		0.85	Unfiltered $K_E = 3.61$ $cm^2 s^{-2}$

Sources a) i)–iii) Nowlin, Pillsbury and Bottero (1981).
      b)   H. Bryden (WHOI) pers. comm.
      c)   Middleton and Foster (1977).

Table 22. West Spitzbergen Current

Identifier sounding (m)	Latitude/ longitude	Sampling depth (m)	Duration (days)	K_E[a] $(cm^2 s^{-2})$	K_M $(cm^2 s^{-2})$	K_E/K_M	Comments
1A	78°58.8'N	98	413	64.1			
1248	06°54.1'E	398	415	32.0			
1B	78°57.0'N	398	416	34.3			
1198	07°38.0'E						
1C	78°57.8'N	92	413	77.0			
1142	07°53.6'E	392	412	50.7			
1D	78°59.2'N	93	398	49.6			
343	08°32.1'E						
2B	78°57.2'N	110	348	124.0			
1210	07°35.6'E	410	288	85.6			
2C	78°58.3'N	100	348	117.9			
1150	07°50.6'E	400	348	71.3			
3C	78°58.8'N	100	307	139.9			
1150	07°50.0'E	400	291	73.5			

Source K. Aagaard and D. Hanzlick (U. Washington) pers. comm.

[a] Filtering carried out by spectral decomposition, deleting all components outside the desired frequency bands and reconstituting the record.

Table 23. Late Data Atlantic (N.B. The statistics are included here for completeness but are neither discussed in the text nor plotted in Fig. 3)

Identifier, sounding (m)	Latitude longitude	Sampling depth (m)	Duration (days)	K_E (cm²s⁻²)	K_M (cm²s⁻²)	K_E/K_M	Comments
a) Biscay Rise (MAFF)							
80-02 4395	47°15.5'N 09°58.2'W	595	347	24.85	7.23	3.44	
		995	347	11.90	3.66	3.25	
		1495	347	4.77	2.03	2.35	
		2995	344	0.88	0.18	4.89	
		3995	347	1.73	0.13	13.31	
80-05 2030	48°05.1'N 09°50.1'W	980	343	19.02	15.64	1.22	
80-06 1640	48°08.3'N 09°45.1'W	790	343	34.74	31.71	1.10	
		1590	343	2.27	1.21	1.88	
80-07 640	48°11.5'N 09°39.8'W	330	286	16.12	1.70	9.48	
		590	286	8.23	21.85	0.38	
80-08 1465	48°07.4'N 09°17.0'W	715	344	11.06	12.37	0.89	
		1415	344	6.30	19.59	0.32	
b) Topographic Study Array							
80-10 4025	45°54.8'N 16°31.4'W	3975	(185)	5.50	16.11	0.34	
80-11 4349	45°50.1'N 16°35.8'W	3299	(185)	1.60	5.01	0.32	
		4299	(185)	3.09	2.53	1.22	
80-12 4280	45°54.4'N 16°37.2'W	2730	257	1.75	4.66	0.38	*3480 m record:* suspect intermittent encoder fault after 135 d. Statistics for the full 257 d record are 1.38, 4.47 and 0.31 respectively
		3480	(135)	0.82	3.69	0.22	
		4230	257	2.32	4.83	0.48	
c) Kings Trough (MAFF)							
80-14 4108	42°25.5'N 20°35.1'W	3058	370	1.71	0.20	8.55	
		4058	370	5.03	0.38	13.24	
80-15 3840	41°44.9'N 21°57.0'W	600	369	12.39	1.32	9.39	
		1500	369	4.71	0.53	8.89	
		3000	369	3.32	1.68	1.98	
		3790	369	3.16	3.20	0.99	
80-16 3568	41°38.6'N 21°08.7'W	3518	367	3.71	3.31	1.12	

Table 23. (continued)

Identifier, sounding (m)	Latitude/ longitude	Sampling depth (m)	Duration (days)	K_E (cm²s⁻²)	K_M (cm²s⁻²)	K_E/K_M	Comments
d) East of Mid Atlantic Ridge (MAFF)							
81-03 3991	46°49.0'N 23°46.2'W	2869 3941	(113) (113)	1.97 2.72	0.33 5.37	5.97 0.51	
81-04 4530	47°07.6'N 21°43.6'W	3369 4480	(111) (111)	1.03 1.37	2.20 2.27	0.47 0.60	
81-05 4540	47°26.9'N 20°11.1'W	3353 4490	(111) (111)	1.86 2.85	3.35 4.72	0.56 0.60	
81-06 4505	47°55.8'N 18°32.5'W	3566 4455	(110) (110)	1.47 3.57	1.60 6.28	0.92 0.57	
e) NEADS-6 (MAFF)							
80-01 4187	52°25.1'N 17°44.8'W	1637 3137 4137	(182) 365 365	9.78 21.05 42.64	4.60 0.03 0.58	2.13 701.67 73.52	Main buoyancy lost after 185 d. Lower instruments supported by backup buoyancy for remainder of deployment
f) South West of Azores (IOS)							
294 4250	32°00.4'N 31°30.0'W	200 700 1500	230 230 230	122.09 33.17 3.16	3.82 0.32 0.44	31.96 103.66 7.18	
295 3737	33°01.4'N 31°47.8'W	250 750 1550	228 228 228	126.20 33.49 2.35	22.91 3.31 0.51	5.51 10.12 4.61	
g) NEADS 1 Site (IOS)							
296 5283	33°05'N 21°58.4'W	4630 5271	225 225	2.08 2.65	0.02 0.00	128.40 ∞	Since these 6 moorings form a tight cluster, grouped averages for two depths are given below:
298 5286	33°10'N 21°56.8'W	5273	225	2.76	0.03	100.73	
299 5300	33°13'N 22°00'W	4642 5287	225 225	2.28 2.75	0.11 0.04	21.02 63.63	
300 5300	33°10'N 22°06'W	4643 5288	225 225	2.57 3.04	0.14 0.17	18.46 17.54	
301 5275	32°54'N 22°00'W	5263	225	3.14	0.04	77.72	
302 5333	33°10'N 22°17.5'W	5321	225	3.43	0.09	39.65	

Grouped averages (g):

	4638 m	5284 m
K_E	2.31	2.96
K_M	0.09	0.06
K_E/K_M	25.67	48.03

h) Antilles Current (University of Miami)

125	25°29.1'N	200	349	210.2	18.22	11.54	Low passed by 30 h
5470	70°44.4'W	400	349	187.3	18.23	10.27	Half-period Lanczos filter
		700	349	65.9	2.02	32.62	throughout
		1000	349	20.6	1.40	14.71	
		1500	349	12.7	1.62	7.84	
		3000	349	16.0	4.25	3.76	
126	23°38.2'N	200	342	184.5	19.04	9.69	
5280	72°27.8'W	400	342	103.8	10.00	10.38	
		700	342	35.2	9.49	3.71	
		1000	342	37.2	0.63	59.05	
		1500	342	111.8	1.73	64.62	
		3000	342	60.5	5.31	11.39	
127	22°59.0'N	200	349	148.9	20.60	7.23	
4760	73°00.8'W	400	349	118.4	11.24	10.53	
		700	349	29.8	6.58	4.53	
		1000	349	20.6	0.87	23.68	
		1500	349	45.7	1.81	25.25	
128	26°12.1'N	200	361	405.3	77.18	5.25	
4670	75°42.8'W	400	361	253.8	59.63	4.26	
		700	361	61.4	22.87	2.68	
		1000	361	29.0	11.11	2.61	
		1500	361	51.2	7.80	6.56	

i) Maxwell Fracture Zone (IFMK, Kiel)

?	48°33'N	184	~365	78	27	2.89	Recordings smoothed using a filter with
3725	26°05'W	389	~410	67	31	2.16	a 30-h half power period
		794	~410	24	9	2.67	
		2515	~410	7	1	7.00	

Sources a)–e) RR Dickson (MAFF, Lowestoft).
    f) WJ Gould (IOS, Wormley) pers. comm.
    g) P Saunders (IOS, Wormley) pers. comm.
    h) F Schott (Univ. Miami) pers. comm.
    i) J Meinke (IFMK, Kiel) pers. comm.

16. Global Summary: Review of Eddy Phenomena as Expressed in Temperature Measurements

W.J. Emery

16.1 Introduction

The purpose of this chapter is to compare and contrast observations of eddy phenomena, as expressed primarily in temperature sections, from various oceanic regions. Some of the material will be drawn from the preceding regional chapters, while other pertinent presentations will be added. An effort will be made to distinguish some of the more subtle differences in regional eddy behaviour in addition to a description of the more obvious similarities. The acute non-uniformity in the coverage of data, with adequate resolution, will often be mentioned and will be frequently invoked in qualifying any of the conclusions drawn. Most discussions will be concerned with the Northern and Equatorial Atlantic and Pacific Oceans.

Any review of the many different studies of oceanic eddies must necessarily preface its discussion with a recognition of the different meanings applied to the term "eddy". Prior to MODE-1 many observations suggested that mesoscale circulation feature made energetic contributions to the ocean circulation. In MODE-1 (MODE Group 1978) a closed mesoscale circulation cell was examined in some detail and shown to exhibit kinetic and potential energy levels greater than the mean flow. The belief that such features may be ubiquitous in the ocean led to the practice of interpreting many two-dimensional isotherm deflection features as expressions of mesoscale eddies. This was consistent with the belief, fostered by turbulence theory, that fluctuations in velocity and density were due to the presence of vortex circulations called "eddies".

Thus today one may use the term eddy to refer to individual closed circulation cells, or to time/space velocity and property fluctuations caused by a wide variety of circulation features. Eddy time and length scales derived in different studies may apply to closed eddies or to all mesoscale fluctuations. Sometimes the dual nature of the term eddy may be used in a single study where spatial thermal observations of a closed eddy are compared with velocity fluctuations at current meter moorings. In the regional chapters this dichotomy of usage is again well represented, and the reader is cautioned to consider carefully the context in which the term "eddy" is used. Some authors specifically mentioned this problem and turn to terms such as "discrete" or "isolated" to signify eddies which appear as closed cells, while applying the word eddy to the description of all mesoscale perturbations from the long-term mean. This distinction will also be used in the present chapter.

Perhaps the dual nature of this term is due, in part, to our inability to observe routinely the three-dimensional mesoscale ocean structure. Infrared sa-

Eddies in Marine Science
(ed. by A.R. Robinson)
© Springer-Verlag Berlin Heidelberg 1983

tellite images, a limited number of spatial surveys and the tracks of drogued buoys offer tantalizing evidence of the variety of mesoscale phenomena that abound in the ocean. The majority of our observations, however, are restricted to two-dimensional sections or time series at a single or limited array of moorings. Neither of these observational approaches resolves the ocean structure in the way satellites, ground stations and upper air measurements display atmospheric structure and circulations. Thus our understanding of oceanic eddies is based on the undersampling of a complex environment.

In experiments where higher sampling resolution has been achieved, usually at the expense of total spatial coverage, new types of eddy behaviour have been discovered. In the POLYMODE Local Dynamic Experiment (Chap. 6) intense space-time measurements revealed the presence of small-scale (~ 30 km in diameter), subsurface eddies with apparently long lifetimes. Exhibiting large Rossby numbers, comparable to even the most energetic currents, these features are an addition to the eddy spectrum. This experience confirms the suspicion that the more closely we look, the wider the variety of phenomena we find. Hence many of the regions of the ocean, now thought to be devoid of certain types of eddies, may merely lack adequate observations to resolve such features.

Generally available oceanographic observations with mesoscale resolution are concentrated in regions known, or suspected, to be dominated by mesoscale fluctuations. Although some efforts have been made to observe regions of lower mesoscale activity, results do not yet provide the wide horizontal range of data needed to characterize even qualitatively some regional aspects of the geographic distribution of mesoscale phenomena. Thus a global summary must depend on comparisons between well-sampled regions and regions with little or no data. Often the conclusion that a region contains low mesoscale activity is more a statement that few mesoscale features have been observed and closer study may reveal fluctuations passed over with previous sampling schemes.

Finally it must be pointed out that the eddy expressions, discussed in this chapter, appear largely in temperature data and can thus reflect only those eddies with a baroclinic structure. Eddies, as determined from direct current observations, include both these baroclinic features and those of a barotropic nature. Thus results for the spatial structure and distribution of baroclinic of eddies which has been most widely studied through XBT observations require careful interpretation when compared with results of studies with current meter measurements which include also the barotropic component.

16.2 Standard Deviation of Temperature

As a global background for the discussion of individual XBT sections the historical XBT file was used to compute the means and standard deviations of temperature at a variety of levels for a variety of grid patterns in the northern hemisphere and equatorial oceans. Data coverage south of 10°S was far too sparse to permit the calculation of standard deviations. Since most of the available mesoscale XBT sections were collected in the North Atlantic and North Pacific the restriction to latitudes north of 10°S does not hamper the comparison.

The computation of XBT temperature standard deviation follows the landmark work by Dantzler (1977) who used similar calculations, in concert with estimates of the Brunt-Vaisala frequency profile from hydrographic data, to make a map of eddy potential energy in the North Atlantic. In his analysis Dantzler employed a 5° square as an averaging interval, noting regions where data coverage was low. In an effort to take advantage of the significantly greater data coverage in many areas, a variable grid (Fig. 1) was selected rang-

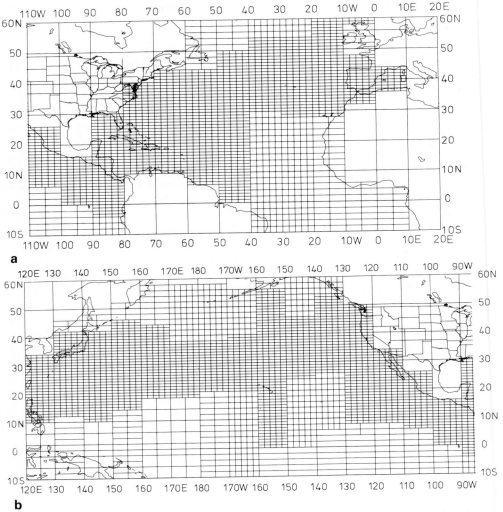

Fig. 1. Variable grid scheme for the XBT historical file analysis. **a** North Atlantic, **b** North Pacific

ing from 1° latitude by 2° longitude squares in data dense regions to 4° latitude by 5° longitude squares in data sparse regions. In the selection of this variable grid an attempt was made to have no fewer than 10 observations per grid area. Unfortunately this was not always possible and even some 4° × 5° squares fall below this limit.

The temperature standard deviations, at 260 m from this computation, are shown in Fig. 2 for the North Atlantic and Fig. 3 for the North Pacific. Before discussing these distributions it should be noted that in his study Dantzler (1977) showed a relatively close correspondence between the spatial pattern of eddy potential energy in the North Atlantic and the distribution of the deviations in the depth of the 15°C isotherm from which it was calculated. Thus there is some reason to believe that a map of mid-thermocline temperature variability (standard deviation) can be related to the level of mesoscale activity. In some regions the 260 m depth does not correspond well to the mean thermocline depth, e.g., in the western North Atlantic and the Sargasso Sea the presence of 18°C water (Worthington 1959) results in an average thermocline depth below 260 m. Thus similar maps (Figs. 4 and 5) of the standard deviation in temperature at 460 m will also be examined.

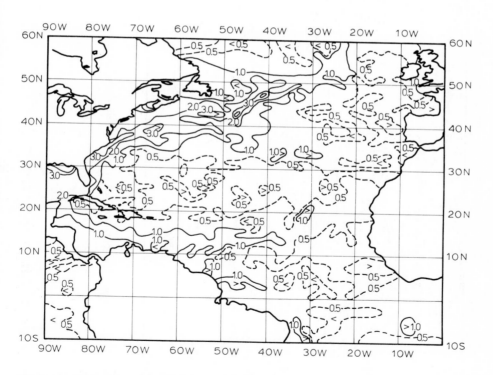

Fig. 2. Standard deviation of temperature (°C) at 260 m based on the variable grid analysis of the XBT file for the North Atlantic should be dropped

W.J. Emery

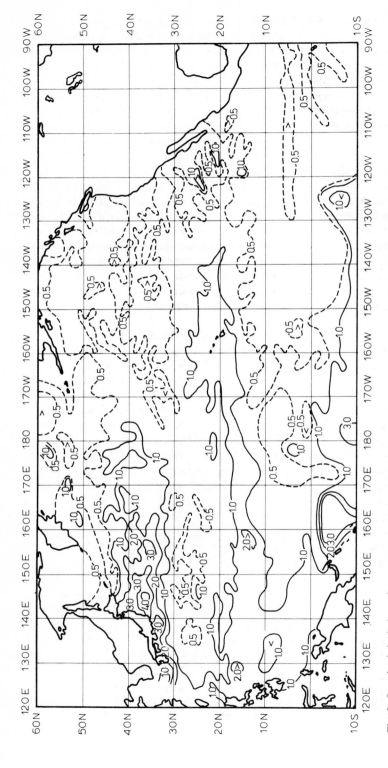

Fig. 3. Standard deviation of temperature (°C) at 260 m based on the variable grid analysis of the XBT file for the North Pacific

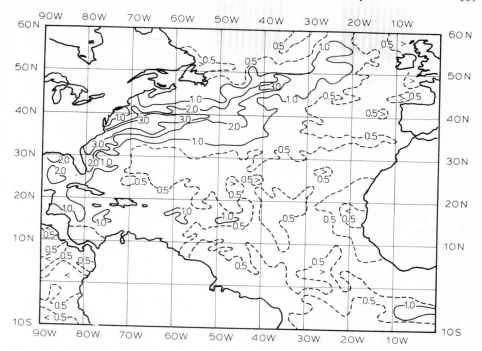

Fig. 4. Standard deviation of temperature (°C) at 460 m based on the variable grid analysis of the XBT file for the North Atlantic

16.3 The North and Equatorial Atlantic

In terms of data sampling with mesoscale resolution the North Atlantic is the best studied of the world's oceans. The presence of the Gulf Stream as a source of mesoscale meanders, rings and eddies, has led to a variety of studies focusing on these expressions of variability. Major co-operative study programs (MODE, POLYGON, POLYMODE, etc.) have been conducted to evaluate the role of eddies in the ocean circulation.

The well-known map of eddy potential energy (Fig. 6), prepared by Dantzler (1977), reveals the essential spatial distribution of this quantity. Maximum values are associated with the path of the Gulf Stream and its eastward extension out into the North Atlantic. There is a marked general decrease in eddy potential energy just west of the Mid-Atlantic Ridge. A broad minimum dominates the zonal band between 20° and 30°N with a narrow zone of slightly higher values just south of 20°N in the west stretching to bracket 20°N in the east.

There are many problems in interpreting this calculation as a strict measure of eddy potential energy. As pointed out by Gould (Chap. 7) this analysis includes all thermal variability thus combining internal mesoscale energetics with

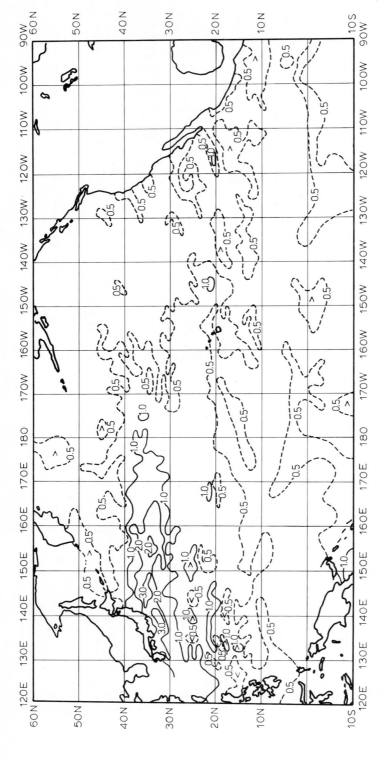

Fig. 5. Standard deviation of temperature (°C) at 460 m based on the variable grid analysis of the XBT file for the North Pacific

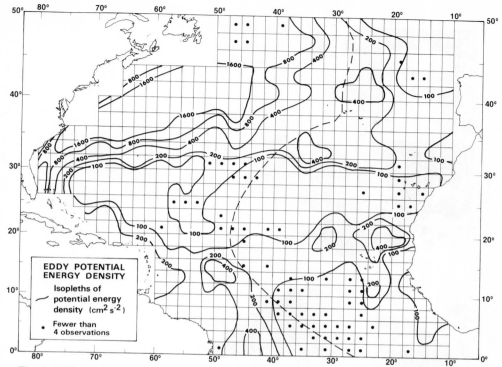

Fig. 6. Eddy Potential Energy Density (Dantzler 1977)

variations due to seasonal heating and cooling. Other criticisms have focused on large-scale variations in the depth of the chosen 15 °C isotherm, the large 5° square averaging area and the use of mean stratification profiles to compute the potential energy of fluctuations. In spite of these, and other difficulties, this approach offers a unique view of geographic variations in ocean variability.

It is encouraging that the general pattern of temperature variance at 260 m (Fig. 2) compares fairly well with Dantzler's (1977) eddy energy map (Fig. 6). Highest values are again in the northwest and are associated with the Gulf Stream. The band of minimum values at mid-latitudes is bordered to the south by a narrow maximum that stretches eastward from South America along about 12 °N. This maximum tongue is absent in the temperature deviation at 460 m (Fig. 4); otherwise the 460 m map is very similar to that at 260 m (Fig. 2).

The same general distribution also appears in the map of eddy kinetic energy (Fig. 7) computed from the historical ship drift file by Wyrtki et al. (1976). The large values in the northwestern Atlantic do not extend as far eastward, stopping at about 50 °W, far west of the Mid-Atlantic Ridge. The southern maximum band is located farther south, being near the equator and extending all the way to the coast of Africa.

This latter summary map of eddy kinetic energy has been invoked by almost all the authors of the regional chapters as the only available global sum-

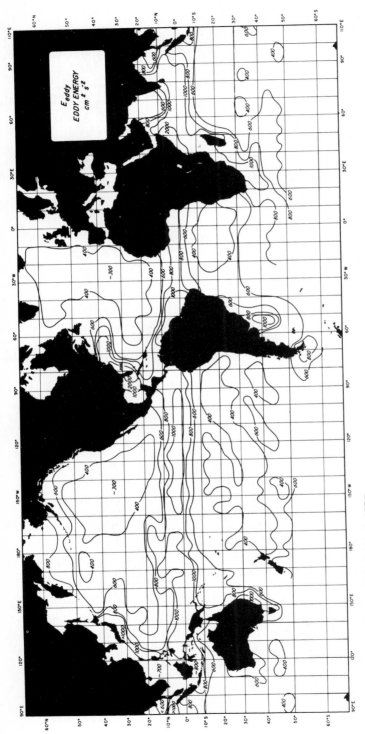

Fig. 7. Eddy Kinetic Energy (Wyrtki et al. 1976)

mary of eddy variability. In his review of current meter records Dickson (Chap. 16) concludes that the overall kinetic energy distribution is qualitatively consistent with the ship-drift analysis. This, coupled with the favourable agreement with Dantzler's eddy potential energy, bespeaks the reality that the dominant mesoscale variability is associated with the strong current regions such as the Gulf Stream. In fact in Chapter 4, Wunsch suggests that the Gulf Stream may be the source of most, if not all, the eddy energy in the North Atlantic. He notes the observation, common to all chapters on the western North Atlantic, that the energy level of mesoscale variability increases with proximity to the Gulf Stream. In Chapter 3 Richardson echoes this philosophy by interpreting almost all mesoscale features in the Atlantic as rings formed by the separation of Gulf Stream meanders.

Approaching this descriptive problem with an entirely different posture Kim and Rossby (1979) argue that rings should be regarded as a special part of the mesoscale spectrum and studied separately. They suggest that Gulf Stream rings are larger in amplitude and have a specified behaviour (namely westward propagation) in contrast to the assumed mesoscale eddy field with its presumed Gaussian statistics. While remarking that the centers of young rings are relatively easy to identify (in XBT sections), they acknowledge that the edges of rings are difficult to distinguish from other mesoscale isotherm deflection features. They add a word of caution that one must carefully define the appropriate mean field from which mesoscale perturbations are computed. In areas known to be dominated by large features such as rings, the mean field may express structure as an artifact of averaging in a number of rings. In this discussion we will follow the design of the volume and continue to regard rings as mesoscale phenomena.

Whether rings, meanders or other eddy forms, mesoscale phenomena in the western North Atlantic appear to share common characteristics. Downstream of Cape Hatteras the eddy scales range between 200 and 300 km horizontally, and between 50 and 150 days temporally. In the high shear region of the Gulf Stream, near the continental margin, both time and space scales tend to be smaller with periods ranging from 4 to 20 days over spatial scales from 25 to 100 km. Most of the eddies, observed in this strong-current region, exhibited a strong baroclinic structure as evidenced in thermal sections and the analysis of current meter data in Chapter 4.

By sharp contrast Gould (Chap. 7) reports that relatively fewer eddy phenomena are observed in the eastern North Atlantic (east of the Mid-Atlantic ridge). This is predicted, in general, by the lower levels of eddy potential energy in Fig. 6. Gould argues, however, that the interpretation of the variance may be different for this region due to deep wintertime convection carrying the seasonal temperature signal down below 600 m. From the available observations he suggests three specific areas where eddies are likely to be found and two regions of little eddy activity. As stated earlier these limits may reflect the relative absence of data adequate to resolve the mesoscale. It is interesting to note that in both Figs. 2 and 4 while there are alternating regions of high and low temperature deviation, there are no sharp gradients, as observed in the western North Atlantic.

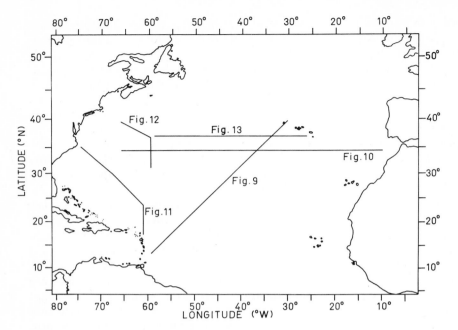

Fig. 8. Locations of Atlantic XBT sections

A comparison between the maps of three general indications of eddy activity (Figs. 2, 4, 6 and 7) and individual XBT sections emphasizes this zonal contrast while pointing out some interesting features. A section (Fig. 8) between Barbados and the Azores (Gould 1976 Polymode News 16) contains only a few expressions of mesoscale fluctuations (Fig. 9). The strongest of these is a cold feature to the south and west of the Azores. Interestingly both Dantzler's (1977) map (Fig. 6) and the temperature deviation at 260 m (Fig. 2) contain a small maximum at about this location. Surprisingly, however, the southwestern portion of this section exhibits no mesoscale activity even thought it crosses into a similar maximum. In fact the most impressive character of this section (Fig. 9) is the total lack of mesoscale fluctuations in the central region (crossing the energy minimum).

By contrast a long zonal section (Fig. 10) along 34°30′N (Seaver 1975 MODE Hot-Line News 84) exhibits a lot of eddy activity west of 30°W. East of this longitude, corresponding to the Mid-Atlantic ridge, smaller scale (~30 km) vertical isotherm excursions dominate the thermal structure. These smaller fluctuations are coherent over a limited portion of the water column and are possibly expressions of internal waves. The larger mesoscale features, in the west, range from 150 to 250 km and are primarily cold cyclonic features. Just west of 30°W is a warm eddy about 150 km wide.

A section (Fig. 11) running southeast from Cape Hatteras (Fig. 8) displays the rapid descent of the thermal structure associated with the Gulf Stream. Cold mesoscale eddies are seen south and east down to about 29°N, 68°W

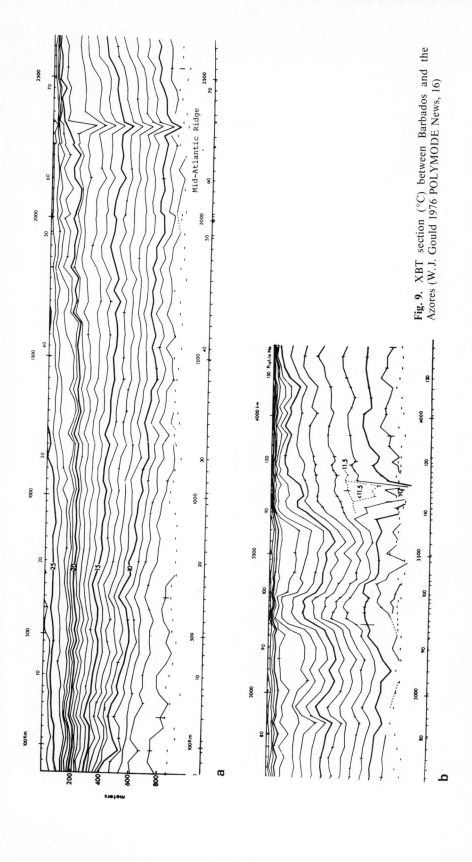

Fig. 9. XBT section (°C) between Barbados and the Azores (W.J. Gould 1976 POLYMODE News, 16)

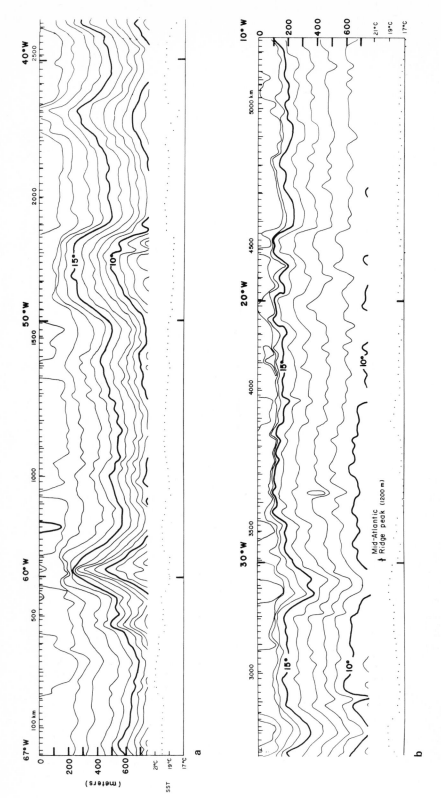

Fig. 10. XBT section (°C) along 34°30′N between 67°W and 10°W (G. Seaver 1975 MODE Hot-Line News, 84)

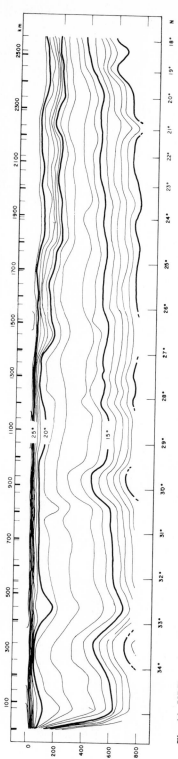

Fig. 11. XBT section (°C) between Norfolk and Antingua (G. Volkmann MODE Hot-Line News, 84)

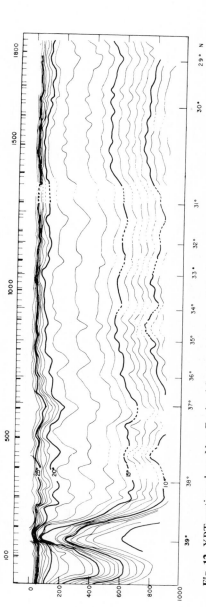

Fig. 12. XBT section along New England Seamounts down 60°W from 37°N to 30°30'N and southeast to about 29°N, 55°W (G. Volkmann 1975 MODE Hot-Line News, 84)

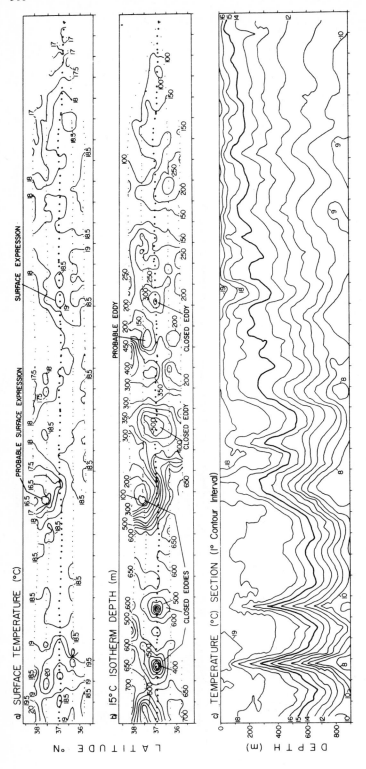

Fig. 13. North Atlantic Multi-Ship Survey: **a** sea-surface temperature (°C), **b** depth of the 15°C isotherm (m) and **c** temperature section (°C) along central ship track

after which the relatively flat thermal structure expresses the lack of mesoscale eddies in the low energy region of the mid-latitudes.

Farther north (Fig. 8) another XBT section (Fig. 12) also confirms the sharp drop in mesoscale features as one crosses out of the maxima in the maps of Figs. 2 and 6. Extending southeast out from the seaward edge of the Gulf Stream a series of pronounced mesoscale isotherm deflection features dominate the western end of the section. Southeastward, along this section, the features are much weaker in amplitude but similar in horizontal scale. Crossing out of the eddy activity maximum, at about 35°N the deflection features become substantially smaller in scale and look more like the internal wave expressions discussed earlier. As the section turns south (Fig. 8) it crosses the eddy energy minimum in agreement with the absence of vertically coherent isotherm deflection features.

Four multi-ship XBT surveys in the North Atlantic (Emery et al. 1980) also support the distribution of eddy energy revealed by the temperature deviations in Fig. 2. All four surveys are north of 30°N and the majority of eddy features confirmed in the three-dimensional surveys occur west of 40°W in agreement with the higher values in Fig. 2. For example the five-ship swath between 30° and 39°N is shown in Fig. 13. The features, identified in the 15°C isotherm depth map as closed eddies, include both isolated rings and more closely packed eddies. The XBT section, from the central ship, exhibits the marked east-west change in mesoscale expression introduced earlier. West of 38°W (the approximate longitude of the Mid-Atlantic ridge) stronger, vertically coherent eddies dominate while east of this position fewer, weaker and shallower eddies are found.

The basic agreement between the individual expressions of mesoscale eddies, as seen in Atlantic XBT sections, and the summary eddy energy plots of Dantzler (1977) and Wyrtki et al. (1976), (along with the temperature standard deviations in Fig. 2) confirms the validity of these maps as general guides to where mesoscale phenomena may or may not be expected. Their use must be tempered, however, by a proper awareness of the limits of the data bases. It should also be recognized that a more detailed understanding of the distribution of eddy energy must await the addition of many more regional studies of mesoscale phenomena.

16.4 The North and Equatorial Pacific

As there is no eddy potential energy map for the North Pacific comparison can only be made between the maps of the standard deviation in temperature and the kinetic energy analysis by Wyrtki et al. (1976). The temperature deviation distribution at 260 m (Fig. 3) is quite different from the kinetic energy map (Fig. 7). Both maps have high values to the east of Japan; the eddy kinetic energy has a maximum just north of the equator, while Fig. 3 has a relative minimum along this latitude. The temperature deviation map has a tongue of high values just south of and along 20°N, extending eastward to 135°W. This tongue is not in the temperature deviation map at 460 m (Fig. 5) which instead exhibits a minimum. In general the level of mesoscale activity and amount of pattern is much less at 460 m than at 260 m. Highest values, at both 260 and 460 m, are concentrated in a tongue stretching east from Japan to about the dateline. In the kinetic energy analysis of Wyrtki et al. (1976) the tongue of relatively high values extends eastward from Japan only to about 165°E. The next zonal change in activity occurs at 170°W bordering a broad minimum in the eastern North Pacific.

Unfortunately no systematic efforts have been undertaken to study mesoscale phenomena in the North Pacific. Some individual efforts have yielded XBT sections with a mesoscale resolution. As discussed in Bernstein and White (1977) a zonal XBT section along 38°N (Fig. 14) exhibits strong mesoscale features west of 175°W about the limit of the tongue of maximum values in Fig. 3. Bernstein and White suggested that this change in eddy regimes was due to the presence of the Emperor seamount chain. East of this longitude the section contains only one eddy at about 150°W. As discussed by Bernstein in Chap. 8 a similar XBT section, along 35°N, also reveals this marked east-west discontinuity in the expressions of mesoscale features. Two multi-ship sections from the North Pacific (Emery et al. 1980, Wilson and Dugan 1978) also exhibit mesoscale eddies west of 180°W. In agreement with Fig. 3, these eddies are predominantly at latitudes north of 30°N and are probably generated by fluctuations of the Kuroshio.

These swaths and sections combined with the maps of Figs. 3, 5 and 7 suggest that as in the North Atlantic, most of the mesoscale activity in the North Pacific is associated with the strong western boundary current, the Kuroshio, and its eastward extension. Isolated high values along the west coast of North America in Fig. 3 suggest the presence of mesoscale phenomena in this area, as mentioned in Chapter 8. These features appear to be related to the strong baroclinic flow in this region and may be spawned by baroclinic instability.

Another region of increased mesoscale activity far from strong boundary currents is the region around Hawaii. The 260 m temperature deviations (Fig. 3) show a maximum around the Islands while the eddy kinetic energy analysis of Wyrtki et al. (1976) also shows relatively higher values in this region. It is therefore not surprising that various surveys have revealed a rich variety of mesoscale structures. An air expendable bathythermography study by Wollard et al. (1969) reveals many warm and cold eddies (Fig. 15) in a region just north of Hawaii. Slightly farther east a more recent multi-ship XBT survey (Fig. 16) demonstrates the existence of a very strong warm eddy most obvious in the temperature at 200 m. There is no sign of the relatively small ($\sim$ 50 km) warm eddy in sea surface temperature and only a limited surface salinity expression. A vertical temperature section from the central ship (not shown) indicated that the eddy was strongest in the thermocline and weakened quite rapidly below 300 m. Somewhat farther south a time series of temperature sections taken along 25°N are interpreted by Bernstein and White (1974) to show the westward propagation of mesoscale eddies. Most of the eddies in the region around Hawaii are relatively small having length scales less than 100 km.

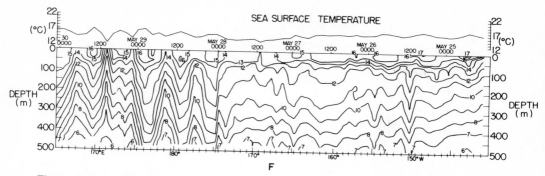

Fig. 14. XBT section along 38°N in North Pacific. (Bernstein and White 1977)

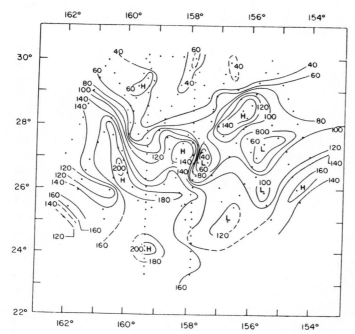

Fig. 15. Depth (m) of the 19°C isotherm 18–26 April 1969 from an AXBT survey. (Wollard et al. 1969)

The area around Hawaii is at the northern boundary of the North Equatorial Current but the eddies are probably not generated by current meanders as have been observed for the Gulf Stream. Eddies observed south of Hawaii (Patzert pers. comm.) in the tracks of drifting buoys are generally larger (∼ 200 km) than those to the north. It is interesting to note that in both the North Atlantic (Fig. 2) and in the North Pacific (Fig. 3) there is an eastward stretching maximum between 10° and 20°N in the temperature standard deviation at 260 m. Perhaps the greater eastward extension of this tongue, in the Pacific, is due to the presence of the topographic ridge of the Hawaiian Islands.

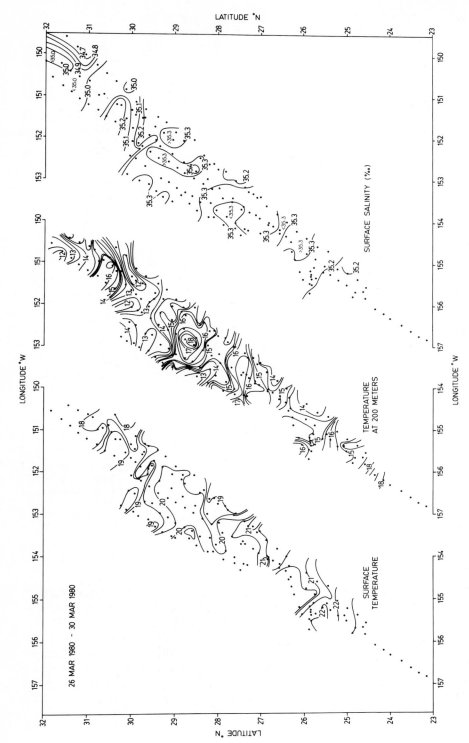

Fig. 16. Surface temperature (°C), temperature at 200 m (°C) and surface salinity (‰) from a multi-ship XBT survey 26–30 March 1980

16.5 Discussion

At present, a truly global summary of mesoscale eddy phenomena is not possible. Our limited sampling of the vast majority of the oceans forces us to focus on the few well sampled regions. The regional bias in the study of eddies to the western North Atlantic is mirrored in this volume which devotes five chapters to the area. One must be careful not to extrapolate the results from this region to all other parts of the ocean. There are indeed similarities between oceans and we see cold eddies form in the southern ocean from meanders of the Antarctic Circumpolar Current (Chap. 15) much in the way cold rings are spawned from the Gulf Stream (Chap. 2). Such an instability process may also be operating off the coast of Labrador and in other areas where a strong current separates two very different water masses.

Marked differences can also be found between various ocean areas. A prime example is the increase of mesoscale activity around Hawaii with no corresponding increase in the North Atlantic. It may be that the topographic influence of the island chain contributes to the generation of eddies or it may mean that the increased number of measurements taken near Hawaii has revealed features that may be found elsewhere if more data were available.

Other evidences for topographic control or influence, in the distribution of eddy activity, are the drop in eddy expressions east of the Emperor Seamounts in the North Pacific, and a similar eddy energy descrease associated with the presence of the mid-ocean ridge in the Atlantic. Also the flow of the Antarctic Circumpolar Current over the topographic ridge south of New Zealand has been shown (Gordon 1972) to generate mesoscale eddies. The increasing evidence of such topographic effects suggests that a future map of eddy energy may resemble a superposition of a map of strong currents with that of bottom topography.

Some regions such as to the East of Australia do not have a western boundary current to spawn eddies; rather the entire circulation field is comprised of a series of warm and cold eddies (Chap. 12). It is suggested in Chap. 13 that the current field off western Australia is also a collection of eddies maintained by energy transfer from larger scale currents. The tracks of drifting buoys, in this region, revealed eddy scales of about 200 km with time scales of about 2 months.

In the same chapter (Chap. 13) Grundlingh discusses the formation of eddies by the Agulhas Current. He concludes that eddies form both as meanders of the main current and as spin-off features on the inshore edge of the mean flow. A somewhat larger, quasi-permanent eddy forms the recirculation of this current back into the Indian Ocean. Thus we see the same mechanisms operating in the Indian Ocean that dominate the North Atlantic and North Pacific.

Unique to the Indian Ocean, however, is the strong seasonal signal associated with the monsoonal wind shift. Swallow, in Chap. 11, concludes that in regions of strong eddy activity (as shown by Wyrtki et al. 1976, Fig. 7) the seasonal signal contributes about half of the kinetic energy fluctuations. In areas of lower energy this ratio drops to between a quarter and a tenth.

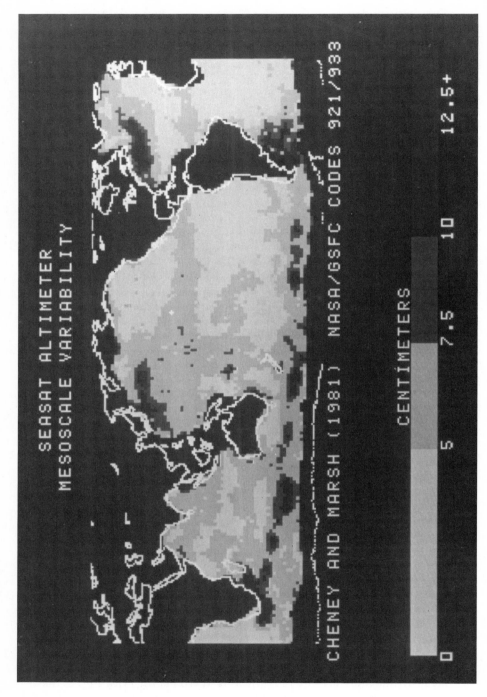

Fig. 17. Mesoscale variability from collinear SEASAT altimeter ground tracks. (Cheney et al. 1981)

In the many other geographic ocean regions, reviewed in this volume, eddies have been encountered. Again detailed comparisons are made difficult by the relative scarcity of appropriate observations in many areas. Any systematic overview which comprehensively summarizes the geographic distribution of eddy energy, eddy scales and the role of eddies in the general circulation must await a significant increase in the data available for analysis. Since resources are not likely to be available for mesoscale surveys throughout the world ocean, the hope for the future lies in imporved regional studies in areas which have, at present, not been well studied.

As a closing comment reference should be made to a new data product which perhaps best represents global mesoscale variability. Figure 17 is a map of mesoscale variability computed from collinear passes of SEASAT from its radar altimeter (Cheney et al. 1981). Along track resolution is 7 km while the spacing between tracks is 700 to 800 km. A total of 25 days of altimetry data were used to formulate this representation of variability. Cheney (pers. comm.) is working on a contour map of the same field which should provide higher resolution in many areas.

This map (Fig. 17) confirms many of the features in the existing eddy energy maps (Figs. 6, 7) while revealing many new areas of high eddy activity. Most significant are the large values in most of the southern ocean and in the region east of southern South. America. It is also interesting to note that the maxima in the northwestern portions of the Atlantic and Pacific, extend somewhat farther eastward in Fig. 17 than they do in the earlier maps (Figs. 6, 7). The equatorial maxima of eddy kinetic energy (Fig. 6) are absent in the map from SEASAT (Fig. 17) and in Dantzler's map for the Atlantic (Fig. 7).

One important and unique facet of Fig. 17 is that it is computed from a data set with almost uniform global coverage and thus does not depend on the availability of data from ships transiting a particular region. Future maps of variability from satellite altimeters may provide the best possible pictures of the geographic distribution of mesoscale variability.

Models

17. Eddy-Resolving Numerical Models of Large-Scale Ocean Circulation

W.R. Holland, D.E. Harrison, and A.J. Semtner, Jr.

17.1 Introduction

During the decade of the 1970's, in association with large field programs (MODE, POLYMODE, ISOS, NORPAX), fine resolution (grid size < 100 km) numerical models of ocean circulation were developed to examine the role of mesoscale eddies in the oceanic general circulation. These models have sought to develop an understanding of dynamics of flow in closed, wind-driven (and in some cases thermally driven) ocean basins, a flow in turbulent equilibrium. The eddy fields in these studies, maintained by instabilities in the currents of the large-scale ocean gyres, radiate energy to distant regions of the basin, interact actively with the mean flow, and ultimately dissipate much of the energy input by wind and heating in bottom or lateral friction.

The mean circulation is intimately tied to the nature of the eddy-mean flow interaction. The mechanisms by which eddies redistribute mean momentum, heat, and potential vorticity have been found to be much more complex than homogeneous diffusion and, although it is as yet unclear whether eddy effects can be adequately parameterized in terms of the mean flow, considerable progress has been made in understanding the details of this interaction.

One important aspect of modeling studies has been a considerable narrowing of the gap between model simulations and observations. For the first time, we have within our grasp the elements necessary to construct valid models of the large-scale circulation, models that agree well with observations in terms of the statistics defining the oceanic general circulation.

Although we will discuss here only the results of eddy-resolving models, it is important to point out the role played by a variety of simpler models. These include stability calculations, models of the interior, models of Gulf Stream Rings, boundary-forced linear waves, and many other models involving a restricted set of processes. In addition, the so-called process models (Chap. 18) which are intended to examine a limited range of processes in full, multi-layered models in turbulent equilibrium, have played a vital intermediate role between the simpler models and models of the complete circulation including boundary currents and interior processes in closed basins. The progress made to date has depended critically upon the interaction of ideas spawned by the whole range of these works.

It is worth emphasizing that turbulent model flows are enormously complex in their behavior, and it is only with rather painstaking analysis that the behavior can be distilled down to a reasonably simple description. The continuing

Eddies in Marine Science
(ed. by A.R. Robinson)
© Springer-Verlag Berlin Heidelberg 1983

task of developing such analyses will be a major factor in future progress toward dynamical understanding of the large-scale oceanic circulation.

In this chapter we shall review results from a number of studies with eddy-resolving general circulation models (EGCM's). Our purpose will be to give a general idea of the nature of these models and the results therefrom without going into great detail. The model studies we shall mention have been carried out by Holland and Lin (1975a, b), Robinson et al. (1977), Semtner and Mintz (1977), Holland (1978), Semtner and Holland (1978), McWilliams et al. (1978), and Semtner and Holland (1980). In addition, we shall look at several analyses of these experiments by Haidvogel and Holland (1978), Harrison and Robinson (1978, 1979), Holland and Rhines (1980), Harrison and Holland (1981), and Schmitz and Holland (1982). All but the last of these are studies to understand various aspects of the dynamical behavior of the solutions; the final study is a first attempt to relate model statistical results directly to the observational data base from the North Atlantic, in order to define better the values of poorly known model parameters that lead to realistic simulations.

All of the papers to be discussed make use of a common numerical technique for solving the highly nonlinear governing equations. Spatial and temporal derivatives in the continuous hydrodynamical equations are replaced by finite-difference approximations involving values of the dependent variables on a regular array of gridpoints. Solutions are obtained by stepping forward in time from an initial resting state until statistical equilibrium in the presence of mesoscale variability is achieved. Time series of data from the period of equilibrium are then analyzed for dynamical information. Complete descriptions of the numerical techniques and analysis procedures, which depend on the exact form of the governing equations to be solved, can be found in the individual papers. The interested reader is directed to the somewhat more technical review papers by Holland (1977, 1979) and by Robinson et al. (1979).

17.2 Review of EGCM Results

In this section we shall discuss results from the eight papers cited above which report basic EGCM experimental results. The experiments described in these papers span a wide range of model processes, from adiabatic two-layer quasi-geostrophic physics in rectangular, flat bottom basins to wind and thermally forced multi-level primitive equation physics in asymmetric basins with bottom topography. Most of these papers have been midlatitude closed basin studies. Semtner and Holland (1980) describe an equatorial closed basin experiment and McWilliams et al. (1978) describe a series of open and partially obstructed zonal channel experiments which are relevant to the Antarctic Circumpolar Current. Although these are all highly idealized models, they have produced a wide range of different flows. Considerable effort has been devoted to formulating analysis methods and simple dynamical hypotheses in order to understand the behavior of these experiments, and a number of analysis papers have been published.

The purpose of this section is to offer a non-technical overview of the model results and the various analysis efforts. In order to provide such an overview, many important aspects of the work will be downplayed; the researcher seriously interested in this area should return to the original papers to become informed about these aspects. Further, in the interest of brevity, not every EGCM experiment and analysis will be described and discussed; rather, experiments will be selected to illustrate the range of model results. Thus, many (for the specialist) very interesting results will not be examined here.

To provide a focus for this overview, it is useful to begin with the questions that EGCM scientists have been trying to answer within the limitations of their idealized models. These questions are, broadly stated,

A. What processes account for the presence of mesoscale variability?
B. Do mesoscale phenomena play a fundamental role in the character and dynamics of the time mean circulation? If so, what is this role and in what regions is it important?
C. Where the effects of mesoscale circulations are important, can they be parameterized in terms of mean field quantities?
D. How do the answers to the above questions change as the model physical processes are changed?
E. What choices of model processes give results similar to oceanic behavior?

In the course of discussing each of these questions, the desired overview will emerge.

17.2.1 What Processes Account for the Presence of Mesoscale Variability?

This question is in many ways the simplest to answer because there is reasonable concensus about analysis methods to answer it. The origin of mesoscale variability in the model flow has been characterized by the energy transformations that maintain the eddy kinetic energy

$$K' \equiv \frac{1}{2}\rho_0(\overline{u'^2} + \overline{v'^2}),\qquad(1)$$

where overbar denotes a long time average, prime denotes instantaneous departure from the long time average, and $\rho_0 = 1$ gm cm^{-3} is the reference density of the fluid [often omitted from Eq. (1) for convenience]. When there is no time-dependent mechanical forcing, K' can only be maintained in the face of dissipation either by the conversion of mean flow kinetic energy,

$$\overline{K} \equiv \frac{1}{2}\rho_0(\overline{u}^2 + \overline{v}^2)\qquad(2)$$

via Reynolds stresses ($\overline{u'v'}$, etc.) or by conversion of mean potential energy $\overline{P}$ via eddy buoyancy fluxes ($\overline{w'T'}$). The natural definition of potential energy varies from one model system to another but the eddy buoyancy conversion is uniquely defined in each case.

Integrated over the model basin, the two energy transformation processes are well defined, and can be denoted $\overline{\langle K \rangle} \rightarrow \langle K' \rangle$ and $\overline{\langle P \rangle} \rightarrow \langle K' \rangle$, respectively, where $\langle \rangle$ indicates a basin integral. When $\overline{\langle K \rangle} \rightarrow \langle K' \rangle$ dominates $\overline{\langle P \rangle} \rightarrow \langle K' \rangle$, the convention has been to speak of the eddies as supported primarily by "barotropic" instability, and in the converse as supported primarily by "baroclinic" instability. When neither process dominates, the instability is referred to as "mixed". The basin-integrated energetics of many of the EGCM experiments discussed in the various papers mentioned above have been discussed in detail by Harrison (1979).

Although convenient, the labels introduced above give only a global breakdown and are not necessarily useful in a local sense. To gain insight into the processes that result in energy transfer to K' locally, it is necessary to examine the model flow fields. The time-averaged flow fields of several EGCM experiments are shown in Fig. 1. All of the mid-latitude closed basin cases have the following general characteristics:

1. strong western boundary currents;
2. strong eastward flowing jets at the mid-basin zero-wind stress curl latitude (multi-gyre wind forcing) or northern boundary current (single gyre wind forcing);
3. weak, broad interior flow, roughly consistent with Sverdrup dynamics.

From a fluid dynamical perspective, any of these three regions of the flow could be unstable. The western boundary current and eastward jet may be unstable if the horizontal curvature of the mean flow is sufficiently strong (i.e., horizontal shear instability) and all three currents may be unstable due to vertical shear in the mean flow (i.e., baroclinic instability). Many of the flows also possess a distinct fourth region, which we will call the westward recirculation. It returns mass from the eastward jet flow without penetrating into the interior. This recirculating flow can also be unstable, either by horizontal shear or baroclinic processes. In fact, because it is a westward current, less vertical shear is required for baroclinic instability than for the eastward jet (Pedlosky 1964a). Which of these currents are unstable and produce the energy to sustain the model eddies?

Maps of K' are suggestive of where the eddy energy sources are to be found. Consider Fig. 2. The single gyre case represented has maximum K' in or near the recirculation region while the multigyre cases generally show maximum K' somewhere in the eastward jet. The multigyre situation varies from case to case; sometimes K' is maximum very near the western boundary and sometimes well offshore. But clearly, maximum K' is associated with strong upper layer currents and not with the interior flow. Efforts to investigate the regional energetics of some of these cases confirm that the energy transformations tend to be maximal in the areas of maximal K' (Semtner and Mintz 1977, Harrison and Robinson 1978, Holland 1978).

These results led Haidvogel and Holland (1978) to examine whether the time and space scales of the model eddies could be estimated by evaluating the most unstable eigenfunction of a meridionally and vertically varying zonal jet. Figure 3 shows some snapshots of different model instantaneous fields. Several

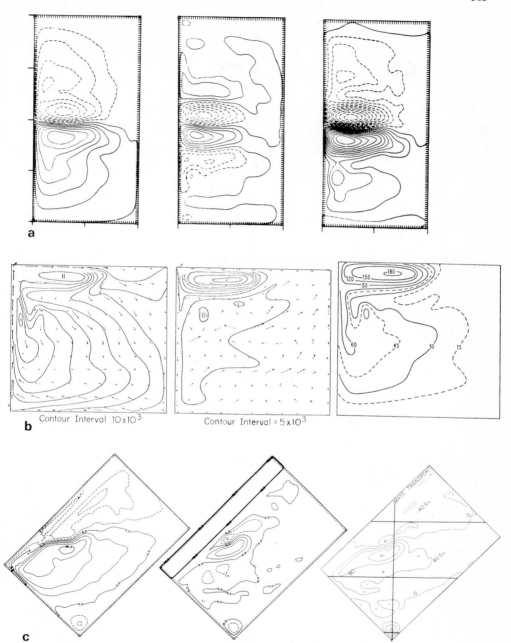

Fig. 1a–c. The time-mean fields of surface flow, deep flow, and total mass transport *(left to right)* for: *(top)* a double-gyre wind driven $1\,000 \times 2\,000$ km^2 ocean (Holland, 1978); *(middle)* a single-gyre wind and thermally driven $1\,000 \times 1\,000$ km^2 ocean (Robinson et al., 1977); *(bottom)* a 2-1/2 gyre wind and thermally driven $2\,000 \times 3\,000$ km^2 ocean with a northwest continental slope (Semtner and Mintz 1977)

384

W.R. Holland et al.

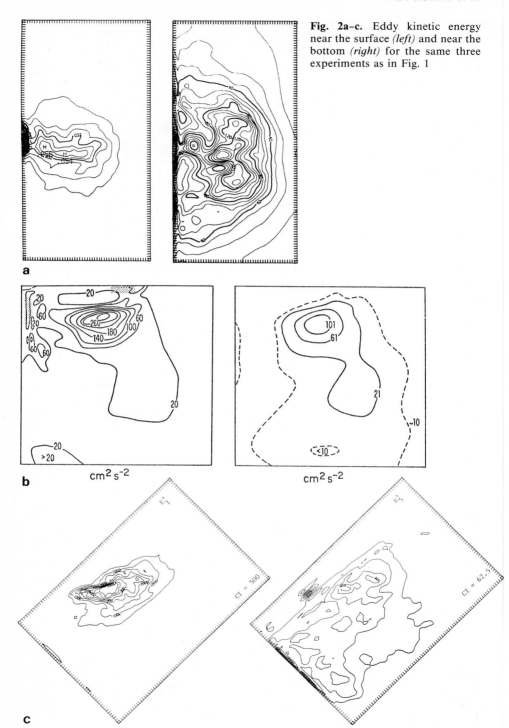

Fig. 2a–c. Eddy kinetic energy near the surface *(left)* and near the bottom *(right)* for the same three experiments as in Fig. 1

different types of behavior can be seen. Sometimes the eddy length scale is comparable to the basin scale and has much the same character over the whole basin (Fig. 3, middle) and sometimes the eddy properties are quite different in different parts of the basin. Haidvogel and Holland had some success predicting the characteristics of the eddies in the combined eastward jet and westward circulation when they did their linear instability analysis of a section of the instantaneous flow. Use of mean flow sections produced less satisfactory results. Harrison and Robinson (1979) tried to account for the behavior away from strong currents, with a linear boundary forcing model that neglects mean currents in the interior. They found that the interior eddy field in several ECGM cases corresponded to a weakly dissipated forced mode that is quite similar in form to a free barotropic basin mode. The excitation of basin modes appears to occur preferentially in very small ocean basins.

The picture that emerges from this work is that the eddy kinetic energy results from some type of instability of either the eastward jet or its westward recirculation or both. Approximately equal number of experiments show "baroclinic", "barotropic," and "mixed" instability as the major source of K′. The characteristics of the eddy field near the strong currents appear to be strongly affected by the instability properties of these currents and may be nearly barotropic or very strongly sheared in the vertical. Holland and Haidvogel (1980) suggest, in fact, that realistic Gulf Stream parameters place the instability in the transition regime between baroclinic and barotropic dominance. In the interior, the eddy field tends to be more barotropic than the strong current eddies and may plausibly result from the radiation of energy outward from the strong current region in these simple models. Baroclinic instability of the interior mean flow does not play a dominant role as an energy source for K′ in any of the studies mentioned above, but several studies (Semtner and Mintz 1977, Semtner and Holland 1978) show a secondary baroclinic instability in the southwestward and westward flow of the Sverdrup gyre. None of the experiments seems to have an important instability of the western boundary current.

17.2.1 Do Mesoscale Phenomena Play a Fundamental Role in the Character and Dynamics of the Time-Mean Circulation?

This second question lies at the heart of the mesoscale problem. It is clear that K′ can be locally much greater than $\overline{K}$ and that $\langle K' \rangle$ can be larger than $\langle \overline{K} \rangle$. Further, the eddy field can completely overwhelm the mean fields in an instantaneous glimpse of the flow, especially in the deep ocean. But neither of these facts establishes that the eddy plays an important role in governing the mean circulation. More detailed analysis of the flow results is necessary.

Several approaches have been tried. The simplest perspective comes from evaluating the individual mean and eddy terms in the time-averaged model equations to see if the eddy terms are ever of leading order in the pointwise budgets. This approach has been taken in a number of EGCM studies (Holland and Lin 1975a, b, Robinson et al. 1977, Semtner and Mintz 1977, Holland

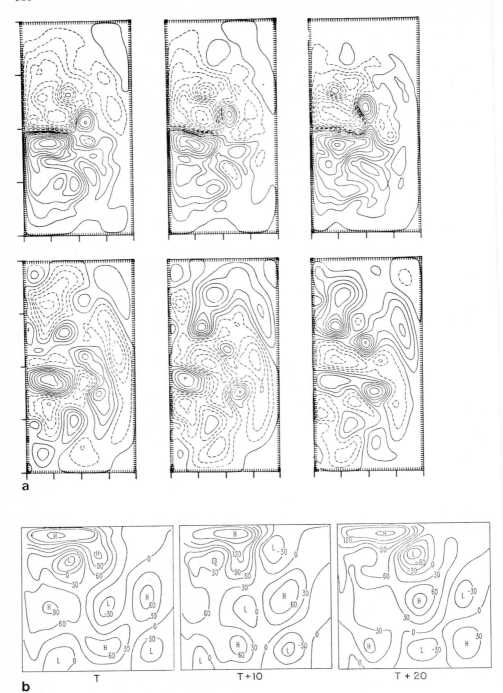

Fig. 3a, b

Fig. 3a–c. Instantaneous maps of surface flow and deep flow at approximately 10 day intervals in the experiments of **a** Holland (1978) and **b** Semtner and Mintz (1977). The **c** *panels* are for mass transport in the experiment of Robinson et al. (1977)

1978, Holland and Rhines 1980). Different authors have examined different equations, and systemmatic efforts to examine all the basic equations (heat, momentum, vorticity) have not been reported. However, in every instance discussed in the literature the eddy term has been part of the lowest order balance in some equation in some part of the circulation.

Interpreting this result is not as straightforward as one would like. In most instances, the spatial distribution of the various eddy terms in the mean equations is complex. Generally these terms have considerable spatial structure and vary importantly on length scales as small as 100 km. As the mean flow patterns are generally (but not always) of larger scale, how should one interpret the result that, for example, the eddy term is order one and positive in a given dynamical balance in a particular part of the mean flow, and order one and negative in the same balance in the same part of the mean flow 100 km away? Figure 14 of Holland and Lin (1975a) provides an illustration of this type of behavior and is reproduced here as Fig. 4. Particularly in the vorticity equation, where the various terms are highly differentiated, one finds considerable small-scale structure in the eddy terms.

A possibility exists that some of the spatial variations result from insufficiently long time-averaging intervals. This has been examined for two experiments and, although the eddy fields continue to adjust in detail with longer averaging intervals, the general character of the variability of the field does not change importantly. The models intrinsically produce small scale structure.

THE RATES OF CHANGE OF VORTICITY AND
THE MEAN VORTICITY

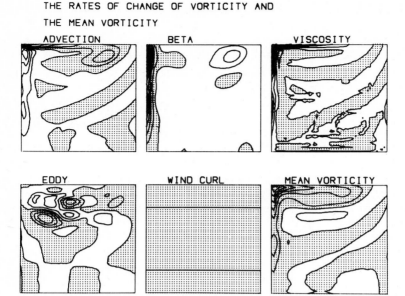

SEE LEGEND FOR CONTOUR INTERVALS

Fig. 4. The maintenance of the mean vorticity in a single-gyre experiment of Holland and Lin (1975a). The contour interval for the upper three maps is 1.0×10^{-7} cm s^{-2}, for eddy and wind curl 0.2×10^{-7} cm s^{-2}, and for the mean vorticity 0.2 cm s^{-1}

This suggests the need to go beyond point balances and to compare regional transports, sources, and sinks resulting from different terms in the mean equations. One can examine the relative importance of different processes in selected regions that correspond to major components of the mean flow.

Regional analysis of vorticity budgets has been used by Holland and Rhines (1980) and Harrison and Holland (1981). These studies clearly illustrate an important aspect of this type of analysis, namely that the selection of analysis regions and analysis techniques can influence the nature of the results. This should not be surprising, since the method consists of evaluating integrals over spatially variable fields. Typically, regional definitions are established by external criteria – for example, Harrison and Robinson (1978) sought to find a region of their flows over which Reynolds stresses provided a pure energy conversion between $\overline{K}$ and K'; Holland and Rhines (1980) were interested in budgets over regions defined by mean flow streamlines in order to use a circulation theorem; and Harrison and Holland (1981) were interested in the vorticity budgets over various subregions of the model mean flow. However, different criteria for selecting regions and different analysis approaches lead to different information and can result in different interpretations.

Within the above context, the regional budget results are nevertheless quite interesting. In the experiment studied by Holland and Rhines (1980) and Harrison and Holland (1981), the eddies clearly play a major role in the mean circulation, because they transport most of the vorticity put into each gyre by the wind stress curl across the boundary between the gyres, thereby allowing equilibration to take place without much need for a frictional vorticity sink. In the absence of mesoscale eddies the antisymmetry of the experiment would require that each gyre attain equilibrium through frictional mechanisms. Further, Holland and Rhines (1980) have shown that the vertical and horizontal fluxes of vorticity by the eddies can be very important in the budgets of regions defined by mean streamlines. In such budgets, where only the wind, model friction and eddy effects can be non-zero, the eddy terms are clearly important. However, Harrison and Holland (1981) found that the zonally averaged net horizontal vorticity transport by eddies seems not to be important except near the boundary between gyres, and that eddy transport is also small over many regions of the mean flow in which the eddy term is pointwise large. The notable exception is in the vicinity of the westward recirculation, where strong eddy heat flux divergence acts as a vorticity source for a lower layer gyre.

Viewed from the perspective of basin-integrated energy budgets, mesoscale phenomena have been found to play a significant role in the dissipation of kinetic energy in a number of experiments. Here the relevant comparison is to examine the energy loss through dissipation of K' versus the energy gain by external forcing of the mean circulation. In those EGCM cases in which a strongly scale-selective operator (∇^4) provides the lateral friction and a bottom friction process is also included, the mesoscale seems always to play an important role in the basin energy budgets (Holland 1978, Harrison 1979). The mesoscale plays a lesser role in those experiments which use less scale-selective horizontal frictional mechanisms.

It has been argued on the basis of scale analysis that, as long as the model mesoscale and mean flows have amplitude and length scale ratios roughly similar to those in the ocean, then the non-dimensional number A_n/RL^n provides an approximate inverse measure of the importance of the mesoscale in model energy budgets (Harrison 1980a). Here, the horizontal friction and bottom friction are modeled by $[A_n(\nabla^n)u]$ and $-Ru$, respectively; and L is the mean flow length scale of the model Gulf Stream system. Existing scale-selective horizontal friction calculations all have $A_4/RL^4 < 0.1$, and indeed the eddies are important in the energy budgets of these experiments. Numerical constraints make it difficult to achieve A_2/RL^2 smaller than 1.0, and "eddy viscosity" friction experiments have not had eddies as important in their budgets. Further experiments are necessary to determine whether other parameters (e.g., the growth rate of strong-current instabilities) may also influence the eddy contribution in the budget energetics. But clearly the choice of the model friction(s) has a strong effect on energy budget results in the existing range of A_2, A_4, R parameters. This is regrettable, since it is not clear which frictional mechanism is more appropriate for these calculations.

To summarize briefly, the model mesoscale is important in the point-by-point dynamical balances throughout the flow of some experiments, but is only significant in parts of the strong current regions in others. Regional transport calculations suggest that the eddy transports can be very important in certain analysis regions and of much less importance in other regions of the same flow field. Finally, much remains to be understood about the implications of different modeling assumptions on the importance of eddies.

17.2.3 Where the Effects of Mesoscale Circulations Are Important, Can They Be Parameterized in Terms of Mean-Field Quantities?

The problem of parameterizing the eddy terms that exist in the mean equations in terms of mean variables is important, given that eddies are in fact important in the dynamics of the mean circulation. If satisfactory parameterizations can be developed, they can be used to improve the relatively coarse spatial resolution ocean calculations, whose purpose is to advance our understanding of the water masses of the ocean and the effect of the ocean on climate, without resolving the mesoscale explicitly. Perhaps because the EGCM results have shown a wide variation in the importance of the mesoscale for the mean circulation, there has been relatively limited effort devoted to the parameterization problem. We shall briefly review the available results.

The traditional assumption has been that mesoscale eddy terms might be describable in "eddy viscosity" form, i.e., a constant times the Laplacian of the mean field. This parameterization is the one which has been used in almost all coarse resolution ocean general circulation studies to describe the effects of motions not resolved by the model. It is thus a natural candidate to compare with the various eddy terms that come out of the analysis of EGCM experiments.

There is a problem in judging the quality of a parameterization. Ideally, one should repeat the numerical experiment, including the hypothesized parameter, and compare the mean fields of this calculation with those of the original calculation. To our knowledge this has never been done with a closed-basin mid-latitude EGCM experiment. Other approaches are to compare regional transports, evaluated from the parameterization and from the eddy terms, or to make point by point comparisons of the two fields.

Generally, the eddy term fields are considerably more complex spatially than are the fields of the Laplacians of the mean fields, so that few detailed comparisons have been made. In one case (Harrison and Robinson, 1978), correlations between the divergence of the Reynolds stress tensor and the Laplacian of $\bar{u}$ and $\bar{v}$, between the divergence of the eddy heat flux and $\bar{\nabla}^2\bar{T}$, and between the Reynolds stress eddy vorticity equation term and the Laplacian of the time-averaged vorticity were sought over various different regions of the model domain (Harrison 1978). No satisfactory correlations were found, but this may be due to the dominance of explicit Laplacian diffusive processes in that case. In an unpublished analysis of the experiment of Semtner and Mintz (1977) with biharmonic diffusions, larger regions of consistent correlation are found, and an effective heat diffusion coefficient ranges from 10^7 to 10^8 cm^2 s^{-1}. However, Holland and Rhines (1980) found in another experiment that there were complex regions of negative and positive effective eddy heat diffusion coefficient. Much more work is needed to understand these early results.

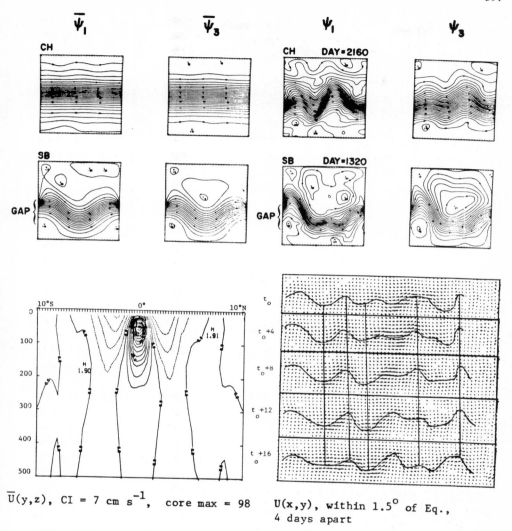

$\overline{U}(y,z)$, CI = 7 cm s^{-1}, core max = 98   U(x,y), within 1.5° of Eq., 4 days apart

Fig. 5. *Above* Time-averaged and instantaneous flow patterns from the idealized Antarctic study of McWilliams et al. (1978), for the cases of an open channel (CH) and a partially blocked channel (SB). *Below* Time-averaged and instantaneous representations of the equatorial undercurrent in the simulation of Semtner and Holland (1980)

The high-latitude periodic-domain experiments of McWilliams et al. (1978) and the equatorial basin experiment of Semtner and Holland (1980) indicate that the parameterization of eddy effects in these contexts may be simpler than in mid-latitude situations (see Fig. 5). In the Antarctic channel experiment, baroclinic instability of the circumpolar flow leads to downgradient lateral heat transfers of order 10^7 cm^2 s^{-1}, upgradient lateral momentum transfers of order -3×10^7 cm^2 s^{-1}, and interfacial form drag between layers of order 10^4 cm^2 s^{-1}. The equatorial experiment of Semtner and Holland has a barotropically

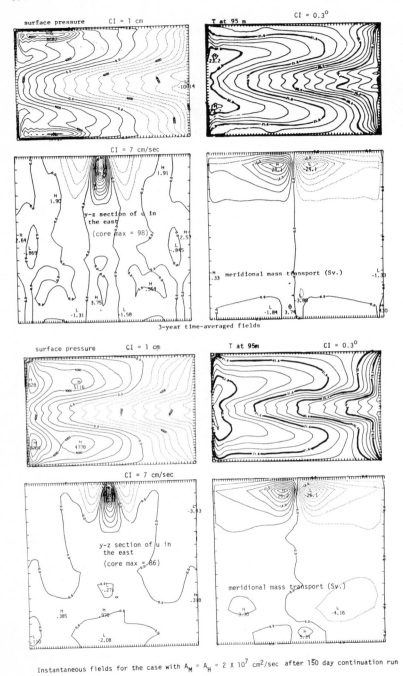

Fig. 6. A comparison of the time-averaged fields of Semtner and Holland (1980) against instantaneous fields for the case with large Laplacian diffusion of heat and momentum

unstable equatorial undercurrent whose time-averaged state can be approximately reproduced by using diffusions of 2×10^7 cm^2 s^{-1} (see Fig. 6).

The most extensive parameterization efforts have been those of Rhines and Holland (1979) and Holland and Rhines (1980). They have discussed a parameterization of eddy effects in terms of transport down the mean potential vorticity gradient, with a variable coefficient of diffusion derived from a Lagrangian dispersion tensor. In the one experiment examined, certain aspects of the eddy terms are rather well reproduced by the parameterization. To be operationally useful, a theoretical relationship between this tensor and the mean potential vorticity field needs to be established.

Parameterization efforts are clearly in their infancy. Much remains simply to describe the general characteristics of model eddy fields, as well as to establish relationships between these and the mean fields. It remains to be demonstrated whether or not adequate parameterization schemes can be devised in order that models of the large-scale circulation need not explicitly resolve the eddy field.

17.2.4 How Do the Answers to Questions About the Role of Mesoscale Eddies Change as the Model Physical Processes Are Changed?

Determining the sensitivity of model results to changes in physics other than parametric changes has not been a major activity in EGCM research. Because EGCM systems display such a wide range of model behavior even when they are highly idealized, it seems sensible to understand first the behavior of the simplest systems. As will be clear from the discussion of the preceding sections, deciding how to analyze the results of even a highly idealized experiment is not a simple task. Limitations on computational resources also encourage the use of simple, efficient models. Thus it is possible to discuss the question of sensitivity of results to model physics primarily in the context of simple EGCM systems; very few experiments with complicated physics have been carried out.

Holland and Lin (1975b) showed that, when horizontal "eddy viscosity" is the model dissipation mechanism, the outcome of experiments is sensitive to the magnitude of the eddy viscosity coefficient and to the nature of the lateral boundary conditions. Changing A_2 by only a factor of 3.3 fundamentally alters the character of the mean flow, the mesoscale variability, and the energetics of the circulation (Fig. 7). Changing the imposed mean stratification also leads to important changes. When quasigeostrophic two-layer equations are used with the "biharmonic" friction ($A_4\nabla^4 u$) and linear bottom drag parameterizations, Holland (1978) has shown that there exists a range of bottom drag amplitudes (2.5×10^{-8} to 1×10^{-7} s^{-1}) over which only modest changes in the model mean flow behavior take place (Fig. 8). While the mean flows are similar, the eddy kinetic energy increases in proportion to the decrease in bottom friction coefficient. Semtner and Holland (1978) have shown that introducing an equivalent surface heat flux at the uppermost interface in two- and three-layer quasigeostrophic systems can produce significant, but not enormous, changes in the model mean circulation and energy budgets. Much more important

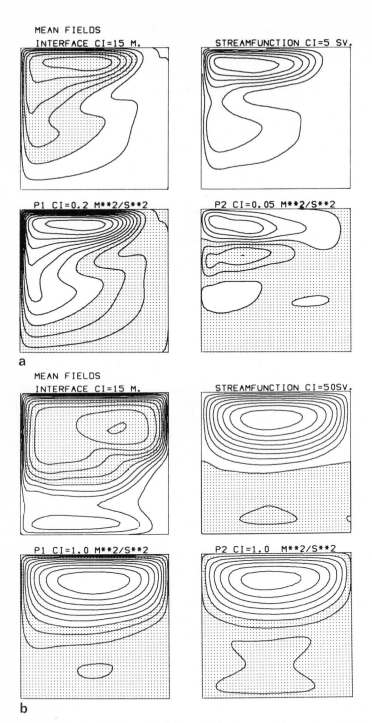

Fig. 7a, b. Mean fields in a $(1000 \text{ km})^2$ ocean with single-gyre wind forcing. *Upper panels* are for lateral viscosity of $3.3 \times 10^6 \text{ cm}^2 \text{ s}^{-1}$ and *lower panels* are for $1 \times 10^6 \text{ cm}^2 \text{ s}^{-1}$. (Holland and Lin 1975b)

changes result when continental shelf/slope topography replaces the flat bottom that has been most widely used. However, many aspects of a primitive equation experiment with topography can be reproduced in a three-layer quasigeostrophic experiment, if the topography and thermal physics of the PE experiment are included in the QG system (Fig. 9).

These various results establish that there is considerable model sensitivity to changes in the model physics. Topography, vertical structure, adiabatic or diabatic thermal physics, form and magnitude of the model dissipation mechanisms, and lateral boundary conditions all make significant differences in model results, in certain areas of "parameter space" (the set of all model physical choices) that have been examined. This variability of model results is itself extremely interesting, because it makes clear the fact that processes which are not well understood and must be parameterized in the models play important roles in establishing the behavior of the mean circulations.

17.3 Comparisons with Observations

From the perspective of geophysical fluid dynamics all EGCM experiments are of interest, since each represents a conceptually possible oceanic system. However, if the experiments are intended to give information about possible behavior of the earth's oceans, only some experiments are directly relevant. Because EGCM modeling is such a young field, rather little attention has been given to trying to "simulate" the ocean. Attention has instead been directed to investigating the effects of different model assumptions in simple systems. Still, a variety of comparisons between model data and ocean data have been made and deserve a brief review.

Comparisons with ocean data are not entirely straightforward, because there are only a few long-time series of direct current measurements in the ocean and because most EGCM experiments have been carried out in rectangular basins that are small compared to the North Atlantic or any other ocean basin. The available North Atlantic data have been largely described in a series of papers (Schmitz 1977, 1978, 1980) and a recent comparison effort, to be described later, has been made by Schmitz and Holland (1982). Most earlier comparisons have been rather casual for several reasons, and these are considered first.

The initial question that must be answered is how to make the comparison. For example, in a rectangular basin, how does one select the model longitudes to compare with Schmitz' data along 55°W and 70°W in the North Atlantic? Then it is necessary to choose the quantities to be compared. Eddy kinetic energy is perhaps the most reliably known ocean quantity, although the mean horizontal flow is reasonably well known at some locations. There are also satisfactory data for estimating $\overline{u'v'}$, $\overline{u'T'}$, and $\overline{v'T'}$ at the POLYMODE Array 2 moorings. Kinematically, one would like the models to give realistic amplitudes of each of these fields as well as comparable patterns of spatial variability. From a dynamical perspective one wants the models to have eddy flux div-

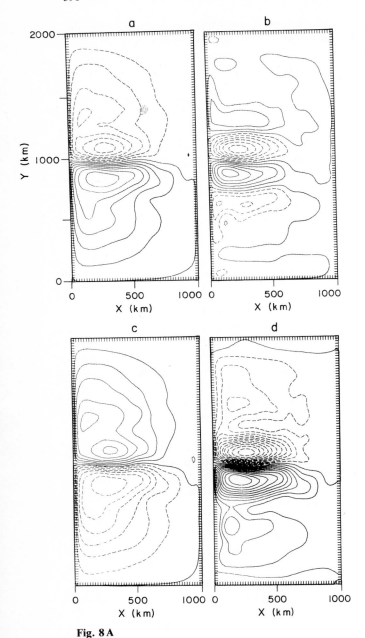

a   b

c   d

Fig. 8 A

ergence fields comparable to those of the ocean. Unfortunately, there is insufficient ocean data for reliable evaluation of ocean eddy flux divergences, so kinematical comparisons are all that have been done at this time (but see Harrison 1980b for some remarks about ocean vs. model divergence results).

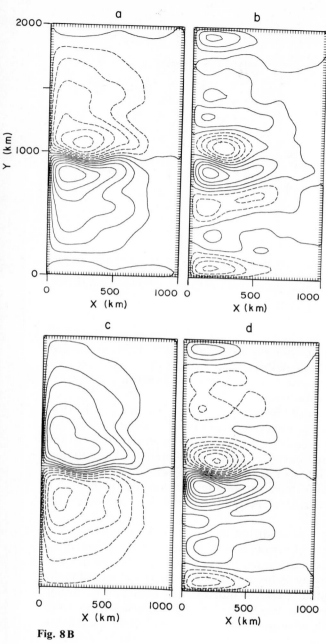

Fig. 8A, B. Mean flows for two experiments with different bottom friction coefficients (Holland, 1978). **Left** The mean fields for experiment 5 **a** ψ_1 (CI = 5000 m^2 s^{-1}; **b** ψ_3 (CI = 1000 m$^{2-1}$); **c** h_2 (CI = 20 m); **d** Transport (CI = 5 × 10^6 m^3 s^{-1}). **Right** The mean fields for experiment 7: **a** ψ_1 (CI = 5000 m^2 s^{-1}); **b** ψ_3 (CI = 2000 m^2 s^{-1}); **c** h_2 (CI = 20 m); **d** Transport (CI = 10 × 10^6 m^3 s^{-1})

Fig. 8 B

The first remark to make is that even the first eddy-resolving studies of Holland and Lin (1975b) showed qualitative correspondences between model and ocean that had not been achieved earlier with steady-state models. The deep Gulf Stream recirculation suggested by Worthington (1976) and later con-

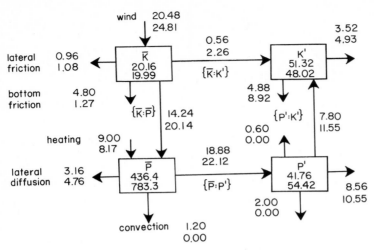

Fig. 9. A comparison *(upper and lower numbers in pairs)* of the box integrated energetics of the primitive-equation experiment of Semtner and Mintz (1977) with a quasigeostrophic experiment using the numerical model of Holland (1978). (From the study of Semtner and Holland 1978)

firmed by long-term current meter moorings (Schmitz 1977, 1978) was reproduced by certain of the model calculations. This feature of the flow was entirely eddy-produced in Holland and Lin (1975a, b) and exists in a qualitative sense in nearly all later EGCM studies.

The second flow feature produced by the eddy models is related to the first: as a consequence of the deep, eddy-driven recirculation, the mean Gulf Stream transport is strongly enhanced over its Sverdrup value. An explanation for this enhancement had been suggested in earlier, steady-state models as due to non-linear recirculation (Veronis 1966) or to topographic-baroclinic interactions (Holland 1973). The eddy models have yet another mechanism: eddy-driven, deep transport. It may well be that all three "mechanisms" are operating in the real ocean.

The third flow feature that was predicted in the earliest eddy models was the qualitative nature of the eddy energy pattern. It now seems obvious to us that the western boundary current extension region should be the site of vigorous eddy activity, but in the early 1970s a number of other explanations for mid-ocean eddies were suggested. These included direct atmospheric forcing, mid-ocean instabilities, topographic-mean current interactions, and so forth. Now, in retrospect, with the observations from Dantzler (1976, 1977), Schmitz (1977, 1978), and others, it would seem that Gulf Stream instability is the primary eddy-generation mechanism in the large-scale circulation and that these other mechanisms are of secondary importance (except locally, of course). This was perhaps the most striking discovery by the early models, found at a time when the observations were largely unavailable and little was known about the processes of eddy generation.

To illustrate the nature of the qualitative comparisons, Fig. 10 presents 1 500 m results from Robinson et al. (1977). The maximum K′ in the model is roughly consistent with the maximum ocean value (125 cm² s⁻² vs. 260 cm² s⁻²), but the interior values are rather large (~25 cm² s⁻² vs. 1 → 10 cm² s⁻²). Further, the model has a rather small zonal extent (~500 km) for the region of K′ > 100 cm² s⁻² (see Fig. 2) while amplitudes > 100 cm² s⁻² are found over at least 1500 km in the ocean (55°W to 70°W). Comparisons of $\overline{u'v'}$ are less favorable. The model maximum value is several times smaller than the ocean maximum value (20 cm² s⁻² vs. 65 cm² s⁻²), and the overall patterns do not compare well. Figure 11 presents comparison results from Holland (1978). Note that zonal averages of model values are compared with 70°W data from Schmitz (1977), that the latitude of the model jet has been taken to be 38°N, and that different amplitude scales have been used for the model and ocean data. K′ data are typically small in the model by a factor of 4, $\overline{u'v'}$ data are

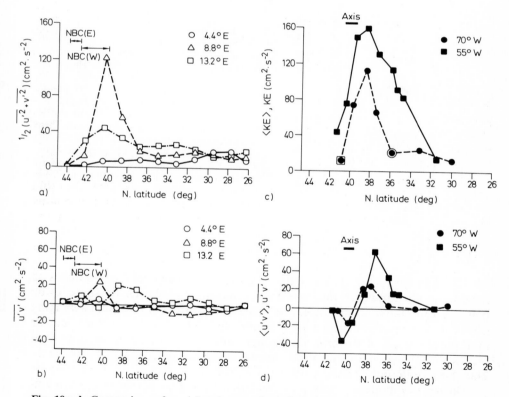

Fig. 10a–d. Comparison of model and ocean from the primitive equation experiment of Robinson et al. (1977). **a** and **c**, $1/2\ \overline{(u'^2 + v'^2)}$ vs. latitude; **b** and **d** $\overline{u'v'}$ vs. latitude. Model data are plotted at three different longitudes, east of the western boundary, with 44 °E through a quiet part of the northern boundary current 8.8 °E through a vigorously eddy-energetic region of that current and 13.2 °E through the separation and turning region. Ocean data are at latitudes 70 °W and 55 °W from Schmitz (1977)

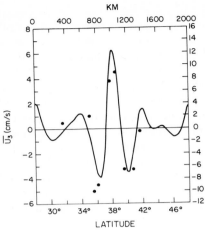

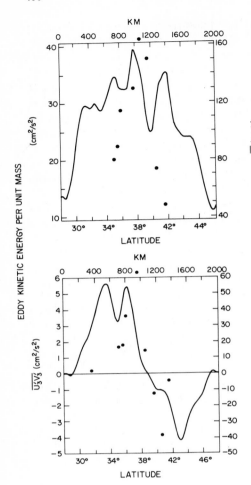

Fig. 11. Comparisons from Holland (1978). *Upper* The meridional distribution of zonally averaged eddy kinetic energy in experiment 3 (*solid line*, scale on the *left*). Also shown are observations presented by Schmitz (1977) along 55 °W (*dots*, scale on the *right*). *Middle* The meridional distribution of the zonally averaged Reynolds stress term $\overline{u_3' v_3'}$ in experiment 3 (*solid line*, scale on the *left*). Also shown are observations presented by Schmitz (1977) along 55 °W (*dots*, scale on the *right*). *Lower* The meridional distribution of zonally averaged eastward deep mean flow u_3 in experiment 3 (*solid line*, scale on the *left*). Also shown are observations (Schmitz 1977) along 55 °W (*dots*, scale on the *right*)

small by a factor of 4 to 10, and $\bar{u}$ data are small by a factor of 2 compared with the ocean data. However, there are some interesting similarities in some of the patterns, especially $\bar{u}$. It is not simple to assess the effect of the zonal averaging of the model results, except that it almost certainly produces values smaller than those typical of the mid-basin longitudes. However, the more important fact is that the model calculations are for a small ocean basin (1 000 km × 2 000 km), quite unlike the North Atlantic in size.

These early models certainly produce fields not grossly inconsistent in amplitude with ocean data, but they do not generally have length scales as large as those suggested by the relatively sparse ocean data. As noted above, if the eddies in the models are to play comparable roles to those in the ocean, the dynamical comparisons must be satisfactory. Different model length scales thus will require different eddy quadratic term amplitudes, to attain comparable eddy flux divergence terms. It is hoped that further ocean data and more realistic model simulations will make these dynamical comparisons possible in the relatively near future.

A rather different problem from how the existing EGCM's compare with the ocean is that of selecting the model physical assumptions and parameters that will lead to satisfactory simulations of the ocean. At present topographic effects, irregular basin geometry, time-dependent wind forcing, and thermal forcing have barely begun to be investigated, but the few results available suggest that each effect may be of some importance in the long-time average dynamics. Further, existing results demonstrate considerable sensitivity to assumptions about how vorticity and energy are removed from the model system. While it is possible that the inclusion of further physical processes will reduce the model sensitivity to the "dissipation" processes, there is no evidence at present that this is the case. Quasigeostrophic dynamics, while well suited to many of the problems now being studied via EGCM experiments, are not likely to remain satisfactory when the range of model physics is expanded. Unfortunately, primitive equation experiments, in which the model establishes its own vertical and horizontal stratification, require much more computational effort than do experiments conducted with quasigeostrophic dynamics. It seem likely that some types of intermediate models will be useful.

The only comprehensive comparison between model results and observations has been carried out by Schmitz and Holland (1982), where a number of new two-layer numerical experiments show the amplitude and spatial dependency of eddy energy and mean currents upon various nondimensional governing parameters. If the basin size and the wind strength are realistic, the strength and pattern of the mean gyre, including the recirculation, are reasonable. The statistics of the eddy field, particularly abyssal eddy kinetic energy, have the correct amplitude and meridional structure (Fig. 12) but the pattern does not extend zonally over as large a range of longitudes in the model studies as in the observations at 55°W and 70°W. An explanation for this discrepancy has been found in ongoing work by Holland, and more realistic patterns of mean currents and eddy kinetic energy are produced in recent numerical experiments that extend the work of Schmitz and Holland.

17.4 Unsolved Problems and Future Prospects

Eddy-resolving general circulation model studies have substantially contributed to our evolving view about the mechanisms responsible for the amplitude and structure of the gyre-scale oceanic circulation. Most of this work has been carried out in a rather idealized framework, that is, rectangular, constant depth basins forced by steady winds. Obviously a number of important mechanisms are not present in such calculations. In addition to understanding internal physical effects, we need to understand the effects of basin shape and topography, the connection of large-scale flow to coastal regions and marginal seas, and indeed the interactions of the various ocean basins. The role of transient forcing

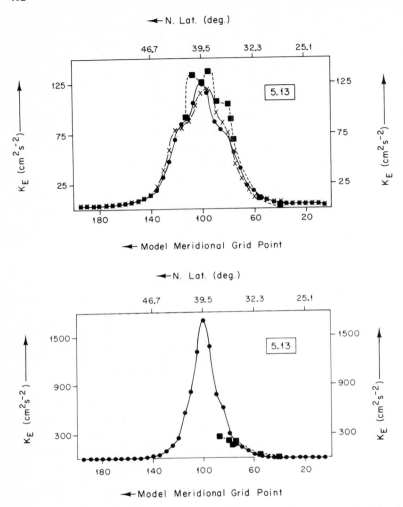

Fig. 12. Eddy kinetic energy as a function of latitude. (Schmitz and Holland 1982). Squares show observed values and dots and crosses show model values for a $4\,000 \times 4\,000$ km^2 basin with realistic wind forcing

by winds has hardly been considered and little is known about the additional role of buoyancy forcing at the sea surface on all time scales.

The difference between midlatitude and equatorial mean and transient flows needs further study. How do midlatitude and equatorial circulations merge? The same questions apply to Southern Hemisphere oceans, where meridionally blocked midlatitude gyres border the channel-like Circumpolar Current.

How do traditional notions about water mass formation fit into this new scheme of strongly turbulent oceans? How does seasonal (and other time scale) forcing interact with the similar time-scale phenomena associated with mesoscale eddying? What are the respective roles of mean advection and eddy mixing in redistributing heat and salt as well as passive tracers? How importantly do the eddies contribute to the net meridional heat flux in the ocean?

This list goes on and on. Given the complexity and diversity of phenomena that may be important, it is clear that the last 6 or 7 years of effort on modeling the eddy field is only the beginning. The framework established by these models, however, is one in which all of these questions can be answered. The use of such models, together with simpler theoretical models with only a few processes operating and with observational programs which continue to elucidate the dynamical nature of the various regions of the world ocean, should make possible a great improvement in our understanding of the turbulent ocean. Such a physical understanding will allow us to solve crucial problems related to climate change, waste disposal, fisheries management, and air/sea interaction.

18. Periodic and Regional Models

D.B. Haidvogel

18.1 Introduction

Periodic (often called process) models and regional models resemble eddy-re-solving general circulation models (EGCM's) in their use of fine horizontal re-solution to achieve a simultaneous explicit representation of both the large-scale and mesoscale components of the oceanic circulation. They differ, how-ever, in that periodic and regional models seek to determine the dynamical equilibrium characterizing the mesoscale variability in a *local* block of ocean, whose interconnection to the remainder of the oceanic gyre is either assumed statistically unimportant (via an assumption of horizontal homogeneity) or pa-rameterized (in terms of some suitable boundary constraints on the local re-gion). The resulting local dynamical models are used to examine the initial de-velopment and subsequent statistical equilibrium which ensues during the evo-lution of the mesoscale eddy field from some specified set of initial conditions. By selective inclusion of physical processes and choice of environmental pa-rameters, the relative importance and dynamical significance of such factors as nonlinear and bottom topographic interactions, planetary vorticity variations, stratification, local viscous effects, and forcing by the large-scale mean circula-tion can be assessed.

Although a few EGCM simulations have utilized primitive equation dy-namics (see Chap. 17), periodic and regional modeling studies have typically been applied to the mid-ocean regions, for which the quasi-geostrophic approx-imation is appropriate (McWilliams 1976). Prompted in part by the initial ob-servational and theoretical interest in describing mesoscale eddy properties in mid-latitude regions, a local β-plane representation of the planetary vorticity gradient has also generally been used. Under these dynamical assumptions and, for definiteness, adopting a two-level vertical structure (Phillips 1951)[1], the associated quasi-geostrophic equations can be written in non-dimensional form as:

$$\frac{\partial q_1}{\partial t} + J(\psi_1, q_1) = -\nu \nabla^6 \psi_1 + \mathscr{F}_1 \tag{1}$$

---

[1] In general, the leveled models to be discussed are strictly equivalent to a layered model only in circumstances for which fluctuations in layer depth in the latter can be ignored. However, we will follow conventional practice and use the terms "multi-level" and "multi-layer" synonymously.

Eddies in Marine Science
(ed. by A.R. Robinson)
© Springer-Verlag Berlin Heidelberg 1983

$$\frac{\partial q_2}{\partial t} + J(\psi_2, q_2) = -\kappa \nabla^2 \psi_2 - \nu \nabla^6 \psi_2 + \mathscr{F}_2 \qquad (2)$$

$$\qquad\qquad\qquad\qquad\; \text{SD} \qquad\;\; \text{LM} \qquad \mathscr{F}$$

where

$$q_1 = \nabla^2 \psi_1 + \beta y + \left(\frac{1}{1+\delta}\right)(\psi_2 - \psi_1) \qquad (3)$$

and

$$q_2 = \nabla^2 \psi_2 + \beta y - \left(\frac{\delta}{1+\delta}\right)(\psi_2 - \psi_1) + \alpha h . \qquad (4)$$

$$\quad\;\; \text{RV} \;\;\; \text{BY} \qquad\qquad\qquad \text{VS} \qquad\quad \text{T}$$

Equations (1) and (2) are statements of conservation of upper and lower level potential vorticity – q_1 and q_2, respectively – following the horizontal fluid motion at each level. As indicated, nonconservative effects, represented by the terms on the righthand sides of Eqs. (1) and (2), are provided by surface drag (SD), lateral mixing (LM)[2], and forcing ($\mathscr{F}$) of the system by external driving agents and/or larger-scale motions. Contributions to the potential vorticities are made by the relative vorticities [RV; $\nabla^2 \psi_i$ ($i = 1, 2$)], planetary vorticity (BY), and vortex stretching due to equivalent "interfacial" displacements between levels 1 and 2 (VS) and topographic roughness (T).

The non-dimensional parameters which enter the physical description of the two-level quasi-geostrophic system are:[3]

$$\beta = \left(\frac{\beta_{\text{dim}} \lambda^2}{U}\right)$$

$$\delta = \left(\frac{H_1}{H_2}\right)$$

$$\alpha = \left(\frac{f_0 h_0 \lambda}{H U}\right)$$

$$\kappa = \left(\frac{\kappa_{\text{dim}} \lambda}{U}\right)$$

$$\nu = \left(\frac{\nu_{\text{dim}}}{U \lambda^3}\right)$$

representing respectively the non-dimensionalized effects of the planetary vorticity and bottom topographic gradients (β and α, respectively), and dissipation (κ, ν), as well as the ratio of the thickness of the fluid layers surrounding levels 1 and 2. The subscript "dim" denotes a dimensional quantity. These two-level equations have been non-dimensionalized by an RMS velocity scale U and the

---

2 The prescribed form of these higher-order viscous terms are discussed, though not fully justified, below.

3 The complete formulation of an initial/boundary value problem based on Eq. (1–4) also involves the specification of other parameters reflecting the character of the initial and boundary conditions, and the details of the selected numerical solution rechnique (see below).

Rossby deformation radius, λ. By suppressing the vortex stretching term in Eq. (4), Eq. (2) describes the dynamics of a single-layer (or barotropic) system, although λ must now be interpreted as some other appropriately chosen length scale. Other dimensional environmental parameters have standard interpretations – see, for instance, Haidvogel and Held (1980).

Note should be taken of the parameterization of lateral mixing processes adopted in Eqs. (1) and (2). Bretherton and Karweit (1975) were the first to implement periodic ocean models analogous to that described by Eqs. (1–4). Their use of the higher-order (often referred to as biharmonic) lateral mixing mechanism was apparently suggested by the work of Leith (1971). In comparison to the traditional (Laplacian or harmonic) form of vorticity diffusion, the higher-order biharmonic operator is much more scale-selective, and hence discriminate, in its removal of vorticity variance (enstrophy) at higher wavenumbers. Despite this heuristic argument for the preferability of the biharmonic form for lateral dissipation, there is little agreement, theoretical or otherwise, on the appropriate parameterization of subgridscale effects in these models. Therefore, although the majority of periodic and regional modeling simulations to date have employed the biharmonic functional form shown in Eqs. (1) and (2), other frictional and filtering mechanisms are also currently in use, including Shapiro-type filters (Shapiro 1971, Robinson and Haidvogel 1980) and higher-order mechanisms suggested by turbulence closure considerations (Holloway 1976, Salmon 1978).

Equations (1–4) are continuous in both horizontal space dimensions (x, y) and in time (t). For situations in which solutions to these equations are not easily found by analytic means, numerical solution techniques must be used. Among other things, the expression of Eqs. (1–4) in a form suitable for numerical solution involves choices of spatial and temporal discretization, as well as the specification of appropriate boundary and initial conditions. It is in relation to these critical modeling choices that periodic and regional models differ most significantly.

The periodic modeling strategy assumes that the statistical dynamical character of the mesoscale eddies in a block of ocean is locally determined and effectively independent of distant eddy generation and/or dissipation regions (except insofar as they may dictate the overall level of eddy energy in a given location). Consistent with this assumed spatial homogeneity of eddy statistics, periodic models invoke truncated Fourier series and (hence) periodic boundary conditions to cast Eqs. (1–4) in a form suitable for numerical solution. The details of this periodic solution technique have been given elsewhere by several authors (e. g., Bretherton and Karweit 1975, Haidvogel and Held 1980). The existence and uniqueness of solutions to the periodic quasi-geostrophic equations have been examined by Bennett and Kloeden (1981), who demonstrate that strong (i. e., smooth) solutions can be shown to exist for a finite time inversely proportional to the initial gradients of vorticity and temperature.

Regional (or open ocean) models are also local dynamical models. In contrast to their periodic analogues, however, regional models are separated from the remainder of the oceanic gyre by arbitrary boundaries across which fluid is free to move in a complicated (non-periodic and time-dependent) manner which reflects not only the internal physics of the local block of ocean but also the state of the exterior environment. The statistical character of the local eddy field is therefore, in general, neither homogeneous nor independent of the exterior flow field. Because of the necessity for parameterizing the region's interconnection with the exterior flow field in which it is embedded, the specification of appropriate boundary conditions on the edges of the open domain is a difficult (and unsolved) problem. The regional models to be discussed in Section 18.7 have been closed by specifying values of streamfunction at all boundary points and values of vorticity at boundary points characterized by inward-directed flow, a prescription first proposed by Charney et al. (1950) for the inviscid barotropic vorticity equation. Despite earlier suggestions to the contrary, the problem as enunciated by Charney, Fjortoft, and von Neumann is actually ill-posed (Bennett and Kloeden 1978), in that it allows for the spurious appearance of vorticity discontinuities (or shocks) in the fluid which are not continuously related to the imposed boundary data. It may be presumed that the addition of a suitable frictional mechanism results in a well-behaved problem; however, this possibility has not been fully investigated. A more complete discussion of the physical, mathematical, and computational questions associated with the choice of open boundary conditions as well as the discrete implementation of regional models with a variety of numerical techniques has been given by Haidvogel, Robinson, and Schulmann (1980).

A few idealized dynamical simulations of periodic model type have also been carried out within closed boundaries (Bretherton and Haidvogel 1976, Haidvogel and Rhines 1983). In this geometric respect, they are therefore akin to the closed-basin EGCM models discussed in Chapter 17. In these simulations, however, the intent is not to achieve a complete description of the general circulation, but rather to investigate isolated physical processes, such as topographic and forced two-dimensional turbulence, in the presence of oceanic boundaries. These papers are included in the subsequent discussion.

18.2 Two-Dimensional Turbulence

By suppressing (1) latitudinal variations in planetary vorticity, (2) vortex stretching due to interfacial and topographic deformations, and (3) all non-conservative effects – i.e., by setting $\beta = VS = \alpha = \kappa = \nu = \mathscr{F}_1 = \mathscr{F}_2 \equiv 0$ – Eqs. (1) and (2) individually reduce to statements of conservation of relative vorticity:

$$\frac{\partial \zeta}{\partial t} + J(\psi, \zeta) = 0 \tag{5}$$

where

$$\zeta = \nabla^2 \psi , \tag{6}$$

and subscripts are unnecessary, each layer being independent of the other. As is now well known, the strict conservation of the single scalar ζ puts powerful constraints on the evolution of flows in an inviscid flat-bottomed, uniformly rotating, barotropic system. The qualitative nature of these constraints on two-dimensional turbulent flow can be seen by first observing that the continuum Eqs. (5) and (6) admit of two conservation statements for kinetic energy and enstrophy:[4]

$$\frac{\partial}{\partial t}[\int \tfrac{1}{2}|\nabla \psi|^2 \, dk] = \frac{\partial}{\partial t}[\int E(k) \, dk] = \frac{\partial}{\partial t} E = 0 \tag{7}$$

and

$$\frac{\partial}{\partial t}[\int \tfrac{1}{2}(\nabla^2 \psi)^2 \, dk] = \frac{\partial}{\partial t}[\int k^2 E(k) \, dk] = \frac{\partial}{\partial t} Z = 0. \tag{8}$$

Here, we use $E(k)$ to denote the kinetic energy, $\tfrac{1}{2}|\nabla \psi|^2$, contained in Fourier components with wavenumbers $|\underline{k}|(=|k,l|)$ between k and $(k+dk)$, and E and Z the total energy and enstrophy contained in all wavenumbers.

The importance of these conserved quadratic properties in dictating the sense of evolution in two-dimensional flow is most easily seen by considering an idealized initial flow state characterized by an energy spectrum peaked within a narrow band of wavenumbers around some k_0. If, during its subsequent evolution according to Eq. (5), the flow undergoes a broadening of its energy spectrum, then this redistribution of energy must favor low wavenumbers $(k < k_0)$ in order to conserve both $\int E \, dk$ and $\int k^2 E \, dk$. At the same time, enstrophy is preferentially transferred to higher wavenumbers $(k > k_0)$. Corresponding to this spectral space description in which the centers of mass of the energy and enstrophy spectra gradually separate is a physical space picture in which the energy-containing eddies gradually increase in size, while vorticity contours increase in length and are sheared out into ever finer laminae. Computer realizations of purely two-dimensional flow within a doubly-periodic domain confirm these tendencies (Fig. 1). Further refinement and discussion of this qualitative picture have been given by Fjortoft (1953), Herring et al. (1974), Merilees and Warn (1975), and Rhines (1979).

Observing that two-dimensional turbulence has both energy and mean-square vorticity as inviscid constants of motions, Kraichnan (1967) demonstrated the formal existence of two inertial ranges:

$$E_\varepsilon(k) \sim \varepsilon^{2/3} k^{-5/3} \tag{9}$$

and

$$E_\eta(k) \sim \eta^{2/3} k^{-3}, \tag{10}$$

where ε is the rate of cascade of kinetic energy (per unit mass) and η is the rate of cascade of enstrophy. (In a later paper, Kraichnan 1971 showed that the second inertial range must be corrected by a logarithmic factor in order that the

---

[4] More formally, it can be shown for two-dimensional flow that energy dissipation does go to zero as $(\kappa, \nu) \to 0$ for finite initial scales of motion. Statements (7) and (8) are therefore correct in this limit (Rhines 1977). This is not true, however, in three dimensions.

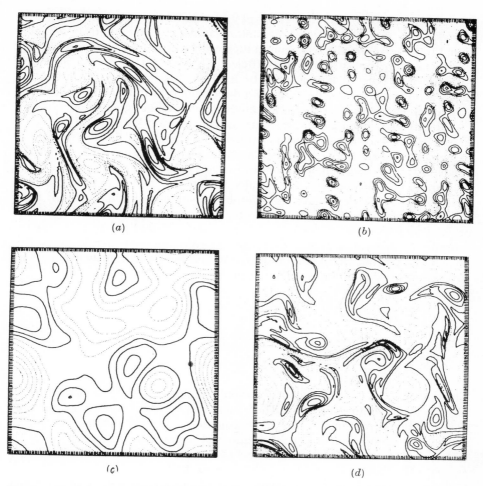

Fig. 1. a Vorticity contours for a typical pure two-dimensional turbulence run, $\beta = 0$, ———, positive; - - -, negative. **b** Streamline field near the beginning of the free-evolution experiment (21st time step). Contour Level (CL) = 0.038. **c** Streamlines at t = 5.8 (725th time step), CL = 0.1. **d** Vorticity contours at t = 5.8, CL = 1.3. (Rhines 1975)

total rate of enstrophy transfer from all wavenumbers be independent of k.) Given the spectral forms (9) and (10), it is easily shown that there is no enstrophy transfer in the first inertial range and no energy transfer in the second. Kraichnan further argued that the $(-5/3)$ range entails energy cascade to low wavenumbers while the (-3) range effects enstrophy transfer to higher k. For a nearly inviscid two-dimensional turbulent fluid in which energy is fed in at a constant rate to a band of wavenumbers near k_0, these inertial range ideas suggest that a quasi-steady state develops in which $E(k) \sim k^{-5/3}$ for $k \ll k_0$ and $E(k) \sim k^{-3}$ for $k \gg k_0$, up to some viscous cutoff.

In contrast to Kraichnan's enstrophy cascade arguments, Saffman (1971) hypothesized that the vorticity distribution in a field of random two-dimensional turbulence develops discontinuities which dictate the spectral shape of the turbulent energy spectrum for sufficiently large k. If this circumstance prevails, Saffman noted that an asymptotic $E(k) \sim k^{-4}$ is expected. Unfortunately, numerical simulations of decaying two-dimensional turbulence have not been able to achieve wavenumber and Reynolds number ranges great enough to verify either the Kraichnan or Saffman inertial range power law dependence (Herring et al. 1974). Nor is experimental evidence conclusive in this regard, although atmospheric energy spectra hint at the existence of a $(-5/3)$ power law behavior for low wavenumber motions (Gage 1979).

Further complicating the numerical verification of turbulent power law behavior is the existence of statistical equilibrium solutions to the spectrally truncated version of the continuous barotropic vorticity Eq. (5). If in analogy to the original partial differential Eq. (5), the ordinary differential equations for the evolution of the retained spectral coefficients conserve total energy and enstrophy in the absence of forcing and dissipation, then the truncated equations have a statistical equilibrium solution of the form

$$E_e(k) = \frac{k}{a + bk^2} \sim k^{-1} \quad \text{(k large)}$$

and   (11)

$$Z_e(k) = \frac{k^3}{a + bk^2} \sim k^{+1} \quad \text{(k large)}$$

where a and b are constants dependent only on the total energy (E) and enstrophy (Z) available to the system. These solutions may be spurious, in that they have no counterpart in the continuous equations. Fox and Orszag (1973) have shown that equilibrium statistical solutions (11) exist for any realizable values of E and Z and that inviscid numerical simulations initiated from a randomly chosen initial state rapidly evolve to approach these equilibrium solutions. (Similar behavior is noted for nearly inviscid turbulence simulations in other geometries, so long as E and Z are constants of motion and an appropriate Liouville's theorem exists for the truncated system. Simulations on a sphere have been described by Frederiksen and Sawford 1980.) Recent numerical simulations by Bennett and Haidvogel (1983) have, in fact, shown that rather large amounts of dissipation may be needed to disturb the tendency of truncated turbulence models to seek steady equilibrium solutions. If weak friction acts, the turbulent system slowly spins down but is never too far away from the equilibrium statistical state appropriate to locally defined values of E and Z. Unfortunately, large amounts of friction, although guaranteeing that a spurious equilibrium solution is not achieved, will nonetheless dictate a spectral shape at large k whose steepness is related directly to the chosen frictional parameterization. Therefore, modelers of turbulent motions in the ocean and atmosphere face a difficult choice: if weak dissipation is invoked, the system will quickly evolve towards an equilibrium statistical solution; with strong dissipation, the high-wavenumber end of the turbulent energy spectrum will be dictated by the (imperfectly known) form of the dissipative mechanism. These considerations underscore the importance of further research in the area of subgridscale parameterization.

18.3 The Effects of β

As discussed in Section 18.2, turbulent fluid motion in a flat-bottomed, inviscid, unforced barotropic system must evolve so as to transfer kinetic energy to ever-larger scales of motion, in what can be referred to as a "red cascade". Simultaneously, enstrophy is transferred to finer spatial scales and wrapped up into ever finer filaments. This physical system is highly idealized, however, relative to the ocean and atmosphere in that, among other things, stratification, gradients of planetary vorticity and topography are missing. The introduction of competing dynamical mechanisms such as these is able to halt or to reverse the two-dimensional cascades of energy and enstrophy for some scales of motion.

For example, with $\beta \neq 0$, Rossby (planetary) waves are possible and serve to redistribute energy by wave radiation. That a competition may thus arise between turbulent and wave-like dynamics was first noted by Rhines (1973) and discussed from a theoretical point of view in succeeding papers (e.g., Rhines 1975). It can be shown that, associated with the restoring force provided by a large-scale planetary vorticity gradient, is a scale-dependent boundary at which wave steepness is of order unity. This spectral boundary divides the energy spectrum into regions in which the mobility of energy is vastly different and at which, depending on the direction and speed of non-linear transfer within these spectral regions, turbulent energy transferred to this dividing scale by the red cascade may pile up. Thus, although the wave-restoring forces associated with β act effectively to disperse energy in physical space, they also act to concentrate the energy spectra of evolving fluids at certain specific wavenumbers.

The scale-dependent spectral transition between turbulence and waves occurs for wave steepness of $0(1)$ – i.e., when $U_{RMS}/C_p \sim \dfrac{2k^2 U_{RMS}}{\beta}$ for waves of average orientation – and therefore corresponds to a dividing wavenumber $k_\beta \simeq (\beta/2 U_{RMS})^{1/2}$. Since β does not alter the gross conservation law [Eq. (7)] and (8) except in the presence of boundaries (Rhines 1977), the energy and enstrophy cascades associated with two-dimensional turbulence are not qualitatively changed, although they may be impeded. For wavenumbers $k \gg k_\beta$, these turbulent processes strongly favor eddies of ever larger scale; upon approaching the transitional scale $(k_\beta)^{-1}$, however, eddies start propagating as Rossby waves as the restoring effects of β increase. When advection has turned the turbulence into a field of steep waves, mutual interactions become weak and selective. Further evolution is greatly slowed.

Periodic simulation of evolving two-dimensional turbulence on a β-plane reproduce the qualitative features of this competitive evolution quite nicely.

Figure 2a, b compares the eddy fields which result from free turbulent evolution of initially small-scale eddies ($k \gg k_\beta$) with $\beta = 0$ (Fig. 2a) and $\beta \neq 0$ (Fig. 2b). The accumulation of energy in eddies of intermediate scale ($k \sim k_\beta$) is a clear feature of the $\beta \neq 0$ case. Time-longitude plots (Fig. 3a, b) further evidence the scale increase of the energy-containing eddies for $\beta = 0$ (Fig. 3a) and the halt of scale expansion in the developing wave field with $\beta \neq 0$ (Fig. 3b). It should be noted that the competition between waves and turbulence is not idiosyncratic of the β-plane approximation; for instance, analogous effects have been obtained in forced turbulent evolution experiments on a sphere (Williams 1978).

In the presence of β, once turbulent evolution has excited scales comparable to $(k_\beta)^{-1}$, further turbulent evolution is slowed and is characterized by growing zonal anisotropy. This tendency to develop zonal currents is a common end-state feature of calculations with β (e.g., Fig. 2b). More generally, contours of mean large-scale potential vorticity $Q(x, y, t)$ – the so-called quasi-geostrophic contours – provide free paths for steady circulation and restoring forces for transient motions. Purely two-dimensional turbulent scale expansion is defeated by these wave forces, and further evolution generates flow along the quasi-geostrophic contours. After passing through a state of propagating waves, the flow tends to alternating zonal jets which are quasi-steady.

A variety of theoretical interpretations have been offered for this observed growing zonal anisotropy. Appealing to weak resonant interaction theory, Rhines (1975) has shown that in the limit of small wave steepness, wave triad energy must be transferred to components of lowest frequency. Since the presence of β may only slow the transfer of energy to lower wavenumbers, evolution must favor lower ω *and* k, a situation which can only be achieved (according to the Rossby wave dispersion relation) by a developing elongation of ed-

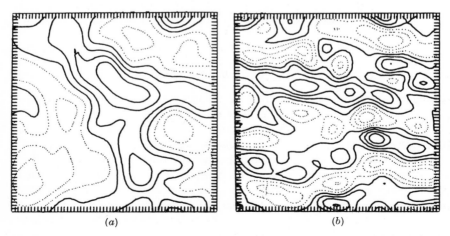

(a)   (b)

Fig. 2a, b. Streamline contours for two-dimensional turbulent flow with and without β. **a** $\beta = 0$, CL = 0.16 **b** Strong beta-effect, CL = 0.06. Besides keeping the scale small, beta acts to produce predominantly zonal flows. (Rhines 1975)

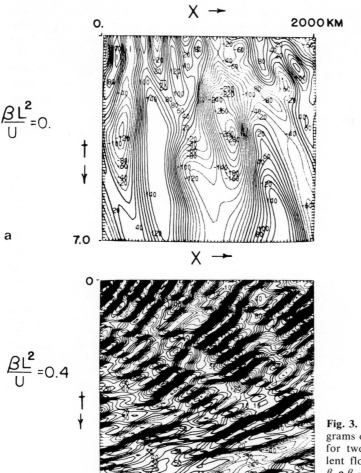

0.   X →   2000 KM

$\dfrac{\beta L^2}{U} = 0.$

t
↓

a   7.0

X →

0

$\dfrac{\beta L^2}{U} = 0.4$

t
↓

Fig. 3. Time-longitude diagrams of the streamfunction for two-dimensional turbulent flow with and without β. **a** $\beta = 0$. **b** $\beta = 0.4$. CL = 0.11. (Rhines 1977)

b   7.0

dies along latitude lines (i.e., quasi-geostrophic contours). Rhines therefore suggests as a likely end-state of the cascade a set of nearly steady, alternating zonal currents which, by stability considerations, must have a width scale exceeding $(2 U_{RMS}/\beta)^{1/2}$. Since developing zonal anisotropy is not strictly a feature of the weak wave limit, however, this theoretical argument must be viewed as suggestive only. As an alternative (or complementary) mechanism, Colin de Verdiere (1979) has explored Rossby wave sideband instabilities and shown that the resulting short-term tendency also favors zonal flow generation. Finally, Holloway and Hendershott (1977) have examined an extension of the turbulence test field model for two-dimensional flow with Rossby waves. In the geophysically interesting case in which long, fast Rossby waves propagate substantially without interaction while short Rossby waves are thoroughly dominated by advection, the turbulence closure model predicts the slowing of the red cascade and the developing zonal anisotropy of flow. In particular, it is shown that the growing anisotropy reaches an equilibrium having a calculable functional dependence on the isotropic part of the flow spectrum, and that the equilibrium arises as a consequence of competition between production, transfer and "loss" of anisotropy. Comparisons of the closure model results with direct numerical simulations of two-dimensional turbulent flows show substantial agreement for β small (Fig. 4).

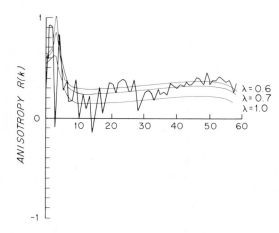

Fig. 4. Anisotropy as observed in numerical simulation of non-linear Rossby waves compared to theoretical predictions for three values of the adjustable test field model parameter λ. (Holloway and Hendershott 1977)

18.4 Mean Flow Generation by Localized Forcing

A particularly simple example of eddy-mean flow interaction and the generation of mean zonal currents for which some theoretical predictions are available is obtained by introducing a localized external driving agent into the periodic barotropic β-plane system of Section 18.3. The mean circulation that can be induced in such a re-entrant β-plane (or channel) model was first noted by Whitehead (1975), whose laboratory experiments exhibited rectified flow generated by forcing localized to a latitudinal circle on a polar β-plane. The sense of the induced jet was prograde (eastwards) at the forced latitudes and westward elsewhere. In Whitehead's experiment, forcing was provided by a small circular disk at mid-depth in the homogeneous fluid. More recent experimental evidence, obtained in a number of experimental configurations and for a variety of forcing mechanisms, confirms the general tendencies noted in these early experiments (Colin de Verdiere 1979, McEwan, Thompson and Plumb 1980).

In a zonally re-entrant geometry, the tendency for retrograde (westwards) motion to develop at unforced latitudes can be anticipated theoretically as follows. Ignoring dissipation and forcing, the zonally averaged u momentum equation can be written

$$\frac{\partial}{\partial t}\hat{u}=\widehat{vq} \tag{12}$$

where a carat ($\hat{\ }$) represents a zonally averaged quantity. If the disturbance induced by the remote forcing is sufficiently weak, the enstrophy equation can be linearized to give.

$$\frac{\partial}{\partial t}\widehat{q^2}=-\beta\widehat{vq}. \tag{13}$$

If initially quiescent, the fluid at unforced latitudes will be energized by the remote forcing; hence, $\widehat{\partial q^2}/\partial t>0$ and $\widehat{vq}<0$ at these latitudes. From Eq. (12), this corresponds to developing westward flow at unforced latitudes and, by conservation of momentum, developing eastward flow in the forced latitudinal band. An associated theoretical connection follows by noting that $\widehat{vq}=-(\widehat{uv})_y<0$, making explicit the net flux of westerly momentum into the region of forcing. Further discussion of these points, and analogous arguments for two-layer systems is given by Held (1975). The vorticity transfer ideas invoked to arrive at Eq. (12) can be generalized to include surface friction and used to obtain an explicit expression for the meridional transport of potential vorticity, $\widehat{vq}$ (Rhines 1977). Comparison of these theoretical expressions with vorticity fluxes observed in the laboratory has been generally favorable, but inconclusive (Colin de Verdiere 1979).

The use of periodic numerical simulations offers an advantageous environment in which to deduce the details of mean flow rectification. In a series of such experiments, Haidvogel and Rhines (1983) have examined the Eulerian and Lagrangian vorticity balances accompanying forcing by a narrow Gaussian plunger (or windstress curl) of e-folding scale 100 km in the geometric center of a $(2000 \text{ km})^2$ periodic domain. In one such experiment, the plunger oscillates in amplitude with a period of 100 days over the range $-4\leq\tau\leq+4$ dynes cm^{-2}; results and further details are shown in Fig. 5.

The localized forcing generates a field of weak Rossby waves of steepness $0(1/2)$. The length scale of these waves and their preferred direction of phase propagation are consistent with the Rossby wave dispersion relation and the properties of the forcing chosen. The resulting transient flow is seen to be weakly non-linear; frequency spectra (not shown) are dominated by the 100-day forcing period and potential vorticity by the planetary variation in f (i.e., $Q\sim\beta y$). Nonetheless, a substantial time-mean flow (Fig. 5a) develops with eastward flow under the plunger and net westward motion elsewhere (as expected from the arguments given above). Representative velocity scales for the eddy and induced mean components of flow are comparable.

Profiles of the time-mean zonal velocity ($\bar{u}$) and Reynolds stress ($\overline{u'v'}$) taken along a meridional line lying to the west of the plunger are shown in Fig. 5b. (Here, the overbar denotes a time-average, and a prime the deviation therefrom.) The time-mean velocity (Fig. 5b) is dominated by easterly flow at

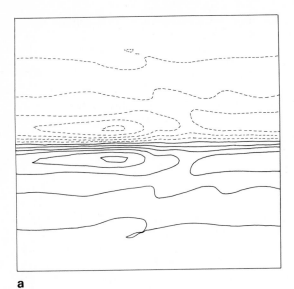

Fig. 5a, b. Periodic barotropic model driven by an oscillating wind stress confined to the center. The periodic domain is 2000 km on a side; the Gaussian wind has an e-folding scale of 100 km, a period of 100 days and a maximum wind stress of 4 dynes cm^{-2}. **a** $\overline{\psi}$, CL = 1.6×10^7 cm^2 s^{-1}. **b** clockwise from upper left: $\overline{u}$ (cm s^{-1}), $\overline{\zeta}$ (3×10^{-8} s^{-1}), $\overline{u'v'}$ (cm^2 s^{-2}), and $\overline{q}$ (3×10^{-8} s^{-1}), evaluated along $\times$ = 0.25. An *overbar* denotes a time-mean quantity (Haidvogel and Rhines 1983)

a

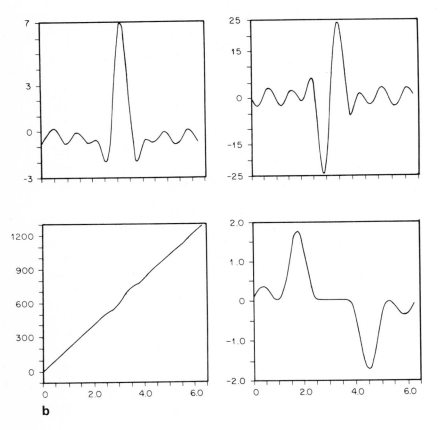

b

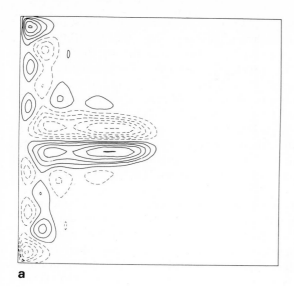

a

Fig. 6a, b. As Fig. 4, except within a closed basin. **a** $\overline{\psi}$, CL $= 7 \times 10^6$. **b** clockwise from upper left: $\bar{u}$ (cm s^{-1}), $\bar{\zeta}$ (s^{-1}), $\overline{u'v'}$ (cm^2 s^{-2}), and $\bar{q}$ (s^{-1}), evaluated along $\times = 0.25$. (Haidvogel and Rhines 1983)

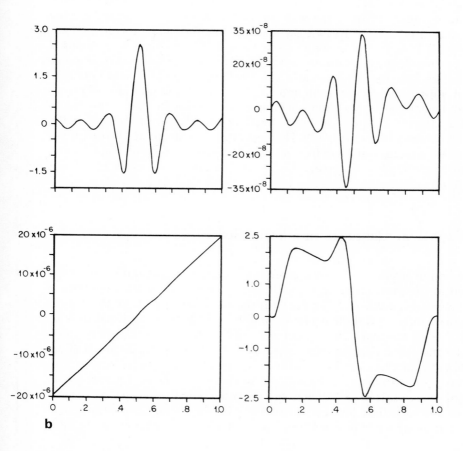

b

mid-latitude; elsewhere, although the net flow is to the west, the mean flow oscillates in strength with occasional bands of eastward flow. The theoretical explanation for the fine-structure observed in the regions of net westward flow is not known at present. A simple relationship between the central and side-lobe jets and the associated time-mean Reynolds stress $\overline{u'v'}$ does not appear to exist. The associated time-mean vorticity balance shows the dominant balance at unforced latitudes to be $\beta\overline{v} \sim -\nabla\cdot\overline{(v'q')}$, the so-called turbulent Sverdrup balance. At forced latitudes, however, this balance is upset, $\beta\overline{v}$ and $\nabla\cdot\overline{(v'q')}$ partially reinforce one another, and dissipation provides the required balance. Further elaboration on these balances, and an interpretation of these results in terms of potential vorticity mixing ideas is given by Haidvogel and Rhines 1983).

Although the theoretical and numerical results discussed thus far have dealt with a periodic or infinite domain, it is important to note the systematic effects induced by meridional boundaries, particularly those lying to the west of the forcing. In such a case, the mean quasi-geostrophic contours characterizing the homogeneous system are no longer closed, but blocked by the western boundary. With the introduction of a western wall, the enstrophy conservation statement (8) for inviscid unforced flow is no longer valid. Large-scale Rossby waves can be reflected from the western wall with higher wavenumber; the western wall can therefore act as a net source of small-scale enstrophy (Rhines 1977). Figure 6 shows the resulting circulation within a completely enclosed domain for the identical localized forcing used above. Because of the presence of the western wall, the induced mean zonal velocity cannot be closed periodically, but must return to the north and south of the forced latitudes (Fig. 6a). The resulting two-gyre circulation is a familiar one; a similar circulation can be induced at deep levels by form drag effects in eddy-resolving models of the general circulation (see Chap. 17). The generally weaker induced flow in the presence of a western wall than in a periodic geometry – compare Figs. 5b and 6b – can be anticipated following the theoretical arguments given by Rhines and Holland (1979).

18.5 Turbulent Cascades in a Stratified Fluid

The addition of stratification has a profound influence on the turbulent system discussed in the previous sections. First, the existence of vortex stretching in three dimensions, and for definiteness in the two-level Eq. (1–4), relaxes the strong constraints on turbulent motion imposed on two-dimensional flow by the existence of the quadratic invariants E and Z. As we will describe below, although some general statements can be made concerning the direction of energy and enstrophy transfer in baroclinic turbulence, in general effective transfer of energy to both larger and smaller scales is now possible. Second, the introduction of stratification supplies the restoring force necessary to support slow and fast baroclinic Rossby waves (the latter in conjunction with bottom slopes). The existence of transient motions on a variety of temporal and spatial scales complicates the interpretation of turbulent flow simulations, as it does observed turbulent motions. For a more complete discussion of the properties of these linear waves and some evidence for their importance in mid-ocean data sets, see Rhines (1977).

Although the two-level equations (1–4) are no longer characterized by only two quadratic invariants, some general conclusions can be drawn from conservation considerations. By recasting Eqs. (1–4) in terms of barotropic and baroclinic streamfunction components (ψ and τ), it can be shown that the two-level system has conserved properties corresponding to total (barotropic plus baroclinic) energy, and sum and difference enstrophy (Salmon 1978, 1980). The general implications of these conservation statements can best be understood by distinguishing between barotropic (ψ_k, ψ_p, ψ_q) and baroclinic (ψ_k, τ_p, τ_q) triads interactions, for each of which the wavenumber selection condition $\underline{k}+\underline{p}+\underline{q}=0$ must be satisfied. For barotropic triads (as in purely two-dimensional flow), the quadratic constraints prevent effective energy transfer in both extremely local ($\underline{k}\simeq\underline{p}$) and non-local ($\underline{k}\ll\underline{p}$) triads; for intermediate (k, p, q), an initial spreading of the energy spectrum favors net transfer to lower wavenumber. For baroclinic triads in the two-level system, two limiting situations emerge. For $(k, p, q)\gg\lambda^{-1}=k_R$ (the Rossby deformation wavenumber), the individual layers act independently as in purely two-dimensional flow; this is the limit of uncoupled layers. For $(k, p, q)\ll k_R$, energy is transferred between the baroclinic components only and the transfer may be either local or non-local.

The general picture that results is shown in schematic form in Fig. 7 (Salmon 1980). If net baroclinic energy is produced at low wavenumbers (say by heating of the atmosphere, or wind forcing of the ocean), this energy will be transferred preferentially towards higher wavenumbers until the Rossby deformation scale is reached. Once wavenumbers near k_R have been excited, efficient transfer of energy between baroclinic and barotropic modes of motion is encouraged. The associated physical space picture is the gradual occlusion or barotropification of upper and lower layer flows, once deformation scale features have been produced. As energy is converted to the barotropic mode, this energy can move back towards lower wavenumbers in more local interactions in the manner of purely two-dimensional flow. The ultimate disposition of en-

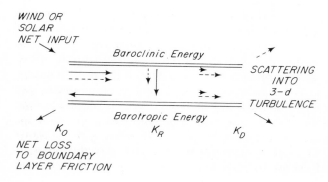

Fig. 7. The wavenumber energy flow diagram for a two-layer system. *Solid arrows* represent energy flow, and *dashed arrows* potential enstrophy flow (Salmon 1980)

ergy at low wavenumbers will of course depend on the specifics of the physical situation, including details of β and surface friction, both of which strongly influence the rate at which energy can be concentrated at ever lower k. Analogous qualitative statements can be made with regard to enstrophy transfer in this system (Fig. 7).

Simulations of evolving turbulence in two-layer systems qualitatively confirm the tendencies suggested above. Figure 8 (Rhines 1977) shows the instability of a single large-scale $(k < k_R)$ Rossby wave in a periodic domain with weak damping (Fig. 8a). Initial baroclinic instabilities quickly distort the single baroclinic wave (Fig. 8b). Triad interactions among the resulting spectrum of baroclinic waves rapidly feeds energy to deformation scale eddies which then lock together in the vertical (Fig. 8c). Scale expansion and preferential elongation of motion in the zonal direction then typifies the barotropic interactions occurring thereafter (Fig. 8d). This turbulent evolution can be schematically represented as $BC(k < k_R) \rightarrow BC(k \sim k_R) \rightarrow BT(k \sim k_R) \rightarrow BT(k < k_R)$. It is important to note that energy transfer to deformation scale eddies and the occlusion of upper and lower layers also accompanies the evolution of an initially small-scale $(k > k_R)$ baroclinic eddy field. Figure 9 shows a numerical realization of this situation; since $\beta = 0$ in this experiment, barotropic energy (once supplied by BC triad interactions near k_R) moves to large scale without the competition afforded by wave propagation. The resulting evolution follows the form $BC(k > k_R) \rightarrow BC(k \sim k_R) \rightarrow BT(k \sim k_R) \rightarrow BT(k < k_R)$.

Figures 8 and 9 have been obtained in the absence of forcing and with weak dissipation (Rhines 1977). Though not able to achieve a long-term mean dynamic equilibrium, such "spin-down" experiments can be used to examine the initial transient and subsequent approximate statistical equilibrium which occur during the slow relaxation from some set of initial conditions. With the addition of some prescribed forcing to the system, a true dynamic equilibrium can be achieved and the effects of parametric changes on dynamical quantities such as energy and heat flux spectra assessed.

A variety of such equilibrium turbulence simulations have been conducted. Haidvogel and Held (1980) have examined the equilibrium dynamics of homogeneous quasi-geostrophic turbulence maintained by an imposed large-scale baroclinic shear field (or temperature gradient). Salmon (1980) has investigated a similar two-layer simulation for environmental parameters and forcing (solar heating) representative of the mid-latitude atmospheric circulation. Finally, McWilliams and Chow (1981) have studied the properties of quasi-geostrophic turbulence in equilibrium with wind-driven flow in a β-plane channel in a parameter range appropriate to the Antarctic Circumpolar Current.

Broadly speaking, the equilibrium character of the quasi-geostrophic turbulence fields achieved in these experiments is independent of the precise nature of the forcing mechanism. The time-averaged turbulent energy spectra are peaked at wavenumbers lower than those at which the energy is supplied to the system, a manifestation of the red cascade associated with barotropic triad interactions. In many cases, the spectral peaks appear to coincide with the transfer arrest wavenumber k_β. The turbulent energy balance is achieved by conversion of available potential energy to eddy kinetic energy at the deformation

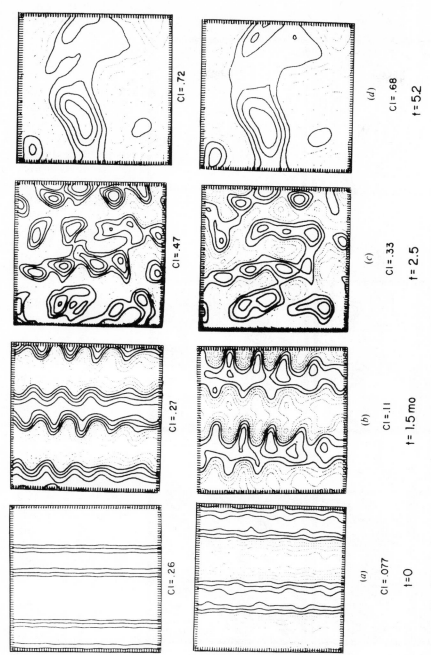

Fig. 8. Instability of a single baroclinic Rossby wave with a weak noise field (Rhines 1977)

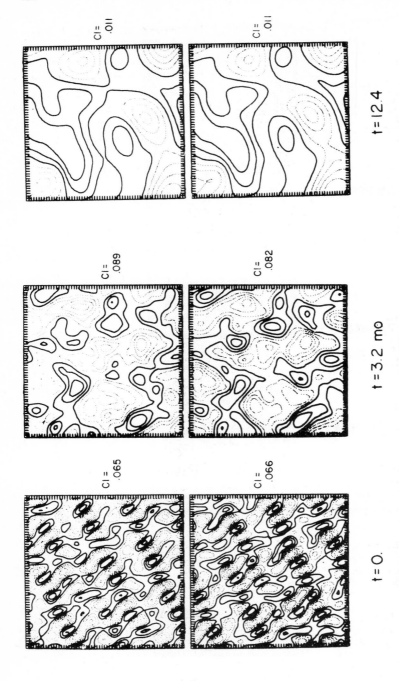

t = 0.   t = 3.2 mo   t = 12.4

Fig. 9. Sequence of two-layer streamfunction fields in a free spindown experiment with $\beta = 0$. The initial field has equal energy in either layer in wavenumber band $7 \leq k \leq 10$ ($k_R = 7$) but with random phase

scale, and its subsequent transfer to increasingly barotropic low-wavenumber motions at which surface frictional effects contribute the dominant energy removal. Despite the importance of non-linear transfer processes in the turbulent equilibrium state, remnant features of the linear wave processes also persist, both as lines in frequency spectra representing propagation of turbulent eddies and in the structural properties of the eddies involved in the baroclinic energy conversion process. A more detailed description of these results – including equilibrium eddy balances of momentum, energy and enstrophy; the observed nonlinear equilibration processes; and Eulerian (pointwise) and Lagrangian (particle) diffusivities – as well as related discussion of the eddy heat flux parameterization question can be found in Haidvogel and Held (1980), Salmon (1980) and McWilliams and Chow (1981).

18.6 Scattering by Topography

Scattering of linear topographic waves is a source of small-scale motion; in particular, bottom roughness can generate relative enstrophy (ζ^2) and thereby effectively counter the cascade of energy to larger scales. For barotropic flow, if total enstrophy (including its topographic contribution) is approximately constant and relative enstrophy increases with time, then

$$\frac{\partial}{\partial t}(q^2) = \frac{\partial}{\partial t}[(\nabla^2 \psi)^2 + (\alpha h + \beta y)\nabla^2 \psi] = 0 \; ;$$

hence

$$\frac{\partial}{\partial t}[(\alpha h + \beta y)\nabla^2 \psi] = -\frac{\partial}{\partial t}(\nabla^2 \psi)^2 < 0$$

and a growing anti-correlation is expected between ζ and the topography $(\alpha h + \beta y)$. In physical terms, this implies the generation of anti-cyclonic (cyclonic) flow above topographic elevations (depressions). In stratified fluids, the scattering by topography of initially large-scale flow has other (related) important effects; among them are a conversion of potential to kinetic energy, and a generation of baroclinic energy in opposition to the tendency towards barotropy accompanying flow over a flat bottom.

Various theories of turbulence above topography have been investigated to make more precise the prediction of growing correlation between ψ and h. Suppose, for instance, that topographic scattering augments the cascade of enstrophy to high wavenumbers (where it is removed by dissipative processes), but that total energy, largely confined to low wavenumbers, is nearly conserved. Then it is relevant to seek the flow field with minimum enstrophy (Z) for a given initial total energy (E) and topographic spectrum [h(k)]. Bretherton and Haidvogel (1976) have examined the associated minimization problem and shown that the desired flow spectrum for $\beta = 0$ is given by

$$\psi(k) = h(k)/(\gamma + k^2)$$

where γ is a function of E and h(k) with units of k^2. This minimum enstrophy solution predicts a circulation about the topographic contours $\psi(k) \sim h(k)/\gamma$ for $k < \gamma^{1/2}$; at smaller scales, $\nabla^2\psi + h \simeq 0$ and fluid particles are swept completely over high-wavenumber bumps. The spectrum of motion, $\psi(k)$, is therefore a smoothed (low-passed) version of the bottom roughness, $h(k)$. For damping weak enough that the theoretical assumptions are approximately valid, numerical simulations of two-dimensional flow above topography within a closed box give the predicted, nearly stationary flow about smoothed topographic contours (Fig. 10). With the addition of β, planetary-scale westward flow is generated in the interior, and the circulation is closed in narrow inertial boundary layers around the edges of the domain (Bretherton and Haidvogel 1976).

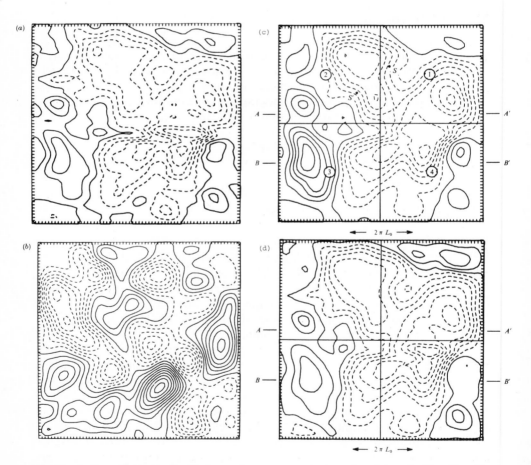

Fig. 10a–d. Two-dimensional turbulence above topography in a closed domain ($\beta = 0$). **a** Topography. **b** Initial streamfunction. **c** Final streamfunction. **d** Final potential vorticity. (Bretherton and Haidvogel 1976)

Statistical models of topographic turbulence have been provided by Holloway (1978), Salmon et al. (1976), and Herring (1977). These have shown that the inviscid equipartition equilibrium spectrum for topographic turbulence is the sum of a two-dimensional turbulent spectrum of form (11) plus the energy associated with a steady geostrophic contour current. For purely barotropic flow, the equilibrium correlation between $\psi(k)$ and $h(k)$ is positive at all wavenumbers. In a two-level system, bottom topography affects the equilibrium upper layer flow only for $k \ll k_R$ and the kinetic energy spectra of the two layers are nearly equal, having the same general form as equilibrium spectra for single layer flows [Eq. (11)]. Smaller-scale topography traps energy preferentially in the lower layer and the layers are effectively uncoupled. On the basis of these theoretical solutions, Salmon et al. (1976) suggest that the statistical trends observed in nonequilibrium flows may be manifestations of a tendency to maximize entropy of the turbulent system, a point of view further elaborated by Holloway (1979).[5] However, direct numerical simulations of turbulent flow above irregular topography are not always close to either the minimum enstrophy or maximum entropy states (Bretherton and Haidvogel 1976; Holloway 1978).

Baroclinic turbulent flow above a rough bottom has been investigated numerically in periodic models by Bretheron and Karweit (1975) and Rhines (1977).[6] Of particular concern has been the interaction of large-scale baroclinic currents with topography, and the modification of baroclinic instability processes accompanying bottom roughness. When the large-scale mean flow is allowed to respond to the presence of topography, it is seen in these simulations that flow from the west must be forced, and produces an energetic barotropic eddy field. Unforced flow, allowed to slowly spindown from an initial eddy field, will naturally respond with flow from the east, for which both free waves and topographic drag are absent (Bretherton and Karweit 1975). Figure 11 shows the flow patterns obtained for rough-bottom baroclinic instability. A visual impression of the streamline patterns in this simulation is one of large-scale flow, quite different than the deformation-scale eddies evidenced in the presence of flat-bottom (see Fig. 7). This is the result of spectral broadening by the topography which circumvents the tendency for baroclinic energy to be transferred to k_R and thence into barotropic motions. Further topographic experiments suggest that the developing vertical structure depends not only on the overall energy level E and topographic spectrum h(k), but also on the manner in which the energy is supplied, e.g., in initially large- or small-scale eddies (Rhines 1977).

---

5 This maximum entropy (equilibrium) solution includes among its family of solutions the minimum potential enstrophy solution.

6 Slightly different formulations are possible depending on whether the streamfunction or the velocity is assumed to be periodic. In the former case, large-scale pressure gradients must exist to balance topographic drag. These are taken to vanish in the latter case, in which large-scale mean flows may develop in response to topographic drag.

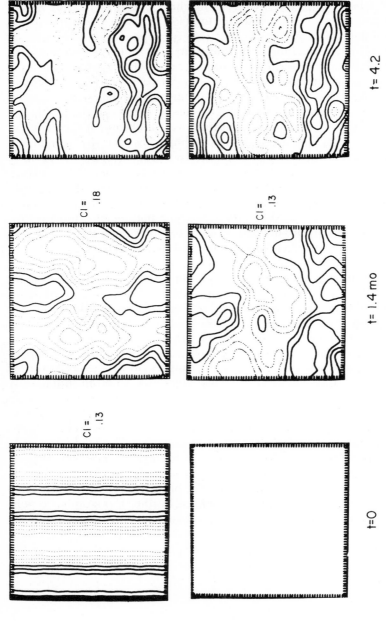

Fig. 11. Flow patterns for rough-bottom baroclinic instability (cf. Fig. 7). Quasi-geostrophic contour flow and a significant amount of deep-water energy develops. (Rhines 1977)

18.7 Regional Models and Non-Local Effects

Although periodic models have been used with some success to reproduce the observed statistical properties of mesoscale motions at a $(500 \text{ km})^2$ site in the Northwest Atlantic Ocean (see Sect. 18.8), significant variations in eddy intensity are known to occur both on larger horizontal scales and in other geographical regions (e.g., Schmitz 1976, Dantzler 1977). In general, therefore, eddy statistics may not be insensitive to the distribution of eddy energy generation and dissipation, nor to the overall detail of the larger-scale mean (general) circulation in which the eddies are embedded. For regions across which eddy properties or their supporting dynamical mechanisms change significantly, or near to which other regions of unique internal physics may lie, the assumption of horizontal homogeneity (and thus the use of periodic boundary conditions) may not be strictly applicable. In such instances, a more general formulation of the mixed initial-boundary value problem for local ocean dynamical studies may be needed.

In regions for which the quasi-geostrophic Eqs. (1-4) are an acceptable dynamical model, the regional modeling problem must be closed by specifying appropriate constraints on the vorticity Eqs. (1-2) and the related Poisson equations (3-4) for the streamfunction. The choice of these boundary conditions involves a variety of physical, mathematical and computational considerations, including the necessity that (1) the boundary conditions properly parameterize the effects of the exterior ocean, (2) the resulting mathematical model be formerly well-posed or well-behaved in some practical sense, and (3) the discrete numerical formulation of the continuum problem be efficiently implementable. Oliger and Sündstrom (1978) give a more detailed discussion of some of these points for a variety of initial-boundary value problems in fluid dynamics.

Despite the important modeling concerns noted above, limited-area models are beginning to be used to investigate general aspects of the initial-boundary value problem for isolated regions of the ocean and, more specifically, to conduct idealized dynamical forecast experiments. For application to a homogeneous mid-ocean region, Haidvogel, Robinson and Schulmann (1980) have investigated three numerical models in which the barotropic vorticity equation is integrated over a limited-area (open) domain. Following the suggestion of Charney, Fjortoft and von Neumann (1950), the specification of the initial-boundary value problem is completed by prescribing initial fields of streamfunction and vorticity, and providing, as a function of time, values of streamfunction everywhere along the boundary and values of vorticity at inflow points on the boundary. For the purely inviscid equations, the resulting continuum problem is ill-posed; spurious vorticity shocks may be generated inside the limited-area region even though the supplied boundary data are smooth (Bennett and Kloeden 1978). In practice, however, bottom and high-order lateral friction are invoked in model simulations for both physical and computational reasons. The inclusion of scale-selective dissipative processes limits the accumulation of enstrophy in high-wavenumber motions. Depending on the precise nature of the dissipation, the resulting damped initial-boundary value problem may be well-posed; however, the damped solutions may be characterized by the un-wanted occurrence of narrow frictional boundary layers adjacent to outflow.

To determine the extent to which the accuracy and efficiency of limited-area calculations depend on the numerical integration scheme, Haidvogel, Ro-binson and Schulmann (1980) solved a variety of test problems – including li-near and non-linear Rossby wave propagation – by three independent methods based respectively on finite-difference, finite-element and pseudospectral ap-proximation techniques. These tests indicated all three models to be capable of delivering stable and accurate solutions to linear and weakly non-linear (Rossby number = $\varepsilon < 0.4$) problems in open domains. The higher-order (fin-ite-element and pseudospectral) codes were shown to be more efficient for specified accuracy, a conclusion entirely consistent with previous evaluations of higher-order numerical methods for use in computational fluid dynamics (e.g., Oszag 1971).

Dynamical forecasting error can be ascribed to three factors: physical er-rors related to the inadequate dynamical representation of resolved and/or un-resolved scales of motion; computational errors which reflect the details of the discrete numerical model and, in particular, its implementation of the open boundary conditions; and observational errors associated with inaccurate ini-tial, boundary condition, and verification data. With the barotropic finite-ele-ment model, Robinson and Haidvogel (1980) have conducted a sequence of numerical simulation experiments to characterize and to quantify the sources of dynamical forecasting error for oceanic motions representative of the MODE and POLYMODE regions. The strategy has been to explore the ability of the model to accurately forecast the evolution of a quasi-turbulent baro-tropic ocean flow ($\varepsilon = 1.5$) when given both "perfect" and systematically de-graded initial and boundary data.

To examine the behavior and properties of the numerical model in the tur-bulent flow limit, an embedding strategy is used in which an exterior numerical calculation in a doubly-sized computational domain (1 000 km square) pro-vides initial condition, boundary and verification data for a forecast experi-ment over a smaller (500 km square) interior region. The simulated flow in the larger domain is therefore taken to represent oceanic "truth", and is con-structed to have a range of scales, amplitudes and phases of motion closely matched to those observed in the mid-ocean. This was accomplished over the exterior domain by use of the best two barotropic Rossby wave fit to the MODE-I 1 500 meter data as initial and boundary data (McWilliams and Flierl 1976, see Fig. 12). Computational forecast error in this type of embedding ex-periment is strictly attributable to computational error associated with the im-plementation of the open boundaries. This minimum (or irreducible) RMS er-ror associated with the finite-element model, the C-F-VN boundary condition prescription and the specific quasi-turbulent flows studied has been shown to be approximately one percent in streamfunction and vorticity per period of forecast integration[7] (Fig. 13a). Some representative forecast results are shown in Figure 15.

---

7 One "period" here is taken to be the shorter of the periods of the two best-fit MODE waves, approximately 128 days.

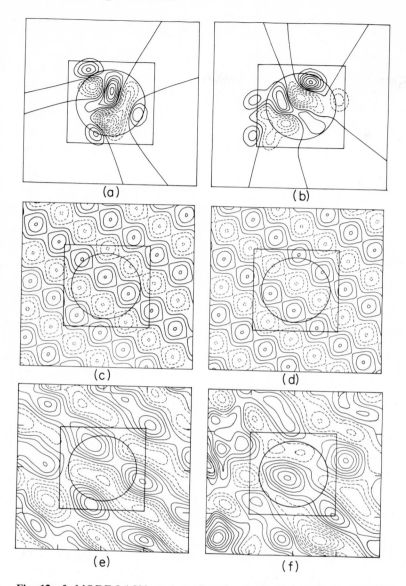

Fig. 12a–f. MODE-I 1500 m streamfunction maps at day 100 **a** and day 165 **b** Best linear Rossby wave fit to MODE-I data; day 100 **c** and day 165 **d**. Simulated (non-linear) stream-function maps at day 240 **e** and day 900 **f**. Outer (inner) square domain is 1000×1000 (500×500) km. (Robinson and Haidvogel 1980)

Forecast quality can be seriously degraded by observational errors associated with inadequate knowledge of required initial, boundary and verification data. Robinson and Haidvogel (1980) have systematically examined the qualitative and quantitative character of these observational errors; in particu-

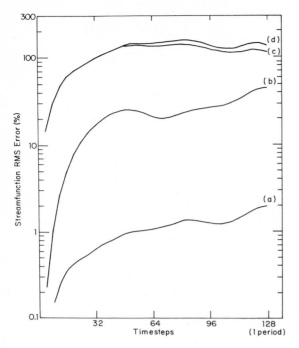

Fig. 13. Global RMS streamfunction error in a barotropic dynamical forecast experiment as a function of the frequency of updating of boundary data (initial fields are correct). **a** Boundary values of ψ and ζ updated every timestep ("perfect" boundary data). **b** ψ updated every timestep, ζ not updated. **c** ψ not updated, ζ updated every timestep. **d** Boundary values not updated (persistence). (Robinson and Haidvogel 1980)

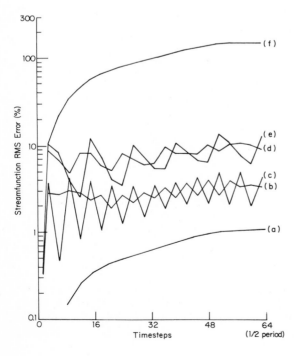

Fig. 14. As Fig. 13. **a** perfect boundary data. **b** boundary values of ψ and ζ updated on average every second time step (two-ship sampling strategy). **c** ψ and ζ updated every second time step (persistence for two time steps). **d** boundary values of ψ and ζ updated on average every fourth time step (one-ship sampling strategy). **e** ψ and ζ updated every fourth time step (persistence for four time steps). **f** boundary values of ψ and ζ not updated (persistence). (Robinson and Haidvogel 1980)

lar, their dependence on (1) the quality of initial and boundary condition data, (2) the strategy used to update boundary information, (3) the method of acquiring vorticity data, and (4) the time and space interpolation procedure used to

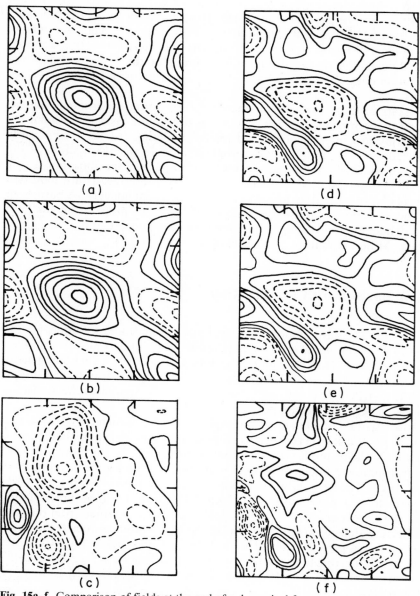

Fig. 15a–f. Comparison of fields at the end of a dynamical forecast experiment (t = 660 days after start) using "perfect" initial and boundary data. **a** Actual ψ, CL = 1, **b** Forecast ψ, CL = 1. **c** ψ error, CL = 0.04. **d** Actual ζ, Cl = 0.9. **e** Forecast ζ, CL = 0.9. **f** ζ error, CL = 0.04. (Robinson and Haidvogel 1980)

obtain missing values of initial and boundary streamfunction and vorticity. These studies indicate that, for forecasts of a few months' duration, the observational error is controlled essentially by the time interval between successive updates of individual points on the boundary (Figs. 13 and 14). (*Global* observational error does not depend strongly on whether the boundary points are updated all at once or piecewise, although the structure of *local* error fields is, of course, related to the updating strategy). Furthermore, taken as a whole, these dynamical forecast experiments support the conclusion that forecast quality, although sensitive to the overall accuracy of the supplied boundary streamfunction data, is relatively insensitive to large errors in boundary values of vorticity.

The results of these barotropic dynamical forecasting experiments are encouraging for the prospects of practical open ocean forecasts. Additional experiments, physical interpretation, and related updating studies are currently being carried out (Fig. 15). These experiments include an investigation of the possibility of updating at interior gridpoints throughout the duration of the forecast simulation (Tu 1981). The extent to which these results are modified by the inclusion of baroclinic processes is also of special concern. A baroclinic regional ocean model is presently under evaluation, in preparation for both idealized dynamical forecasting experiments and for forecast studies using real ocean data (Miller, Robinson and Haidvogel 1983).

18.8 Comparison with Observations

It is hoped that the development of baroclinic regional ocean models for the mid-ocean will allow *direct* intercomparison between models and the growing mid-ocean data base. However, under the assumption of the local homogeneity of eddy properties, the *statistical* character of the eddy fields observed during the MODE and POLYMODE experiments can be compared to the statistical predictions of periodic quasi-geostrophic models.

A statistical intercomparison of this type has been conducted by Owens and Bretherton (1978) using a six-layer quasi-geostrophic model (Fig. 16a). In analogy to the two-layer model described in Section 18.1, the periodic unforced quasi-geostrophic equations with weak bottom and (biharmonic) lateral friction were used. The resting layer thicknesses and densities within each layer were chosen to be representative of the MODE region; the first three internal modes of the Owens and Bretherton model match those obtained from the continuous MODE density profile. Following the assumption of locally homogeneous dynamics, a periodic idealization of the bottom relief in the MODE region was used for the topography underlying the model (Fig. 16b). Other model parameters were all matched a priori to those observed during the MODE experiment.

Initial velocities were confined to the three uppermost layers (above the main thermocline) and given a pattern characteristic of the 10C isotherm displacements observed during MODE (Fig. 17). The magnitude of this input en-

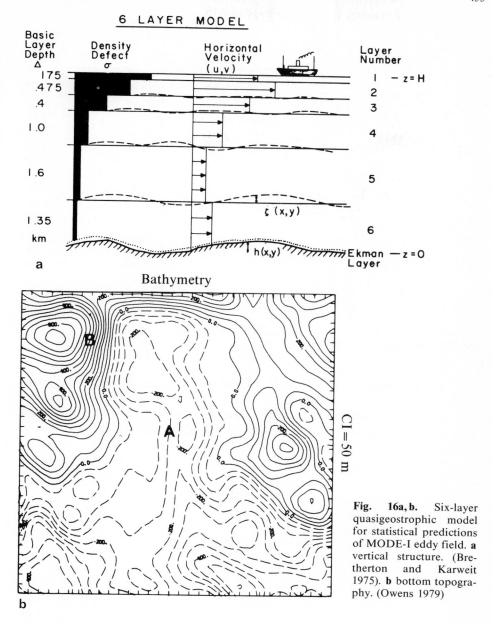

6 LAYER MODEL

Basic Layer Depth Δ · Density Defect σ · Horizontal Velocity (u,v) · Layer Number

175   1 — z = H
.475   2
.4   3
1.0   4
1.6   5

ζ (x,y)

1.35   6
km

h(x,y)   Ekman — z = 0
Layer

a

Bathymetry

CI = 50 m

b

Fig. 16a, b. Six-layer quasigeostrophic model for statistical predictions of MODE-I eddy field. a vertical structure. (Bretherton and Karweit 1975). b bottom topography. (Owens 1979)

ergy was adjusted so that subsequent RMS speeds below the thermocline (layer 4) approximated those of the MODE SOFAR floats. From these initial conditions, the periodic model was run for 200 days, after which a quasi-equilibrium, determined by the internal dynamics, had been reached. During an additional integration of 200 days, eddy spectra slowly decreased in magnitude, but not in shape, due to frictional dissipation. Model statistics generated during this latter

CONTOUR INTERVAL = 10 M

Fig. 17. Typical vertical displacements of the 10C isotherm as observed in the MODE-I experiment and used for initial model velocity profiles (Owens and Bretherton 1978)

quasi-equilibrium period were analyzed and compared to the analogous spectral estimates from the MODE observations. As discussed by Owens and Bretherton (1978), the results are somewhat sensitive to the amplitude of the initial conditions, but not to moderate changes in the initial wavenumber spectrum.

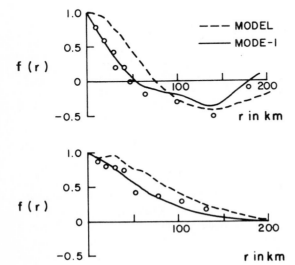

Fig. 18. Eulerian transverse, g(r), and longitudinal, f(r), spatial correlation functions of separation estimated from the MODE SOFAR floats and from the six-layer model at depth of 1500 m. (Owens and Bretherton 1978)

In general, the equilibrium statistical predictions of the periodic model compare favorably with the observed properties of the MODE eddy field, and are consistent with the hypothesis that eddy statistics are locally determined. Figure 18 shows the Eulerian transverse and longitudinal spatial correlation functions as measured in the periodic model and as deduced from the MODE SOFAR floats by Freeland, Rhines and Rossby (1975). If a small-scale noise is added to the model velocities, the predicted and observed correlation functions are nearly identical, indicating that the model correctly reproduces eddy energies at the larger length scales and underpredicts energy at smaller scales.

Eulerian frequency spectra of kinetic energy are shown in Fig. 19. Although the model appears to predict a somewhat larger time scale, the differences between the observed and modeled spectral peaks are not statistically significant. There is also some tendency for high frequency energy content to be lower in the model than that observed. In a careful follow-up study to this MODE simulation, Schmitz and Owens (1979) compared the vertical structure of the observed eddy fields, expressed in terms of temperature and velocity variances, with those of the model. The comparison is highly frequency-dependent (Table 1). For periods of 50–130 days, the agreement between model and observations is good at both deep and shallow levels. At higher frequencies (periods of 5–50 days) the comparison is poor, as suggested by the Eulerian frequency spectra (Fig. 19). The slow spin-down character of the periodic MODE simulation does not permit the prediction of energy levels for periods in excess of 400 days. Detailed intercomparison of these spectral estimates – as well as a discus-

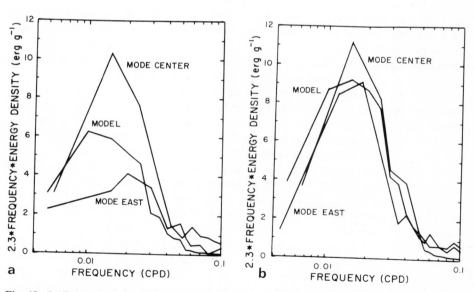

Fig. 19a, b. Eulerian frequency spectra of the kinetic energy the site moorings (MODE center at 28 °N, 69°40′W; MODE east at 28 °N, 68°40′W) and from the model. **a** 1500 m. **b** 4000 m. (Owens and Bretherton 1978)

Table 1. Spectral decompositions for MODE-I site mooring data and six-layer model at selected depths. (Schmitz and Owens 1979)

Period range (days)	K_E $(cm^2 s^{-2})$	MODE CENTER (500 m)	MODE CENTER (4000 m)	Model (Layer 2)	Model (Layer 6)
Band 1	T	20.4	0.3	—	—
(400→1500)	U	18.7	0.2	—	—
	V	1.7	0.1	—	—
Band 2	T	13.5	0.9	9.8	1.7
(133→400)	U	8.5	0.7	5.2	0.8
	V	5.0	0.2	4.6	0.9
Band 3	T	10.4	5.7	9.3	4.2
(44→133)	U	4.7	2.2	5.3	2.0
	V	5.7	3.5	4.0	2.2
Band 4	T	7.2	1.9	1.7	1.5
(5→44)	U	2.4	0.9	1.0	0.8
	V	4.8	1.0	0.7	0.7

sion of spatial variability of eddy energy, westward propagation of eddies, and vertical variation of horizontal wavenumber content – has been given by Bretherton (1975), Owens and Bretherton (1978), and Schmitz and Owens (1979).

The dynamic properties of the simulated MODE eddy field also appear to be consistent with observations made during the MODE experiment. Owens (1979) has examined the equilibrium Eulerian potential vorticity and heat balances in the periodic quasi-geostrophic MODE simulation and shown them to be consistent with the analogous balances inferred from the MODE data (McWilliams 1976). Above the thermocline, the balances are the result of the turbulent cascade of relative vorticity, in a manner analogous to the red cascade of purely two-dimensional flow. The associated vorticity balance is dominated by the passive advection (and shearing out) of the high-wavenumber components of relative vorticity by the larger-scale energy-containing eddies. By transferring to the Lagrangian frame, or focusing on the low-wavenumber energetic motions, these advective effects are suppressed and a nearly linear response by relative vorticity and vertical vortex stretching to changes in planetary vorticity emerges. In the bottom layer, transient oscillations (presumed to be topographic Rossby waves) are superimposed on nearly steady quasi-geostrophic contour flow, a state in close agreement with the theoretical ideas discussed in Section 18.6. Both the scattering of energy into bottom-intensified waves and the tendency to induce low-frequency circulation about topographic features inhibits the strong transfer of energy from baroclinic to barotropic motions (which would otherwise be expected in the absence of topography – Sect. 18.5). Intermediate layers of the MODE simulations are characterized by a balance sharing features of the uppermost and bottommost layers. Further details of these simulated dynamic balances for the MODE mesoscale eddy field are presented by Owens (1979).

18.9 Conclusions and Future Directions

Periodic and regional modeling studies of the mesoscale eddy field, and related theoretical research, have contributed substantially to our present unterstanding of operative dynamical mechanisms in the quieter (mid-) ocean regions. By studying processes in isolation, such process-oriented experiments have identified several idealized interaction and transfer processes which halt, modify or reverse the well-known spectral and physical space evolution accompanying two-dimensional turbulent flow. Examples include: the competition between turbulence and waves on a β-plane, the conversion via non-linear interaction of baroclinic to barotropic energy near the Rossby deformation wavenumber and its eventual migration to larger scales, and the role of bottom topography in impeding the tendency towards purely barotropic flow and in generating mean zonal currents by systematic pressure forces associated with bottom roughness. Taking account of all these dynamical mechanisms in combination, periodic models appear to reproduce successfully the *statistical* dynamical features of the MODE-I eddy field. Regional models, in which the physical and dynamical inhomogeneities associated with the larger oceanic gyre are explicitly retained, offer the future possibility of *deterministic* comparisons of models and data in the mid-ocean regions. Finally, although in need of further quantitative evaluation, potential vorticity transport theories are beginning to provide a useful theoretical framework with which to understand observed and simulated mid-ocean turbulence and waves.

Recent progress in local dynamical modeling of the mid-ocean is only a beginning, however. Many important unsolved problems remain to puzzle ocean modelers and theoreticians. Examples include: (1) a more complete theoretical and/or operational account of the interaction of baroclinic flow with bottom roughness, (2) a Lagrangian characterization of the idealized dynamical processes for which Eulerian (spectral and physical space) signatures are now partly understood, and (3) parameterization schemes for momentum/vorticity/heat mixing by mesoscale eddies and smaller scale (subgridscale) phenomena. Although many more enticing and significant theoretical questions could, of course, be listed, these three serve to indicate some of the complex dynamical issues being actively addressed by ocean scientists at this time and for which at least partial answers may soon be available.

Effects and Applications

19. Eddies in Relation to Climate

A.E. Gill

19.1 Introduction

Eddies in the ocean could be important for climate in two ways. One is through the direct effect of the heat they transport. The other is indirect through the effect of the eddies on the circulation and on air/sea exchange rates. If these are different in the presence of eddies, the heat budget of the ocean and atmosphere will be affected. Evidence about both direct and indirect effects of eddies on the heat balance are considered below.

19.2 Heat Flux Carried by Eddies

The way heat is transported in the *atmosphere* is known quite well from observational studies (e.g., Oort 1971). The traditional method of analysing the data has been in terms of zonal means, with a subdivision of the flux into three parts, one due to the meridionial circulation, one to standing eddies and the third due to transient eddies. Where the heat flux is largest, i.e., in extra-tropical latitudes, the part due to transient eddies (i.e. cyclones and anticyclones) gives the dominant contribution.

The information about heat flux in the *ocean,* whether by eddies or other means, is very sparse, and quantitative estimates are available for very few places. The only available calculations on a global scale are those of Vonder Haar and Oort (1973) and Trenberth (1979) who estimate the ocean heat fluxes as a *residual* after the total flux (atmosphere plus ocean) has been calculated from the Earth's radiation balance, and the atmospheric flux has been subtracted. The result is shown in Fig. 1. The implication is that the ocean carries similar fluxes of heat to those observed in the atmosphere with the largest ocean fluxes at latitudes in the twenties.

Eddy heat fluxes will now be discussed in the light of Fig. 1 for two areas where some information is available, namely the North Atlantic and the Southern Ocean. Before going into detail, it is useful to estimate how large an eddy flux would be considered significant for a given area, and the *North Atlantic* will be discussed first. Bryden and Hall (1980) have used the estimates of atmosphere-to-ocean heat fluxes of Bunker (1976) to find how much heat flux across latitude 25°N is implied. This is the latitude where the flux attains a maximum value of 10^{15} W. To obtain a measure of how large a heat flux might

Eddies in Marine Science
(ed. by A.R. Robinson)
© Springer-Verlag Berlin Heidelberg 1983

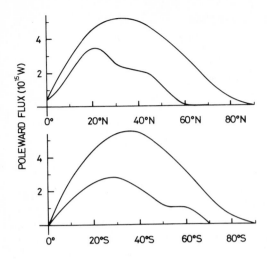

Fig. 1. Poleward flux as a function of latitude in the northern hemisphere (*upper panel* based on Fig. 1 of Vonder Haar and Oort 1973) and in the southern hemisphere (*lower panel* based on Fig. 1 of Trenberth 1979). *Upper curve* is the total flux (atmosphere plus ocean) required to make up the radiative imbalance. *Lower curve* is the oceanic flux implied by subtracting the estimated atmospheric flux

be considered significant, suppose all this flux were carried by eddies with the flux distributed uniformly across the 6 000 km width of the ocean and over the top 1 km of the water column (i.e., mostly above the thermocline). The average flux density would then be

$$1.6 \times 10^5 \text{ Wm}^{-2} = 4 \text{ cal cm}^{-2} \text{ s}^{-1}$$

(i.e., mean perturbation velocity times perturbation temperature = 4°C cm s^{-1}) and this value can be used as a standard of comparison for North Atlantic values. For comparison, measurements in a mooring array located at 15°–17°N, 53°–55°W (Fu et al. 1982) give eddy heat flux magnitudes which are only marginally significant at the 90% level, the values being about 1°C cm s^{-1} and the directions predominantly eastwards. In other words, the meridional flux is weak. Values found by Schmitz (pers. comm.) on the 55°W section also appeared to be weak at latitudes south of 35°N.

From the spot measurements, it would appear that eddies do *not* contribute a great deal to the heat flux in the North Atlantic between 14° and 35°N where the flux is strongest. In fact, calculations by Bryden and Hall (1980) of the flux by the mean flow at 25°N show that this can account for the 10^{15} W through the section without the aid of eddies. The flux is principally due to water being carried northwards through the Florida straits and being returned 13° cooler over the remainder of the section. It is perhaps worth remarking that such a mode of heat transport is not available in the atmosphere because this form of circulation requires east-west pressure differences which can only be established when side boundaries are present.

There are, however, other measurements of eddy heat flux in the North Atlantic which are significant by the above criterion. Observations by Schmitz (pers. comm.) in the energetic recirculating-gyre region south of the Gulf Stream show large eddy heat fluxes *to the south* (i.e., the opposite way to those depicted in Fig. 1). (It is of interest to note that Semtner and Mintz (pers.

comm.) found such a southward flux in the eddy-resolving model reported below.) In particular, at 37.5°N, 55°W the southward flux amounted to about 10°C cm s^{-1} at 600 m, about 4°C cm s^{-1} at 1 000 m, about 1°C cm s^{-1} at 1 500 m and was less than 0.1°C cm s^{-1} at 4 000 m. It follows that eddy heat fluxes in the ocean *are* important, at least in some especially energetic regions. However, other *indirect* effects of eddies on the heat flux in such regions may be more important, as discussed later.

The other region where some measurements of eddy heat flux exist is the Southern Ocean, and these measurements are discussed by Bryden in Chapter 14. He finds that if the results obtained in Drake Passage are extrapolated to the entire circumpolar region, they can account for all of the meridional heat flux. Thus eddies appear to be very important in this region.

19.3 Indirect Effects of Eddies on the Heat Balance

From studies with numerical models of the ocean, it is clear that the presence of eddies can strongly modify the mean circulation of the ocean, and hence the heat flux due to that mean circulation, and can also have strong effects on the rate of exchange of heat with the overlying atmosphere. The subject has not yet advanced to the stage where quantitative estimates of these indirect effects can be made for, say, the North Atlantic, so the best that can be done is to point to numerical experiments which indicate possible effects. These models are also of interest in the results they give for direct effects. Other chapters discuss such experiments in some detail.

Figure 2 (Semtner and Mintz pers. comm.) shows the difference in surface temperature between two experiments with an ocean circulation model. Air-sea exchange rates were determined by standard bulk formulae and meteorological variables were kept fixed at values determined from climatology. The only differences between the two models were that the second had twice the resolution (and so unstable eddies could develop) and a lower lateral eddy viscosity (this has to be larger with the sparse grid to maintain numerical stability). The calculation shows that the whole character of the heat transport in the Gulf Stream region changes when eddies are present, and this is reflected in the sea surface temperature pattern. The eddies in fact produce a fundamental change in the structure of the circulation which looks quite different in the mean as well as at any instant. Such effects on the circulation are discussed in Chapter 17.

Figure 3 (from Mintz 1979) shows the partition of heat transport between the mean circulation and eddies for a simulation in which eddies were resolved. The size of the model basin was small, so magnitudes are low. However, it is interesting that the model shows the significant equatorward eddy heat flux between 25° and 35° latitude, a poleward flux only appearing when the mean flux is added. Separate maps of the heat flux due to mean currents and the heat flux due to eddies in the "Gulf Stream" area show the two fluxes are in opposite directions over most of this region.

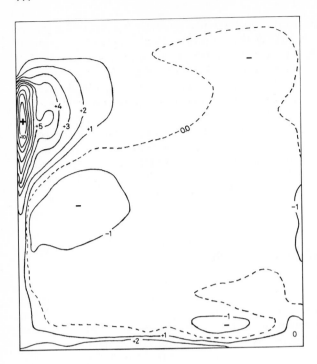

Fig. 2. Difference in °C between the 520-day time averaged temperature at the uppermost level in a simulation with 40 km grid size and the same quantity in a simulation with 120 km grid size (Semtner and Mintz pers. com.). The basin extends from 15°N to 55°N

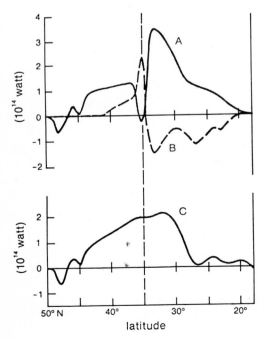

Fig. 3. Poleward heat transport in a model ocean basin (AJ Semtner pers. comm.). **A** Transport by time-averaged circulation; **B** transport by transient eddies; **C** total transport

The implication of these studies appears to be that in the region representing the "Gulf Stream" area, eddies play an important role in the ocean circulation and also in eddy heat transport. These eddy transports are, however, offset by opposing fluxes by mean currents and the dominant transport in the models is still that due to the mean currents.

In summary, the role of eddies in climate is difficult to assess at this stage, as measurements of eddy heat flux and of eddy effects on circulation require very long time series and few exist. The indications are that eddies do play an important role in active regions like the Gulf Stream recirculation region and Drake Passage, but are not directly important in the less active parts of the ocean.

20. Eddies and Coastal Interactions

P.C. Smith

20.1 Introduction

When mesoscale eddies encounter the continental margin, the restoring force associated with the movement of the vortex columns across isobaths supports wavelike motions, known as topographic Rossby waves, over the sloping bottom. For typical depth profiles $h(x,y)$, the ratio of the topographic to planetary restoring forces $\beta L_T/f$ (where L_T is the topographic length scale, $H/\nabla h$) is much smaller than one, so the dispersion properties of these waves are governed primarily by the topography. Strong lateral variations in the bottom *slope* also provide an important mechanism for refraction, reflection and scattering of topographic wave energy. Other important dynamical effects include:

a) density stratification, which weakens the tendency for columnar motion in a rotating fluid and supports baroclinic oscillations,
b) bottom friction, which may cause dissipation in shallow shelf areas, and
c) nonlinear advection and shear in longshore coastal currents.

The proximity of western boundary currents to eastern continental margins provides a convenient energy source for topographic waves. Meanders and eddies shed by these strong currents are likely to shoal on the outer portion of the continental rise, where their deep velocity fields are forced to accommodate the condition of no normal flow through the sloping bottom. The adjustment to this constraint involves radiation of topographic waves to other areas of the continental rise, slope and shelf. However, the penetration of topographic wave energy onto the shelf in particular may be severely limited by topographic constraints (refraction, reflection, scattering into other modes), dissipation by bottom friction, and/or the effects of shear in the longshore coastal current (refraction, critical-layer absorption).

Much of the early theoretical analysis of mesoscale eddy interactions with the continental margin has been formulated in terms of the linear dynamics of topographic waves over small bottom slopes (e.g., Rhines 1969, Suarez 1971). More recently, however, the effects of finite depth changes, bottom friction, and longshore advection with weak shear have been treated (e.g., Kroll and Niiler 1976, Ou and Beardsley 1980, Collings and Grimshaw 1980). In these models, the forcing is approximated by a monochromatic wave which is uniform along the outer boundary. For realistic topography, most of the incident topographic wave energy would be reflected from the steep continental slope (Rhines 1971). However, due to drastic reductions in depth, significant levels

Eddies in Marine Science
(ed. by A.R. Robinson)
© Springer-Verlag Berlin Heidelberg 1983

of kinetic energy are possible over the outer portion of the shelf, where the effects of bottom friction and sometimes longshore current must be taken into account. In addition, some of the incident energy may be fed into trapped baroclinic modes at abrupt changes in the topography (Suarez 1971, Ou and Beardsley 1980). A survey of various modelling efforts for topographic interactions is given in Sect. 20.2.

To date, very little attention has been given to the generation of the topographic waves by a mesoscale feature incident on the continental rise. Analysis of this process is essential since it determines the character (e.g., amplitude, frequency and length scales) of the incident waves. Linear initial value problems for a impulsive vorticity disturbances in homogeneous (Longuet-Higgins 1965, Louis and Smith 1982) and two-layer (Tang 1979) fluids have been solved analytically, and, in the former case, applied to determine the radiation field of a Gulf Stream Ring on the Scotian Rise (Louis and Smith 1982). Kroll (1979) has used harmonic forcing from a distributed source to model the generation of topographic waves by remote weather systems.

To observe mesoscale eddy interactions with the continental margin requires long-term measurements of horizontal velocity and density (or at least temperature) throughout the water column. The vertical spacing of the measurements must resolve the most energetic (low) baroclinic modes through the thermocline as well as the evanescent bottom-trapped mode. In the horizontal plane, a coherent array across the continental rise and slope and onto the shelf requires dense coverage in the cross-isobath direction (consistent with local topographic length scales), but widely spaced instruments in the along-isobath direction to detect the much larger longshore scales. Unfortunately, very few of the existing data sets satisfy all of these requirements. A summary of conventional observations (e.g., data from moored instruments) and their analysis in terms of topographic waves and eddy interactions is given in Sect. 20.3.

With the rapid development of satellite technology and its oceanic applications, a new tool for synoptic monitoring of the mesoscale eddy field in the vicinity of the continental margin has become available. In particular, time series of infrared images of the sea surface, contrasting the meanders and eddies shed by warm western boundary currents with the colder coastal waters, provide the means to monitor length scales, drift and growth rates of, and interactions among these mesoscale features. Furthermore, with regard to the generation process, it is sometimes possible to detect the influence of "fast" barotropic waves as the deformation of surface isotherms in the far field of a radiating eddy (Louis and Smith 1982). The potential for combining the synoptic sea surface data with direct conventional observations has not been fully exploited, but an example of one such attempt is described in Sect. 20.4.

Finally, with regard to the role of topographic waves in the shelf circulation, Garrett (1979) has shown that the strong longshore currents produced at the shelf break (by topographic amplification and focussing of the incident waves) are capable of inducing upwelling onto the shelf via the bottom Ekman layer. The vertical isopycnal displacement resulting on a uniformly stratified shelf is of order V/N, where V is the current magnitude and N is the Brunt Vaisälä frequency. In addition, Garrett showed that on the East Australian Shelf, the onshore divergence of the topographic wave momentum flux makes a significant contribution to balancing the observed longshore pressure gradient.

On the other hand, Smith et al. (1978) estimate that the topographic wave momentum flux is negligible in the dynamical balances on the Scotian Shelf. In that region, however, it appears that mesoscale interactions produce substantial onshore eddy fluxes which make significant contributions to the shelf-wide budgets for heat, salt and nutrients (Houghton et al. 1978, Smith 1978).

20.2 Theoretical Considerations

20.2.1 Stratification, β-Effect

The properties of linear, inviscid, nondivergent topographic Rossby waves in a uniformly stratified fluid over a gently sloping bottom have been discussed by Rhines (1970, 1977). Invoking the Boussinesq and "traditional" (neglect horizontal Coriolis force) approximations, conservation of potential vorticity on a β-plane requires

$$\frac{\partial}{\partial t}\left(\nabla^2 \psi + \frac{f_0^2}{N^2}\frac{\partial^2 \psi}{\partial z^2}\right) + \beta \frac{\partial \psi}{\partial x} = 0, \qquad \nabla^2 = \frac{\partial^2}{\partial x^2} + \frac{\partial^2}{\partial y^2}, \tag{1}$$

where ψ is the streamfunction for horizontal velocities and the Coriolis parameter is given by $f = f_0 + \beta y$. A gentle north–south slope α is required to make the problem separable and leads to the following linearized boundary conditions,

a) $\quad \dfrac{\partial^2 \psi}{\partial z \, \partial t} = 0 \qquad\qquad$ at $z = 0$, and

$$\tag{2}$$

b) $\quad \dfrac{\partial^2 \psi}{\partial z \, \partial t} = -\dfrac{N^2 \alpha}{f_0}\dfrac{\partial \psi}{\partial x} \qquad$ at $z = -H + \alpha y$.

With free horizontal boundary conditions, these equations may be scaled and expanded in the small parameter, $\varepsilon = \dfrac{\alpha L}{H}$, which represents the fractional depth change over a characteristic horizontal length scale, L, to give plane wave solutions of the form

$$\psi \propto e^{i(kx + ly - \omega t)} \begin{array}{l} \cosh \mu z \\ \cos m z \end{array}$$

$$\begin{array}{r}(3\,a)\\(3\,b)\end{array}$$

where the $0(\varepsilon$ frequency is

$$\omega = \frac{-\beta k}{K^2}\left\{\begin{array}{l}\left(1 - \dfrac{\mu^2 H^2}{S^2}\right)^{-1}\\[2ex]\left(1 + \dfrac{m^2 H^2}{S^2}\right)^{-1}\end{array}\right. = -\frac{\alpha N^2 k}{f_0}\left\{\begin{array}{l}(\mu \tanh \mu H)^{-1}\\[2ex]-(m \tan m H)^{-1},\end{array}\right.$$

$$\begin{array}{r}(4\,a)\\[2ex](4\,b)\end{array}$$

$K^2 = k^2 + l^2 \equiv L^{-2}$ and $S = \dfrac{NH}{fL}$ is the Burger number.

The vertical structures of the waves are defined by

$$\mu H \tanh \mu H = \frac{\alpha f_0}{\beta H}[S^2 - \mu^2 H^2] = -\frac{N^2 \alpha H}{f_0 c_1} \tag{5a}$$

and

$$mH \tan mH = -\frac{\alpha f_0}{\beta H}[S^2 + m^2 H^2] = \frac{N^2 \alpha H}{f_0 c_1}, \tag{5b}$$

where $c_1 = \omega/k$ is the eastward component of the phase velocity. According to Rhines (1977), these solutions fall into three classes depending on the character of the vertical shear:

1. "Fast Barotropic" waves have vanishing vertical shear in the limit $S^2 \ll 1$ for Eq. (5a). This implies that the scale of the wave is much greater than the internal deformation radius, NH/f, and the approximate dispersion relation

$$\omega \approx -\frac{k}{K^2}\left(\frac{\alpha f_0}{H} + \beta\right) \tag{6}$$

is that for familiar barotropic waves. For typical bottom slopes near the continental margin, $\alpha = 0(10^{-2})$, the β-effect is negligible i.e., $\dfrac{\beta H}{\alpha f_0} = 0(10^{-2}) \ll 1$.

2. "Fast Baroclinic" waves represent the bottom-trapped mode in the opposite limit for Eq. (5a), $S^2 \gg 1$. In this case, the dispersion relation is

$$\omega \approx -\frac{\alpha N k}{K} \equiv -\alpha N \sin\theta, \tag{7}$$

where θ represents the rotation of the rectilinear motion with respect to the longshore direction. The limit, $\theta = \pi/2$, represents a buoyancy oscillation, trapped to the bottom on a scale of

$$\mu^{-1} \approx \frac{f_0}{NK} \ll H.$$

3. "Slow Baroclinic" waves are the oscillatory modes which exist as free modes only for $\beta \neq 0$. For small slopes $\left(\dfrac{\alpha f_0}{\beta H}S^2 \ll 1\right)$, the vertical structure approaches that of simple baroclinic Rossby waves, i.e.,

$$mH \approx n\pi \quad n = 1, 2, \ldots, \tag{8a}$$

whereas for larger slopes $\left(\dfrac{\alpha f_0}{\beta H}S^2 \gg 1\right)$, a node develops at the bottom i.e.,

$$mH \approx (n + \tfrac{1}{2})\pi \quad n = 0, 1, 2, \ldots \tag{8b}$$

for the low mode numbers. For realistic open ocean parameters, the energy in these waves is concentrated near the surface, so that they have also been dubbed "thermocline eddies".

Recently, Ou (1980) has extended Rhines' analysis to include the effects of finite depth and bottom slope. Considering the propagation of free waves in an infinite uniformly-stratified wedge ($\beta=0$), he found that the problem could be transformed into the corresponding surface gravity wave problem with the roles of surface and bottom boundaries reversed. Thus analytical surface wave solutions are available to describe waves that are either progressive or trapped in the crosswedge direction. For progressive waves, the far field (deep water) solution corresponds to an incident fast baroclinic wave, which is then refracted and transformed to a barotropic wave in the near field around the apex (see Sect. 20.2.3).

20.2.2 Abrupt Topography: Scattering and Reflection

Suarez (1971) has considered the problem of Rossby waves incident on a discontinuity in the bottom slope. In the linear case, when a long, fast incident wave

$$\psi_i = \Psi_i(y,z)e^{i(kx-\omega t)} \quad (k<0) \tag{9}$$

encounters a slope, $\alpha\left(\varepsilon = \dfrac{\alpha L}{H} \ll 1\right)$, at $y=0$, its energy is partially reflected, transmitted as a fast barotropic or baroclinic wave, and scattered into baroclinic "fringe" modes (Ou and Beardsley 1980) which are horizontally trapped to the discontinuity. The matching conditions across the break require that the tangential wave number, k, and frequency, ω, be preserved, so that the structure of the solution in the sloping region is governed by the tangential phase speed, $c_1 = \omega/k$, and the external parameters (α, N, f_0, H).

The vertical trapping scale of the transmitted wave, governed by (5a), has the same limits as described in the last subsection, i.e., a) barotropic for large phase speed or small slope, which implies

$$\mu H \approx \left(-\frac{N^2\alpha H}{f_0 c_1}\right)^{1/2} \ll 1 \tag{10a}$$

b) baroclinic for small phase speed or large slope, i.e.

$$\mu H \approx -\frac{N^2\alpha H}{f_0 c_1} \gg 1. \tag{10b}$$

Similarly, the structure of the baroclinic modes is determined by Eq. (5b), where the above limits for the vertical wave number are given by Eqs. (8a) and (8b). The horizontal trapping scales of the baroclinic modes ($\propto e^{\gamma_n y}$) are given by

$$\gamma_n^2 = \frac{f_0^2 m_n^2}{N^2} + \frac{\beta}{c_1} + k^2 > 0 \tag{11}$$

for $c_1 < -\beta N^2 (f_0^2 m_n^2 + N^2 k^2)^{-1} < 0$. Hence, for realistic parameters at the base of the continental slope ($c_1 = -10^{-1} \mathrm{m\,s}^{-1}$, $H = 10^3$ m, $N/f_0 = 10$, $k^2 = 10^{-9} \mathrm{m}^{-2}$, $\beta = 10^{-11} \mathrm{m}^{-1} \mathrm{s}^{-1}$) the maximum upslope penetration of these modes is the internal deformation scale, $\lambda \equiv NH/f_0$,

$$\gamma_1^{-1} \approx \frac{N}{f_0 m_1} = 0(10 \text{ km}). \tag{12}$$

For arbitrary orientation of the discontinuity, the longshore component of the incident phase speed may have either sign. In this case, the addition of a complementary set of trapped baroclinic modes for $c_1 > 0$ forms a complete orthogonal basis about which to expand the vertical structure of any incoming wave (Suarez 1971).

The effects of finite depth changes on the kinematics and energetics of barotropic waves have been analyzed by Kroll and Niiler (1976). Again taking north as the upslope direction on one dimensional topography (i.e., scale of variation infinite in one horizontal direction), the governing equation is

$$\frac{\partial}{\partial t} \left[\nabla^2 \phi + \frac{d}{dy} \left(\frac{dh}{dy} \Big/ 2h \right) - \left(\frac{dh}{dy} \Big/ 2h \right)^2 \phi \right] + \left(\beta - f \frac{dh}{dy} \right) \frac{\partial \phi}{\partial x} = 0, \tag{13}$$

where the dependent variable ϕ represents a transformation of the transport streamfunction such that,

$$u = -h^{-1/2} \left[\frac{\partial \phi}{\partial y} + \left(\frac{dh}{dy} \Big/ 2h \right) \phi \right],$$

$$v = h^{-1/2} \frac{\partial \phi}{\partial x}. \tag{14}$$

The continental margin is then modelled as piecewise continuous exponential topography of the form,

$$h(y) = H e^{-2ay}, \tag{15}$$

which renders the coefficients in Eq. (13) constant. For barotropic plane wave solutions, the dispersion relation is simply

$$\omega = \frac{-(2af_0 + \beta)k}{K^2 + a^2} = \frac{-\beta^* k}{K^2 + a^2}, \tag{16a}$$

which is similar to Eq. (6) in the short wave limit $K^2/a^2 \to \infty$. The components of the group velocity $\underline{c}_g = (c_{g1}, c_{g2})$ are then given by,

$$c_{g1} = \frac{\partial \omega}{\partial k} = \frac{\beta^* (k^2 - l^2 - a^2)}{(K^2 + a^2)^2}$$

and

$$c_{g2} = \frac{\partial \omega}{\partial l} = \frac{2\beta^* kl}{(K^2 + a^2)^2}, \tag{16b}$$

where $\beta^* = \beta + 2af_0$.

The form of Eq. (14) implies that the mean energy density on exponential topography,

$$\overline{E} = \frac{\rho}{2} (\bar{u}^2 + \bar{v}^2) \propto h^{-1} \tag{17}$$

is inversely proportional to the local depth. This central result of the Kroll and Niiler analysis implies that although the bulk of the incident energy may be reflected from the slope discontinuity, large energy densities are possible over the steep continental slope due to finite depth variations. Furthermore, for a model in which three segments are used to represent the shelf ($a = a_I$), slope (a_S), and rise (a_R) regions, the flux of this energy onto the shelf may be enhanced for selected waves which satisfy the criterion

$$l_S W = n\pi \qquad n = 1, 2, \dots \tag{18}$$

where l_S and W are the offshore wavenumber and width of the slope region. When $a_I = a_R$ this corresponds to the Ramshauer effect (Rhines 1969) in which half wavelengths normal to the shelf fit perfectly between the breaks and there is no reflected wave over the rise. On the other hand, when the amplitude of the reflected wave is significant, it interferes with the incident wave and produces a modulated energy distribution in the offshore region.

Ou and Beardsley (1980) have recently generalized both the barotropic and small slope results using a numerical model for a viscous stratified fluid over realistic topography. The offshore boundary conditions consist of an ensemble of incident and reflected quasi-geostrophic waves. At the inshore boundary, on the other hand, only transmitted waves are permitted under the assumption that transmitted energy is fully dissipated on the shallow portions of the shelf. In their example of three-piece exponential topography, the bulk of the incident energy is again reflected by the steep continental slope leading to a standing wave and modulated energy distribution over the slope and rise. In addition, however, baroclinic "fringe" modes are excited at the junction between the continental rise and slope in order to match the different structures of the propagating waves on either side. At relatively high frequencies, these modes produce an amphidrome in the pressure field which results in reversals of the usual phase propagation and momentum and heat fluxes in the upper part of the water column.

The absence of a reflected wave component on the shelf causes a sharp drop in the kinetic energy across the shelf break suggesting that topographic waves would be difficult to observe on the shelf proper. Furthermore, varying stratification affects the maximum onshore energy transmission coefficients associated with the Ramshauer effect [Eq. (18)] by shifting the peaks (for a given frequency) as a function of the alongshore wavenumber. This difference may be understood, in light of the quasi-geostrophic dispersion relation [Eq. (4)], as variations of l_s with N (or S).

20.2.3 Refractive Effects

When topographic variations are gradual rather than abrupt, that is, when the length scale for changes in the topographic restoring force far exceeds that of the wave itself, ray theory may be applied to study the refraction of topographic Rossby waves. For a given topography, the validity of this approximation is enhanced in the low-frequency limit, since wavelength decreases with frequency. Smith (1971) derives and discusses the equations for the ray paths and amplitude variations in this limit for barotropic waves over slowly-varying topography, $h(x, y)$. He finds that ray trajectories are governed by the geostrophic vector,

$$\underline{G} \equiv h \nabla (f/h), \tag{19}$$

and that rays tend toward regons where $|\underline{G}|$ is large. For instance, if $\underline{G}$ has a pole at the shoreline, the ray paths are circles tangential to the coast. In this case, free wave energy is refracted toward the shore where enhanced dissipation (as $h \to 0$) is expected to provide a partial sink for the incident energy.

In the absence of dissipation, Smith (1971) also demonstrates that conservation of energy between two adjacent rays implies that the kinetic energy density, $\bar{E}$, is inversely proportional to the local depth and the separation of the ray paths, i.e.,

$$\bar{E} \propto [h J(n, t)]^{-1} \tag{20a}$$

where

$$J(n, t) = |\underline{c}_g| J(n, \underline{s}) = c_{g1} \int_0^t \frac{\partial c_{g2}}{\partial n} dt' - c_{g2} \int_0^t \frac{\partial c_{g1}}{\partial n} dt' \tag{20b}$$

is the Jacobian of the transformation to ray coordinates,

$$\underline{x}(n, t) = r_0(n) + \int_0^t \underline{c}_g(n, t') dt' \tag{20c}$$

$$= r_0(n) + \underline{s}(n, t),$$

$\underline{c}_g = (c_{g1}, c_{g2})$ ist the group velocity [Eq. (16b)], n is the ray index, r_0 is the initial position vector of the ray, and $\underline{s}$ is the distance measured along the ray. Louis and Smith (1982) have exploited these formulae to determine the energy distribution over parabolic topography as described in Sect. 20.4.

In a stratified fluid, qualitatively similar effects will govern the refraction of incident baroclinic waves, but as the depth shoals, their vertical structure becomes progressively more barotropic according to the limit (10a) of the dispersion relation. This behavior is exhibited by Ou's (1980) solution for free waves propagating in a uniformly stratified wedge. In the deep water, the solution reduces asymptotically to an incident bottom-trapped wave, but near the apex the amplitude is uniform vertically and the local offshore wavenumber is proportional to $h^{-1/2}$, consistent with the barotropic result.

20.2.4 Bottom Friction

As the topographic wave energy propagates shoreward, the decreasing depth and possible amplification due to other geometric effects leads to strong dissipation in the shallow regions. For barotropic waves, these effects have been analyzed using the linearized form of the bottom stress law (e.g., Csanady 1976)

$$\frac{\tau_b}{\rho h} = R(u, v), \qquad R = \frac{C_D V_b}{h} \qquad (21)$$

where τ_b is the bottom stress, C_D is the drag coefficient, V_b is the bottom velocity scale, and R^{-1} is the dissipation time scale (Garrett 1979). Considering the $h^{-1/2}$ dependence for the velocity components of the wave itself [Eq. (14)], Kroll and Niiler (1976) chose $R \propto h^{-3/2}$ and found various solutions for free waves with friction. As expected, dissipation limits the penetration of topographic wave energy in the upslope direction. On a simple exponential slope, this produces a single energy maximum from which the energy decays rapidly shoreward. However, the position and magnitude of this peak are sensitive to the characteristics of the incident wave (i.e., frequency and longshore wavelength), with the greatest penetration achieved by waves with the largest upslope component of the group velocity. Also according to this solution, total decay of the incident wave energy occurs for $h < h_m/4$ where h_m is the depth at the energy density maximum. For the more realistic three-segment topography, the effect of friction was to reduce the transmission of energy onto the shelf, particularly for the shorter waves, leaving a strong concentration over the continental slope. In the stratified version of this model (Ou and Beardsley 1980), Ekman friction produces anomalous eddy heat fluxes near the bottom.

Using the simplest possible formulation, $R = const$, Collings and Grimshaw (1980) have computed an upslope frictional decay scale, δ, on exponential topography in the presence of a mean longshore current. When the current vanishes, this scale is obtained by solving Eq. (16 a) for a complex-valued onshore wavenumber, $l = l_0 - i\delta$, with l_0, ω, δ, k all real. In the weak dissipation limit $(R/\omega \ll 1)$, the decay is given approximately by the resistance coefficient divided by the onshore component of the group velocity, i.e.,

$$\delta = R/c_{g2} \qquad (22)$$

and corrections to the real part of l occur at higher order (R^2/ω^2).

20.2.5 Longshore Currents

Garrett (1979) and Collings and Grimshaw (1980) have explored the effects of mean longshore current and linear transverse shear on topographic waves over exponential topography with $\beta = 0$. Garrett discusses the kinematics of waves imbedded in a uniform flow, U_0. The longshore component of the trace speed, i.e., the apparent propagation speed of the wave, is given by

$$c_{t_1} = \omega/k + U_0 \tag{23}$$

where ω is the intrinsic frequency relative to the mean flow. Then specifying U_0 and c_{t_1}, the dispersion relation (16a) yields

$$l = \left[\frac{2 a f_0}{(V_0 - c_{t_1})} - k^2 - a^2 \right]^{1/2}. \tag{24}$$

Hence of the offshore trace speed,

$$c_{t_2} = \omega/l + U_0 k/l, \tag{25}$$

consists of an intrinsic speed relative to the mean current and a component resulting from the advection of the sloping phase lines.

Collings and Grimshaw (1980) introduce a linear transverse mean shear, $U = U_0 - A y$, in the nearshore zone matched to a uniform exterior flow. This leads to the possibility of a critical layer at $y = y_c$ [$U(y_c) = c_{t_1}$] which would absorb the incident wave momentum flux. Offshore from this singularity, the solution is expressed in terms of Whittaker functions which describe the refractive effects of the shear.

20.2.6 Generating Mechanisms and Initial Value Problems

Hydrodynamic instability is generally regarded as the primary cause of meandering and eddy formation by strong western boundary currents (e.g., Niiler and Mysak 1971, Orlanski 1969). When the flow is close to the coast, meanders and eddies impinge directly on the continental shelf and instability theory may be used to describe their initial transient development. However, because of the rapid growth rates for the infinitesimal disturbances (typical e-folding time ≤ 7 days), the nonlinear interactions of the unstable waves are important in limiting wave growth and modifying the forms of the disturbances (Orlanski and Cox 1973). This renders the determination of the complete evolution of a mesoscale feature and its interaction with the continental margin hopelessly complex. In many cases, a more enlightening approach is to utilize observed temporal and spatial scales to represent an impinging eddy or meander as impulsive or harmonic vorticity disturbances. Analytically, weighted sums of elementary (linear) wave solutions for point source disturbances may be used to map the response to an observed feature. On the other hand, numerical solutions to initial value problems allow the incorporation of nonlinear effects. Other generating mechanisms for mesoscale meanders and eddies, such as flow over isolated topography (e.g., Huppert and Bryan 1976, Chao and Janowitz 1979), require similar techniques. A summary of some relevant results for topographic interactions is given below.

Longuet-Higgins (1965) derived the linear, barotropic planetary wave response to a point source vorticity impulse [forcing term $\propto \delta(\mathbf{x}) \delta(t)$] on a β-

plane. In the stationary phase approximation, two different wave systems meet at a caustic, which expands at a rate proportional to $\beta \Lambda^2$, where

$$\Lambda^2 = \frac{gH}{f_0^2} \tag{26}$$

is the external deformation radius. The asymptotic solution ($t \to \infty$) at a fixed distance from the source consists of isotropic short-wave radiation representing the slowest of the "fast barotropic" waves. In this limit, both wavenumber and period of oscillation increase as $t^{1/2}$ i.e.,

$$K = \frac{2\pi}{L} \approx \left(\frac{\beta t}{r}\right)^{1/2}$$

and

$$P = \frac{2\pi}{\omega} = 2\pi \left(\frac{2t}{\beta r}\right)^{1/2}, \tag{27}$$

where r is the radial distance from the source. Flierl (1977 b) used this model to study linear dispersion of radially symmetric disturbances on a β-plane.

Louis and Smith (1982) applied Longuet-Higgins' formalism to the problem of topographic wave radiation on constant exponential topography [Eq. (15)] with no planetary effects. In this case, the β-parameter and deformation radius are replaced by $2 a f_0$ and a^{-1} respectively. Hence the expansion rate for the barotropic wave field is scaled by $2 f_0/a$ ($\approx 80 \text{ m s}^{-1}$ on typical mid-latitude continental rise, $a \approx 2.5 \times 10^{-6} \text{ m}^{-1}$). Using the point source impulse solution as a Green's function, they derive an analytical solution for the radiation field of an isolated circular vortex,

$$v_0(r) = \frac{v_m r}{r_m} H(r_m - r), \tag{28}$$

where v_0 is the azimuthal velocity with a maximum v_m at $r = r_m$ and $H(x)$ is the Heaviside function. In the asymptotic limit [Eq. (27)], the solution for the transformed streamfunction is of the form,

$$\phi \approx -\frac{h_0^{1/2} K}{4\beta t} v_m r_m J_2(K r_m) \cos\theta \tag{29}$$

where J_2 is the 2nd order Bessel function and the phase is given by

$$\theta = [2\beta t(x+r)]^{1/2}. \tag{30}$$

The parameters (v_m, r_m) in Eq. (29) govern, respectively, the maximum amplitude and modulation of the radiating wave field, but do not affect its dispersion characteristics.

Furthermore, the approximate form of the mean kinetic energy density,

$$\bar{E} \approx \rho \frac{v_m^2 r_m^2}{16 r^2} e^{2ay} J_2^2(K r_m) \propto (h r^2)^{-1}, \tag{31}$$

contains factor representing the competing effects of upslope amplification and radial decay [since the Jacobian, Eq. (20 b), is proportional to r^2].

Tang (1979) considered the linear response to a barotropic point source impulse of a two-layer β-plane model with mean shear and small bottom slope. The far field is populated by mixed (topographic-planetary) barotropic Rossby waves which disperse rapidly. The near field consists of baroclinic modes and expands at a very slow rate of

$$\beta \lambda^2,$$

where $\lambda^2 = \dfrac{\Delta \rho\, g\, H_1}{\rho\, f_0^2} \left(1 + \dfrac{H_1}{H_2}\right)^{-1}$ \hfill (32)

defines the baroclinic deformation scale and H_i $(i = 1, 2)$ are the layer depths. The character of the baroclinic modes is strongly affected by the hydrodynamic stability of the mean shear, but the effects on the barotropic field are mainly kinematic.

Using a somewhat different approach, Kroll (1979) derives a linear, barotropic, Green's function for harmonic forcing at a point $[\propto \delta(\underline{x}) e^{i \omega t}]$ on the outer portion of two-piece exponential topography. Then he maps the response on the inner ("shelf") region to an offshore axisymmetric wind stress curl distribution. Here again the dominant longshore scale of the forcing $(L_m \approx 500 \text{ km})$ determines the most energetic frequency, ω_m, and local wavenumber, k_m, at any point on the shelf. With no β-effect, characteristic scales for the directly incident energy are estimated by

$$\omega_m = \frac{a_0 f_0 L_m}{\sqrt{7}},$$

and \hfill (33)

$$k_m = \frac{\sqrt{7}}{L_m}$$

where a_0 is the topographic parameter for the offshore region [Eq. (15)].

In the stratified case, Suarez (1971) and Rhines (1977) considered the initial imposition of a current with arbitrary vertical structure, $F(z)$. The elementary solution for a single Fourier component in the horizontal decomposition consists of a fast barotropic or baroclinic wave plus a steady geostrophic flow with a node at the bottom, i.e.,

$$\psi \propto e^{i(kx + ly)} \left[(e^{-i\omega t} - 1) \frac{\cosh \mu z}{\cosh \mu H} F(-H) + F(z) \right] \hfill (34a)$$

where ω and μ are given by Eqs. (4 a) and (5 a) respectively. Summing such solutions over azimuth for a fixed total wavenumber, $K = (k^2 + l^2)^{1/2}$, gives a circularly symmetric response

$$\psi \propto J_0(K \hat{R}) \frac{\cosh \mu z}{\cosh \mu H} + \left(1 - \frac{\cosh \mu z}{\cosh \mu H}\right) J_0(K r) \hfill (34b)$$

where

$$\hat{R}^2 = \left(x - \frac{\omega t}{k}\right)^2 + y^2$$

and $r^2 = x^2 + y^2$ is the radial distance from the initial disturbance. These Bessel solutions for the horizontal structure have the property that they propagate through a uniform medium without changing shape and provide a convenient basis about which to expand an arbitrary symmetric disturbance (Flierl 1977 b). However, composite solutions of this type are dispersive since waves of different K will travel at different speeds. At large distances from the centers of both the fixed ($Kr \gg 1$) or moving ($K\hat{R} \gg 1$) portions of the disturbance, the waves are locally plane.

McWilliams and Flierl (1979) have carried out a comprehensive numerical study of the evolution of isolated, nonlinear vortices on a baroclinic β-plane of uniform depth. For an initial Gaussian vortex consisting of a single (equivalent) barotropic mode, Rossby wave dispersion is inhibited by nonlinearities which are scaled by $E = V_0/\beta L^2$, the ratio of the particle velocity to the linear barotropic wave speed. However, when the initial condition consists of a mixture of the barotropic and first baroclinic modes, far-field eddies are efficiently generated by energy transfer to the barotropic mode and Rossby wave dispersion. This process requires large values of E since the modal coupling terms are nonlinear. The result is an energetic field of barotropic eddies surrounding the main vortex, which approaches a state of deep compensation where the lower-layer velocities vanish.

Topographic irregularities are also capable of generating mesoscale eddies and meanders by interacting with the large scale current field and redistributing vorticity in the vicinity of the disturbance. To investigate effects of variable currents, Huppert and Bryan (1976) considered the initial value problem for forced flow, V_0, over an isolated topographic feature on a stratified f-plane of constant depth. For large values of the Burger number ($S = NH/f_0 R_m$), a region of closed streamlines is formed over the feature (R_m, h_m = horizontal, vertical scales) provided

$$\frac{N R_m}{V_0} \geq 0(1) \tag{35a}$$

and an eddy generated by the initiation process is shed and advected downstream if

$$\frac{N h_m}{V_0} \leq 0(10). \tag{35b}$$

For weaker mean flows, the eddy remains trapped to the topographic irregularity as part of a double vortex structure. The amplitude of the disturbance is also sensitive to the inverse Froude number, $N h_m/V_0$, which governs what portion of the incident flow passes around rather than over the obstacle.

For steady flow, Chao and Janowitz (1979) developed a model for horizontally-sheared barotropic current over a small isolated bump on sloping topography. For exponential topography [Eq. (15)] and special forms of the upstream current profile, they find a pair of lee waves in the wake of the obstacle plus a set of decaying solutions trapped to it. In the downstream region, the maximum seaward deflection of the current occurs 1/4 wavelength from the disturbance and closed eddies may form in the shallow and deep water in the lee wave troughs.

Finally, Brooks (1979) has suggested that short-period oscillations observed in the South Atlantic Bight may be generated by backscattering of long continental shelf waves by variable topography and the Gulf Stream.

20.3 Observations and Analysis

Ideally, the observation and analysis of eddy interactions with the continental margin require simultaneous measurements of both the forcing mesoscale feature and the response of the surrounding coastal waters. Unfortunately there have been very few successful attempts at this formidable task. Topographic Rossby waves have been identified in measurements from several coastal areas, but usually the generating mesoscale feature has gone undetected. On the other hand, satellite imagery of sea surface temperature fronts along the continental margins have revealed intense interactions of eddies and meanders with surrounding surface waters, but except for some gross estimates of the temporal and spatial scales involved, the observations are mostly qualitative. In the following summary of observations and analysis from various regions, the need for an integrated approach is apparent.

20.3.1 The New England Continental Rise

Undoubtedly, the longest series of moored current and temperature measurements from the continental margin have been collected on the New England continental rise near Site D (39°10'N, 70°W). Thompson (1971) used "gappy" records from 3 years of deep water (≥ 500 m) data to identify certain characteristics of topographic Rossby waves with periods from 2 days to 2 weeks. He found that the vertical structure of the low-frequency current field was coherent and in phase and that the oscillations at each level were rectilinear and oriented at a small angle to local isobaths to give a significant downslope momentum flux. For both the barotropic and fast baroclinic waves, this implies a significant onshore component of the group velocity, which led him to suggest the Gulf Stream (100 km to the south) as a likely energy source. His numerical model for random forcing at the deep end of a wedge of homogeneous fluid produced a reasonable velocity field at early times with negative correlation coefficients ($R_{uv} \approx -0.5$) near Site D consistent with the downslope momentum flux. However, this behavior is found only during the transient development stages, as further integration leads to refracted or reflected (when a vertical barrier is placed in the shallow end) waves which substantially reduce the correlation and modulate the offshore energy distribution [Rhines (1971a)].

Thompson (1977) concludes that shallow-water frictional dissipation sufficient to reduce the amplitude of the reflected wave to two-thirds of the incident wave is required to produce the observed negative correlation at Site D.

Using 10 months of data from mooring just south of Site D, Thompson and Luyten (1976) demonstrated that the vertical structure of the kinetic energy at periods from 8 days (corresponding to the high-frequency cutoff for a buoyancy oscillation, $2\pi/\alpha N$) to 16 days is bottom-intensified. Using Eq. (5a), the observed vertical shear implies longshore wavelengths in the range of 100 to 200 km. Furthermore, they showed that the principle axes of the band-passed current ellipses were rotated toward the upslope direction with increasing frequency, consistent with the trend toward the buoyancy oscillation at the cutoff, $\theta = \pi/2$, in Eq. (7). Thompson (1977) provided a solid statistical basis for these and earlier observations by demonstrating that:

a) the momentum flux is downslope over the range of periods from 8 to 64 days,
b) currents are in phase and highly coherent in the vertical at 8 to 32 d,
c) measured wavenumber components fit the dispersion relation (4a) at 8 to 16 d,
d) phase propagation is downslope and westward at 8 to 32 d, and
e) the kinetic energy density at fixed distances off the bottom (200, 1 000 m) decays exponentially in the upslope direction to the 2 500 m isobath.

He also pointed out a remarkable concentration of kinetic energy in the 16-day band at Site D.

Further upslope to the 1 000-m isobath, Schmitz (1974) used 3 months of near-bottom records to examine the low-frequency current field. He observed distinct 10-day oscillations in the longshore velocity component and gave marginally significant estimates of 600 km and 40 cm s^{-1} for the longshore wavelength and phase speed. Unfortunately, cross-isobath coherences at 25 km separation were too low to make meaningful estimates of the offshore component of phase speed.

Shoreward from the 2 500 m isobath, Schmitz found that the deep kinetic energy density no longer decreases monotonically, but achieves a maximum near the slope/rise junction ($\approx 2 000$ m isobath), which he suggests might be related to the trapped baroclinic modes described in Section 20.2.2. Kroll and Niiler (1976), on the other hand, applied their barotropic model for three-piece exponential topography to the New England shelf and found that, for a 14-day wave with a 600 km longshore wavelength, the observed magnification near the slope/rise junction is achieved. However, even for the highest reasonable levels of dissipation, their results greatly overpredict the observed amplification near the 1 000 m isobath.

Recently, Ou and Beardsley (1980) used the results of their numerical model (incorporating stratification, friction, and realistic topography) to interpret measurements from a 5-month mooring experiment on the New England Shelf and Slope. Examining current ellipses and eddy heat fluxes in the period range from 3 to 10 days, they find some evidence for a predicted pressure amphidrome caused by baroclinic "fringe" modes excited at the slope/rise junction. However, a denser array of measurements is required before the dominant physics of the interaction may be resolved.

To date, no conclusive evidence of the dominant energy source(s) for topographic waves on the New England Continental Rise has been presented. However, Hogg (1982) has recently used empirical orthogonal function analysis of a 15-mooring array along 69°30′ and 70°W to describe the low-frequency circulation at periods from 8 to 108 days. Elaborating on Thompson's (1977) conclusions, he finds that the properties of the dominant modes in each of three low-frequency bands are consistent with topographic waves refracted by the rise topography. Then assigning a characteristic frequency, wavenumber and (spatial) energy maximum to each band, he projects the wave energy backward along predicted ray trajectories to a common source which lies east of the array. In the source region, two possible generation mechanisms are suggested:

1. the adjustment of the deep circulation in response to a large meander of the surface Gulf Stream during the first few months of the experiment, and
2. baroclinic instability of the deep circulation on the lower rise.

Gulf Stream rings and meanders are often observed to impinge on the outer rise (e.g., Saunders, 1971) and Rhines (1977, Fig. 13) points out that the deep velocity fields beneath such surface features are dominated by the relatively rapid, rectilinear oscillations which characterize the upslope records. Thus topographic wave energy from several of these isolated sources, as well as the Gulf Stream itself, is likely to radiate onto the continental margin.

20.3.2 The East Australian Shelf

Measurements of the low-frequency longshore surface current field on the East Australian Shelf have been made using observations of ships' drift in the primary coastal shipping lanes which are located 6.5 km (northward passage) and 19 km (southward passage at the shelf break) offshore (Hamon et al. 1975). In two years of data, oscillations with a characteristic period of 120 days were found to exhibit a lag of 7 to 10 days (offshore leading) between the two tracks.

Garrett (1979) has analyzed these data in terms of barotropic inviscid topographic Rossby waves imbedded in a uniform longshore current on an exponential shelf. Using the observed longshore trace speed and period for the spectral energy peak, he computes a lag of 20 days. Furthermore for reasonable parametrization of friction, the opposing effects of dissipation and topographic amplification [Eq. (17)] lead to a net decrease (by a factor of 0.2) in energy between the offshore and inshore tracks, which is consistent with observed levels. The introduction of a linear shear into the mean advective terms decreases the onshore trace speed and thereby increases the discrepancy between observed and theoretical lags (Collings and Grimshaw 1980). However, this trend is counteracted by friction and the net result (19 days) is similar to Garrett's estimate.

Garrett also suggests that the observed topographic waves are generated by meanders and eddies of the East Australian Current (e.g., Nilsson et al. 1977) as they encounter the continental slope, but no direct connection has been established.

20.3.3 The Southeastern Coast of the U.S.

Between the Florida Straits and Cape Hatteras, the Gulf Stream flows close to shore over the Blake Plateau and its lateral migrations impinge directly on the continental shelf. The proximity of the current to the coastline complicates the analysis of the observed fluctuations on the shelf because of the possibility of mutual interactions between wind-driven continental shelf waves and inherent hydrodynamic instabilities of the Stream itself. One essential difference is that the phase of the stable shelf modes propagate southward against the current (e.g., Brooks and Mooers 1977), whereas barotropic (Niiler and Mysak 1971) or baroclinic (Orlanski 1969) instabilities propagate northward. However, the situation is further confused by significant topographic variations which, in conjunction with the Gulf Stream, are capable of (1) backscattering long continental shelf wave energy from remote sources into high-frequency modes (Brooks 1979), or (2) generating localized vorticity disturbances (Chao and Janowitz 1979, see below).

Early observations of the meander patterns of the Gulf Stream off Onslow Bay, N.C. (Webster 1961, Oort 1964) reveal counter-gradient eddy heat and momentum fluxes in the surface layers, which is consistent with baroclinic instability theory (Orlanski 1969). Webster also demonstrated that the fluctuations were coherent with offshore wind forcing at the dominant meteorological time scales (4,7 days), but pointed out that the energy transfer implied by the momentum fluxes far exceeded the wind input. Brooks and Bane (1981) have recently confirmed Webster's conclusions about eddy fluxes and the implied energy conversions, but they find no significant correlation with either wind stress or coastal sea level. Their moored measurements extend the earlier results to at least mid-depth.

Düing et al. (1977) also found a countergradient momentum flux in current meter measurements from the Florida Straits. At the dominant time scale of 12 days, the eddy heat flux was down-gradient in the deep layer and countergradient near the surface, consistent with the extraction of potential energy from the mean density field in the lower layer which characterizes baroclinic instability. However, they were unable to resolve the dominant physics of this 12-day wave since the velocity fluctuations were also coherent with the curl of the wind-stress, again suggesting meteorological forcing. The wave amplitude reaches a maximum on the western side of the Strait, near the Miami Shelf, and a 180° phase change across the current was observed, consistent with both unstable baroclinic and stable barotropic shelf waves. Using current measurements from the 300 m isobath, Schott and Düing (1976) give evidence for southward-propagating fluctuations in the 10- to 13-day band, which they interpret as barotropic continental shelf waves.

At the shelf break in the Florida Straits, Lee (1975) and Lee and Mayer (1977) have observed reversals of the northward current component related to cyclonic "spin-off" eddies of the Florida Current. These fluctuations, which do not appear to be directly induced by wind, tide, or topographic irregularities, are coherent across the Florida Current (Düing et al. 1977) and propagate northward at roughly 35 km d^{-1}. The circulation associated with these features results in the intrusion of warm, salty tongues of Florida Current surface water onto the shelf and upwelling of cold nutrient-rich deep water in the core of the eddy (Lee et al. 1980). Hence they promote an efficient exchange between coastal and offshore waters. Lee and Mayer (1977) have given a loose comparison of their observations with the theory of barotropic shear instability (Niiler and Mysak, 1971), whereas Düing et al. (1977) suggest that these and other Gulf Stream instabilities observed downstream may originate as continental shelf waves in the Florida Straits.

"Spin-off" eddies in the Florida Current have been observed extensively using satellite infrared imagery (e.g., Legeckis, 1975; Vukovich et al. 1979). Vukovich et al. (1979) have noted three stages in the development of these mesoscale features as they propagate northward along the shelf break. Following their initial detection off northern Florida (27°–28°N), they undergo a 30% to 40% increase in size to a mature stage in which their amplitude is relatively stable. The average major and minor axes for the mature elliptical eddies are 136 and 36 km respectively and they propagate northward at 30 km d^{-1}. A decay stage sometimes follows in the vicinity of 30° to 32°N, during which the cold, fresh offshore tongue merges again with shelf water, thus trapping a core of warm Florida Current water on the shelf. The average period (9 days) of these fluctuations corresponds to that of the forcing by passing atmospheric cold fronts and there are indications of seasonal influences in the eddy characteristics (e.g., largest eddies occur in winter). Lee et al. (1980) have recently synthesized satellite, hydrographic and moored current meter data to describe the "spin-off" (renamed "frontal") eddy phenomenon and its implications for biological productivity on the Georgian Shelf.

Further north, Legeckis (1979) has noted a persistent seaward deflection of the surface isotherms representing the inshore edge of the Gulf Stream in the vicinity of Charleston, S.C. The steady component of this deflection has been interpreted by Chao and Janowitz (1979) in terms of lee waves generated by sheared, barotropic flow over the localized topographic irregularity known as the "Charleston bump". Legeckis, on the other hand, calculates the wavelike characteristics of the time-dependent meandering: 150 km wavelength and 40 km d^{-1} phase speed on average. He also tracks some of the small-scale "spin-off" eddies continuously past the bump from the upstream region and finds their amplitudes to increase downstream from the deflection.

20.3.4 The West Florida Shelf

Using 8 months of records from the 100 and 150 m isobaths, Niiler (1976) has described the low-frequency current field on the West Florida Shelf. The response to low-frequency forcing appears to be highly dependent on the local depth (i.e., banded) with wind-driven motions dominating the flow inshore from the 100 m isobath and topographic waves forced by offshore meanders and eddies evident in the deeper portions of the continental slope. At the 150 m isobath, he finds vertically coherent fluctuations, with characteristic period and wavelength of 12 days and 600 km respectively, propagating northward. The variance in a broad spectral peak (periods 10 to 25 days) for the dominant longshore current component is virtually constant over the records as compared to the strong seasonal variations at the 100 m isobath, suggesting off-shore forcing by meanders and eddies of the Loop Current which populate the cyclonic shear zone in the deeper Gulf of Mexico.

Kroll and Niiler (1976) apply their barotropic model for topographic Rossby waves to the West Florida Shelf and conclude that, for periods of 10 to 12 days, waves with lengths of 400 to 600 km may be effectively transmitted up the continental slope. With reasonable levels of frictional dissipation, the predicted ratio of kinetic energy densities on the 100 and 150 m isobaths agrees with that observed during the summer period, but the winter comparison is degraded by the wind-driven circulation.

Satellite observations of the Loop Current and its associated eddy field have been made by Maul et al. (1974) and Vukovich et al. (1979), but no direct comparison with the current field on the continental slope has been attempted.

20.3.5 The Scotian Shelf

Early current measurements from the Scotian Shelf and Slope suggest the presence of topographic Rossby waves similar to those observed off New England (Sect. 20.3.1). Using relatively short records, Petrie and Smith (1977) demonstrated that the low-frequency (periods greater than 10 days) near-bottom currents on the Scotian Slope were rectilinear and oriented at a small angle to local isobaths. In spectra for the summer period, they also found significant concentrations of kinetic energy at periods longer than the meteorological time scales of 2 to 10 days. Using a single wintertime mooring at the shelf break, Smith (1978) pointed out that the eddy momentum flux at periods greater than 10 days is offshore (i.e., downslope), consistent with a shoreward flux of topographic wave energy. The onshore eddy fluxes of heat, salt, and nutrients are also concentrated in these lowest-frequency bands, which has important implications for their respective budgets on the shelf. The suggested forcing mechanism for this low-frequency circulation is the interaction of Gulf Stream rings and meanders with the coastal waters.

Recently, a long-term mooring experiment was carried out to define the low-frequency circulation at the shelf break south of Halifax. A description of the use of these data to define the interaction of a mesoscale feature with the continental margin is presented in the next section.

20.4 Anatomy of a Warm-Core Ring Interaction with the Continental Margin

The Scotian Shelf is a broad (width ≈ 200 km), rugged region of the continental margin off the coast of Nova Scotia. Beyond the shelf are the steep continental slope (width ≈ 15 km, slope ≈ 0.05) and the gentler continental rise. The mean axis of the Gulf Stream in this region lies over the flat abyssal plain at the base of the rise (i.e., 300 to 400 km from the shelf break), but meanders and "pinched-off" rings of the Gulf Stream frequently penetrate northward onto the rise. The region between $63°$ and $65°W$ longitude appears to be a preferred site for large amplitude meanders and the formation of warm-core rings, possibly due to the line of New England seamounts near $67°W$.

In December 1975, scientists from the Bedford Institute began an experiment to study the low-frequency circulation at the shelf break south of Halifax, (Smith and Petrie 1982).

Elements of the experiment included:

1. an array of 11 moorings at 8 sites,
2. seasonal hydrographic surveys,
3. collection of meteorological data from nearby Sable Island, and
4. collection of weekly frontal analyses of sea surface temperature entitled *Experimental Ocean Frontal Analysis* or EOFA (US Naval Oceanographic Office).

The main line of the mooring array (Fig. 1a) lies along $63°30'W$ from $42°$ (2 500 m isobath) to $43°N$ on the shelf. An auxiliary line of two moorings lies along $64°W$. The full array was in place from July 1976 to July 1977 after which a skeletal array of four moorings remained until January 1978. To focus on topographic wave oscillations, the horizontal velocity data were resolved into eastward (u, longshore) and northward (v, onshore) components, low-passed to remove tidal and inertial motions, and subsampled at 6-h intervals. The data discussed below were taken during the spring (April/July, 1977) and summer seasons (July/October 1976 and 1977) when the weakness of the meteorological forcing permits unambiguous interpretation of the response to fluctuations in the deep offshore regions.

The major frontal features in the sea surface temperature field off the Scotian Shelf are the Gulf Stream, warm-core Gulf Stream rings, and the shelf/slope water boundary (abreviated SSB). To monitor the low-frequency offshore forcing, the weekly positions of the SSB and warm-core rings were digitized on a 10×10 km grid oriented to the shelf break (Fig. 1b). In the region between $60°$ and $65°W$, the mean position of the SSB over three years observations is 120 ± 70 km from the shelf break. (For a detailed analysis of the space-time structure of the SSB, see Smith and Petrie (1982) or Halliwell and Mooers, (1979)).

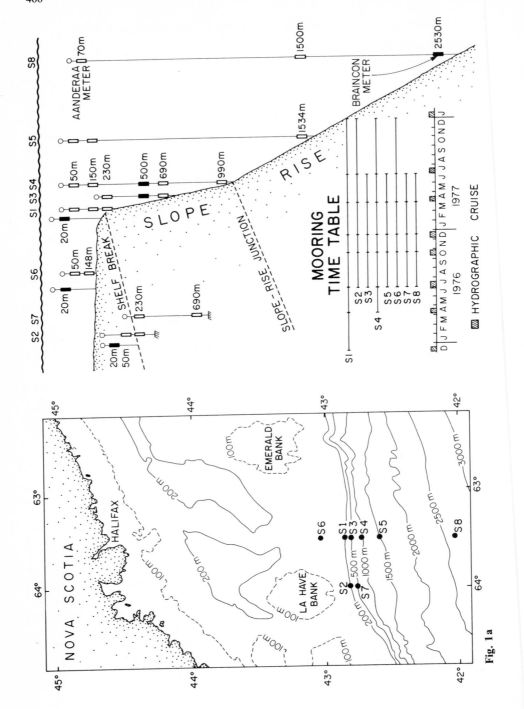

Fig. 1a

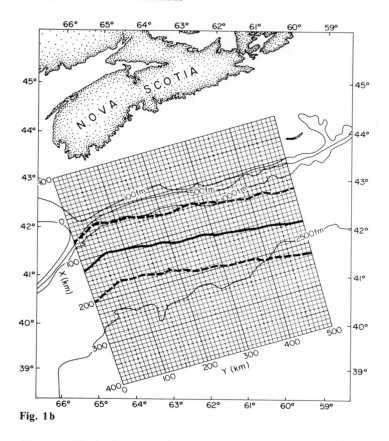

Fig. 1 b

Fig. 1a, b. Monitoring networks for Shelf Break Experiment. **a** Mooring array and time table. **b** 10 × 10 km grid used for digitizing position of the shelf/slope water boundary and Gulf Stream rings: January 1975 to February 1978. Mean position *(solid)* and standard deviation *(dashed)* of the shelf/slope water boundary are shown for the period of the experiment

20.4.1 Low-Frequency Current Data, Summer 1976

Low-passed records of longshore current from deep instruments deployed during July/October 1976 (Fig. 2) reveal a distinct wavelike oscillation, with an average period of 21 days and amplitude of 5 to 10 cm s^{-1}, which lasts for 4 or 5 cycles. Some familar characteristics of the onshore flux of topographic Rossby wave energy, such as offshore phase propagation, are evident by inspection of these smoothed signals. Louis et al. (1982) give spectral estimates of the average phase speed (9.5 km d^{-1}) and wavelength (200 km) of this 21 d wave, which apply not only to the deep but also the near surface (50 m) data in the offshore region. One unfamiliar feature of these observations is the gradual increase in the period of oscillation through the record, which is especially evident at the S4 mooring on the 1 000 m isobath.

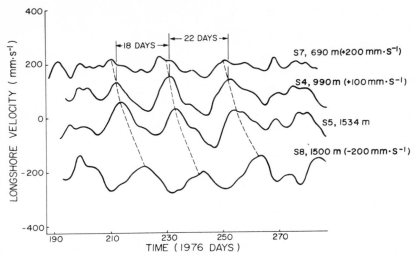

Fig. 2. Low-passed records of longshore (eastward) current at the deep instruments in the off-shore region during July/October 1976. *Dashed lines* indicate offshore propagation and period variations are shown for the S4 record

Inspection of all the currents on mooring S4 (Fig. 3) reveals that the oscillations are vertically coherent and in phase at the base of the continental slope. The variable phase differences, between 50 and 990 m for example, may be accounted for by slow variations in the onshore current at 50 m, which affect the net phase speed. The similarity in amplitude of the 21 d oscillations at 50, 150 m versus those at 690, 990 m shows little evidence for bottom trapping. As an indication of the baroclinic effects in the records, the strong vertical shear, which develops between the near surface (50, 150 m) and deeper instruments near day 250, coincides with the observation of an eddy (labelled Eddy K on the EOFA) in the sea surface temperature data. The apparent location of the

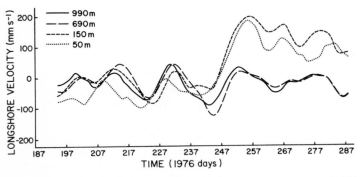

Fig. 3. Low-passed records of longshore current at S4 (50, 150, 690, 990 m) during July/October 1976

center of this feature between moorings S5 and S8 is confirmed by the simultaneous occurrence of strong pulses of longshore current (eastward at S5, westward at S8) on these moorings. As with the measurements at S4, these effects are confined primarily to the surface layers above the main thermocline.

Although the vertical uniformity in phase observed at S4 appears to be a ubiquitous feature, at least in the deep waters of the slope and rise, the same cannot be said for the amplitude of the 21-d oscillation. On the 700 m isobath, for instance, the longshore current records at S7 (Fig. 4) reveal a distinct amplification of the wave at 230 m (relative to the oscillations at S4), accompanied by a severely reduced signal near the bottom. Further up the slope, a dense set of measurements during summer (July/October) 1977, (Fig. 5) indicates a

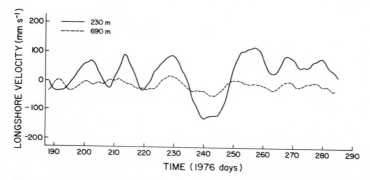

Fig. 4. Low-passed records of longshore current at S7 (230, 690 m) during July/October 1976

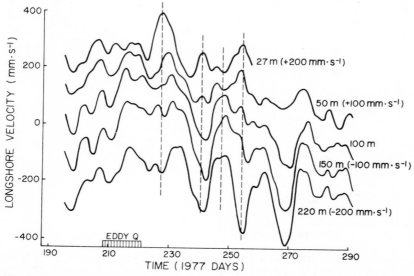

Fig. 5. Low-passed records of longshore current at S1 (27, 50, 100, 150, 220 m) during July/October 1977. Antiphase of near-surface and bottom oscillations following the passage of Eddy Q are indicated

breakdown of the vertical phase relationship as well in a strong 14-day oscillation at the shelf break mooring, S1. From the start of the event near day 220,
1977, the peaks and troughs of the signal at 220 m appear to be exactly out of
phase with those near the surface (27, 50 m) with the transition occurring be-

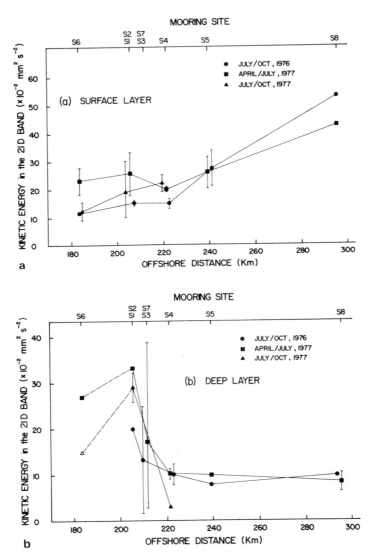

Fig. 6. a Average of spectral estimates for 21d kinetic energy in the surface layers (50–150 m)
from the shelf break array during summer 1976, spring and summer 1977. *Error bars* represent
the range of estimates at 50, 150 m. **b** Average of spectral estimates for 21 d kinetic energy in
the deep water layers (≥ 230 m) from the shelf break array during summer 1976, spring and
summer 1977. *Error bars* represent the range of estimates from 230 m to the bottom

tween 50 and 100 m. Unfortunately, the deep instrument on S4 failed during this period and there were no moorings deployed on the continental slope.

Comparisons of vertically averaged kinetic energy density in the 21-day spectral band for the spring and summer mooring periods (Fig. 6) reveal a different but consistent pattern of upslope variations in the deep (≥ 230 m) and shallow (50–150 m) layers. The near-surface baroclinic energy generally decreases in the onshore direction over the rise, reaching a minimum at the base of the continental slope (S4). Onto the shelf, this decline generally continues, except in spring when it appears there is some wind-driven energy present.

The horizontal distribution of deep (≥ 230 m) kinetic energy over the rise, on the other hand, is relatively uniform. In particular, there appears to be little evidence for offshore structure, such as the spatial modulation caused by the standing component associated with a significant reflected wave, or the upslope amplification (inversely proportional to depth) suggested by the barotropic theory of Kroll and Niiler (1976). On the continental slope, however, the deep energy is strongly baroclinic (e.g., Fig. 4), as indicated by the large "error bars" on the S3 and S7 estimates. The amplification at the 230 m level is also present at the shelf break. The deep measurements at S3 and S7 consistently show very low energy levels, indicating a node of the 21-d oscillation on the topography at the 700 m isobath.

20.4.2 Offshore Forcing

Louis et al. (1982) discovered that the bursts of low-frequency energy in the longshore current component were not only ubiquitous over continental rise and slope, but also appeared to be directly related to offshore fluctuations in the mesoscale eddy field associated with the Gulf Stream. Figure 7 illustrates how the frontal analysis data may be used to detect interactions between the Gulf Stream and slope waters. During late June and early July 1976 (weeks 25 to 28), a meander of the Gulf Stream penetrated northward onto the continental rise, causing a large deflection of the SSB. By week 28, Eddy I had formed at a point due south of the shelf break mooring array. Its diameter is roughly 160 to 180 km and its centre is located 200 km from the shelf break. Using emprical orthogonal function analysis, Smith and Petrie (1982) have shown that this event is associated with a large-scale onshore translation of the entire frontal boundary in this region and may be monitored with the smoothed amplitude of the dominant empirical mode of SSB variance. Louis et al. (1982) also point out that this anomaly and others like it in 1977 are associated with enhanced topographic wave oscillations with periods ranging from 10 to 25 days. Thus, a correlation appears to exist between the formation and shoreward movement of eddies and the presence of topographic Rossby waves at the shelf break.

The interaction of Eddy Q with the coastal waters in summer 1977, (Fig. 8) is considerably more complex and therefore difficult to analyze from the digitized frontal boundary along. Between the weeks of July 27 (day 208) and August 10 (day 220), the surface anomaly associated with Eddy Q is observed to

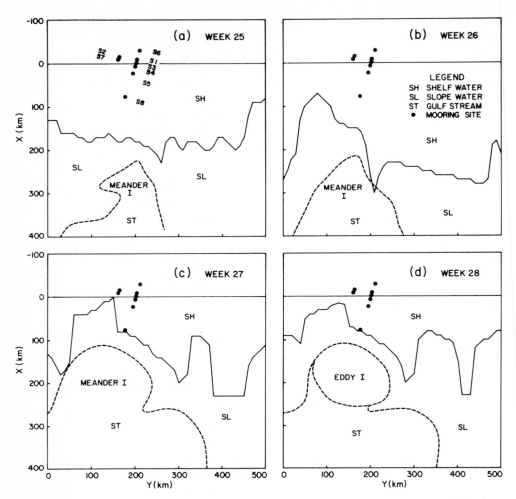

Fig. 7a–d. Frontal analysis of sea surface temperature depicting shelf/slope water boundary *(solid)* and north wall of Gulf Stream *(dashed)* on a grid oriented to the shelf break, south of Nova Scotia (Fig. 1a). The Y-axis is the mean position of the shelf break and the *solid dots* are mooring sites for the Shelf Break Experiment (S1 to S8). Panels **a–d** illustrate the formation and development of Eddy I during weeks 25 to 28, 1976

split into two parts, one of which (Q_w) moves westward to the south of the (skeletal) shelf break array. After interacting with a tongue of cold shelf water, this feature becomes the dominant signature of the eddy (day 236) and subsequently drifts slowly out of the region before encountering the Gulf Stream (day 271). Evidently this sequence of offshore events is associated with the burst of low-frequency oscillations detected at the shelf break during this period (Fig. 5).

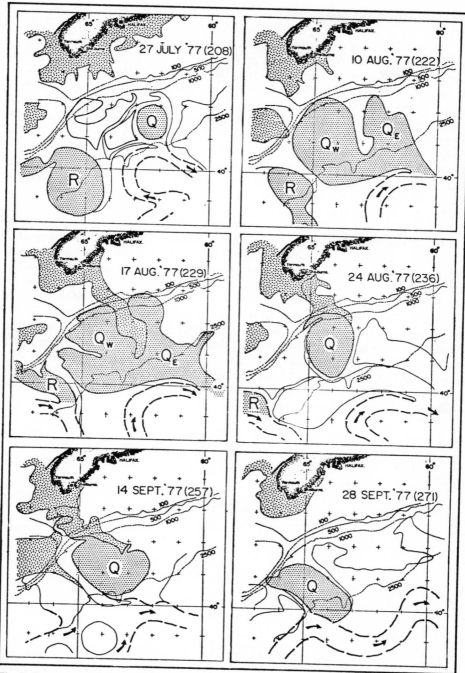

Fig. 8. Sequence of frontal analyses (EOFA, see text) of sea surface temperature depicting the interaction of Eddy Q with coastal waters of the Scotian Shelf and the Gulf Stream. Warm cores of eddies in the region are *labelled*. *Stippled area* near the Nova Scotian coast represents "coldest" shelf water

20.4.3 Analysis and Interpretation

In order to interpret these observations, Louis and Smith (1982) have modelled the formation of Eddy I (Fig. 7) as an isolated circular vortex [Eq. (28)] generated impulsively on the continental rise, 200 km south of the shelf break. For a homogeneous fluid on constant exponential topography [Eq. (15)] the asymptotic solution for "short", isotropic radiation, characterized by [Eq. (27)] with β replaced by $2af_0$, is achieved rapidly at a fixed distance (order 100 km) from the source due to the high expansion rate $(2f_0a^{-1} \approx 6900 \text{ km d}^{-1})$ of the barotropic wave field. The range of predicted offshore wavelengths at a point 200 km from the source (Fig. 9) is consistent with the observed statistical average of 180 to 200 km. Even more revealing is a comparison of the observed and theoretical variations in the period of oscillation (Fig. 10). These data not only support the model physics, but also serve to confirm the temporal origin of the disturbance at week 27 to 1976, i.e., coincident with the formation of Eddy I. In addition, the diameter (70 km) of the forcing vortex may be inferred from the observed amplitude modulations (Fig. 11) to produce a quite reasonable comparison between observed and modelled oscillations in longshore current at the base of the S4 mooring (Fig. 12a).

Further support for this theory also appears in the wave-like deformation of the shelf/slope water boundary to the east of Eddy I during the weeks 27 and 28 following the initial disturbance. The theoretical wave field (Louis and Smith (1982), Fig. 5.2b) indicates that, due east of the disturbance, the wave crests are aligned topographically north-south leading to the observed displacement in the SSB. The 150-km wavelength in the boundary deformation is consistent with the model wave field that would exist during the first few weeks following the initial disturbance (Fig. 9).

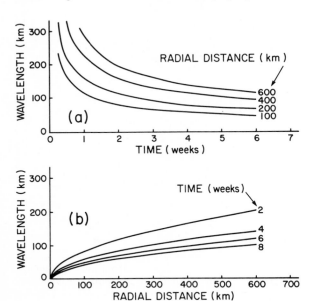

Fig. 9. a Total model wavelength verses time. **b** Total model wavelength verses radial distance from the source

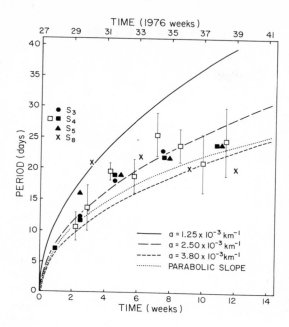

Fig. 10. Observed and model wave period versus time for oscillations topographically north (200 km) of the initial disturbance. *Solid symbols* are visual estimates; *open symbols* represent linear trend in phase of 0.05 cpd complex demodulation at S4 (900 m). *Curves* are given for various values of the exponential slope parameter and for the parabolic topography (see text)

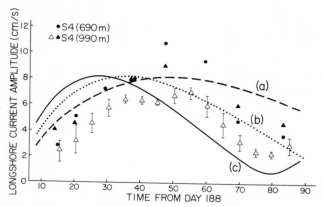

Fig. 11. Wave amplitude of longshore current at S4 mooring site. *Solid symbols* are visual estimates; *open symbols* represent the amplitude of the 0.05 cpd complex demodulation at S4 (990 m). Continuous *curves* are wave amplitude envelopes for the isolated vortex model (28) with parameters, (v_m, r_m): **a** $(-1.9 \text{ m s}^{-1}, 30 \text{ km})$, **b** $(-1.7 \text{ m s}^{-1}, 35 \text{ km})$, **c** $(-1.4 \text{ m s}^{-1}, 40 \text{ km})$

In spite of the success of the simple model, several important questions regarding the lack of evidence for a strong reflected wave in the observations, the upslope distribution of kinetic energy, the orientation of the rectilinear oscillations and the roles of baroclinic, nonlinear and frictional effects remain unanswered. To address some of these questions, Louis and Smith (1982) formulated a linear barotropic WKB model over realistic parabolic topography

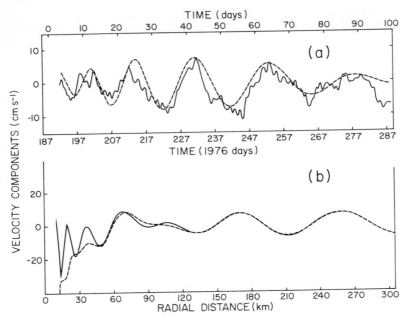

Fig. 12. a Comparison between observed longshore velocity *(solid)* at S4 (990 m) and model result *(dashed)* for isolated vortex (28) with $(v_m, r_m) = (-1.7\ m\ s^{-1}, 35\ km)$. **b** Spatial variations of model velocity components (longshore *dashed*, onshore *solid*) vs. upslope distance

(Fig. 13a). The "slowly varying" parameter in this analysis, proportional to the slope parameter, $a(y)$, is the magnitude of the geostrophic vector [Eq. (19)], i.e.,

$$|\underline{G}| \equiv 2\,a\,f_0 = -\frac{f_0}{2\,y}, \tag{36}$$

where $y(<0)$ is measured from the "shelf break" $(y = h = 0)$ and the disturbance is located at $(0, -y_0)$. In the asymptotic limit $(t \rightarrow \infty)$, the WKB approximation is valid over most of the region, in spite of strong topographic variations, because K increases as $t^{1/2}$. The initial conditions for the ray solutions, consistent with asymptotic state of isotropic radiation on exponential topography, specify a constant total wavenumber, K, on a circular locus $r = r_0 \ll y_0$.

The ray patterns for monochromatic (100 km) radiation over exponential and parabolic topography are compared in Fig. 14. With a constant exponential slope parameter, $a = (4\,y_0)^{-1}$, the phase lines are distorted near the origin due to excessive phase speeds, but the wave pattern in the far field agrees qualitatively with the true dispersive result. With parabolic topography, the principal effects of varying the slope parameter are to refract the wave crests into the observed longshore orientation, decrease the wavelength, and guide the energy along the steep continental slope region. As described by Smith (1971), this behavior is expected near the simple pole singularity $(y = 0)$ in the geostrophic vector magnitude. Considering the location of the shelf break array with re-

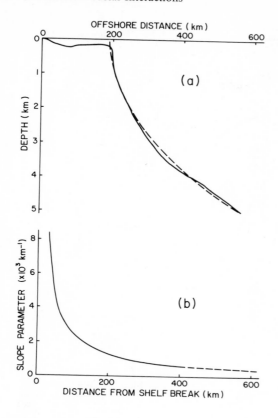

spect to the center of the disturbing eddy, the strong refractive effects may help to explain the absence of a reflected wave in the data set.

In order to define the average kinetic energy distribution associated with barotropic radiation over the parabolic slope, a discrete analogue of the full dispersive wave field was constructed from elementary ray solutions of differing K. Components of the group velocity and their cross-ray derivatives were computed from the dispersion relation and used to construct the scaled energy distribution in the upslope direction via Eq. (20). For comparison with the shelf break data, the theoretical distributions were normalized by the value of E at $y = -100$ km, which represents the mooring S8. Figure 15 compares some observed ratios of deep-water kinetic energy to three model distributions:

a) $\quad \dfrac{E_a}{E_8} = \dfrac{h_8}{h}$,

b) $\quad \dfrac{E_b}{E_8} = \dfrac{(h\,r^2)_8}{h\,r^2}$, and $\hspace{4cm}$ (37)

c) $\quad \dfrac{E_c}{E_8} = \dfrac{(h\,J)_8}{h\,J}$.

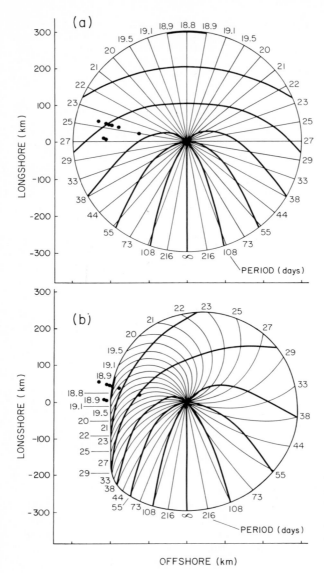

Fig. 14. a The WKB wave pattern for monochromatic radiation (100 km) on constant exponential topography ($a = (4\,y_0)^{-1} = 1.25 \times 10^{-3}\,km^{-1}$). **b** The WKB wave pattern for parabolic bottom topography. *Lighter lines* are rays on which the period of oscillation is constant. *Heavier lines* have constant phase. The *outer perimeter* represents the locus of wave energy at t = 48 d after the initial impulse

OFFSHORE (km)

Case a), for upslope amplification alone, applies to uniform offshore forcing (e.g., Kroll and Niiler, 1976), b) includes the radial spreading factor, and c) includes ray focusing on the parabolic slope. The data from July/October, 1976 generally lie between the latter two curves, but are not well-represented by the first. Hence the competition between focusing and upslope amplification versus decay by radial spreading leads to a nearly uniform distribution of deep kinetic energy over the continental rise and a modest increase over the slope, where the topographic factors dominate. It should be recalled, however, that, in contrast to theory, the slope measurements at S3 and S7 are strongly baro-clinic as indicated by the large "error bars" on those estimates.

To demonstrate that these results are not spurious, scaled energy data from another mooring period (April/July 1977) are included in Fig. 15. At this time, the surface temperature analysis indicated that offshore eddy activity was also present in roughly the same location as Eddy I, but the interpretation is not as clear as that for Fig. 7. Nevertheless, the energy distribution exhibits the same features of offshore uniformity and amplified baroclinic levels over the slope.

To date no detailed investigation of the baroclinic effects in the data set has been undertaken, but a few qualitative results are noteworthy. Over the continental rise, the enhancement of surface layer energies by nonlinear, baroclinic noise has been observed by Thompson (1971) off New England. Conversely, the tendency toward vertical uniformity with increasing distance from the source is a feature of the transient radiation field described by Tang (1979), in which the expansion rate [Eq. (32)] for baroclinic modes is very low (≈ 1.5 km d^{-1}), and hence the "fast barotropic" waves dominate the far field. The sudden enhancement of the vertical shear at the S4 mooring near day 250 (Fig. 3) may signal the arrival of a nonlinear, baroclinic mode which was generated in conjunction with the formation of Eddy I.

Over the slope, the strong amplification at the 230 m level coupled with reduced energy near the bottom (Fig. 4) suggests that part of the incoming baro-

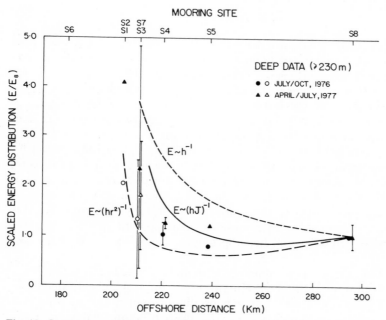

Fig. 15. Comparison of deep-water kinetic energy levels in the 21 d spectral band, normalized to mooring S8, with theoretical curves representing various combinations of upslope amplification ($\propto h^{-1}$), radial spreading ($\propto r^{-2}$) and ray focusing ($\propto J^{-1}$) (see text). Data from two mooring periods, July/October 1976 and April/July 1977, are included. *Error bars* represent the observed variations in energy density over the depth range, ≥ 230 m

tropic energy may be scattered into baroclinic "fringe" modes as described by
Suarez (1971) and Ou and Beardsley (1980). The low values of the alongshore
phase speed component coupled with the large bottom slope suggest that the
limit [Eq. (8b)], with a node on the topography, applies to this region. Furthermore, the horizontal trapping scale implied by the observation of a mode-1 oscillation at the shelf break during summer 1977 (Fig. 5) is roughly the internal
deformation radius (≈ 35 km). Thus a baroclinic mode trapped to the slope/
rise junction would be evident at the shelf break (16 km away), but much reduced further on the shelf (e.g., at S6, 38 km away).

As for frictional effects, Garrett's (1979) simple estimate for the dissipation
time scale is,

$$R^{-1} = \left(\frac{C_D V_b}{h}\right)^{-1} \approx 10^6 \text{ s} = 8.6 \text{ days}$$

where C_D is 2×10^{-3}, $V_b (\approx 0.1 \text{ m s}^{-1})$ is the bottom velocity and $h (\approx 200 \text{ m})$ is
the local depth. Multiplying this value by a characteristic longshore component
of the group velocity ($|c_{g1}| \approx 0.1 \text{ m s}^{-1}$) yields a spatial decay scale of 100 km.
Thus the wave energy which reaches the shelf break is localized in the along-
shore direction to that extent.

20.5 Conclusions

Louis and Smith (1982) have used the results of the barotropic radiation models in conjunction with observed kinetic energy levels to assess the decay rates
for warm-core Gulf Stream rings in the continental rise. These estimates, in the
range of 10^{13} J d^{-1}, are smaller than but generally consistent with those of Barrett (1971) and Olson (1980) (2 and 3×10^{13} J d^{-1} respectively) for the decay of
available potential energy (APE) in cold-core cyclonic Gulf Stream rings. Unfortunately, the hydrographic measurements of Eddy I were insufficient to determine the magnitude of its APE and hence estimate its life time. Nevertheless, linear topographic wave dispersion appears to be an important process for
spinning down anticyclonic, warm-core Gulf Stream eddies, as suggested by
Flierl (1977b) for their cold-core counterparts. The mechanism is enhanced by
the strong topographic slopes of the continental margin, which effectively "li-
nearize" the eddy dynamics by increasing the characteristic phase speed relative to the particle velocities and hence promote dispersion (McWilliams and
Flierl 1979).

Regarding the role of topographic waves in maintaining the circulation and
water mass distribution on the shelf, a major unsolved problem is to determine
mechanism(s) by which the low-frequency oscillations appear to promote eddy
fluxes of heat, salt and nutrients from the deep offshore waters onto the
shelf.

21. Eddy-Induced Dispersion and Mixing

D.B. Haidvogel, A.R. Robinson, and C.G.H. Rooth

21.1 Introduction

One of the most wide-ranging problems facing oceanographers today, which has important applications to both basic scientific and societal concerns, is the description of the processes by which dynamically inactive dissolved or suspended materials are redistributed and dispersed by the ocean circulation. Such substances, often referred to as tracers, include naturally occurring and anthropogenic chemical tracers (oxygen, tritium, etc.), biological nutrients and organisms (phytoplankton, fish larvae), and pollutants or waste materials either accidentally or intentionally released into the oceans (oil, chemical and nuclear wastes). Particularly in the mid-ocean, the energetic mesoscale eddy field is of probable importance to the dispersal of these tracers. However, despite recent observational and theoretical interest in eddy-induced transport processes, a qualitative picture of eddy-related dispersal effects is only now beginning to emerge. In particular, many of the results summarized here represent formally unpublished, preliminary results from ongoing research efforts.

As emphasized in the previous chapter, eddy variability occurs on a wide range of spatial and temporal scales, and assumes a wide range of phenomenological forms including Gulf Stream rings, mid-ocean eddies and smaller scale coherent vorticies ("bullets" or "lenses"). The transport effects associated with these eddy phenomena and other scales of oceanic motion are therefore also diverse. Initially material released in small patches may disperse according to classical fluid dynamical concepts (Csanady 1973) but upon reaching the scale of geostrophic or quasigeostrophic eddies novel behavior is expected.

Simply, but conveniently, eddy transport can be said to occur as either "stirring" or "mixing". Oceanographers frequently illustrate the distinction between these processes visually by considering the fate of the spot of red dye in the ocean. As the dye is redistributed by oceanic motions, it may be stirred into a number of undiluted red streaks or filaments of dye, separated by clear water. Alternately, it may be mixed, resulting in a pink cloud of reduced dye concentration. Stirring is thus transport without mixing. In reality of course both processes occur simultaneously with stirring generally enhancing mixing, but for some time and space scales mixing is negligible.

The ability of eddies to transport dynamically passive (as well as active) substances in these ways has important scientific consequences in a range of fields. Biological productivity in many areas of the world's oceans is closely linked to eddy-related transport of nutrients or of the organisms themselves

Eddies in Marine Science
(ed. by A.R. Robinson)
© Springer-Verlag Berlin Heidelberg 1983

(Chap. 22). Observing the distribution and redistribution of geochemical tracers such as tritium, helium and freon can be used to infer the oceanic general
circulation (see Sect. 21.6). Global-scale transports of heat and momentum by
the eddy field are also thought to play a role in the detailed mechanics of the
earth's ocean-atmosphere climate system (Chap. 19).

In addition to these scientific applications, eddy stirring and mixing in the
ocean relate directly to a variety of planetary-scale resource management issues. For instance, the advisability of, and preferred means for, chemical waste
disposal in the oceans depends importantly on knowledge of the anticipated
eddy dispersal effects in the disposal regions (Goldberg 1979). Practical deci-
sions related to seabed and subseabed disposal of nuclear waste materials must
also carefully consider the likely effects of eddy-induced transport (Marietta
and Robinson 1980). Finally, the degree to which, and the time-scale on which,
the ocean acts as a reservoir for anthropogenic CO_2 has a significant impact on
world climate modification due to human industrial activity (Climate Research
Board 1979).

Eddy stirring and mixing in the ocean are presently under intense observational and theoretical examination. Regional and global eddy-resolving ocean
circulation models are being used to explicitly model eddy transport effects
(see Sects. 21.1–21.5). The growing database of geochemical measurements
from the world's oceans now offers direct estimates of large-scale dispersion
(Sect. 21.6). It is also presently feasible to purposefully release localized distributions of a known tracer, and to monitor their subsequent dispersal by small-
scale motions and the mesoscale eddy field (Ewart and Bendiner 1981). Several
such planned-release experiments are presently under consideration. If carried
out, they promise to yield accurate in situ measurements of horizontal and vertical tracer dispersal on both the small and large oceanic scales.

21.2 Two-Dimensional Dispersal in the Mid-Ocean

Recent numerical studies have examined the stirring and mixing of a passive
scalar in two-dimensional β-plane turbulence as a prototype for scalar dispersal within a single isopycnal layer in the mid-ocean. The modeling strategy is
very similar to that associated with periodic process models of mid-ocean dynamics (see Chap. 18). Briefly, specific realizations of a dynamically evolving
mid-ocean velocity field are generated by direct integration of the forced/
damped barotropic vorticity equation over a spatially periodic domain and for
representative mid-ocean values of the environmental parameters such as β and
RMS velocity. An initial distribution of tracer is then released in mid-domain.
Typically the source is a localized Gaussian distribution or other form of
"spot", and is relased as a delta function in time (during one time step) or as a
step function (continuous release after initiation). The subsequent evolution of
the tracer field is then determined by numerical integration of the advection-
diffusion equation:

$$\frac{\partial\theta}{\partial t} + \underline{v}\cdot\nabla\theta = \nu\nabla^2\theta$$

where $\theta(x, y, t)$ represents the concentration of the tracer, and $\underline{v}(x, y, t)$ is the dynamically simulated velocity field. Here, ν is the "explicit diffusivity"; the associated diffusive term is assumed to properly parameterize mixing processes occurring on scales smaller than the grid scale.

Tracer dispersal simulations of this homogeneous sort (e.g., Haidvogel, Keffer and Quinn 1981), clearly demonstrate the rapid transfer of tracer variance (θ^2) from low to high wavenumbers, and the associated rapid sharpening of the θ gradients (Fig. 1). The continued narrowing of the θ contours, associated with "stirring" by the explicitly resolved eddy velocity field, is eventually halted as diffusive effects increase. Thereafter, a quasi-steady situation is attained in which low-wavenumber (large-scale) tracer variance is continually transferred to the smaller scale, high-gradient θ features. While tracer variance is being depleted at the large scale, at the highest wavenumbers an approximate balance exists between θ variance input from larger scales, and diffusive loss to the explicit subgridscale effects. Much of the qualitative picture is compatible with the predictions of turbulence transfer theories (see, for instance, Batchelor 1951).

Because of the rapid transfer of tracer variance to small-scale features associated with advective stirring, diffusive loss ("mixing") of tracer variance is much more rapid than in the absence of the eddy field. Therefore, the turbulent velocity field greatly enhances tracer dispersal. For a short interval after the release of the tracer, the "effective diffusivity", as measured by the time rate of change of the second moment of the tracer distribution, is little more than the explicit (subgridscale) diffusivity. It quickly increases, however, reaching an approximately constant value of order the product of the RMS velocity and representative length scale of the advecting eddies, as would be given by mixing length ideas.

21.3 Streakiness

Garrett (1983) has advanced a simple model to predict the spreading behavior of an instantaneous point release of tracer along an isopycnal in a mesoscale eddy field. Garrett's model assumes that the mesoscale eddy field acts to strain out the tracer into ever longer and exponentially growing "streaks" of length $L_D = e^{\alpha\gamma t}$ and constant width $L_S = (\eta/\gamma)^{1/2}$, where η is an explicit diffusivity associated with small-scale ocean mixing processes (Young, Rhines and Garrett 1982), $\gamma = (\overline{u_x^2 + v_y^2})^{1/2}$ is the r.m.s. rate of strain, and α is a dimensionless constant of order unity. The area occupied by the tracer will then be $A_T \simeq L_S L_D$, which will also increase exponentially. Obviously, such a scenario is unlikely to continue forever; eventually, a well-mixed state will be approached.

How long, then, will streakiness persist? Garrett argues as follows: In an ensemble of releases, on average, the tracer will fill out a "domain of occupa-

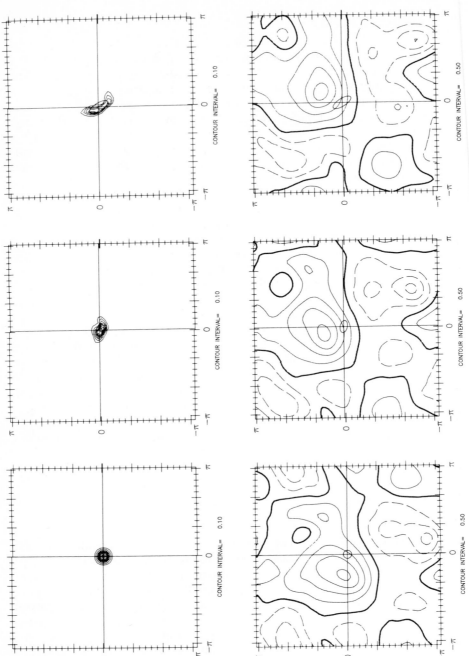

Fig. 1.1

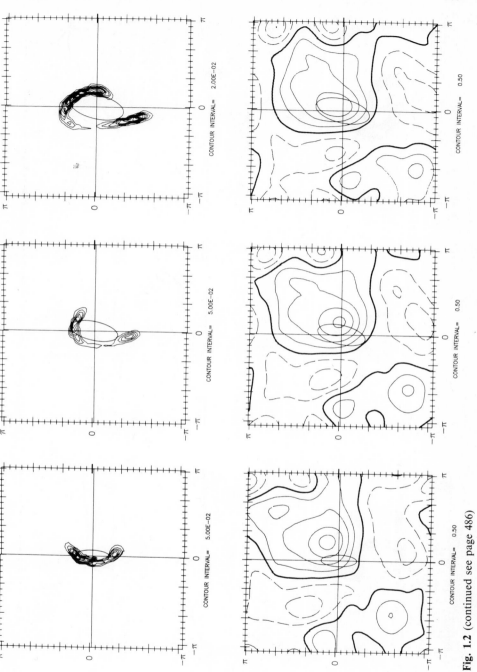

CONTOUR INTERVAL= 2.00E-02

CONTOUR INTERVAL= 5.00E-02

CONTOUR INTERVAL= 5.00E-02

CONTOUR INTERVAL= 0.50

CONTOUR INTERVAL= 0.50

CONTOUR INTERVAL= 0.50

Fig. 1.2 (continued see page 486)

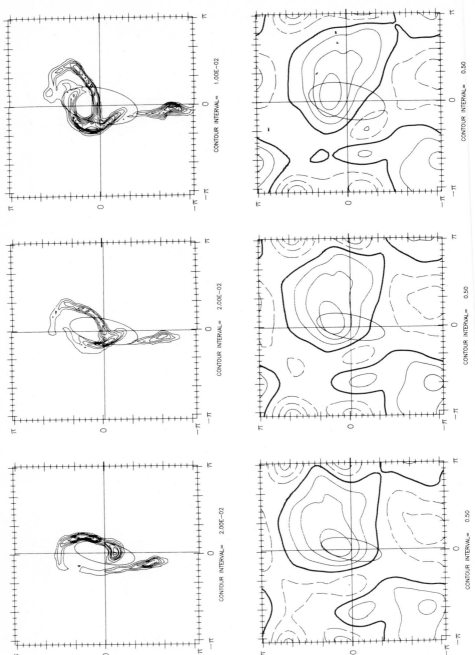

Fig. 1.3

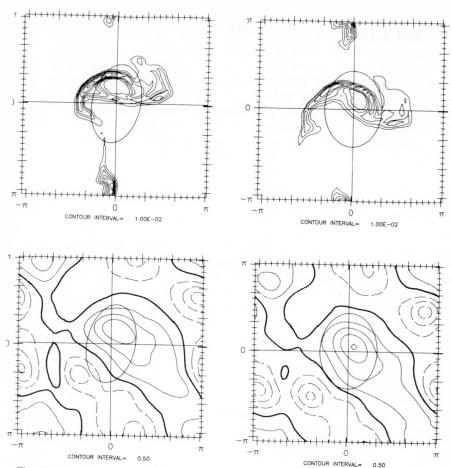

Fig. 1.4

Fig. 1. 1–4. Simulated time series of contour maps showing the temporal evolution of the barotropic streamfunction and tracer fields (*lower* and *upper diagrams,* respectively) in a 600-square-kilometer region of the mid-ocean. A small blob of tracer, approximately 30 km in diameter, is initially released in the center of the simulated ocean. Subsequent maps, separated in time by about 1 week, show how the tracer is advected, dispersed and fragmented by the simulated circulation field. The cross-hairs and inscribed ellipse show the location of the center of mass, and the second movement of the tracer distribution. The stream-function contour interval is 2.5×10^{7} cm^2 s^{-1}; the tracer contour interval is indicated below each map

tion" (Kupferman and Moore 1981) whose area A_0 expands linearly in time after some initial adjustment period. Streakiness, then, can only persist until a time, t_{mix}, at which the area occupied by the exponentially elongating tracer streaks completely fills the domain of occupation, i.e., when $A_T = A_0$. For a typical oceanic environment, t_{mix}, is estimated to be about 1 year, and the associated radius of the domain of occupation about 360 km (Garrett 1983).

Numerical simulations of tracer streakiness (Keffer and Haidvogel 1982) are in rough agreement with these theoretical ideas. In the numerical simulations, however, the exponential growth of contour area begins to slow somewhat earlier than t_{mix}. It appears that this happens when individual streaks have been elongated enough to wrap completely around an eddy, and therefore to interfere with further rapid growth. Due to self-interference, streaks are likely to stop growing exponentially well before their cumulative area completely fills the domain of occupation. Holloway (1982) has also argued for the cessation of exponential growth once the streaks reach the dominant eddy scale.

21.4 Particle Motions

The preceding discussion of the evolving $\theta(x, y, t)$ field has not dealt with the related question of how individual tracer particles are dispersed. The trajectories of particles in two-dimensional model flows typifying oceanic eddies have been examined by Flierl (1981). In this study, the eddy motions are not dynamically generated, but are large-amplitude waves kinematically similar to observed mesoscale eddies. The resulting analysis is applied to Lagrangian motions in both a periodic channel and an isolated circular eddy. The results indicate that mean Lagrangian particle displacement in the channel geometry is quite sensitive to initial particle location; particles may be displaced in either direction relative to the propagation of the finite-amplitude wave. Large volumes of fluid may also be effectively trapped, and move with the wave. In an isolated eddy, some particles are carried with the eddy, while others are left quickly behind. For a typical warm core ring of radius 50 km and translation rate ~ 5 cm s^{-1}, Flierl estimates that particle displacements may be eastwards and as much as 200 km. There also appear to be preferential locations at which particles can enter or leave eddies by whatever small-scale diffusive processes are present (Flierl and Dewar 1982).

Dynamical numerical models of particle motions in the turbulent midocean are also presently being explored. Although it is feasible to construct fully Lagrangian dynamical models of the mid-ocean eddy field and the gyrescale circulation, Lagrangian particle trajectories and statistics can also be recovered from existing Eulerian (grid-point) ocean models by simple interpolation/integration techniques. Haidvogel (1982) has explored the degree to which the accuracy of such particle trajectories, and related Lagrangian dynamic balances, are influenced by details of the Eulerian circulation model. In particular, it is found that particles tracked within an Eulerian ocean model do not precisely conserve vorticity and tracer concentration. The spurious nonconservative effects are typically quite large, leading to a rapid loss of "memory" of initial vorticity and tracer values. The effects of these errors on ensemble-averaged particle statistics are being examined, and an intercomparison of dynamically simulated mid-ocean particle tracks with the LDE SOFAR float dataset is under way.

21.5 Gyre-Scale Dispersal Effects

The majority of dynamical models of tracer dispersal by the mesoscale eddy field have been carried out under homogeneous circumstances, that is, for an eddy field with spatially invariant statistics such as energy level and dominant spatial scales. As is well known, eddies in the real ocean are inhomogeneously distributed. The resulting variations in eddy diffusivity are therefore significant (see Chap. 5 for a discussion of available eddy diffusivity estimates).

The hypothesis that variations in eddy diffusivity may account for some aspects of the observed distributions of ocean tracers has been examined by Armi and Haidvogel (1982). To examine the effects of spatial variability of eddy intensity, solutions are sought to a simple steady-state tracer diffusion model with spatially variable and/or anisotropic eddy diffusivity. The solutions, in which gradients in the eddy diffusivity act much like an imposed advective flow, demonstrate that a purely diffusive field can generate tongue-like property distributions. Although an actual prescription for the oceanic eddy diffusivity field is difficult, estimates based on the variability observed in the eddy potential energy field suggest that a gradient diffusivity velocity of 5 mm s^{-1} may be applicable, even in the relatively low energy region of the Sargasso Sea. It is suggested therefore that tongue-like property distributions in the ocean may not always be associated with purely advective effects.

The long time scale dispersal of particles within closed-gyre domains will almost certainly be strongly modified by the recirculating character of the motion field, and in particular by beta-induced asymmetries including westward intensification and enhanced zonal flow. The spatial gradients across the stream line field are enhanced in the boundary layer in the ratio of the interior gyre scale to the boundary layer scale, while the fraction of time spent there is the inverse ratio. A uniform eddy diffusivity model would accordingly predict comparable cross stream line dispersion to occur within the brief transition periods in the western boundary layer and during the much longer interior transit time. Since theory as well as oceanic observations indicate much higher eddy energy levels to be associated with the boundary regime, it appears reasonable to conclude that cross stream line dispersion of particles will be dominated by the latter, and hence that the cross stream line dispersion time is upper bounded by the gross gyre recirculation time.

21.6 Tracer Evidence for Dispersal Processes

Classical water mass analysis methods, considering temperature-salinity-nutrient correlations, have played an important role in determining the origin and stability of "strong" eddies, such as warm and cold core Gulf Stream rings and mid-thermocline anticyclonic boluses or "bullets" (cf. Chap. 2 and 5). More recently, and particularly because of the urgency of the problem of carbon dioxide uptake by the oceans, substantial attention has been given to a number of

man-made tracers with an injection history appropriate to the study of dispersion processes on decadal time scales. Outstanding among these are the radioactive hydrogen isotope, ^{3}H, with its daughter product, ^{3}He, and the industrially ubiquitous freons.

Tritium stocks presently existing in the oceans derive almost exclusively
from massive injections into the atmospheric water vapor reservoir during the
years 1962–1964. The first attempts to undertake systematic tritium measurements with the aim of quantitative characterization of oceanic dispersal processes were made in the late 1960's (Rooth and Ostlund 1972). Samples collected within the Eighteen Degree Water (EDW) layer in 1968 showed this
layer to be laterally well stirred within only four years of the injection transient.
This set the stage for an effort to determine the rate of downward mixing from
the EDW. The result of a series of vertical profile measurements was an upper
bound estimate of 2×10^{-5} m^2 s^{-1} for the vertical (diapycnic) tracer diffusivity, and also a clear indication, based on irregularities in the vertical profiles of
tritium concentration, that the deeper thermocline layers were at that time in
the early stage of irregular lateral stirring.

A major effort to map the North Atlantic tritium distribution occurred in
the early nineteen seventies, centered around the GEOSECS North Atlantic expedition in 1972 with significant complementary work by groups at the University of Miami and the University of Heidelberg (Sarmiento, Roether and
Rooth, (in press). The distribution of tritium on a set of density surfaces (spanning the range with significant winter time exposure south of the Atlantic Polar
Front) was found to vary with the main advection patterns from zones of exposure, as determined with a large-scale diagnostic circulation model (Sarmiento
and Bryan 1982), provided these advection effects were significantly augmented by lateral eddy mixing across recirculating stream lines.

Rooth (1974) studied the tritium variability in one of the sections on which
the above mentioned mapping effort was based, stretching along 29°N latitude
from the Blake Escarpment to the Mid Atlantic Ridge (the Western Atlantic
basin). He found that while in the deep main thermocline tritium concentrations decrease monotonically at this latitude, and the dissolved oxygen concentration exhibits a pronounced minimum (at about the 10°C potential temperature level), both tracers correlate positively below, as well as above, the minimum oxygen level. The r.m.s. variability for each tracer, evaluated just above
and below the oxygen minimum, and normalized by the NS gradients at constant potential density as derived from the GEOSECS sections, gave an apparent displacement length of 300 km within 10% for all four estimates. This surprisingly large value, accepted with due concern for the representativeness of a
single section, is strongly suggestive of the kind of incomplete stirring effects illustrated in the stirring model experiments in previous sections.

Perhaps the most exciting prospect for further development of our understanding of transport processes based on tracer work is presented by the decay
product of tritium, ^{3}He. As a nobel gas isotope with extremely low water solu-
bility, its kinetic behavior at the sea surface is well defined, and consists essentially of complete escape of the excess above the concentration corresponding
to equilibrium with the average atmospheric composition. The evolution of the

ratio of $^3H/^3He$ in an isolated water parcel with tritium concentrations typical of North Atlantic surface waters provides a Lagrangian time clock with a resolution of about 5×10^6 (Jenkins 1980). Thus, as the tritium distribution gradually approaches a well-stirred state, we have a large-scale ^3H source in the interior, with a Lagrangian clock of resolution approaching eddy time scales in the geostrophic turbulence regime. A major impediment to the effective use of this tool in eddy process studies is the long time needed by shore-based laboratories to process the data. This problem may soon be circumvented by the parallel use of freon analysis, which can be readily performed on ship-board. The freons have been introduced into the atmosphere at rapidly accelerating rates, although recently the concentration growth rates appear to have stabilized. Like helium, they are virtually insoluble in seawater, and hence may equilibrate quickly at the air-sea interface. Extensive freon concentration mapping, supplemented by the more cumbersome tritium/helium technique where the shipboard freon data warrant, promises to provide a powerful new approach to testing the realistic performance of large-scale oceanic stirring models.

In the absence now of such detailed and definitive experiments, we must remain content with the fact that eddy stirring of intensity commensurate with the MODE/POLYMODE physical data is not only consistent with, but actually required for, a reasonable rationalization of existing tracer data. Similar conclusions can be drawn from radiocarbon (Kuo and Veronis 1970).

22. Eddies and Biological Processes

M.V. Angel and M.J.R. Fasham

22.1 Introduction

Surprisingly, there has been little biological sampling done at space or time scales which will detect changes associated with mesoscale features in the open ocean (by mesoscale we mean scales of tens to hundreds of km and 1 to 3 years). Repeated sampling has either been in a localised area (e.g., Angel 1969, Angel et al. 1982), or in labelled parcels of water (e.g., Climax experiment, see McGowan and Walker 1980, for full references), or in highly advective regions (e.g., CALCOFI investigations), or the data are not in a form amenable to such analysis (e.g., Continuous Plankton Recorder Programme, Colebrook, pers. comm.), or have too gross a scale to detect the features (e.g., many zoogeographical surveys).

The observations on biological phenomena at scales pertinent to mesoscale features fall into three main categories (1) the long latitude transects worked by Scripps investigators in the North Pacific, (2) the Gulf Stream Ring investigations carried out from Woods Hole, (3) the data emerging from the recently iniated programme on East Australian warm core eddies. Otherwise mesoscale eddies have tended to be blamed without any hard supporting evidence for unexpected incongruities in data (e.g., Angel 1977). It will be noted that most of the good observational data comes from ring structures which are (1) being actively studied by the physicists, (2) tracked routinely by other agencies, and (3) readily identified in real-time.

This chapter is divided into a series of sections. Initially an attempt will be made to investigate the consequences of present physical theories of the dynamics of rings and mesoscale eddies for biological processes. This theoretical approach will be followed by a review of the published observational data. No attempt will be made to totally integrate these two approaches, mainly because the observational information is so inadequate that any attempt at an overall synthesis would be premature. The final section deals with the sampling methods that are potentially the most useful in resolving some of the problems associated with the understanding of biological processes in mesoscale features and their implications to biological oceanography in general.

Eddies in Marine Science
(ed. by A.R. Robinson)
© Springer-Verlag Berlin Heidelberg 1983

22.2 Frontal Eddies

The term frontal eddies was used by Koshlyakov and Monin (1978) to describe ring-like features that are formed when a meander of an ocean current is cut-off to form an eddy. It has been suggested by these authors and others (MODE Group 1978, Kim and Rossby 1979) that frontal eddies differ from what they term open ocean eddies in being quasi-solitary bodies in which advective motions are predominant and in having a specific kinetic energy two orders of magnitude higher than open ocean eddies. These qualities, coupled with the fact that the ring encloses a body of water initially very different from the water surrounding it, would suggest that the ecological processes associated with rings are likely to be different from those associated with mid-ocean eddies. In recent years rings spawned by the Gulf Stream and East Australian Currents have been the object of joint physical and biological studies which will be described later.

Our present knowledge of the physical structure and properties of rings is described in Chapter 3, this Volume by Richardson. In this section we will merely extract some key information that has a bearing on the study of the biology of these features.

22.2.1 Formation

Rings have been observed wherever swift and narrow currents are found but most of the observations come from cold-core and warm-core rings spawned by the Gulf Stream. Gulf Stream cold-core rings are formed to the right of the Gulf Stream as it travels north-east and enclose colder, less saline slope water (see Fig. 1 of Chap. 3). The diameter of rings, estimated from the cold water anomaly, are typically between 200 and 300 km (Hagan, Olson, Schmitz and Vastano 1978). After formation the rings drift typically westward to south westward with speeds between 1 and 5 km d^{-1} (Lai and Richardson 1977). The ring lifetime has been estimated between 2 and 3 years (Cheney and Richardson 1976, Lai and Richardson 1977) by which time the ring will have decayed into the Sargasso Sea or been re-absorbed into the Florida Current. However, some rings have been observed to be temporarily absorbed back into the Gulf Stream a few months after formation, advected downstream for a while and then reformed (Richardson, Maillard and Stanford 1979, Fuglister 1977, Vastano, Schmitz and Hagan 1980). Recently what appear to be cold-core eddies have also been observed in the mid-Atlantic SW of the Azores (Gould 1981) associated with a south-eastwards flowing current which divided NE Atlantic from NW Atlantic water.

Rotational velocities in the top 100 m of rings show a zone of maximum velocity between 40 and 50 km from the centre where speeds of up to 1.5 m s^{-1} have been observed (Fuglister 1977). Measurements of velocity profiles (Richardson Maillard and Stanford 1979) showed that these high velocities persisted down to 200–300 m below the surface, after which the velocity declined rapidly to values of ~ 7 cm s^{-1} at 1 000 m. Thereafter the rotational velocity declined more slowly, although the depth to which the ring circulation extends is at present in dispute (Richardson Chap. 3, this Vol.).

Warm-core anticyclonic rings are formed to the left of the Gulf Stream and enclose warm, saline Sargasso Sea water. Lai and Richardson (1977) found that typically five warm rings per year were formed and that they were smaller than cold core rings with diameters of ~ 100 km. Saunders (1971) observed that the rotational velocities in one warm core ring were at least one third the magnitude of velocities observed in a typical cold core ring. Warm core rings also form to the south of the East Australian Current (Andrews and Scully-Power 1976, Nilsson and Creswell 1980). Over a period of 18 months Nilsson and Cresswell (1980) observed the formation of three warm core rings; they estimated that about two rings per year were formed, of which most coalesced with the East Australian Current rather than escaped to the south. The rotation speed of these rings was similar to cold core Gulf Stream rings.

After formation the rings will enclose a biological population that will show many specific differences from that of the surrounding water. However, physical exchange processes will begin to take place between the surface water and the atmosphere and between the water in the ring and the surrounding water, and these exchange processes must be understood if we are to attempt to investigate the subsequent fate of the plants and animals that are initially isolated within the ring.

A full understanding of these processes must await the development of coupled physical and biological models, such as have been recently developed in the study of upwelling (e.g., Wroblewski, 1980). In this short review we will limit ourselves to trying to determine which physical and biological process are likely to be important and attempting to estimate some of the space and time scales involved.

22.2.2 Air-Sea Interaction Within a Ring

The heat budget of a body of water is made up of contributions from vertical exchange with the atmosphere and diffusive-advective exchange with the surrounding water. When a ring is formed the advective exchange is considerably reduced as will be discussed later. Thus if a cold-core Gulf Stream ring is formed during the season of net heat input from the atmosphere the surface water of the ring will begin to heat up. This heating process will be enhanced by the south westwards drift of the ring and by horizontal mixing with the surrounding water, with the result that after a few months the temperature anomaly in the top 100 m disappears (Richardson, Maillard and Stanford 1979, Fuglister 1977). If the ring is formed during the season of net heat loss to the atmosphere then presumably the temperature anomaly will persist for a longer period.

In the long term, this heating will have a physiological effect on any plants or animals which can only tolerate a small range of ambient temperatures (stenothermal). However, if the ring is formed in the spring-fall period the short-term biological effects will depend on the changes, if any, in the mixed layer depth within the ring, and there appears to be very little information on this for cold core rings. Ortner (1978) stated that the depth of the seasonal thermocline was consistently 25 m shallower in the slope water than in the Sargasso Sea or in the rings. This suggests that the seasonal thermocline deepened after the ring was formed. If this was so then fresh nutrients would be entrained into the mixed layer which in turn could induce phytoplankton growth. Thus, after formation the slope phytoplankton within the ring could experience a short increase in population followed by a decline due to the stenothermal species dying out as the ring heated up. After this the productivity of the ring will partly depend on the rate at which it is colonised by species from outside the ring. This will be discussed in Section 22.3.

The same physical processes will take place within warm-core rings after their formation. In this case however, the surface anomaly of the ring will persist longer if the ring is formed in spring or summer. If the ring is formed in winter then the heat loss from the surface will be greater than in the surrounding water due to the greater temperature difference between the sea and air temperature, leading to higher sensible heat fluxes (Neuman and Pierson 1966). The surface water will soon become denser than the water beneath, causing convective mixing which will deepen the surface mixed layer. For example Andrews (1979) found that a ring formed from the East Australian current had a mixed layer that was 300 m deep in the centre compared to 50–100 m in the surrounding water. The sub-tropical water from which the ring was formed was poor in nutrients in the mixed layer. However, as pointed out by Tranter et al. (1980), the deep convective mixing will penetrate into the depths at which the nutrient levels are high and mix these nutrients through the whole of the new deep mixed layer. Thus next spring, when the new shallow seasonal thermocline forms, the nutrient levels in the mixed layer of the ring will be higher than those of the surrounding water with consequent increased productivity.

22.2.3 Horizontal Exchanges Between the Ring and Its Surroundings

Once a ring has been formed, then the biological processes within the ring will be critically affected by the rate at which the ring mixes with the surrounding water. Part of this mixing will be produced by radial advection and part by turbulent diffusion and it is important to know the relative importance of these processes. Most of the succeeding discussion will focus on cold-core rings as most of the work has been done on them.

It is probably impossible to measure the radial velocities in a ring directly, and so most of the estimates have been obtained from mathematical models which have been adjusted to fit observations of the spin-down of a ring. The model of Molinari (1970) assumed that a ring is initially in geostrophic balance but that frictional forces will upset this balance giving rise to a radial flow along isopycnal surfaces. If the Prandtl number was greater than one (momentum diffuses faster than density) then the radial flow in the top of the ring was inwards with a downwelling at the centre (Fig. 1). It was found that the inward surface radial velocity possessed a maximum which coincided approximately with the zone of maximum rotational velocity. Assuming a value of 10^6 cm^2 s^{-1} for the horizontal momentum diffusion coefficient κ, and a Prandtl number of 2 yielded values of 0.05 cm s^{-1} for the maximum inward radial velocity of a ring aged about 4 months. A time scale for passive contaminants carried by this advective motion can be obtained by taking as the length scale the width of the high velocity frontal zone, say 20 km, which gives a time scale of 1¼ years. In the surface mixed layer, a process as slow as this is unlikely to be important compared to the transport produced across the frontal area by wind induced Ekman drift. Over a period of time this latter process will probably make the largest contribution to the horizontal mixing of the surface layers. If we assume an Ekman drift of 10 cm s^{-1} then it would take a few tens of days to mix surface layers of a ring with its surroundings. Thus if a ring is formed during a period of medium to high winds there will be rapid interchange of surface plankton, or plankton that migrates to the surface at night, between the ring and its surroundings. If the ring is formed in the summer then the phytoplankton will be concentrated mainly in the seasonal pycnocline which can be at a depth of 100 m. As the Ekman velocity falls off exponentially with depth, the horizontal mixing produced by the Ekman drift will be much reduced and the effect of radial advection and diffusion may be more important.

The model of Schmitz and Vastano (1975) ignored the rotational modes of motion and assumed that the time change of temperature in the ring was produced solely by a combination of radial diffusion and advection. From obser-

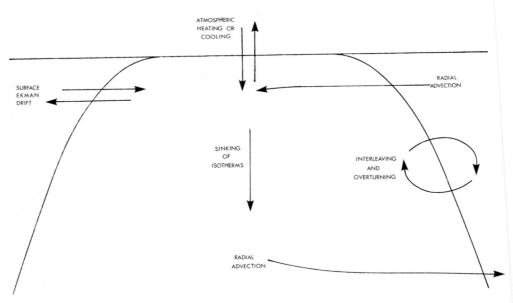

Fig. 1. Diagrammatic representation of exchange processes in a newly formed ring

vations of the change in the temperature structure of a ring over a 52 day period, they calculated the radial stream function for different values of the horizontal diffusion coefficient κ. They found that a best fit to the observations was obtained with a κ value of 10^4 cm^2 s^{-1} giving a maximum inward radial velocity at the surface of around 0.025 cm s^{-1}, about half Molinari's typical value. The return flow took place below 600 m with a maximum velocity of 0.015 cm s^{-1}. The predicted depth of the return flow is in broad agreement with Olsen's analysis of the dynamics of a Gulf Stream ring. Olsen (1980) computed the potential vorticity of the ring along a vertical section and compared the contours of this quantity with that of σ_t. If the two contours for these two quantities are parallel fluid can move in and out of the ring along σ_t surfaces, but where they are not parallel radial flow is inhibited. He found that the contours were only parallel above 200 m and below 800 m and so between these depths the cross-frontal flow was likely to be small. If this is generally true then any zooplankton species whose vertical distribution is confined between these two depths are likely to remain isolated within the ring longer than species that migrate above or below them.

Below 800 m the Schmitz and Vastano model predicted an outward radial velocity of 0.015 cm s^{-1} which gives a time scale of four years for crossing a 20 km frontal zone. Mixing will also take place across the frontal zone by diffusion and a time scale for this process is given by $L^2/4\kappa$ where L is the width of the front and κ the eddy diffusivity. If we take L=20 km as before and use a value for κ of 10^4 cm s^{-1}, which Lambert (1974) considered more correct than the value of 10^6 cm s^{-1} used by Molinari (1970), we get a time scale of 3 years, which is similar to that for radial advection. These results suggest that below the surface layers the concentrations of a passive contaminant within the ring will only change slowly relative to the life expectancy of the ring. Of course biological populations are not passive contaminants and we shall see later how this fact affects these conclusions.

Recently, towed undulator profiles (from the surface to 300 m) have been carried out across eddies to the south-west of the Azores (IOS, Unpublished data) and these showed that above 200 m in the frontal zone there was considerable interleaving of the water masses from either side of the front. In some profiles there was also evidence of overturning similar to that observed by Herman and Denman (1979) in the Nova Scotia shelf edge front. This overturning would bring up fresh nutrients from below the seasonal pycnocline which could explain the higher chlorophyll biomass observed in both this front and also possibly in the frontal zones of the Gulf Stream rings (The Ring Group 1981). This effect must also facilitate horizontal mixing across the front but just how important this is quantitatively is difficult to say at the present.

The evidence presented so far suggests that between 200 and 800 m the ring will maintain its character for some considerable time. What happens below this depth will depend, as pointed out by Richardson (Chap. 3, this Vol.), on whether rings are advected or self-propelled. If they are advected then their whole volumes are transported with the flow. However, if they are self-propelled this will only be true for depths above which the rotational velocity exceeds the translation velocity.

Flierl (1976) has shown how to calculate the stream function of tracer particles in a rotating ring which is itself advecting with a constant velocity C. The stream function for the ring at rest is approximated by a circular gaussian surface with standard deviation r. Thus the area A contained between the centre and the zone of maximum rotational velocity, U, will be $A = \pi r^2$. Flierl shows that for $U > C$ there will be a pear-shaped "trapped" area within the ring where particles will just oscillate around the ring and never escape. Outside this area particles will simply be disturbed by the ring's passage and then left behind as the ring moves away. Flierl suggests that the area of the trapped zone A_T is given by

$$A_T = 4(1-C/U)^2 A$$

In Table 1 we calculate A_T/A for three different depths for a C value of 6 cm s^{-1}, and typical values of U quoted by Flierl.

Table 1. A_T/A at different depths

Depth (m)	U (cm s^{-1})	A_T/A
0	90	3.5
500	60	3.2
750	30	2.6
1000	10	0.6

It can be seen that the trapped area decreases with depth, but that this decline becomes more rapid below 750 m. Thus the fate of the zooplankton species that live below 1 000 m will be critically affected by whether the ring is self-propelled or advected by the mean flow, and it may be that this is a case where biological sampling might throw some light on the physics.

22.2.4 Horizontal Exchange of Plankton Between a Ring and Its Surrounding Water

When we consider the exchange of biological populations we have to take into account the processes of population growth or decline that will be affecting the concentration of an organism independently of any advective-diffusive processes.

Let us now consider the case of a population of organisms in the water surrounding the ring which are nutrient or food limited. Let us further assume that these organisms find themselves at a competitive advantage vis-a-vis related organisms in the ring. This situation might easily arise for phytoplankton, due to the warming of the surface layers of the ring and the fact that nutrient levels in the ring are 2 to 3 times higher than the Sargasso Sea (Ortner 1978). These organisms will diffuse in from the boundary and begin to increase their numbers until some limiting factors come into play. The result is that a "front" of organisms will move inwards towards the centre of the ring. If we assume the logistic law $f(P) = \alpha P - \beta P^2$ for the population growth rate, where α is the intrinsic rate of growth of the population, β is the resource limitation coefficient and P the population density, then it can be shown that the front will move with minimum velocity $v = 2\sqrt{\alpha\kappa}$ where κ is a constant diffusion coefficient (Fisher 1937, Stokes 1977). (Strictly speaking, the formula for v applies to a one-dimensional front; a radial front will move faster or slower than this value, depending on whether the front is diffusing towards or away from the centre.) Taking $\kappa = 10^4$ cm^2 s^{-1} we obtain v values of 0.4 and 0.2 cm s^{-1} for population division times of 3 and 10 days respectively, which are typical of spring bloom conditions. Thus populations of quick-growing phytoplankton are able to spread into the ring considerably faster than if they were passive contaminants. Furthermore, a front moving with a speed of ~ 0.5 cm s^{-1} should be observable during a normal length cruise using continuous fluorometric sampling techniques. In the case of zooplankton, however, the fastest growing species might have a population doubling time of 100 days and thus the velocity of the population front would be 0.07 cm s^{-1} which is comparable with the ring radial velocities. The conclusion is therefore that the enhancement of diffusion by population growth is only likely to be important for phytoplankton or possibly micro-zooplankton.

In all the discussion so far we have assumed that the organisms are incapable of independent motion. While this is effectively the case for phytoplankton and micro-zooplankton it will not be true for macrozooplankton and nekton. Enright (1977) has studied the vertical motion of a copepod *Metridia pacifica* while carrying out diurnal migration and found speeds as high as 2.5 cm s^{-1}. Weihs (1973) has suggested that there is an optimum cruising speed for fish of 1 body length s^{-1} which would give a speed of ~ 5 cm s^{-1} for myctophids. These speeds are two orders of magnitude greater than the fluid radial velocities discussed in the previous sections and give a time scale for crossing a 20 km zone of between 5 and 10 days. However, it is not known whether zooplankton or small fish are capable of directed horizontal motion for these periods although the consensus of opinion is that they are not.

With regard to vertical migration, it has already been pointed out that zooplankton that migrate into the surface layers are likely over a period of time to be mixed into the water surrounding the ring by the surface Ekman drift. It is to be expected therefore that populations of vertically migrating zooplankton contained within the ring on its formation are likely to disappear from the ring at a far faster rate than the non-vertically migrating species. Conversely the ring is likely to be invaded first by vertically migrating species from the water outside the ring.

22.3 Mid-Ocean Eddies

In the previous section on rings we have seen that, because these features are energetic, advective, semi-isolated phenomena enclosing a water mass having

anomalous water properties it is possible to speculate about the possible biological processes going on within them. In the case of mid-ocean eddies however, the increased complexity of the physical processes makes speculation a
much more risky enterprise and a host of questions spring to mind without obvious answers. Are the eddies wave-like or advective? Can we study the biological processes in a single eddy or do we have to consider a statistical ensemble
of eddies? If the latter, how do we describe the turbulent motions? At this stage
we can only briefly discuss some of these problems and must await further progress in the physical understanding of the eddy field and the availability of
suitable data.

22.3.1 Biological Processes in a Single Eddy

The biological processes in a mid-ocean eddy will depend critically on whether
the eddy is an advective entity or a feature of a wave field. If it is the former
then many of the processes discussed in the previous section might also take
place in mid-ocean eddies, although with different parameter values. However,
if the eddy is wave-like then the biological effects are likely to be more transi-
tory. If we assume that a wave-like eddy has a space scale of L and propagates
with a phase speed of C then an organism will experience the effects of the
eddy for a time $t_e = L/C$. Let us assume for the moment that the passage of the
eddy can alter in some way the growth rate of a population, then this effect will
be most marked on organisms whose population doubling-time is small compared with t_e. Freeland and Gould (1976) calculated objective stream functions
from the MODE current meter and float data and estimated the phase speed as
2 cm s^{-1} above the main thermocline and 5 cm s^{-1} below it. If we take L to be
100 km we obtain t_e values of 60 and 20 days for above and below the thermocline respectively. They also concluded that the motion below the thermocline
was wave-like, while above the thermocline a wave-like motion could not be
ruled out. These values for t_e suggest that we should be looking at spatial
changes in the populations of phytoplankton and micro-zooplankton, rather
than zooplankton, for observable effects of wave-like eddies. There are two immediately obvious ways whereby an eddy might alter the productivity of phytoplankton. The first is by altering the depth of the seasonal thermocline thus altering the average light absorbed by the plants above this thermocline. The second is by increasing the vertical mixing through the thermocline bringing fresh
nutrients into the surface mixed layer. The work of Andrews (1979) suggests
that both these processes occur. Andrews found a definite correlation between
eddies in the region of the West Australian Current and the depth of the surface mixed layer. He suggested that when there is net efflux of heat from the
surface the resulting free convection creates turbulence which in turn produces
cross isobar flows. This flow produces a divergence of water away from low
pressures and a convergence over high pressures. The net result is that the
mixed layer deepens in highs, resulting in nutrient entrainment into the surface
layers, and shallows in lows. The effect of both these processes on the phytoplankton will depend on the vertical distribution of the chlorophyll and the

season. However we may speculate that the productivity of the phytoplankton at some depth in the photic zone, though not necessarily at the surface, will be affected by the eddy field. These effects should be observable with an undulating sampler fitted with a CTD and in situ chlorophyll fluorometer.

Other observations have suggested that at least some types of eddies are advective and are capable of transporting anomalous water properties over large distances. McDowell and Rossby (1978) identified a 300 m thick lens of water off the Bahamas at a depth of 1 100 m that had water mass characteristics typical of Mediterranean and eastern Atlantic waters. From observations of neutrally buoyant floats they estimated the phase velocity at 6 cm s^{-1} and concluded that the eddy (dubbed by them a "meddy") had travelled about 6 000 km, over a period of 2½ years, without losing its identity to the surrounding waters. This raises the possibility that biological populations of eastern Atlantic zooplankton might also have been transported with the eddy. However, until more is known of the physics of these "meddies" it is difficult to predict whether these trapped populations could maintain their identity over such long periods. Conversely if it were possible to sample the biological populations of these "meddies" and compare them quantitatively with the surrounding water then the biologist might provide some clues for the physicist.

22.3.2 Spatial Spectrum

It has been mentioned that we now have the technical capability of measuring the mesoscale distribution of chlorophyll, which is a measure of phytoplankton biomass, and that this distribution may be correlated in some way with the eddy field. If the spatial spectrum of some parameter of this eddy field, such as mixed layer depth, was known, can we say anything about the spatial spectrum of chlorophyll? A full discussion of this problem must probably await the merging of population dynamics with recent developments in two-dimensional oceanic turbulence (Rhines 1977, Haidvogel Chap. 18, this Vol.) and we can do no more than discuss two previous attempts to derive the spatial spectrum of chlorophyll.

Denman and Platt (1976) used dimensional arguments to determine the slope of the chlorophyll wavenumber spectrum. They considered that eddies were characterized by a length scale L and time scale t_L, the time taken for the eddy of size L to transfer its kinetic energy to eddies of size L/2. The time scale for the biological population is given by the reciprocal of the growth rate α and Denman and Platt postulated two wave number regimes. When $t_L < \alpha^{-1}$ biological reproduction has little effect on the spatial distribution and so the spatial spectrum will be the same as that for the physical variable. However when $t_L > \alpha^{-1}$, reproduction will be an important factor in the production of horizontal pattern and the chlorophyll spectrum will decrease less sharply with wavenumber than the spectrum of the physical variables. Using dimensional arguments they determined that the spectrum in the latter region was proportional to k^{-1}, where k is the wavenumber. The boundary between the two zones was determined by a critical wavenumber $k_c = (\alpha^3/\varepsilon)^{1/2}$ where ε is the rate of viscous dissipation. A more rigorous mathematical derivation of the spectrum was obtained by Denman, Okubo and Platt (1977). It should be emphasised however that both these derivations relied heavily on the theory of isotropic turbulence as developed by Kolmogorov (1941) and Corssin (1961) which in the ocean is only strictly applicable at length scales of a few hundred metres or less.

Fasham (1978a,b) used a different approach to this problem based on some early work of Whittle (1962). In this model, the turbulence was parameterised by a constant eddy diffusion coefficient. The stochastic variability was introduced by a stochastic forcing function giving for the one dimensional case the equation

$$\frac{\partial P}{\partial t} = \kappa \frac{\partial^2 P}{\partial x^2} + \alpha P + F(x,t) \tag{1}$$

where $P(x,t)$ is the concentration of phytoplankton, κ is the horizontal diffusion coefficient, $F(x,t)$ the forcing function and α the net growth rate of plankton. The function F can be regarded as an added population growth which can vary stochastically in space and time but which is independent of population size. Equation (1) can be transformed to an ordinary differential equation by taking a spatial Fourier transform and from the solution of this transformed equation we can calculate the time dependent spatial spectrum $E(k,t)$ of P. In order to simplify the situation it was assumed that the function F was random in time which yielded the spectrum

$$E(k,t) = \phi(k)[1 - \exp\{-2(4\pi^2\kappa k^2 - \alpha)t\}/2(4\pi^2\kappa k^2 - \alpha)] \tag{2}$$

where $\phi(k)$ is the spatial spectrum of $F(x)$.

If we define a critical wavenumber

$$k_c = (\sqrt{|\alpha|/\kappa})/2\pi \tag{3}$$

then the behaviour of the spectrum for both positive and negative α will be different on either side of k_c. This behaviour is best demonstrated by plotting the spectral ratio $R(k) = E(k,t)/\phi(k)$ as a function of normalised wavenumber $k_n = k/k_c$. This function for $\alpha = \pm 0.5$ and $t = 2$ and 5 is plotted in Fig. 2.

For $k > k_c$ all the curves tend to the same time-independent form with $R(k) \propto k^{-2}$. This means that for high wavenumbers the population spectrum will always fall off more sharply than the forcing function. For low wavenumbers the spectra for both positive and negative α will have the same form as the forcing function. However, for positive α the magnitude of the spectrum will increase with time for $k < 1$, while for negative α it will approach a limiting value given by

$$E(k) = \frac{\phi(k)}{2(4\pi^2\kappa k^2 + |\alpha|)} \tag{4}$$

It will also be noted that for k values in the region of k_c the spectral ratio $R(k)$ for positive α falls off very sharply with wavenumber and, judging by eye,

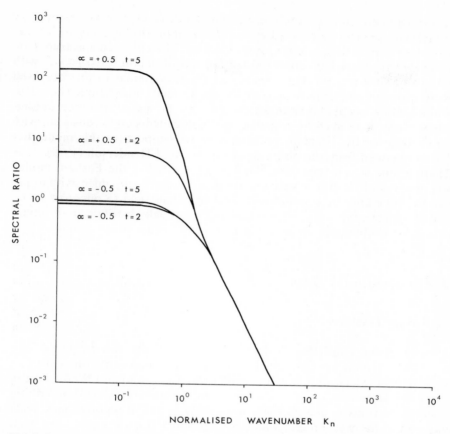

Fig. 2. Spectral ratio for the spectral function of Fasham (1978a) for positive and negative α ($\alpha = 0.5$) and times $t = 2$ and $t = 5$

is proportional to k^{-3}. As time progresses this zone of more rapid fall-off will extend to lower wavenumbers.

It is clear that this spectrum differs substantially from that of Denman and Platt (1976) in that it predicts that the chlorophyll spectrum will have the same slope as $\phi(k)$ for wavenumbers less than k_c while the reverse was true for the Denman-Platt spectrum. However we have already pointed out that the Denman-Platt spectrum is only applicable to isotropic turbulence at small space scales.

The critical wavenumbers k_c depends on the growth rate α and will thus vary with season. In the spring bloom in temperate regions net growth rates of the order of $0.5 \, d^{-1}$ are observed (Fasham, unpublished data) and so for a κ value of $10^4 \, cm^2 \, s^{-1}$ we obtain a k_c of 2.5 km. We thus expect that for wavelengths longer than 2.5 km the chlorophyll spectrum would have the same slope as the eddy spectrum. Recently, Gower et al. (1980) published a chlorophyll spectrum obtained from the LANDSAT multispectral scanner which supports this observation. The area covered was in the Atlantic Ocean south of Iceland and the data were obtained in the month of June. The observed spectrum covered wavelengths from 1 km up to 100 km and was of the form $E(k) \propto k^{-n}$ with a mean value of n of 2.92. This sort of slope is very similar to that observed from airborne radiometer measurements of sea surface temperature. Their conclusion therefore was that over these space scales the spatial spectrum of phytoplankton was unaffected by growth rate, which conclusion is consistent with the predictions of the Fasham spectrum. Further support for this model has also been obtained from the North Sea by Steele and Henderson (1979).

There is one serious criticism that can be levelled at the Fasham model. This is that it is difficult to justify the use of a constant eddy diffusivity to parameterise the eddies in a two-dimensional turbulent field (Kraichnan 1976). Now that we have the ability to measure the mesoscale distribution of phytoplankton this is obviously an area that will repay further theoretical work.

22.4 Observational Data

22.4.1 Scripps Transects

Early work in the North Pacific had suggested that the eastern part of the North Pacific central gyre provided a rather monotonous environment (e.g., McGowan 1974, McGowan and Hayward 1978, McGowan and Walker 1980). Thus it came as something of a surprise when Shulenberger (1978) describing the results of a transect along 28°N from 163°W to 146°E reported mesoscale activity, not at the western end where a Kuroshio influence might have been expected, but in the vicinity of 180° in a region of rough bottom topography over the Hawaiian ridge. Hydrographic observations were made at 1° longitudinal intervals. Both the salinity and temperature profiles showed evidence of strong eddy activity around 180°. Despite sampling being reduced to 2° intervals of longitude, similar structure was evident from the nitrate profiles, but not quite so clearly in the nitrite and phosphate profiles. However, the phosphate profile also indicated eddy-like features at the western end of the transect. The depths of the chlorophyll maximum responded to the variations in the depth of the seasonal thermocline and the nutricline.

On the zoogeographic scale the mesopelagic fish assemblages and the structure of the plankton communities showed strong differences between the ends of the transect. No sampling was conducted specifically to investigate where the change took place or indeed to see if the change was clinal. Shulenberger (1978) pointed out that at his station at 160°E, diel migrants would experience much greater environmental heterogeneity than similar migrants in the Eastern Pacific Gyre, both because of the greater steepness of the physical gradients and the greater horizontal variability of the chemico-physical variables.

In an XBT section across the North Pacific from 122°W to 141°E in March/April 1976, Kenyon (1978) showed substantial mesoscale activity in the form of cold core rings west of 170°E and much quieter conditions to the east. Venrick (1979) has analysed the chlorophyll and nutrient data taken at 2° longitudinal intervals at the same time and on several subsequent cruises. In the western half of the transect there was marked horizontal heterogeneity in chlorophyll distribution, and integration of the vertical profiles showed the standing crop was higher from which Venrick inferred primary production was greater. In the east the chlorophyll standing crop (Fig. 3) tended to have maxima in the warm features rather than the cold rings (e.g., stations 95 and 87) a result of the greater winter mixing of the surface layers leading to greater nutrient enrichment (Sect. 22.3.A and Andrews 1979). One station (67) with a high standing crop appeared to coincide with the edge of a cold feature. Venrick used a principal components analysis to reduce the physical and chemical data to four axes along which she identified five environments, including (a) the Kuroshio, (b) cold core stations, (c) western North Pacific, (d) eastern North Pacific and (e) California Current. Spatial heterogeneity was greatest in the

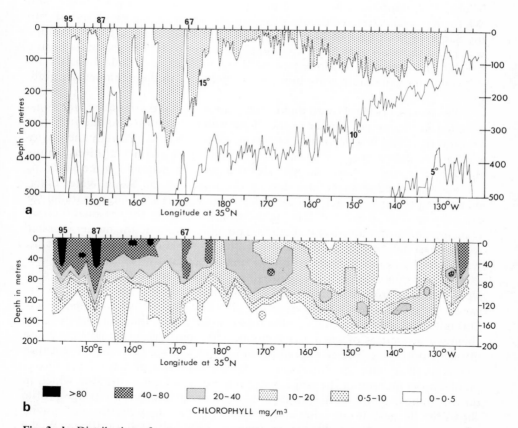

Fig. 3a, b. Distribution of **a** temperature and **b** chlorophyll (mg m^{-3}) across the Pacific at 35°N. (Kenyon 1978 and Venrick 1979)

California Current, and minimum in the eastern Pacific. Thus in areas like the eastern Pacific gyre where there is an absence of mesoscale activity there is a corresponding consistency in the environmental parameters and the biological variables, (e.g., McGowan and Hayward 1978). In the western North Pacific where mesoscale activity is greater, biological heterogeneity is greater. However, because the programme was not designed to investigate mesoscale activity, the sampling spacing was too wide to assess the patterns of the biological processes across eddy and ring structures.

22.4.2 Gulf Stream Ring Structures

The biology of Gulf Stream Rings, mostly cold core rings, has been the subject of a series of papers resulting from Woods Hole study programme (Wiebe 1976, Wiebe, Hulburt, Carpenter, Jahn, Knapp, Boyd, Ortner and Cox 1976, Jahn 1976, Ortner 1978, Ortner, Wiebe, Haury and Boyd 1978, Wiebe and Boyd 1978, Boyd, Wiebe and Cox 1978, Cox and Wiebe 1979, Ortner, Hulburt and Wiebe 1979, Fairbanks, Wiebe and Bé 1980, The Ring Group 1981, Haury and Wiebe in press). These cold core rings are generated by the pinching off into the Sargasso Sea of a meander of the Gulf Stream enclosing a parcel of slope water and so result in the advection of slope water communities into a hostile oceanic environment. In Mode-type eddies generated in oceanic areas the inner and outer communities are more evenly balanced in their competitive ability.

Wiebe et al. (1976) showed that initially the seasonal thermocline and its associated nutricline are similar to the Slope water conditions. As the ring ages and decays, there is a deepening of the nutricline. Not surprisingly the phytoplankton standing crop, as measured by chlorophyll a concentrations, shows that slope > ring > Sargasso Sea, and that primary production measured by ^{14}C shows a similar trend. The starting conditions in the Slope Water determines the precise sequence and rates of the changes. The changes parallel the normal sequence of events following the establishment of the seasonal thermocline in temperate waters, so it should be possible to extend the models applied to the seasonal development and deepening of the thermocline in temperate waters, to the events in these cold core rings (e.g., Pingree 1977).

The phytoplankton cell abundance profiles are interesting (Fig. 4). The data were collected from a 3-month-old ring in March 1974. The differences between the profiles result both from the disparity in the time response of the Spring bloom inside and outside the ring resulting from the differing physico-chemical properties of the water column, and from mixing by advective and diffusive processes. The cell count profiles show a sharp drop in cell densities which coincide with the nitrate-clines. So the centre of gravity of the phytoplankton community in the slope water is shallower than in the ring which in turn is shallower than in the Sargasso Sea Water. In contrast the fringe profile shows the deepest nitrate-cline and consequently the deepest centre of gravity of phytoplankton despite the elevated levels of phosphate at 75–100 m (Wiebe et al. 1976 Fig. 5). Such an observation would be consistent with increased mix-

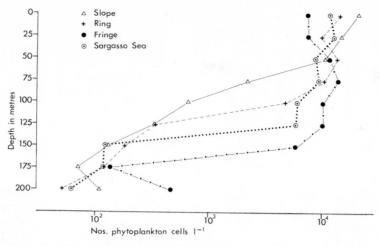

Fig. 4. Vertical profiles of phytoplankton cell concentrations associated with a Gulf Stream Ring. (Based on data from Ortner et al. 1979)

ing in the fringe and a possible mechanism for this was suggested in Section 22.2.3. It is unfortunate that more data are not available from ring fringes.

Ortner (1978) and Ortner et al. (1979) compared the phytoplankton populations inside and outside rings using correspondence analysis; a statistical technique that utilizes abundance data to look for floral discontinuities. They found the differences were not clearly related to ring age, implying that they are as much a local response to the biotic conditions as the result of ring decay. Average population properties such as standing crop and community diversity were generally similar between the ring and the Sargasso Sea communities but were different in the Slope Water community. The correspondence analysis gave similar results for the autumn (fall) samples. However, in the spring and summer samples it was the slope and Sargasso Sea communities that were similar and the Ring community appeared to be distinctive according to the correspondence analysis. Ortner et al. (1979) offer no explanation for these rather surprising observations and without better knowledge of the ways in which seasonal successions of phytoplankton communities proceed and develop, there appear to be no easy explanations. In general diatoms were less abundant in the rings, while a dinoflagellate *Prorocentrum obtusidens* and six coccolithophorid species were more abundant.

These variations in phytoplankton community structure would seem to be determined by complex interactions between seasonal succession, physical mixing processes, grazing pressure and nutrient fluxes supplemented by animal excretion. Ecosystem modelling may produce a better understanding of how these environmental factors interact, once more information is available on the ecology of the species involved, and the successions are better described. The change from Slope to Ring to Sargasso Sea conditions represents a gradation

M.V. Angel et al.

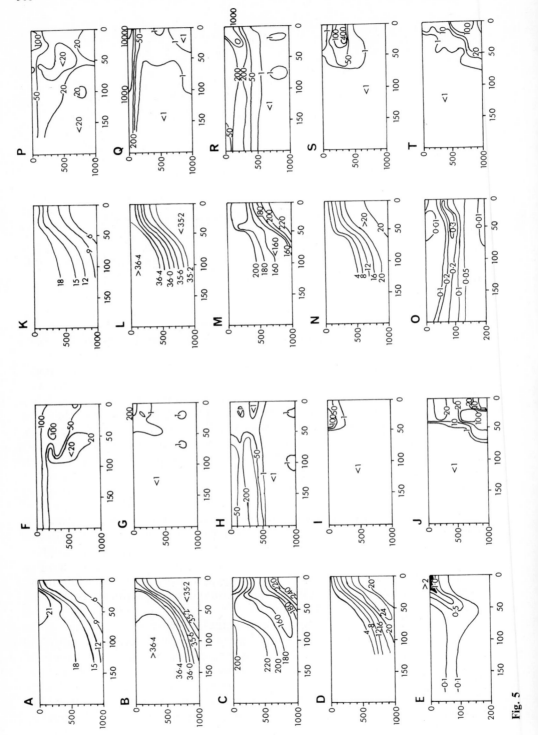

Fig. 5

from conditions where there is a relatively species poor community composed of larger-celled species and having a higher standing crop concentrated into a narrower band of the water column, to a species-rich community of small-celled species with its lower standing crop more thinly spread through a deeper layer of the water column. Ortner et al. (1979) calculate that a herbivore would require to filter an order of magnitude more water and process five times the numbers of cells to obtain the same ration in the Sargasso Sea Water as compared with the Slope Water. Similarly the concentration of cells just above the seasonal thermocline is much greater on the Slope, than in a ring or the Sargasso Sea Water, although this difference is less marked in the Spring.

Thus for a herbivore from slope conditions the rings will represent a deteriorating environment, in which it is harder for a specialist feeder to survive because the standing crop is more evely distributed amongst the greater number of phytoplankton species. For a herbivore from oceanic conditions the rings would provide a richer environment than their normal Sargasso Sea habitat. Wiebe et al. (1976) showed that zooplankton biomass was higher inside rings of all ages examined (up to 12 months) than in the surrounding Sargasso Sea. The increase was mostly accounted for by the enrichment of the zooplankton community at 300–800 m, in comparison the near surface biomass was more similar between the inside and outside of the rings. Note that the enrichment occurred at depths where, in Section 22.3 it was suggested that cross-frontal flow was likely to be small. Non-migrant species would be carried in or out of the ring with equal difficulty. However, the ring might function as a physiological trap for vertical migrants, if the ranges of the migrations are altered by the prevailing conditions either in the front or inside the eddy itself. Isaacs, Tont and Wick (1974) showed how the response of migrators to changes in light profiles between areas of high and low productivity can lead to the aggregation of organisms in the richer zone, resulting from the tendency for current shear to increase with increasing depth. Some preliminary evidence both for vertical migrations being modified and for aggregations of nekton to occur in a frontal zone is presented below. Warm core rings with lower levels of productivity and so with deeper penetration of the isolumes may lose migrating organisms because the ranges of their migrations are extended.

The envelope of the oxygen concentration profiles (Wiebe et al., 1976) show oxygen minima at about 200 m in Slope Water, 300–600 m in the rings and near 800 m in the Sargasso Sea Water, implying a deepening of the main concentrations of midwater organisms. Ortner (1978) and Ortner et al., (1978) give some

Fig. 5. Sections through a ring, the vertical scale in metres and the horizontal scale in km from the ring centre. A–J Ring Bob in April 1977. **A** Temperature, °C; **B** Salinity, ‰; **C** Oxygen μmol kg^{-1}; **D** Nitrate, μmol kg^{-1}; **E** Chlorophyll a, μgl^{-1}; **F** Zooplankton biomass ml 1 000 m^{-3}; **G** *Limacina retroversa* No. 1 000 m^{-3}; **H** *Limacina inflata*, No. 1 000 m^{-3}; **I** *Nematoscelis megalops*, No. 1 000 m^{-3}; **J** *Paraeuchaeta norvegica* ♀ No. 1 000 m^{-3}. **K–T** Ring Bob in August 1977. **K** Temperature, °C; **L** Salinity, ‰; **M** Oxygen, μmol kg^{-1}; **N** Nitrate, μmol kg^{-1}; **O** Chlorophyll a, μg l^{-1}; **P** Zooplankton biomass ml 1 000 m^{-3}; **Q** *Limacina inflata* Night, No. 1 000 m^{-3}; **R** *Limacina inflata* Day No. 1 000 m^{-3}; **S** *Nematoscelis megalops*, No. 1 000 m^{-3}; **T** *Pareucheata norvegica* ♀ No. 1 000 m^{-3}. (After Ring Group 1981)

data, based on oblique hauls of vertical migration behaviour in and outside the same ring in August and November. By expressing the zooplankton biomass in the top 200 m as a percentage of the biomass in the top 800 m of the water column, comparison of the percentage by day and by night gave a rough measure of the intensity of migration. In August the day:night ratio was 15:50 inside the ring and 35:60 outside the ring in the Sargasso Sea Water. Thus although similar proportions of the total population were migrating, the population inside the ring tended to be deeper in its vertical distribution. In November the ratios were 20:30 inside the ring and 30:60 outside, showing a reduction of migratory activity within the ring.

Ortner (1978) found that the ratio between the zooplankton biomass in rings and outside in Sargasso Sea Water decreased as the ring aged. However, the ratio between Slope and Ring biomass did not show a corresponding increase, nor did Slope: Sargasso Sea biomass ratios. This implies there may be a decoupling of the seasonal cycle between the Slope Water regime and the Ring and the Sargasso ecosystem, which may be open to investigation by modelling.

The early disadvantages to Slope Water herbivores inside rings result in Sargasso Sea herbivores moving in rapidly; the pteropods inside the rings were the same as outside. In contrast, carnivores and omnivores move in more slowly. Chaetognaths, for example, are initially an order of magnitude less abundant inside than out. Euphausiids like *Stylocheiron suhmii* and *S. abbreviatus* (both non-migrants) are initially absent. Wiebe et al. (1976) showed how a number of cold water euphausiid species like *Nematoscelis megalops, Euphausia krohnii, Meganyctiphanes norvegica, Thysanoessa gregaria* and *T. longicaudata,* which are characteristic of Slope Water, gradually disappear as a ring ages. Warm water species such as *Euphausia brevis, Nematoscelis microps, N. tenella, Stylocheiron suhmii* and *S. affine* which are rare in Slope Water, gradually become more abundant as the ring ages.

Wiebe and Boyd (1978) and Boyd et al. (1978) investigated the disappearance of *Nematoscelis megalops* in detail. This species is associated with the 10°C isotherm and its vertical range deepened as the 10°C isotherm deepened. In slope water this species is a non-migrant, but in the rings it is possible that it becomes a diel vertical migrant, although Wiebe and Boyd (1978) specifically state it remains a non-migrant. Their interpretation of their data is complicated by the day/night disparity in the catches produced by avoidance. Discovery data (e.g., Roe in press) suggest that although this species is extremely patchy in its distribution and far less abundant in oceanic than in slope and neritic environments, at some stages of its life cycle it is a vertical migrant. If this is the case in rings, then the profiles presented by Wiebe and Boyd (1978) can be interpreted to imply that the upward migration of the animals is halted by the seasonal thermocline, which deepens as the ring ages. Wiebe and Boyd (1978) recorded the capture of a gravid female in the wake of a ring, possibly an indication of spin out. However, they also found that they caught greater numbers of *N. megalops* in the ring when it was 9 months old than when it was 6 months old. This they ascribe to the population having become more moribund and so more susceptible to capture. It could also be the result of lower mortality caused by the lower numbers of predators present. After 17 months the species was no longer present in the ring. Populations sampled on the slope had bimodal size distributions both in summer and in the fall, and during August when the ring was 6 months old, the size frequency distribution of the ring population was similarly bimodal with peaks at carapace lengths of 3–4 mm and 5–6 mm but with fewer in the larger size class. In Discovery material from the northeastern Atlantic the smaller size group corresponds to adolescents which cannot be accurately sexed as the pentasma is not yet fully developed in the males (James pers. comm.). In August large animals occurred in both the ring and the Slope water but the $\female : \male$ ratios were high, 114:1 and 51:1 respectively, probably the result of adult males having much higher mortality in the post-spawning period in the summer. By November when the ring was nine months old the apparent ratio in the ring was $\infty:1$ as no large animals occurred, whereas in the Slope Water where maturation had begun to take place the ratio was 1.5:1. Hence ring conditions appear to either delay or totally inhibit the maturation of this species.

The Ring Group (1981) illustrated two vertical profiles of *N. megalops* sampled in ring "Bob" in 1977. This ring formed in February/March 1977, interacted with the Gulf Stream in April/May and then moved southwestwards through the Sargasso Sea until September when it coalesced back into the Gulf Stream off Cape Hatteras. The ring-core was quite distinct in its potential temperature/oxygen relationship in April from both Gulf Stream and Sargasso Sea water. The oxygen data indicated active interleaving in the ring boundary between the core water and Sargasso Sea Water. In April there was a surface chlorophyll maximum in the ring with a subsurface maximum at 50–100 m in the surrounding water (Fig. 5E). In August the chlorophyll standing crop was higher in the subsurface maximum inside the ring than outside. There were differences in chlorophyll distribution within the ring; the inner core, with a radius of about 20 km, had lower standing chlorophyll levels in the subsurface maximum but higher levels in the surface 20–40 m than the outer zone of the ring (Fig. 5O).

The April zooplankton biomass sections show high shallow-mesopelagic standing crop about 50 km from the ring centre that coincides with a doming in the oxygen levels in the surface waters. Further out there is a marked decrease in standing crop at depths >500 m, which could either be produced by the ring "scavenging" zooplankton or by elevated predation levels in the ring fringe zones. The August profile also shows a hint of zooplankton enrichment away from the centre of the ring and an impoverishment of the deeper standing crop at the ring edge, however, such an interpretation may be overstretching the precision of the data.

The centre of the *Nematoscelis megalops* population deepened from the 0–200 m zone in April to 200–400 m in August, with a decrease according to the Ring Group in its abundance, although in their sections it looks more like an increase. In the carnivorous copepod *Pareuchaeta norvegica* there seems to have been a withdrawal from the flanks of the ring in towards its centre. The high concentration between 100–300 m seemed to have moved down the water column. However, the data as presented do not rule out the possibility that this is a normal pattern of seasonal migration and not necessarily a consequence of the environmental effects of the ring. The data presented on two pteropod species are interesting. *Limacina retroversa* occurred in small numbers in the ring in April but almost completely disappeared by August. *Limacina inflata* was more abundant outside the ring in April (Fig. 5H), but had become much more abundant inside the ring by August; it increased in water column abundance 300 times inside while there was the normal seasonal 30% decrease outside (Fig. 5Q, R). In August this species showed marked diel vertical migration from 200–400 m by day to 0–100 m at night outside the ring and in its outer zone, but the range of movement was reduced in the inner core. Again such an interpretation may be straining the credibility of the data, but it does indicate that the modification of vertical migratory behaviour may occur in at least some species inside rings as compared with outside rings. As discussed above, the existence of such changes whatever the causes could result in the aggregation within or loss from the ring of the species involved. In the example of *L. inflata* in Ring Bob the data do not allow the cause of the dramatic increase in population within the ring between April and August to be elucidated. The choice between alternatives such as the in situ growth of the population within the ring through high reproduction stimulated by better grazing and lower predation, or whether the ring "scavenges" animals from the surrounding waters, cannot be made.

The Ring Group (1981) strongly emphasised the probable influence of temperature both on the survival of species and in regulating their vertical distributions. This interpretation needs to be tested; in a recent Discovery cruise (114) sampling conducted to examine the influence of temperature on regulating the vertical distribution of species across the front bounding the 18°C Sargasso Sea Water, where isotherms changed depth by 100–200 m, has given initial results that suggest that for some species temperature does not influence vertical distribution. However, these Discovery samples need much fuller analysis before their full relevance to this problem can be assessed.

Boyd et al. (1978) studied the physiological and biochemical condition of *Nematoscelis megalops* from aging rings. Individuals from rings had lower total body lipids, carbon and nitrogen, but a higher water content than those from Slope Water. The ring animals had respiration rates as low as $70 \mu l O_2 \ g^{-1}$ wet weight h^{-1} as compared with a corresponding rate of $680 \mu l O_2 \ g^{-1}$ wet weight h^{-1} in slope animals. The Ring Group (1981) suggest that the average depression of respiration is 5%–20%. The same physiological state can be induced in slope animals by 4 days of starvation. Once again the data as presented do not exclude the possibility that these effects are produced by an acceleration of the normal seasonal cycle of changes in the animal rather than the direct influence of ring conditions.

Vastano and Hagan (1977) suggest that the sinking of the population of *N. megalops* observed by Wiebe and Boyd (1978) is consistent with the animals staying within a zone of vertical stability where ring Western North Atlantic Water is being formed by the mixing along isopycnal surfaces of warm saline water from above the mid-thermocline region outside the ring with the cooler lower salinity ring water.

Fairbanks, Wiebe and Bé (1980) considered another aspect of the advection of cold water slope faunas into warmer oceanic areas, the blurring of the paleoceanographic record of planktonic Foraminifera in bottom sediments. In a 9-month-old ring there was the expected reduction in advected slope species and an increase both in the relative and absolute abundances of warm water species. The standing stock of foraminiferans in the top 200 m was an order of magnitude greater than in the surrounding oceanic water. So, because rings occupy 6%–13% of the total area of the Sargasso Sea at any one time, at least half the foraminiferan standing crop occurs in rings.

Wiebe et al. (1976) give some preliminary data on the comparison of micronekton biomass inside and outside rings and in Slope Water; the concentration of biomass showed the same pattern as for plankton of Slope > ring > Sargasso Sea.

Fish in the Gulf Stream cold core ring were studied by Jahn (1976). Fish are relatively long-lived and mobile, and so population and community changes over the time scale of a few months are likely to be dominated by the fluxes of animals into and out of the rings. Jahn made observations on four rings; two were around 3 months old and were observed in November/December and February; and two were 10–12 months old and were observed in March/April and September/October. Species diversity and equitability were lower in the young rings and higher in the older rings. Apart from during September/October when Slope Water, a ring and the North Sargasso Sea all contained fish faunas of similar diversity, the ring communities were intermediate between the lower diversity Slope Water and high diversity North Sargasso Sea communities.

The application of a correspondence analysis grouped the Slope Water hauls and North Sargasso Sea hauls separately as elongate clusters on the first two axes. The ring samples joined the two cluster forming a horse-shoe. The young ring samples lay close to the Slope Water cluster, the March/April year-old ring samples were intermediate, and the September/October year-old ring samples were closest to the North Sargasso Sea observations.

Jahn found that there were few fish species specific to either Slope Water or North Sargasso Sea water and so the species were considered under groups showing marked, moderate or no preference for the two water types. Three species were only caught in Slope Water, *Maurolicus muelleri, Lampanyctus macdonaldi* and *Myctophum affine* and these species were absent from all the rings sampled. Two species abundant in slope water but rare in North Sargasso Sea Water, *Benthosema glaciale* and *Ceratoscopelus maderensis* were quite abundant in young rings, but were rare in the older rings. All these species are active vertical migrants.

Jahn considered several species to be more abundant in rings than in either slope water or North Sargasso Sea Water. Two of these species *Chauliodus sloani* and *Scopeloberyx opisthopterus* were normally considered to prefer Slope Water, and were generally more abundant in all the rings. *Diaphus rafinesquei* another Slope Water species was absent in the young ring sampled in November, but very abundant with high numbers of small individuals (<28 mm) in the ring of similar age sampled in February. It was still more abundant in the old ring sampled in March, mostly larger specimes >20 mm, but scarce in the old ring sampled in September. The maintenance of populations of this species in rings seems to be related to when the ring was formed. Juveniles of this species are migrants, whereas adults are not.

Four species, considered to show moderate preferences for North Sargasso Sea Water, *Vinciguerria attenuata, Valenciennellus tripunctulatus, Hygophum benoiti* and *Lampanyctus pusillus* were abundant in rings. The first species, a non-migrant, was not notably abundant in the young rings, nor in the 12-month-old rings sampled in March, but was very abundant in the old ring sampled in September. The other three species were no more abundant in the young ring sampled in November than in Slope Water, but were exceptionally abundant in the young ring sampled in February. *Valenciennellus* a non-migrant, was still abundant in both the old rings sampled, but the other two species were scarce in the old ring sampled in March, but abundant in the old ring sampled in September.

Jahn used correspondence analysis to try and simplify the complexity of the interrelationships of the fish communities. Although the habitat preferences of the various species were reflected in some of the clustering of the species, the underlying factors controlling many of the interactions are still not clear. Nor is there any clear indication of whether the flux of vertical migrant species into or out of rings differs from the flux of non-migrants. The Ring Group (1981) also presented some fish data. They again related the occurrence and abundance of various species to temperature, but in this case using the depth of the 15°C isotherm as a measure of ring age (cf. Parker, 1971). *Benthosema glaciale* is a sub-polar-temperate species that is abundant in Slope Water and, as noted above, in the cores of newly formed rings. As the 15°C isotherm sinks below 150 m, so the vertical range of the fish deepens. When the isotherm reaches 250 m, no *B. glaciale* occur above 500 m, and by the time it reaches 525 m. *B. glaciale* has disappeared. *Hygophum benoiti* and *Lampanyctus pusillus* are ring exploiters. *L. pusillus* occurs in abundance, in the core of rings in which the 15°C isotherm occurs between 150–350 m, which are up to an order of magnitude higher than in other conditions (Fig. 6), i.e., Slope Water, Sargasso Sea Water, ring edges and old rings. Both *Hygophum* species and the *Lampa-*

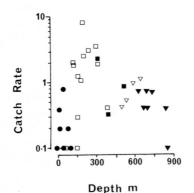

Fig. 6. Catch rates (No. 1000 m^{-3}) of the myctophid fish *Lampanyctus pusillus* plotted against the depth of the 15°C isotherm from Ring Bob ● Slope Water, □ Ring Centre, ■ Ring flank, ▽ Ring fringe, ▼ Sargasso Sea. Five hauls all in slope water contained no specimens and have been omitted from the plot. (After Ring Group 1981)

nyctus species are small with life cycles of a year or less, and which have protracted or continuous spawning seasons. From Parker's estimate of the sinking rate of the 17°C isotherm of 0.6 m d^{-1}, the duration of the abundance peak of *L. pusillus* is less than a year, less than the periodicity of the life cycle. Furthermore, the increase in abundance takes place very soon after ring formation, consequently this increase is more likely to be the result of movement into and retention within the ring, or an enhancement of larval survival in the ring rather than a direct breeding response of the fish stimulated by the ring environment.

There have been relatively few observations on the warm core eddies. Ortner (1978) showed the plankton standing crop of a warm core ring was higher than that of a cold core ring but less than for the slope water. Cox and Wiebe (1979) estimate that 8%–16% of the zooplankton populations occurring in the middle Atlantic Bight, are advected onto the shelf by warm core eddies.

22.4.3 East Australian Current Rings

Hamon (1965a) was the first to report on the existence of warm core anticyclonic eddies off the east coast of Australia. The most recent of a series of papers reporting on the physical oceanography of these rings (Nilsson and Cresswell 1980) shows that the formation of these warm core rings by the pinching off of meanders of the Australian Current (EAC) is analogous to the formation of Gulf Stream Rings. In winter these rings develop deep homogeneous cores as a result of convective cooling (see Sect. 22.2) and this core is then capped by summer warming. The EAC eddies decay with a time constant of 690 ± 50 days as a result of upwelling below the seasonal thermocline.

Scott (1978) used apparent oxygen utilization (AOU) as an indication of biological activity, in a ring in which the summer "capping" was well established. The central core of the eddy at depths of 200–400 m had unusually low AOU values. The coastal zone had a lower productivity than the surrounding ocean, and it was expected that the concentration of nutrients and AOU would show a consistent relationship. Data points (i.e. for nitrate and AOU) from the ring edges did indeed lie on a smooth curve, but those from the core were scattered. Since the silicate/nitrate relationship followed a smooth curve throughout, Scott interpreted the data as evidence of deep nutrient-rich water from outside the eddy mixing up with shallow nutrient poor water inside the eddy. Further data on nutrient distributions in the eddies has been collected (Gardner in prep.).

Tranter, Parker and Cresswell (1980) investigated the sequence of events involved in primary productivity in the rings. Initially the source for these warm rings is low in surface nutrients and is generally unproductive. A recently formed ring studied in September (locally early spring) had a lower productivity inside than outside. By November the situation had reversed. Inside the ring the mixed layer had shoaled from 215–320 m in September to 60–65 m in November as a result of the summer capping, while outside the ring it deepened on average from 37–88 m to 60 m. In September nitrate concentrations were slightly higher inside than out, 2.9–3.2 compared to 1.1–3.3 μmol Nl^{-1}. As discussed in Section 22.2 these elevated nutrient levels inside the ring were produced by surface cooling of the warm core ring causing deeper convective mixing and so stirring nutrient-rich water up into the surface layers. Presumably the greater vertical instability inside the ring caused a reduction in the productivity by mixing phytoplankton down below the photic zone. However, once the summer capping created vertical stability conditions a bloom occurred and productivity in the ring rose. Nitrate concentrations fell to 0.3–0.8 μmol Nl^{-1} compared with 0.6 μmol Nl^{-1} outside.

One interesting contrast with the Gulf Stream rings is that Tranter, Parker and Vaudrey (1979) found on the southern edge of one eddy that there was a marked fourfold increase in the in vivo surface chlorophyll fluorescence, where the phytoplankton was dominated by diatoms. Inside the warm core rings average cell size appeared to be smaller, which is consistent with the observations of Ortner et al., (1979) for cold core rings. Jeffrey and Hallegraeff (1980) investigated the phytoplankton populations within eddy F during December 1978, 3 months after its separation from the EAC. They found that surface chlorophyll levels were similar at all positions sampled but that higher concentrations occurred in the subsurface maximum in the centre of the eddy as compared to the eddy edge or half-way positions. Their two suggested mechanisms for this increase were that either convective overturn occurred inside the eddy, or that there was some concentrating mechanism resulting from "hydrodynamic forces and the centre of the rotating eddy." The importance of the nanoplankton chlorophyll fraction declined from 72%–76% at the eddy edge to 45%–59% at the centre. This was combined with a change in the relative abundance of the two codominant diatoms, *Nitzschia seriata* taking over as the dominant species in the ring centre from *Rhizosolenia alata* which dominated the edge samples. An alternative explanation to the eddy core enhancement observed by Jeffrey and Hallgraeff is that they witnessed the time when a "front" of organisms (see Sect. 22.4) had arrived at the core centre. Estimates of the distance of their central stations from the edge of the eddy suggest the organism "front" would need to move in at a rate of 0.6–0.8 cm s^{-1}, which is sufficiently close to the estimates given in Section 22.2.4 not to rule it out as a possibility, as it would merely require phytoplankton population division times of 1–2 days. The contrast between the observation of Jeffrey and Hallegraeff (1980) and Tranter et al. (1979) again illustrates that the time of formation of the eddy and the season are important factors in the sequence of biological development within eddies.

The unfavourable conditions for herbivores observed in the cold-core rings may also occur in warm-core rings prior to the summer capping, with the lower phytoplankton standing crop combined with smaller average cell size making it much harder for herbivores to survive. Thus when conditions improve for the phytoplankton following the summer capping of the warm core, the herbivorous grazing plankton will be lower in the ring than outside. However, this would lower the regeneration rate of nutrients by zooplankton excretion, and so the ring might subsequently revert to being of poorer productivity than the surrounding water as seen by Scott (1978). This depression in herbivore abundance may also influence the abundance of nekton. Brandt, Parker and Vaudrey (1981) gave an initial report on the biological characteristics of another eddy (J) in September–October 1979. The phytoplankton was relatively sparse within the eddy and increased rapidly at the edges. Maximum levels occurred to the northeast and to the south of the eddy. Micronekton abundance showed a similar pattern, with low levels in the centres of the cores and maximum concentrations occurring just outside the eddy. The average number of myctophid species taken was low inside the eddy compared with the number caught outside the eddy.

Brandt (1981) has just recently published the results for midwater fishes of the trawling conducted in "eddy F" in December 1978. 109 species were identified out of the total collection of 14,602 specimens, but only 46 species contributed > 1% of the catch (i.e., > 15 specimens) and 39 of these were myctophids. He observed highest species numbers at 250 m at the eddy edge and lowest counts at 50 m both outside the eddy and at the edge. There was no biomass increase in the edge region. Five species were caught primarily in the eddy and another eight occurred at greater depths inside. A number of species including both migrants and non-migrants occurred in equal abundance at all positions sampled. One myctophid, *Scopelopsis multipunctatus*, occurred equally abundantly both inside and outside the eddy, but large specimens were restricted to the waters outside the eddy. Stomach content analysis showed that these large specimens fed predominantly on the salp *Thalia democratica* which only occurred outside the eddy. Brandt emphasised the stirring up of normal zoogeographic distribution patterns caused by the advection of warm water species into high latitudes. However, there was also a marked difference in the pattern of dominance. Outside the eddy 60% of the specimens belonged to one species. At the edge two species contributed 20½% and 21½% of the specimens while inside the eddy five species contributed 18%, 17%, 12½%, 12½%, and 11½% respectively.

22.4.4 Incidental Observations

A number of rather assorted observations have been made which relate to this problem. The eddies reported as a more or less regular feature of the surface circulation off the west coast of Australia (Cresswell, Golding and Boland 1979) are probably relevant to the recruitment of commercial rock lobsters *Panulurus longipes cygnus* (Williamson, 1967). Phillips, Rimmer and Reid (1978) have shown that between 20°–32°S settlement in early spring of puerulus larvae of the lobsters is high to the north, maximum in the centre and very low to the south. The larvae spend 9–11 months in the plankton and there is great year-to-year variability in recruitment. Recruitment is probably greatly effected by the stability of the eddy which will keep the larvae close to the Australian coast. Satellite-tracked buoys that spin out of the eddy, have been followed across the Indian Ocean, to close to the North tip of Madagascar and up the East African coast. The same species of rock lobster occurs on these coasts and the stock could well originate from Western Australia. So eddies of the quasi-stable character may well be important in determining the zoogeographic distribution of species, particularly those with long lived planktonic larvae.

Fornshell (1979) used a continuous plankton recorder to examine the patch size of 11 species of microplankton in the region of the Grand Banks of Newfoundland. He observed patches with length scales of the order of 10 km. One tow was parallel to the axis of the Labrador Current and so could have reflected meanders in the current. The value of the paper is more in indicating a technique for studying mesoscale features than in the results themselves.

Angel (1977) studied the vertical distribution of plankton and micronekton in the vicinity of 44°N 13°W in the northeastern Atlantic during early April 1974. The analysis of the samples revealed large and unexpected disparities between the day-time and night-time samples collected between 300–500 m. The logistics of the sampling programme had resulted in samples being collected from over a relatively large area so that these day and night samples were geographically separated by 20–40 km. There were no similar disparities either between samples taken at other depths in the same series of samples, or in the profiles from eight other stations, therefore, it seemed reasonable to look for alternative causes for the variability other than glibly attributing it to sampling error. Several further series of samples were collected in the area within the following weeks, and from these it became clear that the daytime community sampled initially were a week or two more advanced in the seasonal succession than the night-time community sampled. Thus there is evidence that the onset of the Spring bloom had not been synchronous over the whole area but had occurred patchily, possibly as a consequence of mesoscale activity in the area (e.g., Madelain and Kerut 1978).

Observations on the variability of zooplankton at 1000 m near 42°N 17°W specifically designed to look for the existence of vertical migration by species at such depths (Angel, Hargreaves, Kirkpatrick and Domanski in press) showed that for many species the standard deviation was about 25% of the mean (n = 24), an unusually low variability for zooplankton observations. The samples were all collected within an area 50×30 km. Whereas the variability of the various species was low, a geographical pattern was distinguishable which was consistent for a number of species. Rank correlations between hauls for the eleven species of medusae revealed the same geographical pattern. Thus there is some preliminary evidence of the occurrence of zooplankton distribution patterns with length scales of order of tens of kilometres at depths of 1000 m.

During two recent Discovery Cruises (120 and 121 May-June 1981) an attempt was made to investigate the changes in the biological communities between the Western Altantic water mass (WAW) typified by the broad layer of 18°C water and the Eastern Altantic Water (EAW) at the front which lies to the south-west of the Azores in the vicinity of the CRUISER, PLATO and ATLANTIS Seamounts (Gould 1981). This front probably demarcates an edge of one of the return branches of the Gulf Stream which flows into the Eastern Atlantic. It meanders and evidence was found of a meander pinching off and forming a cold core eddy which moved westwards into the Western Atlantic

Water. Using a fluorometer mounted in a batfish, chlorophyll fluorescence profiles were obtained which showed that the phytoplankton standing crop was higher in the front than in EAW which in turn was higher than in WAW. Day and night profiles of plankton and micronekton down to 1 200 m (Fig. 7)

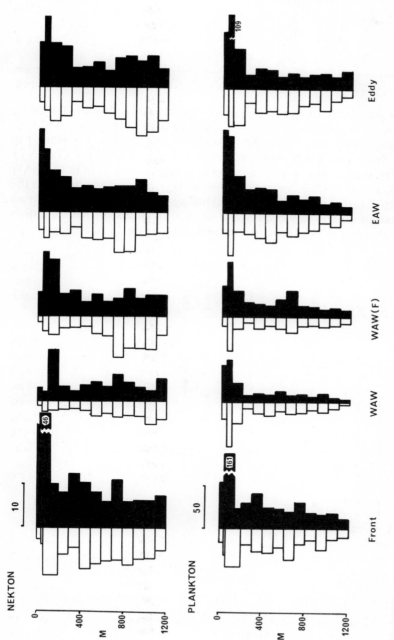

Fig. 7. Day and night profiles of standing crop of plankton and micronekton in the top 1 200 m at five *Discovery* stations from the central North Atlantic to the south-west of the Azores, showing the effect on the populations of the front between the Western Atlantic Water (WAW) and Eastern Atlantic Water (EAW), and a newly formed eddy. Standing crop is expressed as the displacement volume of samples per 1 000 m³ of water filtered

Table 2. Biomass (Displacement volume m^{-2}) in the top 1200 m of the water column at five stations to the south-west of the Azores

Station No.	10380	10378	10376	10379	10382
Water type	WAW	WAW near front	Front	EAW	Newly formed eddy
Nekton	$45\frac{1}{2}$	61	120	$75\frac{1}{2}$	72
Plankton	141	199	305	291	255
Ratio nekton:plankton	1:3.1	1:3.3	1:2.5	1:3.9	1:3.5

at five stations across the front and in a newly formed eddy showed changes in the distribution of biomass and in diel migration activity. The integrated estimates of standing crop suggest that the abundance of plankton was similar in the eddy, the front and EAW, but was much lower in the WAW (Table 2). However, the standing crop of nekton in the front was higher than in the EAW and eddy and at least twice as great as in the WAW. The marked change in the plankton: nekton ratio suggests that the nekton aggregated in the front. In an attempt to examine the possible influence of modifications of diel migration patterns in the vicinity of the front, the vertical shifts of the 25%, 50%, and 75%

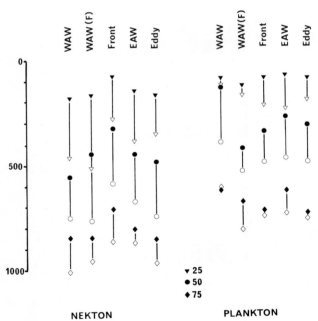

Fig. 8. The day and night changes in the 25%, 50% and 75% population levels of plankton and micronekton at the five *Discovery* stations from the central North Atlantic: 10380 Western Atlantic Water (WAW), 10378 Western Atlantic Water near front [WAW(F)], 10376 Front, 10379 Eastern Atlantic Water (EAW), and 10382 Eddy

levels of the population were examined. These data (Fig. 8) are distorted by two factors (a) the incidence of diel migration by nektonic species such as the myctophid *Ceratoscopelus* from daytime depths > 1 200 m into the region of the pycnocline at night; and (b) daytime avoidance of the nets resulting in day catches being on average 25% lower than night catches. The pattern of plankton migration changed little in the move from EAW to the front including the eddy station. In WAW near the front mean vertical movement was reduced, but deep into WAW there was maximum movement of the 50% level, and an apparent slight reverse movement of the 75%. The nekton data show large changes in the region of the front, and the levels of the quartiles tend to be shallower in the EAW than in the WAW. The difference between the EAW values and the eddy do not show any clear pattern. However, it is clear from these data that diel migration patterns of nekton do alter in the vicinity of fronts and that this can result in a modification of the horizontal pattern of nekton abundance. The increase in nektonic predation in the front may have masked similar changes in the planktonic populations.

22.5 Sampling Procedures

One of the major problems of biological sampling is matching the time/space scale of the phenomena to be studied with the time/space characteristics of the sampling programme. It is clear from the theoretical considerations above that different elements of the biological communities have very different response times to variations in the environment. Mesoscale features are far more likely to have a measurable signature in phytoplankton, bacterial and microplankton communities, than macroplankton communities with their longer population growth times. Nekton, with population growth times of months or years, may have a mesoscale signature resulting from behavioural responses, but not from population growth.

It would seem to us that the region fundamental to the understanding of many of the biological processes associated with eddies is the ring or eddy boundary. At the boundary the chemicophysical gradients change most rapidly horizontally. So these boundaries offer the best conditions to study the responses of organisms to changes in their abiotic environment. It is across the boundary that the flux of organisms, both passive and active, occurs which is so important in determining the succession of the communities within the feature.

The key approach would seem to be to make successive transects across the boundary. The required frequency of sampling will be determined by time scales of the events to be studied. It is clearly essential to have as comprehensive knowledge of the physical processes as possible, and one of the major problems for the biologist is the incompatibility of biological sampling programmes with many of the demands of the basic requirements for understanding the physics. Cruises to carry out mesoscale studies will have to be multidisciplinary if not multivessel.

For rapid surveys of phytoplankton standing crops an in situ fluorometer mounted in an undulating batfish fitted with a CTD, (Herman and Denman 1976) offers the best technique for making the rapid surveys needed. Even with an underway surveying technique it is not possible to traverse across a 200 km feature and back again without the comparison between the start and finish observations being blurred by the speed of temporal succession. Addition of an oxygen sensor to the batfish will give data useful in providing an insight into mixing processes (e.g., Lambert, 1974). Vertical pump profiles can then supplement the batfish transects to provide nutrient profiles and samples for floristic and particle spectrum analysis.

Multiple-serial sampling is the only effective technique for studying zooplankton. The biological situations are likely to be too complex for simple techniques of sampling using vertically or obliquely hauled nets. The spatial dimensions of the phenomena being studied will determine the type of sampler used. For spatial patterns with dimensions of about 10^2–10^3 m, the Longhurst Hardy Plankton Recorder (Longhurst, Reith, Bower and Siebert 1966) is the only technique in which a sufficiently large volume of water is filtered for each subsample to allow adequate analysis of the most abundant species, especially if the modifications recommended by Haury, Wiebe and Boyd, (1976) are employed. Analysis of organisms with mean abundances of less than one per subsample is very likely to give misleading results. Furthermore series of less than 50 subsamples are unlikely to be adequate for statistical analysis except in exceptional circumstances.

For spatial patterns $> 10^3$ m, several multiple serial samplers have been devised e.g. MOCNESS (Wiebe, Boyd, Burt and Morton 1976), RMT $1+8$M (Roe and Shale 1979), and the Canadian system (Sameoto, Jaroszynski and Fraser 1977). With in situ sensors individual environmental parameters such as temperature, depth and light can be monitored in real time and kept constant during the tow. If a conductivity sensor is used in combination with a temperature sensor, or if the T–S relationship is regular, density surfaces can be followed so that the effects of internal waves on the distribution of the zooplankton can be removed. In this way the zooplankton may provide useful clues as to how the physical mixing processes occur.

For species in the middle range of abundance, i.e., occurring in numbers of 5–100/1 000 m^3, at least 2 000 m^3 of water need to be filtered to obtain mean abundance estimates with low enough variability to be interpretable. For less dense organisms considerably greater volumes need to be filtered. Community analysis is more likely to give useful information than studies of individual species or possibly even of single taxonomic groups. For community analysis, too much information is likely to be lost if only the few very abundant species are examined, but logistic limitations and redundancy of effort rule out total community analysis involving all the rarer species.

The study of nektonic species in mesoscale eddies is more of a problem. The only multiple serial net system developed for micronketon are the RMT $1+8$M, the large version of the MOCNESS, and the cod-end system developed by Pearcy, Krygier, Mesecar and Ramsey (1977). In an hour's tow the RMT 8 filters 25–30,000 m^3 of water; while this is adequate for the smaller micronektonic species like euphausiids, decapods and mysids, the results of repeated tows (Angel unpubl.) suggest that adequate quantitative sampling of fish needs about an order of magnitude more water to be filtered. Rather surprisingly, no attempts appear to have been made to date to examine acoustic scattering in relation to mesoscale features, which might be a useful initial indication of the behaviour of fish relative to them.

The descriptive basis is still far too incomplete to be able to make firm recommendations as to the best sampling strategy to adopt when studying eddies and which statistical techniques are likely to prove to be the most useful. Without knowing information such as which physical parameters are of major importance in determining the vertical distribution of species, and whether vertical migratory activity is more or less extensive in warm or cold eddies than in the surrounding water, sample programme design is a matter of intuition, although it is intuition based on a steadily improving background of physical theory.

22.6 Conclusions

1. The biological effects of eddies will be determined by the response time of the organisms, the starting conditions during eddy development, and the degree to which eddies and their component parts are advective or propagative.

2. Theoretical considerations suggest that phytoplankton and possibly microplankton have population response times that will enable them to maintain themselves within mesoscale eddies by their net population growth rates.

3. Zooplankton and nekton population response times are probably too long to enable the populations to be maintained within eddies against diffusion.

4. Diurnal vertical migration is observed to vary within a single species inside and outside ring structures and at fronts. It could be a major factor in determining whether or not a population is maintained or aggregated within an eddy.

5. Feeding conditions may change significantly across ring boundaries for herbivorous plankton not only as a result of changes in primary productivity, but also via changes in the phytoplankton population structure and cell size, the evenness of the partitioning of the standing crop between species, and the thickness of the mixed layer either concentrating or diluting the densities of the cells.

6. Biological sampling of eddies needs to be accompanied by a thorough description of their physical and chemical structure, if there is to be any hope of establishing any sort of understanding of the processes involved.

7. Continuous profiling across eddies with in situ fluorometers, physical sensors (e.g., batfish), and multiple serial samplers (e.g., LHPR, CPR, MOCNESS or RMT 1+8M), and even scattering layer observations are still needed to de-

scribe the biological characteristics of ring structures and eddies before the many uncertainties alluded to in this chapter can even start to be resolved.

8. Long transects of samples within areas rich in MODE-type eddies are needed before sensible sampling programmes can be designed to explore the influence of typical oceanic eddies, as compared with frontal eddies, on the oceanic biological processes.

23. Eddies and Acoustics

R.C. Spindel and Y.J.-F. Desaubies

23.1 Introduction

The ocean is a fluid waveguide through which sound propagates according to the scalar wave equation

$$D^2 p = c^2 \nabla^2 p \tag{1}$$

In this form p is the acoustic pressure perturbation, c is the speed of sound, and the operator $D \equiv \frac{\partial}{\partial t} + \underline{u} \cdot \nabla$ represents the material derivative in the presence of a large-scale (eddy) velocity field $\underline{u}$. In general c and $\underline{u}$ are functions of three spatial coordinates and time which determine, with appropriate boundary conditions, the propagating acoustic field. The ocean surface acts nearly as a reflector scattering almost all of the sound energy in space and dispersing it in frequency. The ocean bottom has highly variable acoustic properties that can span the range from an almost perfect reflector to an almost perfect attenuator. Sea floor roughness and volume variability of c and $\underline{u}$ can also cause significant scattering. Although the sound speed usually varies only weakly with range, in the case of propagation through ocean eddies it is the eddy associated horizontal sound speed variations that are most significant in altering the pattern of acoustic energy distribution. In the case of warm or cold core rings horizontal variations in c can be quite large.

The range to which sound can propagate efficiently is primarily a function of acoustic frequency, f. The attenuation of sound in seawater is roughly proportional to f^2. Practical constraints limit achievable ranges to hundreds of meters for $50\,\text{kHz} < f < 500\,\text{kHz}$; kilometers for $5\,\text{kHz} < f < 50\,\text{kHz}$; tens of kilometers for $1\,\text{kHz} < f < 5\,\text{kHz}$ and hundreds of kilometers for $f < 1\,\text{kHz}$. Eddy spatial scales are such that we are most concerned with rather long range transmission and therefore relatively low acoustic frequency, $1\,\text{Hz} < f < 1\,\text{kHz}$. Acoustic wavelengths at these frequencies range from 1500 m to 1.5 m respectively, much less than the horizontal extent of a mesoscale eddy.

The relationship between temperature, pressure, salinity and sound speed is expressed empirically. The classic relationship was given by Wilson (1960) with more recent versions supplied by Frye and Pugh (1971), Del Grosso (1974) and Chen and Miller (1977). Simplified to first order terms the Del Grosso formula is

$$c = 1402.392 + 5.011\,T + 0.156\,P + 1.329\,S \tag{2}$$

Eddies in Marine Science
(ed. by A.R. Robinson)
© Springer-Verlag Berlin Heidelberg 1983

where c is in m s^{-1}, T in °C, P in kg cm^{-2} and S in parts per thousand. Sound speed depends strongly on T and P and only weakly on S since salinity variations are small.

At mid-latitudes $c = c(z)$ typically has a minimum value about 2% to 3% less than its value at the surface at a depth of about 1 000 m (see Fig. 1). This is due to the opposing effects of temperature, which dominates near the surface and causes a rapid decrease in $c(z)$, and pressure whose increase eventually results in an increasing $c(z)$. Sound is trapped or channeled by zones of minimum speed since energy is alternatively refracted downward when propagating in the region of negative sound speed gradient above the minimum, and upward when propagating in the region of positive gradient below it. Acoustic energy thus cycles through the water column in a sound channel called the SOFAR channel (Sound Fixing and Ranging) by its discoverers (Ewing and Worzel 1948). The wavelength for a complete cycle consisting of one upward and one downward loop is a function of the sound velocity profile and is typically of the order of 50 km; this region of energy focusing is known as a convergence zone.

Since the acoustic wavelength is much smaller than eddy scales one can think of the propagation in terms of rays (geometric acoustics).

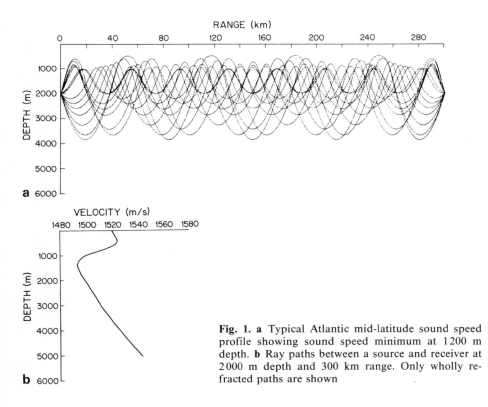

Fig. 1. a Typical Atlantic mid-latitude sound speed profile showing sound speed minimum at 1200 m depth. **b** Ray paths between a source and receiver at 2000 m depth and 300 km range. Only wholly refracted paths are shown

The major effect of SOFAR channel propagation is that sound can propagate over great distances along wholly refracted paths where it suffers only attenuation and geometric spreading losses; scattering and attenuation losses associated with boundary interactions are avoided. The geometric spreading is essentially cylindrical, with energy density decreasing with range R as $1/R$, as opposed to spherical spreading where the decrease is proportional to $1/R^2$. At ranges beyond several hundred kilometers most bottom and surface reflected paths are severely attenuated with respect to refracted paths.

Figure 1b is a ray diagram of typical SOFAR channel propagation. It shows the loci of normals to the propagating wavefronts for wholly refracted paths. A consequence of the SOFAR channel is that long range sound transmissions invariably exhibit multipath propagation characteristics. There are numerous distinct paths connecting a source to a receiver. Acoustic lenses or antennas that would confine energy to within the angular range of a single path are, because of their physical size, difficult to implement. Thus almost every form of long-range transmission is characterized by multipath reception. Since the multiple paths are each somewhat different in length and cycle vertically through portions of the water column where the sound speed is highly variable, the travel time from source-to-receiver differs from path-to-path. The result is severe temporal dispersion; a very rough rule of thumb calls for 1 s of dispersion for each 100 km range.

Superimposed on the mean sound speed profile are variations in $c(x,y,z,t)$ that influence the character of acoustic propagation. Such variations are caused by most forms of oceanic dynamics ranging in scale from microstructure to 1000 km tides. They produce changes in the direction of propagating wavefronts resulting in a redistribution of acoustic energy and an alteration of the multipath arrival pattern. Ocean eddies cause significant variations in $c(x,y,z,t)$ primarily because of their anomalous temperature structure.

23.2 Dynamic Ocean

Any dynamical process that alters the mean ocean state has an effect on sound propagation through perturbations in the index of refraction field (the ratio c_0/c of some reference speed c_0 to the local $c(\underline{x},t)$) and the generation of currents; surface waves scatter incident sound waves. The ocean fluctuates in space and time, with the time scales involved always much longer than any acoustic period. Therefore during a transmission the ocean can be considered frozen and the time variability is important only when repeated transmissions are considered.

In contrast, the spatial inhomogeneities produce significant scattering whose characteristics depend on various parameters: the acoustic wavelength, the length scale and amplitude of the inhomogeneities, the propagation range.

For typical frequencies of 50 to 200 Hz the acoustic wave lengths are 30 to 7.5 m, respectively. At those scales internal waves are the preponderant scattering features. A thorough discussion of the scattering induced by internal waves is given in a monograph edited by Flatté (1979) where it is shown that the fluctuations in acoustic phase and amplitude result in several effects: fluctuations in travel time, intensity (with occasional fades), and arrival angle, loss of spatial and temporal coherence, and spectral broadening of the transmitted signal. Quantitative estimates of these perturbations are given in several distinct parametric regimes; in general the degradation increases with range and frequency.

Similarly, barotropic and baroclinic tides affect the propagation of sound through changes in the refraction, the diffraction being negligible in view of the long length scales involved.

For long range propagation through eddies, internal waves and tides constitute a high frequency noise, some of whose effects can be eliminated from the data through proper filtering and averaging. The remaining noise limits the accuracy with which travel time and arrival angle of individual paths can be measured, and the resolution of the multipaths.

Significant index of refraction perturbations are also associated with such oceanographic features as fronts (Roden 1979) and intrusions, but since no appropriate model exists for them, they can be considered only on a deterministic, case by case, basis. Mesoscale disturbances (eddies, rings) are characterized by large perturbations of the temperature and velocity field which affect sound propagation significantly.

The sound speed in a vertical plane through a cold Gulf Stream Ring is shown on Fig. 2. The eddy acts on the sound rays as a lens on light, a lens with significant aberrations. The depth of the sound channel axis is decreased by about 500 m and the sound velocity gradients are most severely perturbed

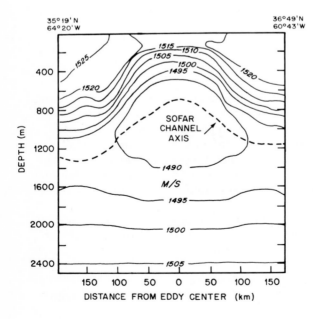

Fig. 2. Sound speed contours through a Gulf Stream ring. (W. Gemmel and E. Khedouri, NAVOCEANO Tech. Note 6 150-21-74, March, 1974, Naval Oceanographic Office)

above 1000 m. Some of the eddy effects to be discussed in detail below can be anticipated qualitatively from this figure. The position of the convergence zone (the "focus length") is modified, and, since the deep cycling rays are more affected than the near axis ones, the multipath arrival pattern is distorted. For given source-receiver configuration some rays might disappear, some new ones become possible, or they might change type, from R-R to RSR for instance, where R-R stands for refracted-refracted, i.e., for a ray that is wholly refracted in the ocean interior and never reaches the surface or the bottom. RSR signifies refracted-surface reflected, i.e., refracted at the lower turning point, reflected at the surface. The vertical and horizontal arrival angles are also perturbed. The question of the possible influence of the eddy currents must also be considered.

23.3 Sound Propagation in a Range-Dependent Environment

Assuming an harmonically dependent pressure disturbance $p \approx e^{i\omega t}$, and neglecting the velocity field $\underset{\sim}{u}$, the elliptic Helmholtz equation is obtained from Eq. (1)

$$\nabla^2 p + k_0^2 \mu^2 (\underset{\sim}{r}) p = 0 \tag{3}$$

where the wavenumber $k_0 = \omega/c_0$ and the index of refraction $\mu = c_0/c$ are defined with respect to some reference velocity c_0. Several method of solution of Eqs. (1) and (3) are available; for a stratified ocean, where $\mu = \mu(z)$ depends only on depth, they are considerably simpler than in the general three dimensional case. Some of these methods are briefly outlined below. More complete accounts and reference can be found in Keller and Papadakis (1977) and De-Santo (1979).

23.3.1 Geometrical Acoustics

The ray approximation can be used when the scale of variation of $\mu(\underset{\sim}{r})$ is much smaller than the acoustic wavelength. The ray equations can be derived either from Eq. (3) by substituting $p = A \exp(i k_0 S)$ and solving, WKB fashion, for the amplitude A and phase S, or from Fermat's principle which states that a ray is the path minimizing the travel time

$$\tau = \int c^{-1} ds = c_0^{-1} \int \mu \, ds \tag{4}$$

(Details are given in Born and Wolf 1975, Witham 1974 and Landau and Lifchitz 1971). The resulting ray equation,

$$\frac{d}{ds} (\mu \underset{\sim}{n}) = \nabla \mu \tag{5}$$

($\underset{\sim}{n}$ is the unit vector tangent to the ray, ds its length element) is quite general.

The conservation of the energy density flux in a ray tube implies that the intensity varies in inverse proportion to its cross section.

In a moving medium, such as the ocean, the ray equations are more complicated (Ugincius, 1972); some effects of currents are discussed below.

For a stratified ocean Eq. (5) reduces to Snell's law by projection on the horizontal axis (x, say): $\mu \cos \theta =$ constant ($\cos \theta$ is the horizontal component of $\underline{n}$). In this case a ray propagating in a sound channel is characterized by its launch angle $\cos \theta_0$; it cycles between two turning depths ($\cos \theta = 1$) and it is periodic in range. This fact is used in numerical ray tracing where the ray must be calculated only over a period (of the order of 50 km).

When the sound velocity profile is range-dependent, Eq. (5) must be integrated numerically over the whole propagation path, a time consuming process. If the departure from a stratified ocean is small (small range variations) Fermat's stationary principle implies that the changes in ray geometry are of second order in the sound velocity perturbation (Hamilton et al. 1980), so that the rays can be calculated as in the stratified case by choosing an appropriate mean profile $\bar{\mu}(z)$. (Milder 1969) exploits the analogy between Fermat's principle

$$\delta \int \mu \, ds = \delta \int \mu [1 + (dz/dx)^2]^{1/2} \, dx = 0 \tag{6}$$

and the principle of least action in classical mechanics $\delta \int L \, dt = 0$ to derive the existence of an adiabatic invariant (Landau and Lifchitz 1968). The Lagrangian function is $L(z, \dot{z}, x) = \mu [1 + (dz/dx)^2]^{1/2} = \mu / \cos \theta$ the range variable x being analogous to the time variable. The Hamiltonian is easily shown to be $H = -\mu (1 + \dot{z}^2)^{1/2} = -\mu \cos \theta$; if μ is not a function of x, Hamilton's equation $dH/dx = \partial H/\partial x$ reduces to Snell's law. In a range dependent environment "energy" is not conserved, but the adiabatic invariant $I = (2\pi)^{-1} \oint p \, dz = (2\pi)^{-1} \int \mu \sin \theta \, dz$ is approximately conserved over a period $\underline{X}$ (Here $p = \partial H/\partial \dot{z} = \mu \sin \theta$ is the momentum conjugate to z). The use of adiabatic invariants requires that the changes in the medium over a ray period be small. In this case the ray parameters (turning depths, period) vary slowly over each period.

Unfortunately it is difficult to estimate the errors involved in using the undisturbed ray geometry or the adiabatic invariants, and there are perturbations in the ocean such that the range variations in sound are large and the resulting errors inadmissible. The full ray Eq. (5) must be solved.

But there are other limitations inherent to the ray approximation itself. In regions where rays converge, i.e., caustics, the amplitude becomes infinite and the approximation breaks down. Geometrical acoustics is inadequate to calculate propagation losses at long range. Ray theory does not account for the leaky modes nor for the energy in the shadow zones.

23.3.2 Normal Modes

If the index of refraction is range independent it is natural to seek solutions of (3) by separation of variables. In a cylindrical coordinate system with a point source at $\rho = 0$, $z = z_0$ the solution is (Graves et al. 1975)

$$p(\rho, z) = -(i/4) \sum_{n=0}^{\infty} H_0^{(1)}(k_n \rho) u_n(z_0) u_n(z) \tag{7}$$

where the vertical modes $u_n(z)$ satisfy

$$\frac{d^2}{dz^2} u_n + K_n^2(z) u_n = 0 , \qquad K_n^2 \equiv k_0^2 \mu^2 - k_n^2 \tag{8}$$

with $\quad u_n = 0$ at the bottom $z = o$
$\qquad \partial_z(u_n) = 0$ at the surface $z = h$

Equation (8) constitutes a classical eigenvalue problem yielding an infinite set of eigenvalues k_n and eigenfuntions u_n. The modes for which k_n is positive imaginary are non-propagating, or evanescent, because then the Hankel function of zero order of the first kind in (7) decays exponentially with range ρ (Keller and Papadakis 1977); there are only a finite number of propagating modes (k_n positive real). (This can be seen immediately in the trivial case $\mu = $ constant.)

When the sound velocity varies horizontally as well as vertically, a plausible method of solution consists in using the modes [Eq. (8)] where now the eigenvalues vary with range: $k_n = k_n(\rho)$,

$$K_n = K_n(z, \rho) = k_0 \mu^2(z, \rho) - k_n^2(\rho), \qquad h = h(\rho)$$

A solution of (3) is then sought in the form

$$p = \sum_{n=0}^{\infty} \varphi_n(\rho) u_n(z, \rho) \tag{9}$$

Substitution into Eq. (3) and use of the normalization $\int u_n u_m \, dz = \delta_{nm}$ gives the governing equations for φ_n

$$[\nabla_h^2 + k_n^2(\rho)] \varphi_n = \delta(\rho) u_n(z_0, 0) - 2 \sum_m A_{nm} \nabla_h \varphi_m - \sum_m \varphi_m B_{nm} \tag{10}$$

with the coupling coefficients

$$A_{nm} = \int_0^{h(\rho)} u_n \nabla_h u_m \, dz , \qquad B_{nm} = \int_0^{h(\rho)} u_n \nabla_h^2 u_m \, dz \tag{11}$$

This set of coupled equations is rather complicated and the technique is useful only in those situations where the horizontal gradients are small enough for the coupling coefficients A_{nm} and B_{nm} to be negligible. A discussion of the conditions for the modes to be invariants is given by Milder (1969): he shows that the effective change in index of refraction μ^2 over a classical ray period must be small compared to the energy level spacing $K_n^2 - K_m^2$. It can be seen from Eq. (11) that if for some mode $K_m \approx K_n$, then $k_n \approx k_m$ and the "forcing terms" on the right hand side induce a near resonance. Milder also points out that the coupling coefficients depend on the vertical structure of the horizontal gradients $\nabla_h^2 \mu$; if they are uniform with depth there is no coupling.

Neglecting the coupling of the modes constitutes the adiabatic approximation. Here the adiabatic invariant is the mode number; each mode propagates
independently of the others, as in the stratified case, with a slowly changing
amplitude $\varphi_n(\rho)$ governed by

$$[\nabla_h^2 + k_n^2(\rho)]\,\varphi_n = \delta(\rho)\,u_n(z_0, 0) \tag{12}$$

Pierce (1965) gives an example of solution of this equation by ray techniques.

23.3.3 The Parabolic Approximation

The parabolic approximation method is widely used in theoretical and numerical studies. A very complete discussion is given by Tappert (in Keller and Papadakis, 1977) and only the basic principle is presented here in the crudest
fashion.

One of the difficulties involved in solving Eq. (3) is that it is elliptic and requires boundary conditions over a closed domain; this difficulty is circumvented in the expansion (7) by imposing a radiation condition. Similarly in the
parabolic method one assumes that as the sound wave propagates through the
medium it is progressively distorted (scattered) but that the backscattered wave
is negligible.

Consider then that the pressure, in Eq. (3), can be represented as a plane
wave propagating in the x direction modified by an envelope ψ:

$$p(\underline{r}) = \psi(\underline{r})\,e^{ik_0x}. \tag{13}$$

Then

$$\frac{\partial^2\psi}{\partial x^2} + 2ik_0\frac{\partial\psi}{\partial x} + \nabla_T^2\psi + k_0^2(1-\mu^2)\,\psi = 0. \tag{14}$$

The parabolic approximation consists in neglecting the term $\partial^2\psi/\delta x^2$ with respect to $2ik_0\,\partial\psi/\partial x$. This requires that the envelope vary slowly over a wavelength: $\partial\psi/\partial x \ll k_0\psi$. In the ocean one also has, the transverse deriva-
tive $\nabla_T^2\psi = \left(\dfrac{\partial^2}{\partial y^2} + \dfrac{\partial^2}{\partial z^2}\right)\psi \approx \dfrac{\partial^2\psi}{\partial z^2} \gg \dfrac{\partial^2\psi}{\partial x^2}$. Equation (14) now becomes

$$2ik_0\frac{\partial\psi}{\partial x} + \frac{\partial^2\psi}{\partial z^2} + k_0^2(1-\mu^2)\,\psi = 0 \tag{15}$$

To consider further the approximations involved, substitute $\psi \approx e^{i\alpha r}$ into the
operator $\left[\dfrac{\partial^2}{\partial x^2} + 2ik_0\dfrac{\partial}{\partial x} + \nabla_T^2\right]\psi = 0$

$$\alpha_1^2 + 2k_0\alpha_1 + \beta^2 = 0, \qquad \beta^2 \equiv \alpha_2^2 + \alpha_3^2$$

$$\alpha_1 = -k_0 \pm (k_0^2 - \beta^2)^{1/2} \approx k_0 \pm k_0(1 - \beta^2/2k_0^2).$$

The root $\alpha_1 = -2k_0 + \beta^2/2k_0$ corresponds to the backscattered wave, while $\alpha_1 = -\beta^2/2k_0$ corresponds to the operator $\left(2ik_0 \dfrac{\partial}{\partial x} + \nabla_T^2\right) \psi$. Thus the conditions for the validity of the parabolic approximation are that the acoustic wavelength be much smaller than the scale of variation of the medium ($\beta \ll k_0$) and that the scattering be essentially at small angles ($\alpha_1/\beta \ll 1$) in the forward direction (no backscattered wave).

The advantages of Eq. (15) over Eq. (14) are obvious: whereas (14) is elliptic and would require boundary conditions on a closed domain, (15) is of the parabolic type and can be marched out in range, given some initial field at x_0, say. In numerical calculations a range dependent environment is easily handled: typically Eq. (15) is Fourier-transformed in depth, marched one step in range and transformed back. Thus the calculations involve two FFTs at each range step and can be time consuming. In particular the number of range steps and of depth points for the FFT are proportional to the acoustic wavenumber (hence to the frequency) and must be increased with increasing sound speed gradients; bottom interactions can also be difficult to model. For these reasons the parabolic equation method is most appropriate for deep ocean, low frequency propagation.

The three methods, ray tracing, normal modes, and the parabolic equation have been translated into computer codes; a complete review of numerical methods in underwater acoustics is given by DiNapoli and Deavenport (in De-Santo 1979) and will not be discussed here. Instead some of the specific applications to sound propagation through eddies will be presented.

23.4 Analytical and Numerical Studies of Sound Propagation Through Eddies

One of the difficulties in calculations of the sound field through an eddy is the dearth of models for the spatial distribution of the index of refraction. Since Gulf Stream rings are such conspicuous features they lend themselves more easily to an analytical representation. Such a model is proposed by Henrick et al. (1977) who derive an approximate solution of the quasigeostrophic potential vorticity equation and obtain an analytical representation of the sound velocity field in terms of Bessel functions, similar to that of Andrews and Scully-Power (1976) and Csanady (1979). In their model the eddy is an isolated feature with cylindrical symmetry, and limited radial and depth extent, its axis drifting slowly along an arbitrary trajectory. The eddy is characterized by a set of adjustable parameters: its radius, maximum current speed, depth of influence, and drift trajectory. However even this idealized model is intractable, in part because of the complicated form of the equation relating sound speed to temperature and pressure, and in one application Henrick et al. (1980) must simplify it further to the point where its dynamical relevance is in doubt. Nonetheless their model has enough adjustable parameters to describe empirically nearly circular isolated rings.

Henrick et al. (1980), study the effects of such an eddy and its currents on short-range propagation (less than 50 km) for a variety of source-receiver positions with respect to the eddy. They use geometrical acoustics to discuss the changes in ray geometry and travel time; they find that currents are more apt than horizontal index of refraction gradients to induce horizontal refraction, but that such effects are insignificant over the short ranges they consider. Current effects are important and can account for up to 20% of the predicted phase variation along a ray. The phase delay related to the sound speed perturbation varies with each ray so that a relative phase speading of the arrivals is predicted, on the other hand current effects are almost-ray independent, so that they result in a uniform shift of the arrival sequence.

Most numerical calculations of the sound field consider only the refraction in a vertical plane. Vastano and Owens (1973) trace rays through a Cold Gulf Stream ring and calculate the transmission loss as a function of range for a source at the ring center. As opposed to propagation without a ring, in the presence of the ring the rays cycle deeper, the first convergence zone is closer to the source with successive zones correspondingly displaced. It follows that at a given range the depth of maximum intensity can be displaced by up to 1 000 m while at a given receiver there can be differences of up to 20 db in intensity between the ring and no-ring cases. Baer (1980) confirms those results by means of a corrected split-step parabolic equation numerical algorithm. The ring considered is based on an analytic model due to Henrick et al. (1977) which reproduces the data of Vastano and Owens. The use of the parabolic equation allows detailed calculation of the acoustic field even at frequencies as low as 25 Hz. At low frequency more energy is absorbed into the bottom in the presence of the eddy. Baer also emphasizes the very strong focusing property of the eddy by calculating the angular energy distribution at the receiver. The ring concentrates a high percentage of the incoming energy at small angles. As the ring moves from the source towards the receiver fluctuations in intensity of up to 30 db are observed.

Similar conclusions are reached by Weinberg and Zabalgogeazcoa (1977) who consider the motion of a cold core ring transverse to the propagation path. By coherently summing the multipaths they calculate intensity fluctuations of about 20 db as the eddy passes by. The amplitude fluctuation is related to phase variations and interference effects, rather than amplitude changes along each ray. They also note that not only does the eddy perturb the pattern of ray arrivals but also can change the nature of some rays [from refracted-refracted (RR) to refracted-surface reflected (R-SR), etc.].

The horizontal refraction of ray paths by mesoscale eddies, neglecting vertical effects, has also been investigated by Munk (1980) and Weinberg and Clark (1980). Munk derives analytical expressions for the refraction angle based on simple expressions for the radial dependence of the sound field in the eddy (in one case he uses the solution of Csanady, 1979), a maximum angle of 4° is found. He shows that rays can be split into horizontal multipaths only for unrealistic eddy parameters, he gives formulae for the fluctuations in intensity and travel time along a ray, and concludes that the signatures of warm and cold core eddies will differ markedly. Weinberg and Clark (1980) fit the sound speed field at 500 m in an observed ring with an arctan function and resort to numerical ray tracing to estimate the horizontal deflection angle over a range of 1 200 km. They find angles close to 2°. The deflection is sensitive to the location of the eddy with respect to the source and receiver, and is largest when the eddy is closer to the latter. However, in a recent study Baer (1981) shows, by solving numerically the three dimensional wave equation in the parabolic approximation, that the horizontal refraction is at most of the order of 0.5°. Thus models that consider ray propagation in a single horizontal plane overpredict the refraction angle. Baer also discusses the degradation of the performance of horizontal arrays in the presence of a ring.

From those studies the following acoustic properties of rings emerge: the presence of a ring affects significantly all the characteristics of the acoustic field, in particular, at the receiver, the arrival sequence, the amplitude, and the angular distribution of energy. Some rays can change type, the position of the convergence zones is modified, horizontal refraction of a fraction of a degree is expected. Current effects have generally been neglected, but one numerical study indicates that this could lead to large errors in phase calculations.

23.5 Acoustic Measurements in Eddies

Acoustic measurements are conducted with a wide variety of signal types ranging from the transmission of narrowband tones of long duration to wideband energy bursts generated by underwater explosions. Two kinds of measurement are prevalent. The first uses mobile sources and/or receivers and is usually employed to determine acoustic transmission loss as a function of range. This measurement can be performed quickly (with respect to eddy time scales) and can provide a synoptic picture of acoustic energy distribution in the presence of an eddy. Explosive charges are often detonated at opening range from a fixed receiver in order to plot received signal level versus range for selected frequency bands. Alternatively an electroacoustic source can be towed from a ship. Usually such sources are operated at shallow depths and they tend to be highly resonant and therefore narrowband devices. The other class of measurement is called fixed point in that the positions of both source and receiver are fixed. Acoustic signals are transmitted from source-to-receiver in order to characterize the acoustic properties of a specific path or to detect changes in transmission characteristics due to changes in the path itself. Effects of eddy-acoustic interaction have been suggested by both types of measurement. Indeed, as noted above, analytic models have predicted profound and readily observable effects. However, with few notable exceptions there is a lack of unequivocal evidence linking specific acoustic observations with the presence of a mesoscale eddy. In general this is a result of both the rather long time constant associated with the eddy field and the large spatial extent of a mesoscale feature. Fixed end point acoustic transmissions must be continued for many months in order to capture an eddy effect. Simultaneously, detailed oceanographic observations must be made to measure significant eddy features. Unfortunately, there have not been many experiments in which both requirements have been met. On the other hand, conventional transmission loss experiments have only recently been conducted in known eddy fields.

One such experiment in the vicinity of a strong warm core eddy was conducted jointly by Australia, New Zealand and the United States during 1974–1975, (Scully-Power et al. 1975, Nyson and Scully-Power 1978). A well-defined eddy was located in the Tasman Sea on the East coast of Australia. A variety of acoustic measurements were conducted in and around the eddy and for the most part observations were in keeping with analytic predictions. The warm core eddy caused an elongation of the convergence zone range by about 8 km. In addition, the warm water lens produced a shallow (200 m) sound channel that behaved as a leaky wave guide. For the case of near surface sources the shallow channel traps energy that would otherwise couple directly into the deeper SOFAR channel.

Cold core eddies are expected to have opposite effects. The U.S. Navel Research Laboratory sponsored an extensive series of environmental and acoustic measurements in the region of an Atlantic cold core eddy during 1979 (Mosley et al. 1979). The results of acoustic tests have not yet been published, but it is not unreasonable to expect that they will be in keeping with analytic predictions.

An interesting series of observations is provided by Beckerle et al. (1980). The signal strength of acoustic signals emitted by two selected neutrally buoyant SOFAR floats was monitored for about one year. These floats are designed to drift with subsurface currents. Their acoustic signals are intended for position tracking through triangulation. A large intersity change was noted at the time an eddy intersected each acoustic transmission path. Other intensity changes could be attributed to changes in the convergence zone range.

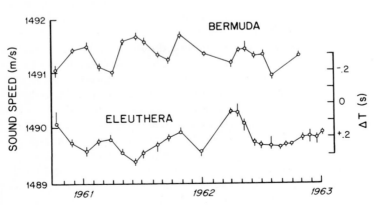

Fig. 3. Variations in mean axial sound speed and equivalent travel time change. (Hamilton 1974). The source is at Antigua

One of the earliest records of possible eddy induced acoustic variations is supplied by a fixed point experiment conducted during 1961–1964 (Hamilton 1974). The experiment was designed to measure the travel time stability of acoustic signals propagating through the SOFAR channel. Explosives were regularly detonated at a precise position near Antigua and received at known locations scattered about the Atlantic. Some pertinent results are summarized in Fig. 3 which shows the measured variation in average sound speed along a path to Bermuda and a path to Eleuthera. Periodic variations in sound speed, or equivalently in arrival time, are clearly evident. Through careful measurement of sound speed at the end points it was concluded that most of the observed variation was due to changes in the intervening water mass. A large eddy (400 km diameter) moving slowly (2 km d^{-1}) would take about 6 months to transit the Antigua to Eleuthera transmission path. A typical eddy might have a sound velocity anomaly at 1 000 m depth of about 5 m s^{-1}. Cycling rays spend about as much time above 1 000 m as below it, but are most affected at shallow depths. Thus a SOFAR ray comes under the influence of such an eddy for about 200 km or roughly 10% of its total path length. This would result in a sound speed change of about 10% of the 5 m s^{-1} anomaly or about 0.5 m s^{-1}. From June to September 1962 a comparable change was observed.

Evidence of eddy-acoustic interaction has also been supplied by simple continuous tone measurements over long paths. For this type of measurement the observed variables are acoustic intensity and phase. Travel time variations, δt, are deduced from phase measurements, $\delta\theta$, through $\delta t = \delta\theta/2\pi f$, where f is the acoustic frequency. For a small 200 km eddy with a 5 m s^{-1} sound velocity anomaly the expected travel time variation is roughly 0.2 s. Measurements at 406 Hz along a 1 250 km path showed phase variations of 100 cycles over a one month period which is equivalent to 0.25 s travel time variation (Clark and Kronengold 1974).

Although fixed point continuous tone experiments are relatively simple to instrument, they suffer a number of drawbacks that have led to more complicated experimental configurations (for example, Steinberg and Birdsall 1966, Spindel 1979). The interpretation of phase variation in terms of an equivalent travel time variation can be ambiguous. Long range transmission geometries result in a received signal which is composed of a (usually) large number of multipaths. The composite reception consists of the vector sum of many sinusoids each with independent amplitude and phase statistics. In the limit of an infinite number of paths the resultant signal is stochastic with Rayleigh distributed amplitude and uniform phase statistics. In practice a relatively small number of paths (four or five) is sufficient to approach a uniform phase distribution. As a result phase variations are due primarily to this random walk effect and cannot be interpreted as travel time variations caused by large scale ocean dynamics. An exception to this occurs when one or two paths are dominant. However, even under these circumstances phase measurements can be ambiguous. During amplitude fades (due to destructively interfering multipaths) the phase becomes uniformly distributed since the observation is then of ambient sea noise rather than signal. Upon exiting from a fade the phase record, and therefore the travel time record, is discontinuous. Furthermore,

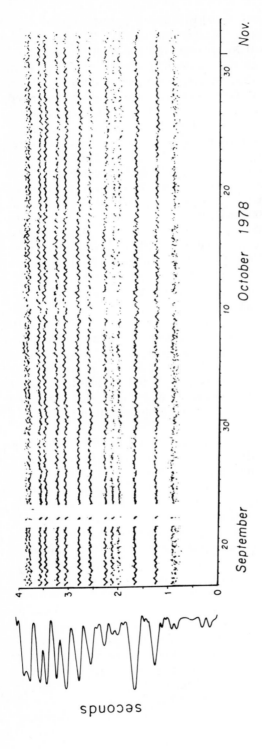

Fig. 4. Multipath arrivals for 48-day period over a 900 km path. This plot shows the last 4 s of total travel time. The *left portion* of the figure shows a single average pulse response. In the remaining portion arrivals whose SNR exceeds 13 db are plotted as a point. A measurement was made every 10 min. (Spiesberger et al. 1980)

phase is measured modulo 2π making it difficult to track travel time over extended time periods.

These considerations have led to experiments that are remarkably similar to the explosive experiment described above. Travel time changes are measured directly through the transmission of pulse-like signals. Electro-acoustic sources are used rather than explosives in order to eliminate uncertainty in travel time determinations due to variations in firing time and position, to mechanize the measurement procedure and to improve the quality of the measurement in other respects. For example, the transmission of coherent signals enables the use of sophisticated signal processing techniques to enhance received signal-to-noise ratios in order to improve travel time precision and resolution.

The possibilities of these techniques for eddy measurements are illustrated by the results of two experiments. Figure 4 shows data from a 2-month fixed point experiment (Spiesberger et al. 1980). A phase-modulated tone was transmitted at ten minute intervals from an acoustic source moored at 2 000 m depth to a bottom mounted acoustic receiver 900 km distant. The transmitted signal

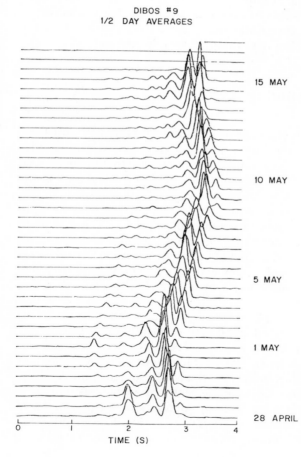

DIBOS #9
1/2 DAY AVERAGES

15 MAY

10 MAY

5 MAY

1 MAY

28 APRIL

TIME (S)

Fig. 5. Multipath arrivals for transmission through a Gulf Stream ring or meander. The last 4 s of total travel time are shown. A measurement was made every 10 min. Each curve is an average (12 h) of 72 measurements

had a sharply peaked autocorrelation function. The received signal was cross-correlated with a replica of the transmitted signal thereby producing the pulse response of the intervening ocean. The cross-correlated signal is equivalent to that which would be received if an actual pulse had been transmitted. Each ten minute transmission produced some fifteen resolvable multipath arrivals spread out over approximately four seconds. An average pulse response is shown on the left of Fig. 4. The remainder of the figure is produced by plotting the peak of each multipath pulse arrival as a point. The 10-min intervals are so close together on this scale that they blend together causing the reception time of a particular multipath to appear as a continuous time series. There is a conspicuous two-cycle-per-day oscillation of about 100 ms due to internal tidal effects. There are also much longer period variations that might be due to eddy influences. A more dramatic example of acoustic travel time perturbations due to ocean effects is provided by a second fixed point experiment in which simultaneous oceanographic observations were taken. A similar signal processing scheme was used and the pulse response for a three week period is shown in Fig. 5. Each plotted curve is the daily average of 72 responses at 10-min intervals. There is a large change in arrival time, amounting to almost 0.6 s, between 1 May and 15 May. Oceanographic measurements showed that it was due to the intrusion of a strong cold core water mass, an eddy-like meander of the Gulf Stream, into the transmission path (Spindel and Spiesberger 1981).

23.6 Acoustic Eddy Monitoring

Both these experiments were preliminary tests in the development of a fixed point system for monitoring the location and strength of mesoscale features over ocean basin areas. It is clear from the above discussion that the eddy field produces acoustic effects and that alteration of the acoustic travel time is one of the most readily observable. A technique known as ocean acoustic tomography seeks to exploit this phenomenon to provide synoptic maps of eddy induced sound speed anomalies and by inference, density (Munk and Wunsch 1979). The procedure consists of measuring travel time variations between multiple fixed sources and receivers scattered about an ocean basin. An image of the interior ocean sound speed field is reconstructed using inverse techniques. In order to resolve 100 km spatial scales in a 1 000 km^2 ocean about five sources and receivers are needed.

A demonstration tomography system was deployed in the Atlantic during 1981. Cooperating laboratories included the Scripps Institution of Oceanography (W. Munk), the Massachusetts Institute of Technology (C. Wunsch), the University of Michigan (T. Birdsall), NOAA Miami (D. Behringer) and the Woods Hole Oceanographic Institution (R. Spindel). Sources and receivers were moored at SOFAR channel depths in the center of a deep basin in order to eliminate the effects of surface and bottom boundary interaction. A set of oceanographic measurements was taken periodically during the 6-month duration of the acoustic experiment. Specific experimental results have not yet been published. This system was designed specifically for the monitoring of eddy scales, but it should be noted that in principle it can be employed on other oceanic scales as well.

Alternative implementations of acoustic tomography are under consideration. For example, a moving source and/or receiver can be employed if positions are well-known and if the time scale of motion is considerably less than that of the eddy field. Continuous tones can be used if large aperture receivers are employed to isolate individual ray paths via arrival angle. The phase then can be converted directly to equivalent travel time. If either source or receiver is in motion the differential Doppler shift of each arriving ray can be used as a distinguishing feature and variations due to eddies can be computed.

23.7 Conclusions

It is clear from the experimental and theoretical evidence available that mesoscale eddies have a very significant effect on all the characteristics of the sound field. There are however very few published detailed studies of the relationship between the various features of the eddy-induced sound speed perturbations and the propagation of acoustic waves.

Experimentally it is difficult to monitor the ocean fluctuations over the required time and length scales while an acoustic transmission experiment is conducted.

On the other hand, analytical methods are of limited use because of the complexity of the representation of the index of refraction field. The governing equations can be solved only in the most idealized situations, leading sometimes to large errors in the predictions (of the horizontal refraction, for instance). Numerical techniques are therefore more appropriate to the calculation of the acoustic field. In many cases the parabolic equation method is particularly suited to such calculations. Numerical solutions have been applied only to the problem of sound propagation through rather strong, albeit not atypical, Gulf Stream rings. For such severe disturbances the corresponding acoustic fluctuations are large. No calculations are available however for more typical, less energetic eddy fields.

Further numerical and experimental work is needed to determine the effects of the flow velocity associated with the eddies and to assess the validity as the eddy strength increases, of various small perturbation approximations (i.e., that the ray geometry is unaffected by the eddy, adiabatic invariants and uncoupling of the modes, propagation in a single vertical plane).

24. Instruments and Methods

R.H. Heinmiller

24.1 Introduction

The instruments and techniques for investigation of mesoscale phenomena in the ocean have undergone drastic changes since the early 1960's. The double impact of new electronics and greatly improved mooring techniques came at the same time as the recognition of the mesoscale as an major factor in the dynamics of the ocean. Baker (1981) has presented a survey of the historical development of instrumentation and methods in the broader context of physical oceanography in general, with an overview of the state of the art in the field.

As in other areas of physical oceanography, technology has been a major factor in the advance of mesoscale ocean science. The Mid-Ocean Dynamics Experiment (MODE) marked the coming-of-age of many new techniques and instruments (MODE Group 1978). One hopes that the advent of microprocessor technology will accelerate our capabilities in ocean engineering in the future as it already done in so many other areas of technology.

This is a rapidly evolving area, and the reader will find that the literature lags somewhat behind the actual state of the art. In many cases, instruments described here have undergone extensive modifications by individual users, and specifications and capabilities may vary depending on when the instruments were produced and who is presently using them. Therefore the scientist or engineer who is interested in becoming familiar with the field must, in many cases, go directly to manufacturers and users for up-to-date information.

Another problem for the interested scientist is that much of the instrumentation used is not commercially available "off-the-shelf". In some cases, only two or three instruments of a given type exist, all built to order. Much of the instrumentation and equipment requires extensive shore-based facilities for proper use, such as calibration tanks, special tape-reading hardware and software, and maintenance shops.

We can divide the techniques and instruments used in mesoscale measurements into four general categories: moored, drifting, profiling, and remote sensing. It is impossible to present here detailed specifications for each of the many instruments available for mesoscale measurements, or even to discuss all instruments in a given area. We will, however, survey the primary instruments used in each category, with an attempt to direct the reader to sources of more detailed information.

Eddies in Marine Science
(ed. by A.R. Robinson)
© Springer-Verlag Berlin Heidelberg 1983

24.2 Moored Instumentation

The mooring and the instruments that are mounted on it must be considered, in any general analysis, as a system. The type of instrument used for a given measurement may depend upon the type of mooring used. The quality of the measurements is dependent on both the instrument and the mooring type. More obviously, the depth at which the measurements are to be made will determine the type of mooring used.

24.2.1 Moorings

Moorings may be considered to be divided into four categories: surface, sub-surface, bottom, and multi-point (Heinmiller and Walden 1973, Berteaux 1975).

24.2.1.a Surface Moorings. Surface moorings are those in which the primary buoyancy is on the surface (Fig. 1). The interaction between the buoyancy package and the surface forcing functions creates several problems. The buoyancy is exposed to the extreme environmental conditions imposed by waves, high surface currents, and wind. In addition, assuming that water temperature is highest at the surface, the corrosion rate will be highest there (Morey 1973).

The engineering problems involved in the survival of surface moorings have largely been solved (Walden and Panicken 1973, Heinmiller 1976a, Berteaux and Walden 1969), although this generally means that a surface mooring requires heavier components than a sub-surface mooring.

A more subtle problem is the introduction of noise into current meter measurements made on surface moorings. The horizontal and vertical motion imparted to the surface float by wave motion will be transmitted to the mooring wire and then to the current meters mounted on that wire. Since the meter measures motion of the water relative to itself, and since most current meter sensors are sensitive to vertical motions, this mooring and instrument motion will appear as contamination in the current meter records (Gould and Sambuco 1974, Pollard 1973, Halpern 1978a, Saunders 1976). This problem can be minimized by careful choice of instrument, sampling rate, and data processing techniques (Halpern 1978b).

24.2.1.b Bottom Moorings. Bottom installations are, in general, the simplest of moorings. They may be moorings which extend, for example, less than 100 m off the bottom (Fig. 1), or they may actually be rigid structures sitting on the bottom. In either case, the mooring is likely to be relatively stable, and involve far less gear than surface or sub-surface moorings. Again, however, precise depth may not be known, due to uncertainties in the measurement of the bottom depth. Distance from the bottom, on the other hand, may be known very accurately.

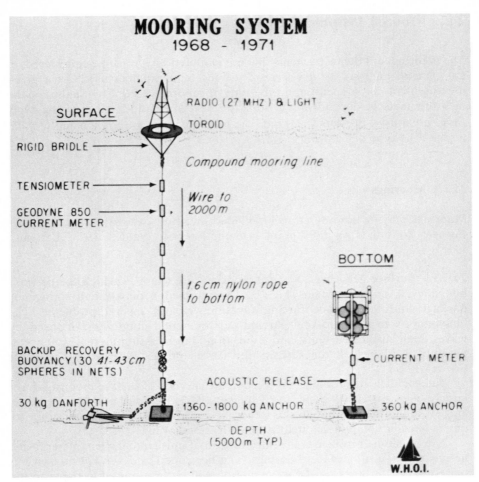

MOORING SYSTEM
1968 - 1971

Fig. 1. Typical Woods Hole Oceanographic Institution Buoy Project surface and bottom moorings. (Courtesy George Tupper)

24.2.1.c Sub-Surface Moorings. Sub-surface moorings are those in which the primary buoyancy is below the surface (Fig. 2). This buoyancy may be in a single package at the top of the mooring, or it may be distributed along the mooring line.

Use of sub-surface buoyancy reduces the destructive forces acting on the mooring. In addition, by reducing or eliminating the action of waves on the mooring, the noise problem is greatly reduced or eliminated, however, a new problem is introduced. Variations in the current profile encountered by the mooring will cause the mooring to "sway" horizontally. Since the mooring is fixed to the bottom and has a fixed length of line, the depths of the instruments attached are variable. Considerable buoyancy may be necessary to withstand

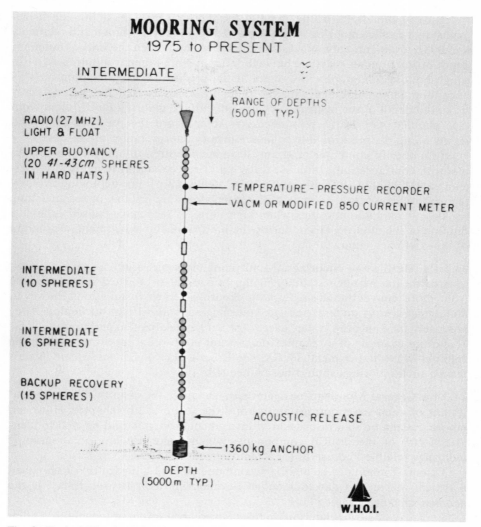

MOORING SYSTEM
1975 to PRESENT

INTERMEDIATE

RANGE OF DEPTHS
(500m TYP.)

RADIO (27 MHZ),
LIGHT & FLOAT

UPPER BUOYANCY
(20 *41-43 cm* SPHERES
IN HARD HATS)

TEMPERATURE - PRESSURE RECORDER

VACM OR MODIFIED 850 CURRENT METER

INTERMEDIATE
(10 SPHERES)

INTERMEDIATE
(6 SPHERES)

BACKUP RECOVERY
(15 SPHERES)

ACOUSTIC RELEASE

1360 kg ANCHOR

DEPTH
(5000m TYP.)

W.H.O.I.

Fig. 2. Typical Woods Hole Oceanographic Institution Buoy Project sub-surface mooring. (Courtesy George Tupper)

the force of strong current profiles and keep instruments at or near the desired depth. This buoyancy may be bulky and expensive. In general, the addition of buoyancy will also increase the drag and the tension on the mooring line. Thus there are limits to the "stiffness" that can be achieved with a sub-surface mooring.

It should be noted that, for a sub-surface mooring with its buoyancy close to the surface, wave-induced motions (and therefore current meter errors) are not completely eliminated. The problem is similar to that of surface moorings, although much less (Halpern and Pillsbury 1976).

Ignoring the effects of the slope of the mooring line, the depths of instruments on a surface mooring may be considered to be well-known and relatively fixed. On a sub-surface mooring, on the other hand, even the initial (nominal) depth of instruments may not be well-defined, since when a mooring is set in rough bottom topography, the depth at the precise mooring location may be uncertain. This, combined with the time variations caused by currents, may make the depth of the instruments on a sub-surface mooring far different from that planned. Generally, it is necessary to measure the pressure at several points along the mooring line to insure proper interpretation of the data.

A number of computer programs have been written to analyze sub-surface mooring configurations, both statically and (less common) dynamically (Berteaux and Chhabra 1973, Albertson 1974, Moller 1976). For deploying sub-surface moorings when the depths of the instruments are critical, or for analyzing the data from such moorings when high currents may have caused excessive dipping of the mooring string during the lifetime of the experiment, a program of this sort is essential.

24.2.1.d Multi-Point Moorings. Multi-point moorings are generally intended to provide the extreme stability in the horizontal or vertical not obtainable from more conventional single-point moorings. Such installations, however, are almost always quite expensive, complicated, and difficult to deploy. They are rarely used in deep water, except for very specialized experiments.

For an example of a complex deep water multi-point mooring, used in the Internal Wave Experiment, IWEX, see Briscoe (1975). For mesoscale experiments, such moorings should not be needed.

24.2.1.e General Mooring Considerations. Detailed discussions of the configurations of deep-sea moorings are beyond the scope of this chapter. However, before passing on to a discussion of instrumentation, it would be well to mention a few of the critical components which make modern oceanographic moorings reliable and useful for scientific measurements.

Foremost among these auxiliary components, and a device that has played a critical role in the development of the mooring as it is used today, is the acoustic anchor release.

Recovery of sub-surface and bottom moorings is impossible without a reliable method of releasing the mooring proper from the anchor when desired. Surface moorings, in theory, can be recovered complete with their anchors still attached. In practice, however, lifting an anchor to the surface on a mooring line which has already withstood many months of exposure to the elements is usually impossible.

Acoustic releases are now widely used for this purpose. The ability to release the anchor when desired, and not at a predetermined time (as in the case of timed releases) is crucial, considering the vagaries of marine weather and ship schedules arranged months beforehand but always subject to change. As in the case of instrumentation, devices made by a few manufacturers dominate the field (Heinmiller 1968, Heinmiller and LaRochelle 1977).

Other auxiliary components which play a role include radios, lights, and acoustic transponders which allow the gear to be located and recovered after the anchor has been released and the mooring has surfaced (Heinmiller and Walden 1973).

Deployment and recovery operations for deep-sea moorings may require considerable operational capability in terms of shore facilities, personnel, and auxiliary equipment (Heinmiller 1976b). This can be a considerable obstacle to the scientist who only wants to deploy a few moorings for a single experiment.

24.2.2 Fixed Depth Moored Current Meters

A straightforward approach to the problem of measuring currents in the ocean might suggest that one can produce an instrument that measures the flow past itself (perhaps adapted from instruments used in industrial flow technology for many years), hang it on a fixed mooring, and measure the ocean current at a fixed geographical point at a single depth. Like the mooring problem itself, experience has shown that the problem is not so simple (Woodward et al. 1978).

Historically, attempts have been made with almost every type of sensor imaginable, in a wide variety of configurations and packages, to measure currents on fixed moorings. The degree of success, first in reliability, and second in quality of the data, has been as varied as the instruments themselves (McCullough 1974).

Presently, only a few current meters are in widespread use on moorings. While the small number of such instrument types in use is partly the result of a desire not to "re-invent the wheel", it is also, given the tendency of scientists and engineers to regard their own measurement needs as unique, partly a reflection of the difficulty of the engineering problem.

24.2.2.a Rotor-Vane Current Meters. The long and honorable history of propellors, rotors, and paddlewheels in both air and water for purposes of measurement and power generation led naturally to many attempts to use such devices in recording instruments in the ocean. The severe environmental conditions, combined with lack of access for maintenance, have reduced the number of such sensors in serious use to a few hardy survivors.

We will discuss briefly current meters which utilize rotating sensors for speed sensing. Since the necessity for direction measurements usually leads to the inclusion of a vane of some sort, these are commonly referred to as "rotor-vane" or "propellor-vane" current meters.

The Vector Averaging Current Meter (VACM) is one of two instruments which dominate the field (the other being the Aanderaa). This instrument was developed at Woods Hole Oceanographic Institution (McCullough 1975) and is now available commerically from EG&G Sea-Link (Fig. 3).

The VACM is a cylindrical package which is designed to be inserted in the mooring line. It senses speed with a Savonius rotor and direction relative to the case with a vane assembly. An internal compass senses the orientation of the case relative to magnetic north. The instrument also records temperature.

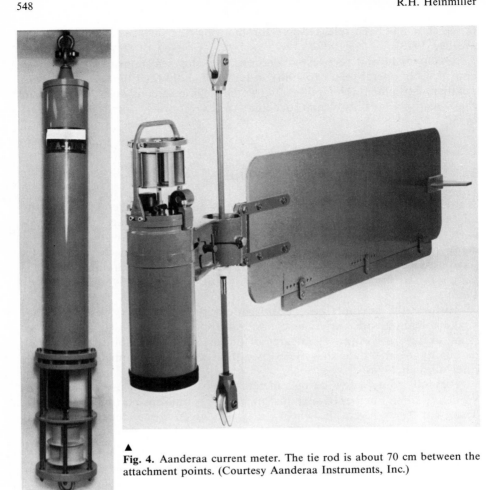

▲
Fig. 4. Aanderaa current meter. The tie rod is about 70 cm between the attachment points. (Courtesy Aanderaa Instruments, Inc.)

▲
Fig. 3. Vector averaging current meter (VACM). The instrument is approximately 2 m long. (Courtesy EG&G Sea-Link Systems)

The COSMOS electronics in the VACM samples and processes the sensor input, allowing a somewhat flexible sampling scheme. The data is recorded as a true vector average. Sampling rate for speed and direction is partly determined by the rotor rotation rate.

The Savonius rotor was originally developed for power generation (Savonius 1931). However, its ruggedness, omni-directionality, and linearity in steady flow have made it, in its present form, an attractive sensor. The Savonius rotor does, however, have some problems with regard to its response to time-varying flow (Fofonoff and Ercan 1967, Kalvaitis 1972). This, plus the mechanical configuration of the sensor mountings, makes the VACM measurements susceptible to contamination by vertical flows, so, like most other current meters, it is not suitable for measurements near the surface, where wave action creates substantial vertical components of velocity.

VACM's were first used extensively in the MODE experiment. A major problem developed at that time. An electrical potential set up by the use of different materials caused carbonates to be deposited in the vane and rotor bearings, resulting in degraded performance (Dexter et al. 1975). This problem was later solved through careful use of materials. But it illustrates the pitfalls of placing sensors in the ocean environment.

The Aanderaa meter also uses a Savonius sensor (Aanderaa 1964, Dahl 1969, Appell et al. 1974). The mounting of the instrument is quite different from that of the VACM. The instrument package proper is free to swivel on gimbals about a vertical tie-rod which is the component actually inserted in the mooring line. A relatively large vane keeps the entire package oriented into the flow, so that only a compass is needed to measure direction of the flow relative to magnetic north. The Aanderaa also records temperature (Fig. 4).

The large vane system and high rotational moment of inertia of the package render the Aanderaa, like the VACM, not suitable for use in the near-surface part of the ocean.

Like all instruments in the ocean, the Aanderaa has been through a series of problems (for instance, see Hendry and Hartling 1979). However, experience indicates that with suitable maintenance and care, they are reliable. The Aanderaa has become, with the VACM, one of the "workhorses" of deep-sea, long-term current measurements.

The Vector Measuring Current Meter (VMCM), also known as the Weller-Davis meter, has proven effective in measuring currents near the surface, where other meters, as pointed out above, are susceptible to contaminated data from vertical currents and high frequency horizontal currents from wave action (Fig. 5).

The VMCM uses two orthogonal propellors to achieve a sensor package which has an excellent cosine response: that is, insensitive to current flow at right angles to the propellor axis (Davis and Weller 1978, Weller 1978, Weller and Davis 1980).

A number of current meters of the rotor-vane type have been used by oceanographers in the U.S.S.R. for some time. The oldest of these is the Alexeev. The original Alexeev meter used a completely mechanical recorder driven by a spring, and recorded on paper tape with ink and type faces. In recent years, these mechanical instruments have been fitted with electric motors.

The TSIT and TSIIT instruments are magnetic type instruments, with simple sampling of a single speed and direction per cycle. The speed sensors in each case are a form of paddlewheel, shielded on one side from the flow.

All of these meters are limited in general to a depth capability of about 1 500 m, since the pressure cases are of brass. A few titanium cases are available for deep use. The meters are suspended from a triangular frame which is clamped to the mooring line. The instrument is free to swing from the frame in three dimensions. It is also free to rotate about the vertical axis, with a vane mounted on the case, and direction is measured with a compass.

Because of the suspension system, these instruments are very susceptible to mooring and wave noise. Comparisons with other current meter moorings and with SOFAR floats have shown that speed is consistently overestimated by a factor of up to 1.4 when these meters are used on surface moorings (SCOR Working Group 21 1974 and 1975).

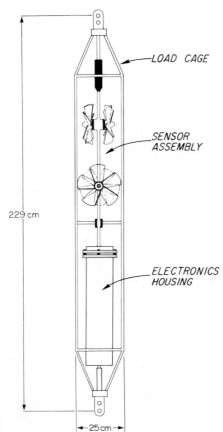

Fig. 5. Vector measuring current meter (VMCM). (Courtesy Russ Davis)

24.2.2.b Acoustic Current Meters. Current meters which measure velocity by means of acoustic techniques fall into two categories: those that rely on measuring the travel time of a pulse, and those that use the doppler effect.

Acoustic travel time meters have two major disadvantages. First, the measurement must be made in the vicinity of the acoustic source and receiver. Therefore, hydrodynamic effects of the transducers and their supports may cause substantial errors in the measurements. Second, very small time intervals must be measured. This implies relatively high power consumption.

Many of the problems of these meters are being overcome. An excellent survey of this type of instrument has been done by McCullough and Graeper (1979).

Doppler current meters in general measure the velocity of the water at a distance from the transducers. Thus, they can avoid the hydrodynamic problems of the travel time current meters. In addition, since the frequency comparison may be made by a beat technique, the time intervals to be measured will be larger (frequencies lower), thus reducing power requirements.

In general, while both doppler and travel time instruments may be obtained from several manufacturers, they are not yet in wide use in oceanographic work. It seems likely that this will change in the near future, due to the advantages such instruments have over other types of moored current meters (no moving parts, fast response, flexibility of sampling).

24.2.2.c Electromagnetic Current Meters. The prospect of measuring water velocity by measuring the electromagnetic field produced when a moving conductor (sea water) cuts a magnetic field has induced several attempts to produce a reliable sensor along these lines. Such sensors are not at present in wide use as moored sensor/recording packages, although they are being used quite successfully in a free-fall current profiling package.

The problems with such sensors fall into two areas: the hydrodynamics of the sensor configuration and fouling of the electrodes (McCullough 1978b, Simpson 1971). An electromagnetic current meter must employ sensors on two orthogonal axes, which can create certain problems in interpreting the flow and the resultant data. The fouling problem, or course, is much less severe in the free-fall profiler which is not in the water for very long periods of time and which can be cleaned as necessary.

24.2.2.d Analysis and Interpretation. For most purposes, and ignoring special needs, a set of more or less standard approaches have been developed for analyzing current meter time series. They encompass the general run of techniques for dealing with time series data. Numerous examples are shown in the companion chapters to this volume.

Displays of current meter data include progressive vector diagrams, geographical vector plots, "stick diagrams", and analogue plots of velocity components as a function of time. These displays are well illustrated in the series of current meter data reports put out by the Woods Hole Buoy Project (see, for instance, Tarbell 1980). The choice of a suitable display is, of course, heavily influenced by personal preference and the intent of the user of the data.

24.2.2.e Calibration and Intercomparison. Traditional laboratory calibrations of current meter sensors are straightforward in principle. The sensor is towed in a tank of water and a calibration factor calculated. However, this deals only with the static calibration and perhaps with the starting threshold for the sensor. More elaborate calibrations must be carried out to investigate the dynamic response of the sensor. Generally, these calibrations have been carried out for most commonly used sensors. The validity and accuracy of such calibrations may be in question, however, depending on the care taken, the purpose of the calibration, and the facilities available for the calibration (see, for example, McCullough 1974, or Kalvaitis 1974).

Intercomparisons for various sensors may, of course, be based on separate laboratory calibrations, but given the variability in calibration techniques, they may or may not be valid. Direct intercomparisons, if done carefully, may be more useful, that is, comparisons with each sensor done in the same tank with identical techniques and parameters. In a variation on this technique, Neilson and Jacobsen (1980) have carried out intercomparisons from fixed masts on the seabed and by towing behind a ship.

In one sense, intercomparisons carried out at sea on moorings should be better, since they reflect actual environmental conditions of variability, dynamics, noise, etc. However, the spatial (and temporal) variability of the ocean works against this, since two instruments even on the same mooring a few meters apart in depth may experience somewhat different inputs both from the ocean and from mooring noise. For instruments on separate moorings, obviously, the problem is even worse.

A number of current meter intercomparisons have been carried out, most notably by SCOR Working Group 21 (SCOR Working Group 21, 1969, 1974 and 1975) and in the course of the U.S./U.S.S.R. POLYMODE program, where an effort was made to compare U.S. and Soviet current meters for the eventual purpose of merging the data sets. A comparison has also been made between the U.S. and Soviet current meters and the SOFAR floats. In general, these comparisons bear out, at least at low frequencies, the results of mooring intercomparisons (H.T. Rossby pers. comm.). Some intercomparisons have also been carried out specifically in the near-surface regime, where the problem is quite different (Halpern et al. 1981).

24.3 Moored Profiling Current Meters

An obvious direction for development of moored instrumentation is to develop an instrument that would move up and down the mooring line, making measurements as it went, rather than staying at a fixed depth, thus collecting a profile of a given parameter as a function of depth. Two instruments have been developed to do this. In each case, the package includes a current meter and a CTD.

The Cyclesonde[R], from Marine Profiles, Inc., is attached to a surface mooring and moves in cycles along the mooring line from about 20 m to as deep as 500 m (van Leer et al. 1974). Depth is controlled by means of an inflatable bladder (Fig. 6).

The Profiling Current Meter (PCM) uses a sub-surface mooring (Dahlen et al. 1977). The instrument moves from a "stop", very near the top of the mooring, to a depth of up to 300 m. The minimum depth of the instrument travel is obviously limited by the minimum depth that the top float of the sub-surface mooring can be placed. This may vary with bottom topography and other factors. The PCM depth is varied by driving a piston to change the buoyancy volume (Fig. 7).

In each case, the number of cycles, depth of travel, sampling, and endurance of the instrument are all inter-related, being at least partially dependent on battery power available.

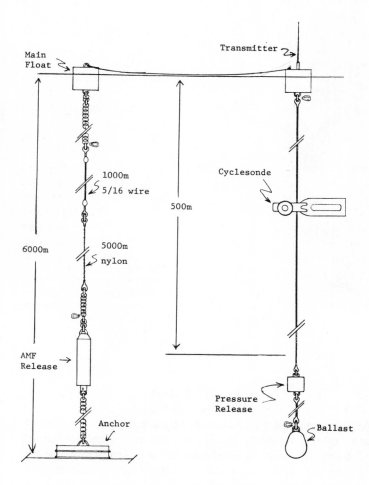

Fig. 6. A deep water Cyclesonde[R] array. (Courtesy John Van Leer)

Analysis and interpretation of data from such instruments may be viewed as a combination of approaches as used for shipboard profiling and moored time series. If several moorings are used at different locations, measurements at a given depth may not be made at all moorings at the same time, unless the clocks in individual instruments are well synchronized.

Fig. 7. Profiling current meter (PCM). The small black sphere is the electromagnetic velocity sensor; the larger sphere is the pressure housing for the electronics. The mooring wire runs through the central tube. (Courtesy John Dahlen)

24.4 Temperature Recorders

Temperature is possibly the easiest parameter to measure on a moored buoy. As noted above, many of the standard current meters in use also routinely measure temperature. Even so, achieving good time series of temperature is not a simple engineering problem. The most serious problems are in the sensors, achieving drift-free, calibrated measurements.

The most widely-used instrument today for moored temperature measurements is the Wunsch-Dahlen Temperature-Pressure Recorder (Fig. 8). This instrument has been used successfully in a wide range of experiments since about 1972 (Wunsch and Dahlen 1974).

The combination of temperature and pressure measurements is particularly useful on the sub-surface moorings on which the recorder is generally de-

Fig. 8. Temperature-pressure recorder (TP). The sphere is a pressure housing. The bar becomes a strength member in the mooring line. (Courtesy John Dahlen)

ployed, since it gives the true depth of the measurement as the mooring dips in the currents. Pressure measurements from the TP's have been particularly useful in evaluating mooring performance for engineering purposes.

24.5 Other Moored Instrumentation

Several instruments have been used to measure deep-sea pressures from bottom installations, for both tidal and mesoscale investigations (Cartwright 1977, Baker 1969). Generally, these measurements are difficult, involving as they do the recording of very small differences in a large number.

Two approaches have been taken to this problem. The first is to measure absolute pressure, that is, the ambient pressure is measured against some mechanical standard (Filloux 1970, Snodgrass 1968, Snodgrass et al. 1974, Caldwell et al. 1969). The second approach is to measure differential pressure; that is, ambient pressure measured against the pressure of an internal gas volume (Baker et al. 1973).

Another bottom moored instrument is the Inverted Echo Sounder, or IES (Watts and Rossby 1977). This instrument measures dynamic height by transmitting from its bottom site a sound signal and measuring the time delay until the surface reflected pulse is received.

An instrument with promise for the future is the "pop-up" profiler under development by Doug Webb at W.H.O.I. This device will store a canister of probes in a bottom-mounted package (up to 50 in the initial version). At timed intervals, a probe will be released to travel to the surface under its own buoyancy. The bottom station will track the probe with phase detection acoustic techniques, thus recording a velocity profile.

24.6 Drifting Instruments

Perhaps the oldest technique for research in the ocean is the drifting marker or drift bottle. Deposited in the water at a known location and date, with a note or message to have it returned to the "owner" if found, the drift bottle provided some of the earliest data on large scale ocean currents. There can be few oceanographers who, at some point in their early careers, have not sealed a note in an empty bottle and thrown it hopefully over the side of a research vessel. Rarer still are those who have had the message answered.

However, the principle survives in modern drifting instruments designed, not to be located and reported by a helpful party, but to be tracked continuously, in some cases telemetering other data as well. Drifting instruments may be divided into surface and deep drifters.

24.6.1 Surface Drifters

Surface drifters come in a wide variety of shapes, sizes, and intended uses. While there are more or less "standard" drift buoys made by manufacturers such as Polar Research Laboratories, Hermes, Ltd., and Telecommunications Enterprises, many investigators and programs have designed and built their own (Fig. 9). Vachon (1978) and Kerut (1980) give overviews of the state of the art in surface drifters.

Most surface drifters for open ocean use are tracked by satellites. Some, intended for near-shore use, may be tracked by VHF or HF radio.

Surface drifters may be undrogued (and assumed to be following the surface currents) or drogued (and assumed to be following the water at a specified near-surface depth).

The use of drift buoys, drogued or undrogued, is still very much of an art. There is little agreement on the effect or efficiency of drogues, for instance, with some evidence that the presence or absence of drogues has little or no effect on buoys in certain regimes. Some work is being done on the optimum configuration of drift buoy hulls for different wave and wind conditions, and a number of different drogue shapes have been tried. Vachon (1978) gives an overview of the state of the surface drift buoy art.

Fig. 9. PRL-Richardson drift buoy being launched. (Courtesy Phil Richardson)

However, the fact remains that much good work is being done with surface drifters, and the state of the art will certainly improve in the 1980's (Cheney et al. 1976, Kirwan et al. 1976, Martin and Gillespie 1978, Richardson et al. 1977 and 1979).

24.6.2 Deep Drifters

More recently, the drift buoy principle has been employed in submerged neutrally buoyant floats. These instruments are compressible packages ballasted to descend to a particular depth to be tracked acoustically. Some also telemeter data such as ambient temperature. All are built by the users; as yet no manufacturer produces them as "off-the-shelf" items.

Swallow floats are small neutrally buoyant floats which have been used in a variety of applications (Swallow 1955, Rossby and Webb 1970). These floats, in their simplest configuration, are tracked acoustically from a vessel in their vicinity. They operate at a medium frequency (5–6.5 KHz), and were used effectively in MODE (Swallow et al. 1974) and other experiments, including the Aries Expedition.

SOFAR floats are larger, low frequency (230–275 Hz) floats which can be tracked acoustically over much greater distance from shore stations by means of the SOFAR sound channel (Webb 1977, Rossby et al. 1975, Voorhis and Webb 1973). Tracking of the SOFAR floats is a more complicated procedure than for the Swallow floats, often requiring the cooperation of the military to use existing listening stations (Walen, 1976, Walden et al. 1973).

The SOFAR floats in general must be close enough in depth to the local SOFAR depth for the acoustics to allow good tracking. However, there appears to be a great deal of leeway in this; SOFAR floats in POLYMODE were used from 700 to 2000 m depth (Fig. 10).

Recently, the potential areas of the ocean in which the SOFAR floats may be used has been considerably expanded by the development of the Autonomous Listening Station, or ALS. This installation consists of a mooring with an acoustic receiver and recording package. Thus, the reliance on shore-based listening stations is reduced. The drawback is that the data is no longer available during the actual experiment, as it may be with the shore-based stations.

Fig. 10. SOFAR floats being loaded aboard ship. (Courtesy Doug Webb)

24.6.3 Problems and Interpretation

The errors and uncertainties in deep drift buoy data are better understood than those in surface drift buoy data, which are susceptible to all the problems caused by wave and wind action.

Uncertainties in the deep float data are largely those inherent in the tracking system, assuming that inertial effects are negligible. The tracking efficiency and errors are a function of distance from the tracking receivers, frequency, and acoustic propagation conditions (Swallow et al. 1974).

Surface buoy data uncertainties are much more complicated. The effects of surface winds and waves are highly dependent on buoy hull and drogue configurations, as well as weather (Kirwan et al. 1979, Nath 1977). "Slippage" must be estimated for each buoy and drogue configuration.

Analysis and interpretation techniques are not as well developed for Lagrangian measurements as they are for Eulerian, although much progress has been made (Freeland et al. 1975, Molinari and Kirwan 1975). In addition, the integration of Lagrangian and Eulerian data is an area which has not yet been adequately addressed. The most common form for display of Lagrangian data is the familiar "spaghetti diagram", seen in various places in the present volume.

24.7 Profiling Instruments

A number of instruments are used to measure one or more parameters as a function of depth, either by being lowered from a ship or by free-falling and being recovered later. In a few cases, the "instrument" is expendable.

24.7.1 Temperature and Salinity Instruments

Some of the very earliest oceanographic measurements were temperatures obtained by lowering thermometers on a wire from a ship. The more sophisticated lowered profilers, and the newer free-falling instruments, are essential to the rapid and efficient collection of large scale oceanographic survey data.

It should be noted that, in addition to the techniques described below, it is routine on many cruises to collect continuous surface salinity and temperature while underway, by sampling a continuous flow of sea water taken in through a hull inlet near the surface.

24.7.2 Nansen Bottles

Nansen bottles are an old technique for collecting salinity and temperature measurements from shipboard. Large-scale Nansen bottle surveys are largely a thing of the past, having been supplanted by the faster and more accurate CTD and STD profilers. The bottles are still used, however, in the so-called "rosette sampler" or as single bottles, to calibrate the conductivity or salinity measurements of the CTD or STD by collecting in-situ samples.

24.7.3 CTD and STD

Several manufacturers make conductivity/temperature/depth (CTD) and salinity/temperature/depth (STD) instrumentation. Two STD instruments are currently in use by Soviet oceanographers.

The Neil Brown CTD is manufactured by Neil Brown Instrument Systems, Inc (Fig. 11). Besides standard capability for measuring temperature (with a combination thermistor and platinum resistance thermometer), depth (with a

Fig. 11. Neil Brown Mark IIIB CTD. (Courtesy Neil Brown Instrument Systems, Inc.)

strain gauge), and conductivity (with a platinum electrode cell), the instrument will accept an oxygen sensor if desired (Brown, 1974).

Soviet oceanographers use the AIST and ISTOK profilers (Paramonov et al. 1979). These instruments are somewhat comparable to western CTD's in specifications, but generally are not as well calibrated. In addition, one area in which they suffer is in sampling rate, which is much lower than in the Neil Brown instrument (Fofonoff and Millard 1976).

Analysis techniques for density measurements in the ocean are well developed, as witness many of the discussions elsewhere in this volume. But see Scarlet (1973 and 1975) for examples of some pitfalls. In addition, merging of measurements from several different types of STD or CTD instruments may require careful intercomparison and intercalibration (Fofonoff et al. 1974, Fofonoff and Millard 1976).

24.8 Expendable Bathythermographs

The Expendable Bathythermograph, or XBT, manufactured by Sippican Ocean Systems, has been one of the most spectacularly successful new instruments in the physical oceanographer's "toolkit". In a few years it has almost completely displaced the earlier mechanical bathythermograph (BT). Several models of the XBT are currently available, the differences being primarily in the depth achievable with the probes.

The T-7 XBT will go to 750 m, the T-10 to 200, the T-6 to 450, and the T-5 to 1830 m, all at ship speeds of up to 15 knots. The T-4 will go to 450 m at speeds of up to 30 knots, and the "Deep Blue" is a 20 knot version of the T-7. The Airborne XBT, or AXBT will go to 750 m in its latest version.

The XBT is an expendable probe, which, once released from the ship, descends at an approximately known rate, sending a temperature signal back along a fine wire which unreels both from the probe and from the shipboard launcher, until the maximum depth is reached (Fig. 12).

A number of attempts have been made to increase the usefulness of the XBT probe data by using digital techniques to process and record the signal from the probe wire, instead of the standard technique of recording the data in analogue form on a strip chart. Currently, packages to do this are available from Sippican Ocean Systems, Bathy Systems, Inc. and from EG&G. In general, the probe is capable of resolving much finer scales than can be recorded on the strip chart. In addition, the magnetic tape format for recording the data is usually more convenient for processing and storage than the rather fragile strip chart paper.

Historically, with oceanographic instrumentation of all types, after an instrument or technique becomes widely used, scientists and engineers begin to look more closely at it, to push its capabilities to their limits, and inevitably to discover more and more subtle errors, which were not of concern when the instrument first became available. The XBT is no exception.

Flierl (1977a) has looked at the problem of using the XBT to infer dynamic heights in regions of large salinity anomalies. Several investigators have carried out comparisons of temperature profiles taken with the XBT with those taken from various types of CTD's and STD's (Flierl 1974, Flierl and Robinson 1977, McDowell 1977, Federov et al. 1978 and 1979, Seaver and Kuleshov 1979).

XBT EXPLODED VIEW

1. AFTERBODY
2. PROBE SPOOL
3. SEA ELECTRODE
4. THERMISTOR
5. ZINC NOSE
6. LABEL
7. SHIPBOARD SPOOL
8. SIGNAL WIRE
9. CANISTER
10. RETAINING PIN
11. SHIPPING CAP

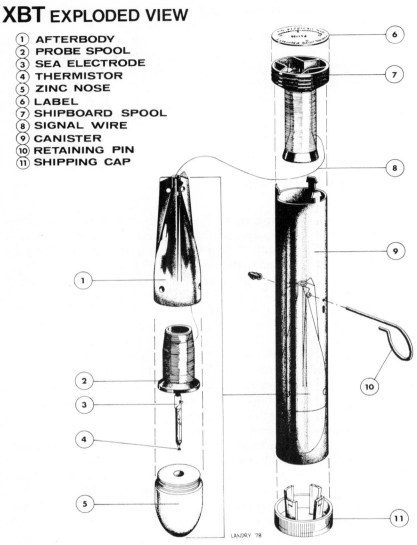

Fig. 12. General configuration of an expendable bathythermograph (XBT). (Courtesy Sippican Corp.)

Recently, an attempt has been made to pull together the various results of intercomparisons of the T-7 XBT with STD's and CTD's (Heinmiller et al. 1982). The results strongly indicate that the T-7 and T-4 XBT's have a systematic error in depth. This error is within the manufacturer's specifications ($\pm 2\%$ in depth); however, since it is systematic it may affect certain types of calculations made with XBT's in bulk. In the model T-7, the error amounts to as much as 15 m at 750 m.

24.9 Current Profilers

In recent years, several instruments have been developed to measure currents by means of free-fall and tethered profilers deployed from vessels. They utilize various techniques for the actual current measurement.

24.9.1 Electromagnetic Velocity Profiler

The Electromagnetic Velocity Profiler, or EMVP, uses a set of electrodes to measure the currents induced in the conductor represented by the sea water moving through the earth's magnetic field plus the local electromotive force. (Drever and Sanford 1970 and 1976, Sanford et al. 1974 and 1978). Processing and analysis of the signals sensed by the electrodes is complex. The data is recorded on magnetic tape for later retrieval, along with temperature, depth, and conductivity.

Since the original version of the profiler only measured velocity relative to a local average profile, a more recent version of the instrument (the AVP, or Absolute Velocity Profiler) adds a doppler acoustic package to measure the absolute velocity of the probe relative to the bottom, for a short period when the probe is near the bottom during its profiling trip (Fig. 13).

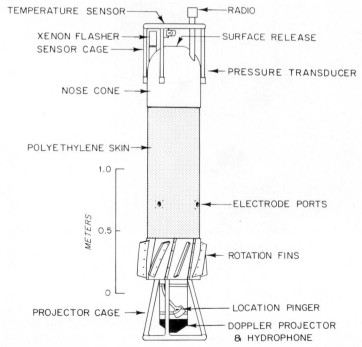

Fig. 13. The Absolute Velocity Profiler (AVP). (Courtesy Thomas Sanford)

24.9.2 The White Horse

The White Horse utilizes a network of bottom-mounted acoustic transponders to track a free-falling probe, which also measures temperature, pressure, and conductivity (Luyten and Swallow 1976). The transponders are on small, simple bottom moorings with acoustic releases, and can be deployed and recovered efficiently. However, they do add to the amount of equipment and ship-time required to use the profiler (Fig. 14).

Fig. 14. White Horse profiler being launched. (Courtesy James Luyten)

24.9.3 Schmitz-Richardson Profiler

Richardson and Schmitz (1965) used a simple probe system for measurements of transport. The probe is released from a vessel, drops to a pre-set depth, releases ballast and returns to the surface. The difference between the drop and recovery points can be used to calculate the transport between the surface and the depth of ballast drop. Accurate navigation is required, and so the technique is probably not suitable for most open-ocean work. This profiler has been used, however, for studies in the Straits of Florida (Niiler and Richardson 1973). A version suitable for deployment from aircraft was used in MODE, but has not been further developed (Richardson et al. 1972).

24.9.4 Other Velocity Profilers

Pochapsky and Malone have used an acoustically tracked probe in the Gulf Stream (Pochapsky and Malone 1972, Pochapsky 1975), and in MODE (Pochapsky 1976). This instrument telemetered the acoustic travel times, as well as temperature and pressure to the ship. Expendable acoustic transponders were used.

The Düing profiler is a tethered instrument which is ballasted slightly negatively buoyant to slide down a hydrowire. It carries an Aanderaa package that records speed, temperature, and depth (Düing and Johnson 1972).

Doppler acoustic profilers can be used from vessels underway. This technique is generally limited to the shallower currents, with an acoustic device on the ship's hull "looking" at a series of depth ranges (Regier 1982, Joyce et al. 1982).

24.9.5 Calibration and Intercomparison

Profiling current instruments can be calibrated against each other or by data collected with techniques such as moored current meters, or currents calculated from geostrophy. Rossby and Sanford (1976) have compared the acoustic and electromagnetic methods, with results indicating good agreement.

24.10 Remote and Inverse Techniques

With the rise of new technologies, such as orbiting satellites, the intriguing possibility presents itself of making measurements in the ocean without venturing out upon the ocean (Born et al. 1979). Some observational techniques have been used from aircraft for several years. In addition, a new technique, acoustic tomography, offers the possibility of determining the density field within a broad region of ocean from measurements made from outside that region.

24.10.1 Satellite and Airborne Measurements

Measurements from satellites and aircraft are generally limited to the surface or (for aircraft) near-surface parameters.

The most common measurement from aircraft is temperature. The Airborne Expendable Bathythermograph (AXBT) has been used for some years for temperature profiles (Sessions et al. 1974, Sessions 1979). These probes are limited to about 350 m depth. Sea surface temperature can be measured from aircraft with an infra-red thermometer.

Availability of aircraft is frequently a limiting factor in these operations. However, this can be a cost effective way to collect data, particularly if an aircraft is already being utilized for other purposes.

In terms of mesoscale oceanography, satellite sea surface temperature measurements are still largely limited to areas where the signal-to-noise ratio is high, such as near the Gulf Stream (Maul et al. 1978). Improving these measurement capabilities requires knowledge of the humidity and other factors, but it would be unwise to rule out the possibility of a more general capability for mesoscale sea surface temperature measurements for the future.

Altimetry, or determination of the height of the sea surface from a satellite, is still in its infancy. However, the results have been promising (Wunsch and Gaposhkin 1980).

Presently, satellite techniques are being used primarily to supplement the more traditional mesoscale approaches. One can assume that, as the technology evolves, they will come into their own as primary tools for eddy research (Barnett et al. 1979, Bernstein et al. 1977).

24.10.2 Tomography

Acoustic tomography is an interesting new technique for mesoscale measurements utilizing an array of acoustic sources and receivers (Munk and Wunsch 1979).

Consider a single source and receiver at a fixed distance in the ocean. If both source and receiver are roughly near the depth of the SOFAR sound channel, then there are a number of possible acoustic paths between them. If a single pulse is emitted by the source, several different pulses will be received at the receiver, due to the varying travel times for each path.

These arrival times and the time of emission of the original source pulse can be used, with some assumptions about the water masses between the source and receiver, to calculate the sound velocity profile as a function of distance from the source to the receiver. Thus a "vertical slice" of the ocean sound velocity field is defined.

If a two-dimensional array of sources and receivers is used, the vertical slices along the various geographical paths form a three-dimensional picture of the sound velocity field within the array.

The first pilot test of this technique was deployed southwest of Bermuda in early 1981 for four months. At the time of writing, the data are still being analyzed. If the technique lives up to its promise, more elaborate experiments will be planned.

Tomography is intended to give a statistical, integrated picture of the field. For the pilot experiment, the resolution horizontally will be of the order of 50 km. The results can be improved by inclusion of a numerical model to dictate considerations of continuity, or by updating with in-situ environmental data.

A related technique is that of reciprocal shooting, in which the difference in travel times between pulses sent in both directions between source/receiver moorings is used to calculate water velocities.

24.11 Intercomparison Between Instrument Types

For a given type of instrument (current meters, density profilers, etc.), comparisons have generally been carried out, either laboratory, at-sea, or both. However, there have been few intercomparisons between instruments in different categories.

One of the few such cross-comparisons was a result of the MODE experiment (MODE-I Intercomparison Group 1974). Since a wide variety of instruments (including instruments of a given type made by different manufacturers) was used in a relatively small area of the ocean at approximately the same time, it was possible to compare the results obtained.

A related problem is that of merging data sets, such as CTD and XBT measurements. This approach has been taken in the processing of the data from the U.S. POLYMODE Local Dynamics Experiment (See Chap. 6. this Vol.).

References

Aagaard K (1980) Moored current measurements over the Lomonosov Ridge, Transactions, Am Geophys Union 61 (17) (abstract)

Aagaard K, Malmberg SA (1978) Observational summary. MONA-5, MONA-6, ICES Overflow Working Group, 2 pp, 52 fig

Aanderaa IR (1964) A recording and telemetering instrument. Tech Rep, Fixed Buoy Proj, NATO Subcomm Oceanogr Res 16, 53 pp

Academy of Sciences of the USSR (1979) POLYMODE Hydrophysical Expedition. Oceanological Researches, 30 (Russian)

Academy of Sciences of the USSR (1980) POLYMODE Hydrophysical Expedition. Oceanological Researches, 31 (Russian)

Adem J (1956) A series solution for the barotropic vorticity equation and its application in the study of atmospheric vortices. Tellus 3:364–372

Albertson ND (1974) A survey of techniques for the analysis and design of submerged mooring systems. Nav Civ Eng Lab, Port Hueneme, Cal, Rep R-815, August

Allen AA (1979) Current variability at the offshore edge of the Labrador Current. MS Thesis, Dalhousie Univ

Amos AF, Gordon AL, Schneider ED (1971) Water masses and circulation patterns in the region of the Blake-Bahama Outer Ridge. Deep Sea Research 18:145–165

Anderson DLT (1980) The Somali Current. Ocean Modelling 34:6–9, Dept Appl Math Theor Phys, Univ Cambridge, England (unpublished manuscript)

Andrews JC (1977) Eddy structure and the West Australian Current. Deep-Sea Res 24:1133–1148

Andrews JC (1979) Eddy structure and the West and East Australian currents. Flinders Inst Atmos Mar Sci, Res Rep No 30

Andrews JC, Scully-Power P (1976) The structure of an East Australian Current Anticyclonic Eddy. J Phys Oceanogr 6 (5):756–765

Andrews JC, Lawrence M, Nilsson CS (1980) Observations of the Tasman Front. J Phys Oceanogr 10:1854–1869

Angel MV (1969) Repeated samples from a deep midwater planktonic ostracod community. J Exp Mar Biol Ecol 3:76–89

Angel MV (1977) Studies on Atlantic halocyprid ostracods: vertical distributions of the species in the top 1000 m in the vicinity of 44°N 13°W. J Mar Biol Ass UK 57:239–252

Angel MV, Hargreaves P, Kirkpatrick P, Domanski P (1982) Low variability and the absence of diel migration in planktonic and micronektonic populations at 1000m in the vicinity of 42°N 17°W. Biol Oceanogr 1:287–319 (in press)

Appell GF, Boyd JF, Woodward W (1974) Evaluation of the Aanderaa Recording Current Meter. NOAA Tech Rep NOASIC-TM-0X5/74-AR04, 25 pp

Armi L (1981) Transient tracers in the ocean preliminary hydrographic data report Leg 3 16 May–14 June 1981, Vol II. Phys Chem Data Facility Scripps Inst Oceanogr, San Diego, USA

Armi L, Haidvogel DB (1982) Effects of variable and anisotropic diffusivities in a steady-state diffusion model. J Phys Oceanogr (in press)

Baer RN (1980) Calculations of sound propagation through an eddy. J Acoust Soc Am 67:1180–1185

Baer RN (1981) Propagation through a three-dimensional eddy including effects on an array. J Acoust Soc Am 69:70–75

Bane JM, Brooks DA (1979) Gulf Stream meanders along the continental margin from the Florida Straits to Cape Hatteras. Geophys Res Lett 6:280–282

Bane JM, Brooks DA, Lorenson KR (1981) Synoptic observations of the three-dimensional structure and propagation of Gulf Stream meanders along the Carolina continental margin. J Geophys Res 86:6411–6425

Bang ND (1970a) Major and frontal structures in the Agulhas Current retroflexion area in March, 1969. Proc symp oceanogr S Afr, Durban, 16 pp

Bang ND (1970b) Dynamic interpretation of a detailed surface temperature chart of the Agulhas Current retroflexion and fragmentation area. S Afr Geogr J 52:67–76

Baker DR Jr (1969) On the history of the high seas tide gauge. WHOI Ref 5-69, Woods Hole Oceanogr Inst

Baker DJ Jr (1979) Polar oceanography II: Southern Ocean. Rev Geophys Space Phys 17:1578–1585

Baker DJ Jr (1981) Ocean instruments and experiment design. In: Warren BA, Wunsch C (ed) Evolution of Physical Oceanography. MIT Press, Cambridge MA, pp 396–433

Baker DJ Jr, Wearn RB Jr, Hill W (1973) Pressure and temperature measurements at the bottom of the Sargasso Sea. Nature 245:25–26

Baker DJ Jr, Nowlin WD Jr, Pillsbury RD, Bryden HL (1977) Antarctic circumpolar current: Space and time fluctuations in the Drake Passage. Nature 268:696–699

Barnett TP, Patzert WC (1980) Scales of thermal variability in the tropical Pacific. J Phys Oceanogr 10:529–540

Barnett TP, Patzert WC, Webb SC, Bean BR (1979) Climatological usefulness of satellite determined sea-surface temperatures in the tropical Pacific. Bull Am Meteorol Soc 60:197–205

Barrett JR (1965) Subsurface currents off Cape Hatteras. Deep Sea Res 12:173–184

Barrett JR (1971) Available potential energy of Gulf Stream rings. Deep Sea Res 18:1221–1231

Barrett JR, Schmitz WJ (1971) Transport float measurements and hydrographic station data from three sections across the Gulf Stream near 67°W. RV Crawford Cruise 168, June–July 1968. WHOI ref No 71-66

Batchelor GK (1949) Diffusion in a field of homogeneous turbulence. Aust J Res 2:437–450

Beckerle J, Baxter L, Porter R, Spindel R (1980) Sound channel propagation through eddies southeast of the Gulf Stream. J Acoust Soc Am 68:1750–1767

Belyakov LN (1972) Triggering mechanism of deep eposodie currents in the Arctic Basin (in Russian). Problemy arktiki i antarktiki, 39:25–32

Bennett AF (1978) Poleward heat fluxes in southern hemisphere oceans. J Phys Oceanogr 8:785–798

Bennet AF, Haidvogel DB (1983) Low resolution numerical simulation of decaying two-dimensional turbulence. J Atmos Sci (in press)

Bennett AF, Kloeden PE (1978) Boundary conditions for limited area forecasts. J Atmos Sci 35:990–996.

Bennett AF, Kloeden PE (1980) The quasi-geostrophic equations: approximation, predictability and equilibrium spectra of solutions. Q J R Meteorol Soc 107:121–136

Bennett AF, Kloeden PE (1981) The ill-posedness of open ocean models. J Phys Oceanogr 11:1027–1029

Bernstein RL (1971) Observations of currents in the Arctic Ocean. PhD Thesis, Columbia Univ, New York, 78 pp

Bernstein RL (1978) Mesoscale eddy activity in the Pacific North Equatorial Current. POLYMODE News, No 45 (unpublished manuscript)

Bernstein RL, White WB (1974) Time and length scales of baroclinic eddies in the central North Pacific Ocean. J Phys Oceanogr 4:613–624

Bernstein RL, White WB (1977) Zonal variability in the distribution of eddy energy in the mid-latitude North Pacific Ocean. J Phys Oceanogr 7:123–126

Bernstein RL, White WB (1981) Stationary and traveling mesoscale perturbations in the Kuroshio Extension Current. J Phys Oceanogr 11:692–704.

Bernstein RL, White WB (1982) Meridional eddy heat flux in the Kuroshio Extension Current. J Phys Oceanogr 12:154–159

Bernstein RL, Breaker L, Whritner R (1977) California current eddy formation: ship, air, and satellite results. Sci 195:353–359

Berteaux HO (1975) Buoy Engineering. Wiley & Sons, New York 319 pp

Berteaux HO, Chhabra NK (1973) Computer programs for the static analysis of single point moored surface and subsurface buoy systems. WHOI Ref 73–22 Woods Hole Oceanogr Inst

Berteaux HO, Walden RG (1969) Analysis and experimental evalution of single point moored buoy systems. WHOI Rep 69–36 Woods Hole Oceanogr Inst

Bisagni JJ (1976) Passage of anticyclonic Gulf Stream eddies through deepwater dumpsite 106 during 1974 and 1975. NOAA Dumpsite Eval Rep 76–1, 39 pp

Boland FM (1979) A time series of expendable bathythermograph sections across the East Australian Current. Aust J Mar Freshwater Res 30:303–313.

Boland FM, Church JA (1981) The East Australian Current, 1978. Deep Sea Res 28:937–958

Boland FM, Hamon BV (1970) The East Australian Current, 1965–1968. Deep Sea Res 17:777–794

Born M, Wolf E (1975) Principles of Optics. Pergamon Press, Oxford

Born GH, Dunne JA, Lame DB (1979) Seasat mission overview. Sci 204:1405–1406

Boyer DL, Guala JR (1972) Model of the Antarctic Circumpolar Current in the vicinity of the Macquarie Ridge. Antarctic Oceanology II: The Australian–New Zealand Sector, Antarct Res Ser 19:79–93

Boyd SH, Wiebe PH, Cox JC (1978) Limits of Nematoscelis megalops in the Northwestern Atlantic in relation to Gulf Stream cold core rings II. Physiological and biochemical effects of expatriation. J Mar Res 36:143–159

Bradley A, Tillier P (1980) A mesoscale Lagrangian measurement system. ICES, Hydrogr Comm CM 1980/C:7

Bradley KF (1981) A summary of Woods Hole Buoy Group Moored Stations for the POLYMODE program. WHOI Tech Rep Ref No 81–15

Brandt SB (1981) Effects of a warm-core eddy on fish distributions in the Tasmann Sea off East Australia. Mar Ecol Prog Ser 6:19–33

Brandt SB, Parker RR, Vaudrey DJ (1981) Physical and biological description of warm-core eddy J during September-October 1979. CSIRO Div Fish Oceanogr, Cronulla 126:1–8

Brekhovskikh LM, Fedorov KN, Fomin LM, Yampolsky, AD (1971) Historical Evolution of Modern Oceanography towards Large-scale Physical Experiments in the Ocean. Proceedings of the Royal Society of Edinburgh (B) 72 (34):351–356

Brekhovskikh LM, Fedorov KN, Fomin LM, Koshlyakov MN, Yampolsky AD (1971) Large-scale multi-buoy experiment in the tropical Atlantic. Deep Sea Res 18:1189–1206

Bretherton FP (1975) Recent developments in dynamical oceanography. Q J R Meteorol Soc 101:705–721

Bretherton FP, Haidvogel DB (1976) Two-dimensional turbulence above topography. J Fluid Mech 78:129–154

Bretherton FP, Karweit MJ (1975) Mid-ocean mesoscale modelling. In: Numerical Models of Ocean Circulation. Ocean Affairs Board Nat Res Counc NAS, Washington, DC, pp 237–249

Briscoe MG (1975) Preliminary results from the tri-moored Internal Wave Experiment (IWEX). J Geophys Res 80:3872–3884

Broida S (1962) A report of data obtained in Florida Straits and off the west coast of Florida. January–June, 1962. Inst Mar Sci, Univ Miami, Tech Rep No 62511, 138 pp

Broida S (1963) A report of data obtained in Florida Straits and off the west coast of Florida. July–December, 1962. Inst Mar Sci, Univ of Miami, Tech Rep No 63-3, 196 pp

Broida S (1964) A report of data obtained in Florida Straits and off the west coast of Florida. January–June, 1963. Inst Mar Sci, Univ of Miami, Tech Rep No 64-1, 179 pp

Broida S (1969) Geostrophy and direct measurement in the Straits of Florida. J Mar Res 27:278–292

Brooks DA (1978) Subtidal sea level fluctuations and their relation to atmospheric forcing along the North Carolina coast. J Phys Oceanogr 8:481–493

Brooks DA (1979) Coupling of the Middle and South Atlantic Bights by forced sea-level observations. J Phys Oceanogr 9:1304–1311

Brooks DA, Bane JM (1981) Gulf Stream fluctuations and meanders over the Onslow Bay Upper Continental Slope. J Phys Oceanogr 11:247–256

Brooks DA, Mooers CNK (1977) Wind forced continental shelf waves in the Florida Current. J Geophys Res 82:2569–2576

Brooks IH, Niiler PP (1977) Energetics of the Florida Current. J Mar Res 35:163–191

Brown ED, Owens WB (1981) Observations of the horizontal interactions between the internal wave field and the mesoscale flow. J Phys Oceanogr 11:1474–1480

Brown NL (1974) A precision CTD microprofiler. In: IEEE Int Conf Eng Ocean Environment Rec, Catalogue No 74 CH0873-0 OCC, pp 270–278

Brown OB, Bruce JG, Evans RH (1980) Evolution of sea surface temperature in the Somali Basin during the southwest monsoon of 1979. Sci 209:595–597

Bruce JG (1968) Comparison of near surface dynamic topography during the two monsoons in the western Indian Ocean. Deep Sea Res 15:665–677

Bruce JG (1979) Eddies off the Somali coast during the southwest monsoon. J Geophys Res 84(C12): 7742–7748

Bruce JG, Volkmann GH (1969) Some measurements of current off the Somali coast during the northeast monsoon. J Geophys Res 74(8):1958–1967

Bruce JG, Quadfasel DR, Swallow JC (1980) Somali eddy formation during the commencement of the southwest monsoon of 1978. J Geophys Res 85 (C11):6654–6660

Brundage W, Dahme A (1969) Temperature and salinity tabulations and profiles with charts from the northern region. MILOC 65 Tech Rep No 143, SACLANTCEN La Spezia, Italy

Bryan KA (1963) A numerical investigation of a nonlinear model of a wind-driven ocean. J Atmos Sci 20:594–606

Bryan K, Cox MD (1968a) A nonlinear model of an ocean driven by winds and differential heating, Part I. Description of the three-dimensional velocity and density fields, J Met 25:945–967

Bryan K, Cox MD (1968b) A nonlinear model of an ocean driven by winds and differential heating, Part II. An analysis of the heat, vorticity and energy balance, J Met 25:968–978

Bryan K, Lewis LJ (1979) A water mass model of the world ocean. J Geophys Res 84:2503–2517

Bryden HL (1979) Poleward heat flux and conversion of available potential energy in Drake Passage. J Mar Res 37:1–22

Bryden HL (1982) Sources of eddy energy in the Gulf Stream recirculation region. J Mar Res (in press)

Bryden HL, Hall MM (1980) Heat transport by ocean currents across 25°N latitude in the Atlantic Ocean. Sci 207:884–886

Bryden HL, Millard R (1980) Spatially averaged Local Dynamics Experiment CTD stations. POLYMODE News 77

Bryden HL, Pillsbury RD (1977) Variability of deep flow in the Drake Passage from year-long current measurements. J Phys Oceanogr 7:803–810

Bubnov VA, Egorikhin VD, Matveeva ZN, Navrotskaya SE, Filippov DI (1979) Graphical presentation of the USSR oceanographic observations in the tropical Atlantic during GATE (June to September, 1974). Univ of Miami, Fl, USA Tech Rep No TR-79-1, 163 pp (unpublished manuscript)

Bubnov VA, Vasilenko VM, Krivelevich LM (1980) The study of low-frequency variability of currents in the tropical Atlantic. Deep Sea Res, 26:199–216 (Gate suppl 1)

Bulgakov NP, Djiganshin GF, and Belous LM (1977) Instrument current measurements in synoptic eddy. In: Nelepo B A (ed) Marine Hydrophysical Research No 4 (79), Mar Hydrophys Inst Ukr SSR Acad of Sci, 273 pp

Bunker AF (1976) Computations of surface energy flux and annual airs-ea interaction cycles of the North Atlantic Ocean. Mon Weath Rev 104:1122–1140.

Caldwell DR, Snodgrass FE, Wimbush MH (1969) Sensors in the deep sea. Phys Today 22:34–42

Cartwright DE (1977) Ocean tides. Rep Prog Phys 40:665–708

Chao SY, Janowitz GS (1979) The effect of a localized topographic irregularity on the flow of a boundary current along the continental margin. J Phys Oceanogr 9:900–910

Charney JG (1955) The generation of ocean currents by wind. J Mar Res 14:477–498

Charney JG (1971) Geostrophic turbulence. J Atmos Sci 28:1087–1095

Charney JG, Stern ME (1962) On the stability of internal baroclinic jets in a rotating atmosphere. J Atmos Sci 19:159–172

Charney JG, Fjortoft R, Neumann J von (1950) Numerical integration of the barotropic vorticity equation. Tellus 2 (4):237–254

Chen TC, Miller FJ (1977) Speed of sound in seawater at high pressures. J Acoust Soc Amer, 62:1129–1135

Cheney RE (1977a) Entrapment of SOFAR floats by Gulf Stream rings. POLYMODE News 31 (unpublished manuscript)

Cheney RE (1977b) Synoptic observations of the oceanic frontal system east of Japan. J Geophys Res 82:(34) 5459–5468

Cheney RE, Marsh JG (1981a) SEASAT altimeter observations of dynamic topography in the Gulf Stream region. J Geophys Res 86:473–483

Cheney RE, Marsh JG (1981b) Oceanic eddy variability measured by GEOS 3 altimeter crossover differences. Trans Am Geophys Union (EOS) 62:743–752

Cheney RE, Richardson PL (1976) Observed decay of a cyclonic Gulf Stream ring. Deep Sea Res 23:143–155

Cheney RE, Gemmill WH, Shank MK, Richardson PL, Webb D (1976) Tracking a Gulf Stream ring with SOFAR floats. J Phys Oceanogr 6:741–749

Cheney RE, Richardson PL, Nagasaka K (1980) Tracking a Kuroshio cold ring with a free-drifting surface buoy. Deep Sea Res 27A:641–654

Cheney RE, Marsh JG, Crano V (1981) Global mesoscale variability from SEASAT collinear altimeter data. EOS Trans, AGU, G2 (17) (abstract)

Chew F (1974) The turning process in a meandering current, a case study. J Phys Oceanogr 4:27–57

Church PE (1932) Progress in the investigation of surface-temperatures of the western North Atlantic. Trans Am Geophys Union pp 244–249

Church PE (1937) Temperatures of the western North Atlantic from thermograph records. Association d'Oceanographie Physique, Union Geod Geophys Int, Publ Sci 4:3–40

Clark RA, Gascard JC (1982) Deep convection in the Labrador Sea. J Phys Oceanogr (submitted)

Clarke RA, Gascard JC (1983) The formation of Labrador Sea water: Part I. large-scale processes (in press)

Clark JG, Kronengold M (1974) Long-period fluctuations of CW Signals in deep and shallow water. J Acoust Soc Am 56:1071–1083

Clarke RA, Hill H, Reiniger RF, Warren BA (1980) Current system south and east of the Grand Banks of Newfoundland. J Phys Oceanogr 10:25–65

Climate Research Boards (1979) Carbon Dioxide and Climate: A Scientific Assessment. US Nat Acad Sci, Washington, DC, 22 pp

Colin de Verdiere A (1979) Mean flow generation by topographic Rossby waves. J Fluid Mech 94 (1):39–64

Colin de Verdiere A (1980) The Tourbillon Experiment: a cruise report. POLYMODE News 73.

Collings IL, Grimshaw R (1980) The effect of current shear on topographic Rossby waves. J Phys Oceanogr 10:363–371

Corrsin S (1961) The reactant concentration spectrum in turbulent fluctuations in an isotropic turbulence. J Fluid Mech 11:407–418

Cox MD (1980) Generation and propagation of 30-day waves in a numerical model of the Pacific. J Phys Oceanogr 10:1168–1186

Cox J, Wiebe PH (1979) Origins of oceanic plankton in the Middle Atlantic Bight. Estuarine Coastal Mar Sci 9:509–527

Cresswell GR (1976) A drifting buoy tracked by satellite in the Tasman Sea. Aust J Mar Freshwater Res 27:251–262

Cresswell GR (1977) The trapping of two drifting buoys by an ocean eddy. Deep Sea Res 24:1204-1209

Cresswell GR (1979) Currents in the northern Mozambique channel. INDEX Occasional Notes No 13, Ocean Sci Center, Nova Uni, USA (unpublished manuscript)

Cresswell GR (1981) The coalescence of two East Australian Current warm-core rings. Sci 215:161-164

Cresswell GR, Vaudrey DJ (1977) Satellite-tracked buoy data Report I. Western Australian releases 1975 and 1976. CSIRO, Div Fish Oceanogr, Rep 86, 49 pp

Creswell GR and Golding TJ (1979) Satellite-tracked buoy data Report III. Indian Ocean 1977, Tasman Sea July-December 1977. CSIRO Div Fish Oceanogr Rep 101:1-44

Cresswell GR, Golding TJ (1980) Observations of a south flowing current in the south eastern Indian Ocean. Deep Sea Res 27:449-466

Cresswell GR, Golding TJ, Boland FM (1978) A buoy and ship examination of the Subtropical Convergence south of Western Australia. J Phys Oceanogr 8:315-320

Csanady GT (1973) Turbulent diffusion in the envirnonment. Reidel Publishing Co, Dordecht, Holland

Csanady GT (1976) Mean circulation in shallow seas. J Geophys Res 81:5389-5399

Csanady GT (1979) The birth and death of a warm core eddy. J Geophys Res 84:777-780

Dahl O (1969) The capability of the Aanderaa recording and telemetering instrument. Prog Oceanogr 5:103-106

Dahlen JM, Chhabra NK, McKenna JF, Scholten JR, Shillingford JF, Siraco FJ, Toth WE (1977) Draper laboratory profiling current and CTD meter. Tech Rep Charles Stark Draper Lab R-1095, 122 pp

Dantzler HL (1976) Geographic variations in intensity of the North Atlantic and North Pacific oceanic eddy fields. Deep Sea Res 23:783-794

Dantzler HL (1977) Potential energy maxima in the tropical and subtropical North Atlantic. J Phys Oceanogr 7:512-519

Darbyshire M (1966) The surface waters near the coast of southern Africa. Deep Sea Res 13:57-81

Das PK (1980) Oceanographic data, MONEX ships, bathy. MONEX data set 5.1. Int MONEX Manage Centre, India Meteorol Dep, New Delhi

Das VK, Gouveia AD, Varma KK (1980) Circulation and water characteristics on isanosteric surfaces in the northern Arabian Sea during February-April. Indian J Mar Sci 9 (3):156-165

Davis RG, Weller R (1978) Propellor current sensors. In: Instruments and Methods in Air-Sea Interactions, preprint vol for NATO School, Ustaoset, Norway, April 1978, NATO Sci Comm

Deacon GER (1937) The hydrology of the Southern Ocean. Discovery Rep, 15:1-124

Deacon, GER, organizer, (1971) A Discussion of Ocean Currents and Their Dynamics. Philosophical Transactions of the Royal Society of London. (A) 270 (1206):349-465

Delecluse P, Philander SGH (1981) The Somali current. Ocean Modelling 36:4-6 (unpublished manuscript)

Del Grosso VA (1974) "New equation for the speed of sound in natural waters (with comparisons to other equations)", J Acoust Soc Am 56:1084-1091

Denham RN, Crook FG (1976) The Tasman Front. NZ Jour Mar Freshwat Res 10:15-30

Denman KL, Platt T (1976) The variance spectrum of phytoplankton in a turbulent ocean. J Mar Res 34:593-601

Denman K, Okubo A, Platt T (1977) The chlorophyll fluctuation spectrum in the sea. Limnol & Oceanogr 22 (6):1033-1038

DeSanto JA (Ed) (1979) Ocean Acoustics. Topics in Current Physics (8), Springer Berlin Heidelberg New York

De Szoeke RA (1977) A model of baroclinic instability in the Southern Ocean EOS 58, 1168 pp (abstract)

De Szoeke RA (1978) Inferred eddy salt fluxes in the Drake Passage EOS 59, 1120 pp (abstract)

De Szoeke RA, Levine MD (1981) The advective flux of heat by mean geostrophic motions in the Southern Ocean. Deep Sea Res 28, 1057-1085

Dexter SC, Milliman JD, Schmitz WJ (1975) Mineral deposition in current meter bearings. Deep Sea Res 22:703–706

Dickson RR, Gurbutt P (1979) A 5500 km XBT section from the northeast Atlantic. Polymode News 64

Dickson RR, Gurbutt P (1980) XBT profiles of the northeast Atlantic in June 1979. Polymode News 74

Dickson RR, Hughes DG (1981) Satellite evidence of mesoscale activity over the Biscay Abyssal Plain. Oceanologica Acta 4:43–46

Dickson RR, Gurbutt PA, Medler KJ (1982) Long-term water movements in the southern trough of the Charlie-Gibbs Fracture Zone. J Mar Res 38:571–583

Dickson RR, Gould WJ, Gurbutt PA, Killworth PD (1982) A seasonal signal in ocean currents to abyssal depths. Nature 295, pp 193–198

Dietrich G (1935) Aufbau und Dynamik des südlichen Agulhasstromgebietes. Veröffentlichungen des Instituts für Meereskunde, Berlin, Reihe A (27) 79 pp

Dietrich G (1969) Atlas of the hydrography of the northern North Atlantic Ocean based on the Polar Front Survey during the International Year winter and summer 1958. Int Counc Explor Sea, Serv Hydrogr Copenhagen, 140 pp

Doblar RA, Cheney RE (1977) Observed formation of a Gulf Stream cold-core ring. J Phys Oceanogr 7 (6):944–946 (R748)

Dow D, Rossby HT, Signorini S (1977) SOFAR floats in MODE, final report of float trajectory data. Univ of Rhode Island, Graduate School Oceanogr, Tech Rep 77-3, 108 pp

Drever RG, Sanford TB (1970) A free-fall electromagnetic current meter – instrumentation. In: Proc IERE Conf Electron Eng Ocean Technol, Inst Electr Radio Engineer, London, pp 353–370

Drever RG, Sanford TB (1976) A velocity profiler based on acoustic doppler principles. WHOI Ref No 76-96 Woods Hole Oceanogr Inst

Düing WO, Mooers CNK, Lee TN (1977) Low-frequency variability in the Florida Current and relations to atmospheric forcing from 1972 to 1974. J Mar Res 35:129–161

Düing WO, Johnson D (1972) High resolution current profiling in the Straits of Florida. Deep Sea Res 19:259–274

Düing WO, Molinari RL, Swallow JC (1980) Somali current: evolution of surface flow. Sci 209:588–590

Düing WO, Ostapoff F, Merle J (1980) Physical oceanography of the tropical Atlantic during GATE. Global Atmos Res Prog (GARP), Atl Trop Exp, Univ Miami, Kingsport Press, USA

Dugan JP, Mied RP, Mignerey PC, Schuetz AF (1982) Compact, intra-thermocline eddies in the Sargasso Sea. J Geophys Res 87:385–393

Duncan CP (1968) An eddy in the Subtropical Convergence southwest of South Africa. J Geophys Res 73:531–534

Duncan CP (1970) The Agulhas Current. PhD Thesis, Univ of Hawaii, 76 pp

Ebbesmeyer CC, Taft BA (1979) Variability of potential energy, dynamic height, and salinity in the main pycnocline of the western North Atlantic. J Phys Oceanogr 9:1073–1089

Elliott BA (1981) Anticyclonic rings in the Gulf of Mexico. J Phys Oceanogr (submitted)

Emery WJ, Ebbesmeyer CC, Dugan JP (1980) The fraction of vertical isotherm deflections associated with eddies: an estimate from multiship XBT surveys. J Phys Oceanogr 10:885–899

Enright JT (1977) Copepods in a hurry: sustained high-speed upward migration. Limnol & Oceanogr 22 (1):118–125

Eriksen CC (1980) Evidence for a spectrum of equatorial waves in the Indian Ocean. J Geophys Res

Erofeeva GS (1972) Geostrophic currents of the northeast Atlantic. Trudy Gosudarstvennyi Okeanogr Ins 114:96–108 (Russian)

Ertel H (1942) Ein neuer hydrodynamischer Wirbelsatz. Meteorol Z 59:277–282

Evenson AJ, Veronis G (1975) Continuous representation of wind stress and wind stress curl over the world ocean. J Mar Res 33:131–144 (Suppl)

Ewart TE, Bendiner WP (1981) An observation of the horizontal and vertical diffusion of a passive tracer in the deep ocean. J Geophys Res 86:10974–10982

Ewing M, Worzel JL (1948) "Long Range Sound Transmission". In: Propagation of Sound in the Ocean. Geol Soc Amer Mem 27

Fairbanks RG, Wiebe PH, Bé AWH (1980) Vertical distribution and isotopic composition of living planktonic Foraminifera in the Western North Atlantic. Sci 207 (4426):61–63

Fandry C (1979) Baroclinic instability of the Antarctic Circumpolar Current in the Drake Passage. Ocean Modelling 22:8–9 (unpublished manuscript)

Fasham MJR (1978a) The statistical and mathematical analysis of plankton patchiness. Oceanogr Mar Biol Ann Rev 16:48–79

Fasham MJR (1978b) The application of some stochastic processes to the study of plankton patchiness. In: Steele JH (ed) Spatial pattern in plankton communities. Plenum Press, New York, pp 131–156

Fedorov KN, Ginsburg AI, Zatsepin AG (1978) Systematic differences in isotherm depths derived from XBT and CTD data. POLYMODE News 50 (unpublished manuscript)

Fedorov KN, Ginsburg AI, Zatsepin AG (1979) On systematic discrepancies of isotherm depth positions by XBT and CTD data. Okeanologicheskie Issledovaniya, No 30 POLYMODE Hydrophys Expedition, Acad Sci USSR, Sov Geophys Comm, Moscow (In Russian)

Findlay AG (1866) A Directory for the navigation of the Indian Ocean, 1st edn. Richard Holmes Laurie, London

Filloux JH (1970) Deep-sea tide gauge with optical readout of bourdon tube rotations. Nature 243:217–221

Fisher A Jr (1972) Entrainment of shelf water by the Gulf Stream north-east of Cape Hatteras. J Geophys Res 77:3248–3255

Fisher RA (1937) The wave of advance of advantageous genes. Ann Euge 7:355–369

Fjortoft R (1953) On the changes in the spectral distribution of kinetic energy for two-dimensional nondivergent flow. Tellus 5:225–230

Flatte SM (ed) (1979) Sound transmission through a fluctuating ocean. Cambridge Univ Press, 299 pp

Flierl G (1974) XBT-CTD intercomparison. In: Instrument Description and Intercomparison Rep of the MODE-I Intercomparison Group, December 1974 (unpublished manuscript)

Flierl G (1976) Motions of tracer particles in Gulf Stream rings. POLYMODE News 8 (1) (unpublished manuscript)

Flierl GR (1977a) Correcting expendable bathythermograph (XBT) data for salinity effects to compute dynamic heights in Gulf Stream rings. Deep Sea Res 25:129–134

Flierl GR (1977b) The application on linear quasigeostrophic dynamics to Gulf Stream Rings. J Phys Oceanogr 7 (3):365–379

Flierl GR (1979) Baroclinic solitary waves with radial symmetry. Dyn Atmos Oceans 3:15–38.

Flierl GR (1981) Particle motions in large-amplitude wave fields. Geophys Astrophys Fluid Dyn, 18:39–74

Flierl GR, Dewar WK (1982) Motion and dispersion of dumped material by large amplitude eddies. (in preparation)

Flierl GR, Robinson AR (1977) XBT measurements of thermal gradients in the MODE eddy. J Phys Oceanogr 7:300–302

Fofonoff NP (1981) The Gulf Stream system. In: Warren BA, Wunsch C (eds), Evolution of Physical Oceanography, chapter 4. MIT Press, Cambridge, MA

Fofonoff NP, Ercan Y (1967) Response characteristics of a Savonius rotor current meter. WHOI Tech Rep 67-33, Woods Hole Oceanogr Inst

Fofonoff NP, Hayes SP, Millard RC Jr (1974) WHOI/Brown CTD Microprofiler: Methods of calibration and data handling. WHOI Ref No 74-89, Woods Hole Oceanogr Inst

Fofonoff NP, Millard B (1967) CTD intercomparisons aboard R/V Akademik Vernadsky. POLYMODE News 19 (unpublished manuscript)

Fofonoff NP, and Tabata S (1966) Variability of oceanographic conditions between Ocean Station P and Swiftsure Bank off the Pacific coast of Canada. JFRB 23 (6):825–868

Fofonoff NP, Webster F (1971) Current measurements in the Western Atlantic. Philos Trans R Soc London, A2/0:423–436

Ford WL, Longard JR, Banks RE (1952) On the nature, occurrence and origin of cold low salinity water along the edge of the Gulf Stream. J Mar Res 11:281–293

Fornshell JA (1979) Microplankton patchiness in the Northwest Atlantic Ocean. J Protozool 26, (2):270–272

Foster LA (1972) Current measurements in Drake Passage. MS Thesis, Dalhousie Univ Halifax, 61 pp

Fox DG, Orszag SA (1973) Inviscid dynamics of two-dimensional turbulence. Phys Fluids 2:169–171

Frankignoul CC (1981) Low frequency temperature fluctuations off Bermuda. J Geophys Res 86:6522–6528

Frederiksen JS, Sawford BL (1980) Statistical dynamics of two-dimensional inviscid flow on a sphere. J Atmos Sci 37:717–732

Freeland HJ, Gould WJ (1976) Objective analysis of mesoscale oceanic circulation features. Deep Sea Res 23:915–923

Freeland HJ, Rhines PB, Rossby HT (1975) Statistical observations of the trajectories of neutrally buoyant floats in the North Atlantic. J Mar Res 33:383–404

Fruchaud-Laparra B, Le Floch F, Le Roy C, Le Tareau JY, Madelain F (1976) Etude hydrologique et variations saisonnieres dans le proche Atlantique en 1974. Publ Centre Nat pour l'Exploitation des Oceans (CNEXO) Ser Rap sci tech Nr 30, 108 pp

Frye HW, Pugh JD (1971) A new equation for the speed of sound in seawater. J Acoust Soc Amer 50:384–386

Fu L-l, Wunsch C (1979) Recovery of POLYMODE Array III Clusters A and B. POLYMODE News 60 (unpublished manuscript)

Fu L-l, Keffer T, Niiler PP, Wunsch C (1982) Observations of mesoscale variability in the western North Atlantic: A comparative study, J Mar Res (in press)

Fuglister FC (1960) Atlantic Ocean Atlas of temperature and salinity profiles and data from the International Geophysical Year of 1957–58. Woods Hole Oceanogr Inst Atlas ser 1, 209 pp

Fuglister FC (1963) Gulf Stream '60. Prog Oceanogr 1:265–383

Fuglister FC (1972) Cyclonic rings formed by the Gulf Stream 1965–1966. In: Gordon A (ed) Studies in Phys Oceanogr: A tribute to George Wust on his 80th birthday. Gordon and Breach, New York, pp 137–168

Fuglister FC (1977) A cyclonic ring formed by the Gulf Stream, 1967. In: Angel M (ed) A Voyage of Discovery, the George Deacon 70th Anniversary Vol. Supp Deep Sea Res

Fuglister FC, Worthington LV (1947) Hydrography of the Western Atlantic; meanders and velocities of the Gulf Stream. Woods Hole Oceanogr Inst Tech Rep No WHOI 47-9

Fuglister FC, Worthington LV (1951) Some results of a multiple ship survey of the Gulf Stream. Tellus 3 (1):1–14

Furlong L (1812) The American Coast Pilot, 7th edn. EM Blunt, New York, 311 pp

Gage KS (1979) Evidence for a $k^{-5/3}$ law intertial range in mesoscale two-dimensional turbulence. J Atmos Sci 36:1950–1954

Galt J (1967) Current measurements in the Canadian basin of the Arctic Ocean, summer 1965. Tech Rep, Dept Oceanogr, Univ Washington, 184, 17 pp

GARP/WMO-ICSU (1973) The First GARP Global Experiment: Objectives and Plan. GARP Publication Series No 11, Secretariat, World Meteorological Organization, Geneva

Garrett C (1979) Topographic waves off East Australia. Identification and role in shelf circulation J Phys Oceanogr 9:244–253

Garrett C (1983) On the initial streakiness of a dispersing tracer in two- and three-dimensional turbulence. Dyn Atm Oceans (submitted for publication)

Garrett C, Munk W (1979) Internal waves in the ocean. Annu Rev Fluid Mech 11:339–369

Garrett JF (1980) Availability of the FGGE drifting buoy system data set. Deep Sea Res 27A:1083–1086

Gascard JC (1978) Mediterranean deep water formation, baroclinic instability and oceanic eddies. Oceanologia Acta 1:315–330

Gatien MG (1976) A study in the slope water region south of Halifax. J Fish Res Board Canada 33:2213–2217

Gent PR, McWilliams JC, Nergaard D (1978) The USNS Bartlett POLYMODE Cruise. PO-LYMODE News 47

Gentili J (1972) Thermal anomalies in the eastern Indian Ocean. Nat Phy Sci 238:93–95

Gill AE (1968) A linear model of the Antarctic circumpolar current. J Fluid Mech 32:465–488

Gill AE (1975) Evidence for mid-ocean eddies in weather ship records. Deep Sea Res 22:647–652

Gill AE (1977) Potential vorticity as a tracer, Appendix to A.F. Pearce. Some features of the upper 500m of the Agulhas Current. J Mar Res 35:752–753

Gill AE, Green JSA, Simmons AJ (1974) Energy partition in the large-scale ocean circulation and the production of mid-ocean eddies. Deep Sea Res 21:499–528

Godfrey JS (1973) Comparison of the East Australian Current with the western boundary flow in Bryan and Cox's (1968) numerical ocean model. Deep Sea Res 20:1059–1076

Godfrey JS, Robinson AR (1971) The East Australian Current as a free inertial jet. J Mar Res 29:256–280

Godrey JS, Cresswell GR, Golding TJ, Pearce AF, Boyd R (1980) The separation of the East Australian Current. J Phys Oceanogr 10:430–440

Goldberg ED (ed) (1979) Proceedings of the Workshop on Assimilative Capacity of U.S. Coastal Waters for Pollutants: 29 July–4 August, 1979. NOAA/ERL Boulder CO 284 pp

Golding TJ, Symonds G (1978) Some surface circulation features off western Australia during 1973–1976. Aust J Mar Freshwater Res 29:187–191

Golding TJ, Cresswell GR, Boland FM (1977) Sea surface current and temperature data report from the "Sprightly" programme off western Australia 1973–1976. CSIRO, Div Fish Oceanogr, Rep No 90, 46 pp

Gonella J, Fieux M, Philander G (1981) Mise en evidence d'ondes de Rossby équatoriales dans l'océan Indien au moyen de bouées dérivantes. CR Acad Sci 292 (11):1397–1399

Gordon AL (1972) On the interaction of the Antarctic Circumpolar Current and the Macquarie Ridge. In: Hayes DE (ed) Antarctic Oceanology II. AGU Washington DC pp 71–78

Gordon AL (1975) General ocean circulation. Numerical models of ocean circulation, N A Sci, Washington, DC, 364 pp

Gordon AL (1978) Deep Antarctic convection west of Maud Rise. J Phys Oceanogr 8:600–612

Gordon AL, Georgi DT, Taylor HW (1977) Antarctic polar front zone in the western Scotia Sea, Summer 1975. J Phys Oceanogr 7:309–328

Gordon AL, Molinelli EJ, Baker T (1978) Large-scale relative dynamic topography of the Southern Ocean. J Geophys Res 83:3023–3032

Gordon AL, Molinelli EJ, Baker T (1982) Southern Ocean Atlas. Columbia Univ Press, New York

Gotthardt, GA (1973) Observed formation of a Gulf Stream anticyclonic eddy. J Phys Oceanogr 3:237–238

Gotthardt GA, Potocsky GJ (1974) Life cycle of a Gulf Stream anticyclonic eddy observed from several oceanographic platforms. J Phys Oceanogr 4:131–134

Gould WJ (1976a) A formation zone for Big Babies near the mid-Atlantic ridge? POLY-MODE News 16

Gould WJ (1976b) Mesoscale monitoring in the eastern North Atlantic (North East Atlantic Dynamics Study) NEADS. POLYMODE News 13

Gould WJ (1981) A front southwest of the Azores. Int Counc Expl Sea Hydrogr Comm CM 1981/C:16

Gould WJ, Cutler AN (1980) Long period variability of currents in the Rockall Trough. Int Counc Expl Sea Hydrogr Comm CM 1980/C:30

Gould WJ, Sambuco E (1975) The effect of mooring type on measured values of ocean currents. Deep Sea Res 22:55–62

Gould WJ, Schmitz WJ Jr, Wunsch C (1974) Preliminary field results for a mid-ocean dynamics experiment (MODE-O). Deep Sea Res 21:911–931

Gower JFR, Deman KL, Holyer RJ (1980) Phytoplankton patchiness indicates the fluctuation spectrum of mesoscale oceanic structure. Nature London, 288:157–159

Grachev YM, Koshlyakov MN (1977) Objective analysis of synoptic eddies in POLYGON-70. POLYMODE News 23 (unpublished manuscript)

Graves RD, Nagl A, Uberall H, Zazur GL (1975) Range dependent normal modes in under-water sound propagation: Application to the wedge-shaped ocean. J Acoust Soc Am 58:1171–1177

Groupe Tourbillon (1983) The Tourbillon Experiment: A Study of Mesoscale Eddy in the Eastern North Atlantic (submitted for publication)

Gründlingh ML (1974) A description of inshore current reversals off Richards Bay based on airborne radiation thermometry. Deep Sea Res 21:47–55

Gründlingh ML (1977) Drift Observations from Nimbus VI satellite-tracked buoys in the southwestern Indian Ocean. Deep Sea Res 24:903–913

Gründlingh ML (1978) Drift of a satellite-tracked buoy in the southern Agulhas Current and Agulhas Return Current. Deep Sea Res 25:1209–1224

Gründlingh ML (1979) Observation of a large meander in the Agulhas Current. J Geophys Res 84:3776–3778

Gründlingh ML (1980) On the volume transport of the Agulhas Current. Deep Sea Res 27:557–563

Gründlingh ML, Lutjeharms JRE (1979) Large scale flow patterns of the Agulhas Current System. S Afr Sci 75:269–270

Gulf Stream Monthly summary 5 (11):1–19. US Nav Oceanogr Off Washington, DC, 1970

Gulf Stream Monthly summary 6 (12):1–16. US Nav Oceanogr Off Washington, DC 1971

Gulf Stream Monthly summary 9 (7):1–12. US Nav Oceanogr Off Washington, DC, 1974

Gunn JT, Watts DR (1982) On the currents and watermasses north of the Antilles/Bahamas Arc. J Mar Res 40 (1)

Hagan DE, Olson DB, Schmitz JE, Vastano AC (1978) A comparison of cyclonic ring struc-tures in the northern Sargasso Sea. J Phys Oceanogr 8 (6):997–1008

Hager JG (1977) Kinetic energy exchange in the Gulf Stream. J Geophys Res 82:1718–1724

Haidvogel DB (1982) On the feasibility of particle tracking in Eulerian ocean models. Ocean Modelling (unpublished manuscript, in press)

Haidvogel DB, Held IM (1980) Homogeneous quasi-geostrophic turbulence driven by a uni-form temperature gradient. J Atmos Sci 37:2644–2660

Haidvogel DB, Holland WR (1978) The stability of ocean currents in eddy-resolving general circulation models. J Phys Oceanogr 8:393–413

Haidvogel DB, Robinson AR, Schulman EE (1980) The accuracy, efficiency, and stability of three numerical models with application to open ocean problems. J Comput Phys 34 (1):1–53

Haidvogel DB, Keffer T, Quinn BJ (1981) Dispersal of a passive scalar in two-dimensional turbulence Effective diffusivity. Ocean Modelling 41:1–4 (unpublished manuscript)

Haidvogel DB, Rhines PB (1983) Waves and circulation driven by oscillatory winds in an ide-alized ocean basin. Geophys and Astrophys Fluid Dyn (in press)

Halliwell GR Jr, Mooers CNK (1979) The space-time structure and variability of the shelf wa-ter-slope water and Gulf Stream surface temperature fronts and associated warm-core ed-dies. J Geophys Res 84 (C12):7707–7725

Halpern D (1978a) Mooring motion influences on current measurements. In: Proc Working Conf Current Measurement, Tech Rep DEL-SG-3-78, Coll Mar Stud, Univ Delaware, Newark pp 69–76

Halpern D (1978b) Moored current measurements in the upper ocean. In: Instruments and methods in air-sea interaction, preprint vol for NATO School, Ustaoset, Norway, April, 1978, NATO Sci Comm

Halpern D (1979) Observations of upper ocean currents at DOMES Sites A, B and C in the tropical central North Pacific Ocean during 1975 and 1976. In: Bischoff JL, Piper DZ (eds) Marine Geology and Oceanography of the Pacific Manganese Nodule Province. Ple-num Press, New York, pp 43–82

Halpern D, Pillsbury RD (1976) Influence of surface waves upon subsurface current measure-ments in shallow water. Limnol Oceanogr 21:611–616

Halpern D, Weller RA, Briscoe MG, Davis RE, McCullough JR (1981) Intercomparison tests of moored current measurements in the upper ocean. J Geophys Res 86:419–428

Hansen B, Meincke J (1979) Eddies and meanders in the Iceland-Faroe Ridge area. Deep Sea Res 26A:1067–1082

Hansen DV (1970) Gulf Stream meanders between Cape Hatteras and the Grand Banks. Deep Sea Res 17:495–511

Hamilton GR (1974) Time variations of sound speed over long paths in the ocean. In: International workshop on low frequency propagation and noise, Woods Hole, MA, October 14–19, pp 7–30

Hamilton KG, Siegmann WL, Jacobson MJ (1980) Simplified calculation of ray-phase perturbations due to ocean environmental variations. J Acoust Soc Am 67:1193–1206

Hamon BV (1961) Structure of the East Australian Current. CSIRO Aust Div Fish Oceanogr Tech Pap No 11

Hamon BV (1965a) The East Australian Current, 1960–1964. Deep Sea Res 12:899–921

Hamon BV (1965b) Geostrophic currents in the south-eastern Indian Ocean. Aust J Mar Freshwater Res 16:255–271

Hamon BV (1968a) Temperature structure in the upper 250 metres in the east Australian current area. Aust J Mar Freshwater Res 19:91–99

Hamon BV (1968b) Spectrum of sea level at Lord Howe Island in relation to circulation. J Geophys Res 73:6925–6927

Hamon BV (1970) Western boundary currents in the South Pacific. In: Wooster WS (ed) Sci Expl S Pac. Nat Acad Sci, Wash, p 50

Hamon BV (1972) Geopotential topographics and currents of West Australia, 1965–1969. CSIRO Division Fish Oceanogr, Tech Paper No 32, 11 pp

Hamon BV, Cresswell GR (1972) Structure functions and intensities of ocean circulation off East and West Australia. Aust J Mar Freshwater Res 23:99–103

Hamon BV, Kerr JD (1968) Time and space scales of variations in the East Australian Current. From Merchant Ship Data, Aust J Mar Freshwater Res 19:101–106

Hamon BV, Godfrey JS, Greig MA (1975) Relation of mean sea level, current and wind stress on the east coast of Australia. Aust J Mar Freshwater Res 26:389–403

Harris TFW (1970a) Planetary-type waves in the Southwest Indian Ocean. Nature 227:1043–1944

Harris TFW (1970b) Features of the surface currents in the South West Indian Ocean. Symp Oceanogr S Afr, Durban, 1970

Harris TFW (1972) Source of the Agulhas Current in the spring of 1964. Deep Sea Res 19:633–650

Harris TFW, van Foreest D (1978) The Agulhas Current in March 1969. Deep Sea 25 (6):549–561

Harris TFW, Stavropoulos CC (1978) Satellite-tracked drifters between Africa and Australia. Bull Am Meteorol Soc 59:51–59

Harris TFW, Legeckis R, van Foreest D (1978) Satellite infrared images in the Agulhas Current system. Deep Sea Res 25 (6):543–548

Harrison DE (1978) On the diffusion parameterization of mesoscale eddy effects from a numerical ocean experiment. J Phys Oceanogr 8:913–918

Harrison DE (1979) Eddies and the general circulation of numerical model gyres: an energetic perspective. Rev Geophys Space Phys 17:969–979

Harrison DE (1980a) Dissipation mechanisms and the importance of eddies in model ocean energy budgets. J Phys Oceanogr 10:900–905

Harrison DE (1980b) Some Eulerian scale analysis results: eddy terms in the mean heat, momentum and vorticity equations. J Phys. Oceanogr 10:1221–1227

Harrison DE, Holland WR (1981) Regional eddy vorticity transport and the equilibrium vorticity budgets of a numerical model ocean circulation. J Phys Oceanogr 11:190–208

Harrison DE, Robinson AR (1978) Energy analysis of open regions of turbulent mean eddy energetics of a numerical ocean eddy experiment. Dyn Atmos Oceans 2:185–211

Harrison DE, Robinson AR (1979) Boundary-forced planetary waves: A simple model mid-ocean response to strong current variability. J Phys Oceanogr 9:919–929

Hart JE (1974) On the mixed stability problem for quasi-geostrophic ocean currents. J Phys Oceanogr 4 (3):349–356

Hart JE, Killworth PD (1976) On open-ocean baroclinic instability in the Arctic. Deep Sea Res 23 (7):637–645

Hartline BK (1979) POLYMODE – Exploring the undersea weather. Sci 205:571–573

Harvey RR, Patzert WC (1976) Deep current measurements suggest long waves in the eastern equatorial Pacific. Sci 193:883–885

Hastenraht S (1980) Heat budget of tropical ocean and atmosphere. J Phys Oceanogr 10:159–170

Hata K (1969) Some problems relating to fluctuation of hydrographic conditions in the sea northeast of Japan (Part II) – Fluctuation of the warm eddy cut off northward from the Kuroshio. The Oceanogr Mag 21 (1):13–29

Hata K (1974) Behavior of a warm eddy detached from the Kuroshio. J Meteor Res 26:295–321 (in Japanese with English abstract)

Haury LR, Wiebe PM (1982) Oceanic zooplankton: distribution in fine scale multispecies aggregations. Nature (in press)

Haury LR, Wiebe PH, Boyd SH (1976) Longhurst-Hardy Plankton Recorders: their design and use to minimize bias. Deep Sea Res 23:1217–1229

Hayes SP (1979) Benthic current observations at DOMES sites A, B and C in the tropical North Pacific Ocean. In: Bischoff JL, Piper DZ, eds. Marine Geology and Oceanography of the Central Pacific Manganese Nodule Province. Plenum Press, New York, pp 83–112

Hayes SP (1982) The bottom boundary layer in the eastern tropical Pacific. J Phys Oceanogr (in press)

Hazelworth JB (1976) Oceanographic variations across the Gulf Stream off Charleston, South Carolina, during 1965 and 1966. NOAA Tech Rep ERL 383-AOML 25, 73 pp

Heath RA (1980) Eastward oceanic flow part northern New Zealand. NZ J Mar Freshwat Res 14:169–182

Heath RA, Bryden HL, Hayes SP (1978) Interaction of the Antarctic Circumpolar Current with topography south of New Zealand. Antarct J U S 13:76–78

Heinmiller RH (1968) Acoustic release systems. WHOI tech rep 68–48, Woods Hole Oceanogr Inst

Heinmiller RH (1976a) The woods hole buoy project moorings – 1960 trough 1974. WHOI ref 76–63, Woods Hole Oceanogr Inst

Heinmiller RH (1976b) Mooring operations techniques of the Buoy Project at the Woods Hole Oceanographic Institution. WHOI Ref 76–79, Woods Hole Oceanogr Inst

Heinmiller RH, LaRochelle RA (1977) Field experience with acoustic releases at the Woods Hole Oceanographic Institution. WHOI Ref No 77–10, Woods Hole Oceanogr Inst

Heinmiller RH, Walden RG (1973) Details of Woods Hole Moorings, WHOI Ref No 73–71, Woods Hole Oceanogr Inst

Heinmiller RH, Ebbesmeyer CC, Taft BA, Olson DB, Nikitin OP (1982) Intercomparisons of XBT and CTD isotherm depths in the nortwestern Atlantic. J Phys Oceanogr (in press)

Held IM (1975) Momentum transport by quasi-geostrophic eddies. J Atmos Sci 32 (7):1494–1497

Helland-Hansen B, Nansen F (1909) The Norwegian Sea. Rep Norw Fish Mar Invest 2 (2):359 pp

Hendry RM (1981) On the structure of the deep Gulf Stream. J Geophys Res (submitted)

Hendry RM, Hartling AJ (1979) A pressure-induced direction error in nickel-coated Aanderaa current meters. Deep Sea Res 26:327–335

Henke M (1978) A ten megameter Atlantic XBT section. POLYMODE News 59 (unpublished manuscript)

Henke M, Zenk W (1980) Megameter Atlantic XBT sections, II. POLYMODE News 77 (unpublished manuscript)

Henrick RF, Siegmann WL, Jacobson MJ (1977) General analysis of ocean eddy effects for sound transmission applications. J Acoust Soc Am 62:860–870

Henrick RF, Jacobson MJ, Siegmann WL (1980) General effects of currents and sound speed variations on short range acoustic transmission in cyclonic eddies. J Acoust Soc Am 67 (1):121–134

Herman AW, Denman KL (1976) Rapid underway profiling of chlorophyll with an in situ fluorometer mounted on a 'Batfish' vehicle. Deep Sea Res 24 (4):385–397

Herman AW, Denman KL (1979) Intrusions and vertical mixing at the shelf/slope water front south of Nova Scotia. J Fish Res Bd Can 36:1445–53

Herring JR (1977) On the statistical theory of two-dimensional topographic turbulence. J Atmos Sci 34:1731–1750

Herring JR, Orszag SA, Kraichnan RH, Fox DG (1974) Decay of two-dimensional homogeneous turbulence. J Fluid Mech 66:417–444

Hogg NG (1976) On spatially growing baroclinic waves in the ocean. J Fluid Mech 78:217–235

Hogg NG (1982) Topographic waves along 70°W on the continental rise. J Mar Res 39:627–649

Hogg NG, Schmitz WJ Jr (1980) A dynamical interpretation of low frequency motions near very rough topography – The Charlie Gibbs Fracture Zone. J Mar Res 38:215–248

Hogg NG, Sanford TB, Katz EJ (1978) Eddies, islands and mixing. J Geophys Res 83:2921–2938

Holland WR (1973) Baroclinic and topographic influences on the transport in western boundary currents. Geophys Fluid Dyn 4:187–210

Holland WR (1977) Ocean general circulation models. In: Goldberg E (ed) The Sea. Vol 6, Wiley & Sons, New York

Holland WR (1978) The role of mesoscale eddies in the general circulation of the ocean: numerical experiments using a wind-driven quasigeostrophic model. J Phys Oceanogr 8:363–392

Holland WR (1979) The general circulation of the ocean and its modelling. Dyn Atmos Oceans 3:111–142

Holland WR (1982) Well-mixed regions of potential vorticity in numerical models of midlatitude ocean circulation. J Phys Oceanogr (submitted)

Holland WR, Haidvogel DB (1980) A parameter study of the mixed instability of idealized ocean currents. Dyn Atmos Oceans 4:185–215

Holland WR, Lin LB (1975a) On the generation of mesoscale eddies and their contribution to the oceanic general circulation. I. A preliminary numerical experiment. J Phys Oceanogr 5:642–657

Holland WR, Lin LB (1975b) On the generation of mesoscale eddies and their contribution to the oceanic general circulation. II. A parameter study. J Phys Oceanogr 5:658–669.

Holland WR, Rhines PB (1980) An example of eddy-induced ocean circulation. J Phys Oceanogr 10:1010–1031

Holloway G (1976) PhD diss, Univ California, San Diego

Holloway G (1978) A spectral theory of non-linear barotropic motion above irregular topography. J Phys Oceanogr 8:414–417

Holloway G (1979) Mechanism and statistical mechanism in ocean circulation. POLYMODE News 61 (unpublished manuscript)

Holloway G (1982) A comment on streakiness. Ocean Modelling 43:5–6 (unpublished manuscript)

Holloway G, Hendershott MC (1977) Stochastic closure for non-linear Rossby waves. J Fluid Mech 82:747–765

Horn W., Schott F (1976) Measurements of stratification and currents at the Norwegian continental slope. Meteorol Forschungsergeb, Reihe A 18:23–63

Houghton RW, Smith PC, Fournier RO (1978) A simple model for cross-shelf mixing on the Scotian Shelf. J Fish Res Board Can 35:414–421

Howe MR, Tait RI (1967) A subsurface cold-core cyclonic eddy. Deep Sea Res 14:373–378

Hunkins KL (1974) Subsurface eddies in the in the Arctic Ocean. Deep Sea Res 21:1017–1033

Hunkins KL (1980) Review of the AIDJEX oceanographic program. In: Sea Ice Processes and Models. Univ Washington Press, Seattle pp 34–45

Huppert HE, Bryan K (1976) Topographically generated eddies. Deep Sea Res 23:655–679

Isaacs JD, Tont SA, Wicks GL (1974) Deep scattering layers: vertical migration as a tactic for finding food. Deep Sea Res 21:651–656

Iselin CO'D (1934) T-S correlation within the Florida Current. Trans Am Geophys Union, pp 208–209

Iselin CO'D (1936) A study of the circulation on the western North Atlantic. Pap Phys Oceanogr Met 4 (4):101 pp

Iselin CO'D (1940) Preliminary report on long-period variations in the transport of the Gulf Stream. Pap Phys Oceanogr 8 (1):40 pp

Iselin CO'D, Fuglister FC (1948) Some recent developments in the study of the Gulf Stream. J Mar Res 7 (3):317–329

ISOS Executive Committee (1977) International southern ocean studies program summary and long-range plans. Nat Sci Found, Washington, DC, 55 pp

Jahn AJ (1976) On the midwater fish faunas of Gulf Stream rings, with respect to habitat differences between Slope Water and northern Sargasso Sea. PhD Thesis, Woods Hole Oceanogr Inst, 172 pp

Jeffrey SW, Hallegraeff GM (1980) Studies of phytoplankton species and photosynthetic pigments in a warm core eddy of the East Australian Current. 1. Summer populations. Mar Ecology – Progress Ser 3:285–294

Jenkins WJ (1980) Tritium and He-3 in the Sargasso Sea J. Mar Res 38:533–569

Jennings JC Jr (1980) Meridional flux of silica in the Drake Passage. EOS 61, p. 263 (abstract)

Johns WE, Watts DR (1982). On the relationship of deep currents to Gulf Stream meanders east of Cape Hatteras. (in preparation)

Joyce TM, Patterson SL (1977) Cyclonic ring formation at the polar front in the Drake Passage. Nature 265:131–133

Joyce TM, Zenk W, Toole JM (1978) The anatomy of the Antarctic polar front in the Drake Passage. J Geophys Res 83:6093–6113

Joyce TM, Patterson SL, Millard RC Jr (1981) Anatomy of a cyclonic ring in the Drake Passage. Deep Sea Res 28:1265–1287

Joyce TM, Bitterman DS, Prada K (1982) Shipboard acoustic profiling of upper ocean currents. Deep Sea Res 29:903–913

Kalvaitis AN (1972) Survey of the Savonius rotor performance characteristics. Mar Tech Soc J 7, 3, May–June, pp 21–22

Kalvaitis AN (1974) Effects of vertical motion on vector averaging (Savonius rotor) and electromagnetic type current meters. NOAA Tech Mem NOAA-TM-NOS, NOIC-3, March

Kao TW, Cheney RE (1982) The Gulf Stream front: a comparison between SEASAT altimeter observations and theory. J Geophys Res 87:539–545

Kartavtseff A, Billant A (1979) Experience NEADS Northeast Atlantic Dynamic Studies Vol 1 Mesures de Courant dans l'Atlantique nord-est (47N 10W) Mai 1977 – Octobre 1978. Resul Campagnes a la Mer No 18, Centre Nat pour l'Exploitation des Oceans, Brest, France

Kartavtseff A, Billant A (1980) Experience NEADS Northeast Atlantic Dynamic Studies Vol 2 Mesures de Courant dans l'Atlantique nord-est (47N 10W) Octobre 1978 – Mai 1979. Resul Campagnes a la Mer No 18, Centre Nat pour l'Exploitation des Oceans, Brest, France

Kawai H (1980) Rings south of the Kuroshio and their possible roles in transport of the intermediate salinity minimum and in formation of the skipjack and albacore fishing grounds. Proc of Fourth Cooperative Studies of the Kuroshio (CSK), Tokyo, February 1979, Saikon, Tokyo, pp 250–273

Keffer T (1981) Time dependent temperature and vorticity balances and the stability of the Atlantic North-Equatorial Current. (submitted)

Keffer T, Haidvogel DB (1982) Numerical simulations of tracer streakiness. Ocean Modelling (unpublished manuscript, in press)

Keffer T, Niiler P (1978) Recovery of Polymode Array III, Cluster C in the North Atlantic equatorial current. POLYMODE News 56

Keller JB, Papadakis JS (1977) Wave propagation and underwater acoustics. Lecture notes in Physics V70, Springer, Berlin Heidelberg New York

Kenyon, KE (1978) The surface layer of the eastern North Pacific in winter. J Geophys Res 83:6115–6122

Kerr RA (1981) Small eddies proliferating in the Atlantic. Sci 213:632–634

Kerut EG (1980) Development of drifting buoy systems for oceanographic and meteorological applications. OCEANS '80 Proc, September, 1980

Killworth PD (1979) On "Chimney" formation in the ocean. JPO 9:531–554

Killworth PD (1980) Barotropic and baroclinic instability in rotating stratified fluids. Dyn Atmos Oceans 4:143–184

Kim K, Rossby T (1979) On the eddy statistics in a ring-rich area: A hypothesis of bimodal structure. J Mar Res 37:201–213

Kinder TH, Coachman LK (1977) Observation of a bathymetrically-trapped current ring. JPO 7:946–952

Kirwan AD Jr, McNally G, Coehlo J (1976) Gulf Stream kinematics inferred from a satellite-tracked drifter. J Phys Oceanogr 6:750–755

Kirwan AD, McNally GJ, Reyna E, Merrell WJ (1978) The near-surface circulation of the eastern North Pacific. J Phys Oceanogr 8:937–945

Kirwan AD Jr, McNally G, Pazan S, Wert R (1979) Analysis of surface current response to wind. J Phys Oceanogr 9:401–412

Kitano K (1975) Some properties of the warm eddies generated in the confluence zone of the Kuroshio and Oyshio currents. J Phys Oceanogr 5:245–252

Knauss JA (1967) The transport of the Gulf Stream. Rpts Plenary Session, 2nd Int Oceanogr Congress, June 1966, UNESCO Special Pub, pp 67–82

Knauss J (1969) A note on the transport of the Gulf Stream. Deep Sea Res 16:117–123 (suppl)

Knox RA (1976) On a long series of measurements of Indian Ocean equatorial currents near Addu Atoll. Deep Sea Res 23:211–221

Knox RA (1981) Time variability of Indian Ocean Equatorial Currents. Deep Sea Res 28:291–295

Knox RA, McPhaden MJ (1980) Profiles of velocity and temperature near the Indian Ocean equator. SIO Ref Ser No 80–1. Scripps Inst Oceanogr, Univ California, USA

Koblinsky CJ, Niiler PP (1982) The relationship between deep ocean currents and winds east of Barbados. J Phys Oceanogr 12(2):144–153

Koblinsky CJ, Keffer T, Niiler PP (1979) A compilation of observations in the Atlantic North equatorial current. Oregon State Univ, Ref 79–12, 119 pp

Kolmogorov AN (1941) The local structure of turbulence in an incompressible viscous fluid for very large Reynolds number. Dokl Akad Nauk SSR 30:299–303

Koninklijk Nederlands Meteorologisch Instituut (1952) Indian Ocean oceanographic and meteorological data. Publ No 135, 2nd edn Royal Netherlands Meteorol Inst, De Bilt

Koshlyakov MN, Monin AS (1978) Synoptic eddies in the ocean. Ann Rev Earth Planet Sci 6:495–523

Koshlyakov MN, Galerkin LI, Truong Dinh Hien (1970) On the mesostructure of geostrophic currents in the open ocean. Okeano 10(5):805–814 (in Russian; English translation in Oceano 10(5):637–646)

Koshlyakov MN, Grachev Yu M, Chyong Din Khiyen (1972) Technique of studying quasi-stationary ocean currents. Okeano 12(4):728–734 (in Russian; English translation in Oceano 12(4):609–614)

Kort VG, Titov VB, Osadchiy AS (1977) Kinematics and structure of currents in a study area in the Norwegian Sea. Oceano 17(15):505–508

Kraichnan RH (1967) Inertial ranges in two-dimensional turbulence. Physics Fluids 10:1417–1423

Kraichnan RH (1971) Inertial-range transfer in two- and three-dimensional turbulence. J Fluid Mech 47:525–535

Kraichnan RH (1976) Eddy viscosity in two and three dimensions. J Atmos Sci 33:1521–1536

Kraus W, Meinke J (1981) Drifting buoy tragectories in the North Atlantic Current. Nature (submitted)

Krishnamurti TN, Ardanuoy P, Ramanathan Y, Pasch R (1980) On the onset-vortex of the summer monsoons. In: Grossman RL (ed) Results of summer MONEX field phase research (part B). FGGE Oper Rep Vol 9, W M O, Geneva, pp 115–166

Kroll J (1979) The kinetic energy on a continental shelf from topographic Rossby waves generated of the shelf. J Phys Oceanogr 9:712–722

Kroll T, Niiler PP (1976) The transmission and decay of barotropic topographic Rossby waves incident on a continental shelf. J Phys Oceanogr 6:432–450

Kuo HH, Veronis G (1971) Distribution of tracers in the deep oceans of the world. Deep Sea Res 20:871–888

Kupferman S, Moore DE (1981) Physical oceanographic characteristics influencing the dispersion of dissolved tracer released at the sea floor in selected deep ocean study areas. Sandia Nat Labor Rep SAND80-2573

Lafond EC (1957) Oceanographic studies in the Bay of Bengal. Proc Indian Acad Sci B 46 (1):1–46

Lafond EC, Lafond KG (1968) Studies of oceanic circulation in the Bay of Bengal. Bull Nat Inst Sci India 38 (1):164–183

Lai DY, Richardson PL (1977) Distribution and movement of Gulf Stream rings. J Phys Oceanogr 7 (5):670–683

Lambert RB (1974) Small-scale dissolved oxygen variations and the dynamics of Gulf Stream eddies. Deep Sea Res 21:529–546

Landau L, Lifchitz E (1968) Mecanique. Editions de la paix, Moscow

Landau L, Lifchitz E (1971) Mecanique des fluides. Editions MIR, Moscow

LaViolette PE (1981) Variations in the frontal structure of the southern Grand Banks. Tech Note 87, April 1981, NORDA, NSTL Station, MS, 48 pp

LaViolette PE, Peteherych S, Gower JFR (1980) Oceanographic implications of features in NOAA satellite visible imagery. Boundary-Layer Meteor 18:159–175

Lazier JRN (1973) The renewal of Labrador Sea water. Deep Sea Res 20:341–353

Leaman KD, Chang S (1980) POLYMODE 1978: Hydrographic observations made aboard the R/V James M. Gillis during the POLYMODE Local Dynamics Experiment. Univ Miami, Tech Rep UMRSMAS-80004, 117 pp

Lee TN (1975) Florida Current spin-off eddies. Deep Sea Res 22:753–765

Lee TN, Mayer DA (1977) Low-frequency current variability and spin-off eddies of southeast Florida. J Mar Res 35:193–220

Lee TN, Mooers CNK (1977) Near-bottom temperature and current variability over the Miami Slope and Terrace, Bull Mar Sci 27:758–775

Lee TN, Atkinson LP, Legeckis R (1980) Observations of a Gulf Stream frontal eddy on the Georgian continental shelf, April 1977. Deep Sea Res 28:347–378

Leetmaa A, Bunker A (1978) Updated charts of the mean annual wind stress, convergences in Ekman layers, and Sverdrup transports in the North Atlantic. J Mar Res 36:311–322

Leetmaa A, Rossby HT, Saunders PM, Wilson D (1980) Subsurface circulation in the Somali current. Sci 209:590–592

Legeckis R (1975) Applications of synchonous meteorological satellite date to the study of time-dependent sea surface temperature changes along the boundary of the Gulf Stream. Geophys Res Lett 2:435–438

Legeckis R (1977a) Oceanic polar front in the Drake Passage-satellite observations during 1976. Deep Sea Res 24:701–704

Legeckis R (1977b) Long waves in the eastern equatorial Pacific Ocean: A view from a geostationary satellite. Sci, pp 1179–1181

Legeckis R (1979) Satellite observations of the influence of bottom topography on the seaward deflection of the Gulf Stream off Charleston, South Carolina. J Phys Oceanogr 9:483–497

Legeckis R, Gordon AL (1982) Satellite observations of the Brazil and Faulkland Current, 1975–1976 and 1978. Deep Sea Res 29:315–401

Leipper DF (1970) A sequence of current patterns in the Gulf of Mexico. J Geophys Res 75:637–657

Leith CE (1971) Atmospheric predictability and two-dimensional turbulence. J Atmos Sci 28:145–161

Levitus S, Oort AH (1977) Global analysis of oceanographic data. Bull Amer Meteor Soc 59:1270–1284

Lighthill MJ (1969) Dynamic response of the Indian Ocean to onset of the southwest monsoon. Phil Trans Royal Soc London, A 265:45–92

Longhurst AR, Reith AD, Bower RE, Seibert DLR (1966) A new system for the collection of multiple serial plankton samples. Deep Sea Res 13 (2):213–222

Longuet-Higgins MS (1965) The response of a stratified ocean to stationary or moving wind-systems. Deep Sea Res 12:923–973

Louis JP, Smith PC (1982) The development of the barotropic radiation field of an eddy over a slope. J Phys Oceanogr 12:56–73

Louis JP, Petrie BD, Smith PC (1982) Observations of topographic Rossby waves on the continental margin off Nova Scotia. J Phys Oceanogr 12:47–55

Lumley JL (1970) Stochastic tools in turbulence. Academic Press, New York, 194 pp

Lutjeharms JRE (1978) Studying the ocean by satellite. S Afr Shipp Fish Ind Rev 33 (10)

Lutjeharms JRE (1980a) Drift of the Venoil in the Agulhas Current. Mariners Weather Log 24 (1):1–6

Lutjeharms JRE (1980b) The influence of the Agulhas Current. CSIR S Afr Res Rep 376, 16 pp (Paper pres at IUGG Symp Canberra, Dec 1979)

Lutjeharms JRE (1981) Features of the southern Agulhas Current circulation from satellite remote sensing. S Afr J Sci 77:231–236

Lutjeharms JRE, Baker D Jr (1980) A statistical analysis of the meso-scale dynamics of the Southern Ocean. Deep Sea Res 27:145–159

Lutjeharms JRE, Fromme GAW, Duncan CP (1980) Large scale motion in the south west Indian Ocean. CSIR S Afr, Res Rep 375, 22 pp (Paper pres at IUGG Symp, Canberra, Dec 1979)

Luyten JR (1977) Scales of motion in the deep Gulf Stream and across the continental rise. J Mar Res 35:49–74

Luyten JR (1981) Recent observations in the equatorial Indian Ocean. In: Lighthill J, Pearce RP (eds), Monsoon dynamics. Cambridge Univ Press, Cambridge London, New York, pp 465–480

Luyten, JR (1982) Equatorial current measurements, 1: moored observations. J Mar Res 40 (1):19–41

Luyten JR, Roemmich DH (1982) Equatorial currents at semiannual period in the Indian Ocean. J Phys Oceanogr 12 (5):406–913

Luyten JR, Swallow JC (1976) Equatorial undercurrents. Deep Sea Res 23:999–1001

Luyten JR, Fieux M, Gonella J (1980) Equatorial currents in the western Indian Ocean. Sci 209:600–602

Mackintosh NA (1946) The Antarctic convergence and the distribution of surface temperatures in Antarctic waters. Discovery Rep 23:177–212.

Madelain F, Kerut E (1978) Evidence of mesoscale eddies in the Northeast Atlantic from a drifting buoy experiment. Oceanologica Acta 1:159–168

Maillard C (1981) Mean circulation in the North East Atlantic from historical data. Int Counc Expl Sea Hydrogr Comm CM 1981/C:41

Malan OG, Schumann EH (1979) Natal shelf circulation revealed by LANDSAT imagery. S Afr J Sci 75:136–137

Manley TO (1981) Eddies of western Arctic Ocean – their characteristics and importance to the energy, heat, and salt balance. PhD Thesis, Columbia Univ, 426 pp

Manley TO, Hunkins KL (1980) Oceanographic measurements at the Fram I Ice Station, etc. Trans Am Geophys Union, 61 (17) (abstract)

Manley TO, Hunkins K, Tiemann W (1980) Arctic ice dynamics joint experiment 1975–1976. Physical oceanography data report. Profiling current meter data, Camp Caribou, Vol 1. Tech Rep 4, Lamont Doherty Geo Observatory, Columbia Univ

Mann CR (1967) The termination of the Gulf Stream and the beginning of the North Atlantic Current. Deep Sea Res 14:337–359

Mann CR (1977) Currents and water masses in the vicinity of Drake Passage. Polar Oceans, Arct Inst North Am Calgary, 681 pp

Marietta MG, Robinson AR (1980) Proceedings of a workshop on physical oceanography related to the subseabed disposal of high-level nuclear waste: 14–16 January 1980. Sandia Nat Labor Doc SAND80-1776, 318 pp

Martin P, Gillespie CR (1978) Arctic Odyssey – five years of data buoys in AIDJEX. AIDJEX
    Bull No. 40:7–14

Masuzawa J (1969) Subtropical mode water. Deep Sea Res 16:463–472

Maul GA (1977) The annual cycle of the Gulf Loop Current. Part I: observations during a
    one-year time series. J Mar Res 35:29–47

Maul GA, Norris DR, Johnson WR (1974) Satellite photography of eddies in the Gulf Loop
    Current. Geophys Res Lett 1:256–258

Maul G, DeWitt PW, Yanaway A, Baig SR (1978) Geo-stationary satellite observations of
    Gulf Stream meanders: infrared measurements and time series analysis. J Geophys Res
    83:6123–6135

McCartney MS (1976) The interaction of zonal currents with topography with applications to
    the Southern Ocean. Deep Sea Res 23:413–427

McCartney MS, Worthington LV, Raymer ME (1980) Anomalous water mass distributions at
    55W in the North Atlantic in 1977. J Mar Res 38:147–172

McCartney MS, Worthington LV, Schmitz WJ Jr (1978) Large cyclonic rings from the north-
    east Sargasso Sea. J Geophys Res 83 (C2):901–914

McCullough JR (1974) In search of moored current sensors. Proc 10th Ann MTS Conf, Wash,
    DC

McCullough JR (1975) Vector averaging current meter speed calibration and recording tech-
    nique. WHOI Ref No 75-44, Woods Hole Oceanogr Inst

McCullough JR (1978a) Near-surface ocean current sensors: problems and performance. In:
    Proc Working Conf on Current Meas, Tech DEL-SG-3-78, College of Marine Studies,
    Univ Delaware, Newark, Delaware

McCullough JR (1978c) Techniques of measuring currents near the ocean surface. In: Instru-
    ments and methods in air-sea interaction, preprint volume for NATO School, Ustaoset,
    Norway, April 1978 NATO Sci Comm

McCullough JR, Graeper W (1979) Moored Acoustic Travel Time (ATT) current meters: Evo-
    lution, performance, and future designs. WHOI Ref No 79-92, Woods Hole Oceanogr
    Inst

McDowell SE (1977) A note on XBT accuracy. POLYMODE News 29 (unpublished manu-
    script)

McDowell SE (1982) Isopycnal hydrography and mixing in the North Atlantic Ocean (sub-
    mitted)

McDowell SE, Rossby HT (1978) Mediterranean water: an intense mesoscale eddy off the Ba-
    hamas. Sci 202:1085–1087

McDowell SE, Rhines P, Keffer T (1982) North Atlantic potential vorticity and its relation to
    the general circulation. J Mar Res (submitted)

McEwan AD, Thomson RORY, Plumb RA (1980) Mean flows driven by weak eddies in rotat-
    ing systems. J Fluid Mech 99:655–672

McGowan JA (1974) The nature of oceanic ecosystems. In: Miller CB (ed) The biology of the
    oceanic Pacific. Oregon State Univ Press, pp 7–28

McGowan JA, Hayward TL (1978) Mixing and oceanic productivity. Deep Sea Res 25:771–
    793

McGowan JA, Walker PW (1980) Structure in the copepod fraction of the community of the
    North Pacific central gyre. Ecol Monogr 49 (2):195–226.

McWilliams JC (1976) Maps from the mid-ocean dynamics experiment. Part II: Potential vor-
    ticity and its conservation. J Phys Oceanogr 6:828–846

McWilliams JC, (1979) A Review of Research on Mesoscale Ocean Currents. Reviews of Geo-
    physics and Space Physics 17 (7): 1548—1558.

McWilliams JC, Chow JHS (1981) Equilibrium geostrophic turbulence: a reference solution
    in a β-plane channel. J Phys Oceanogr 11:921–949

McWilliams JC, Flierl GR (1976) Optimal, quasi-geostrophic analyses of MODE array data.
    Deep Sea Res 23:285–300

McWilliams JC, Flierl GR (1979) On the evolution of isolated, nonlinear vortices. J Phys
    Oceanogr 9:1155–1182

McWilliams JC, Heinmiller RM (1978) The POLYMODE Local Dynamics Experiment: ob-
    jectives location and plan. US POLYMODE Office, MIT, Cambridge, MA

McWilliams JC, Holland WR, Chow JHS (1978) A description of numerical Antarctic Circumpolar Currents. Dyn Atmos Oceans 2:213–291

MEDOC Group (1970) Observation of deep water formation in the Mediterranen Sea. Nature 227:1037–1040

Meinke J (1975). Evidence for atmospheric forcing of Arctic water overflow events. ICES CM 1975/C:29 13 pp (mimeo)

Merilees P, Warn H (1975) On energy and enstrophy exchanges in two-dimensional non-divergent flow. J Fluid Mech 69:625–630

Meyers G (1980) Do Sverdrup transports account for the Pacific North Equatorial Countercurrent? J Geophys Res 85:1073–1075

Middleton J (1979) The analysis of FGGE drifter positional data. Proc Australia-New Zealand GARP Symp, Melbourne (unpublished manuscript)

Middleton JH, Foster TD (1977) Tidal currents in the central Weddell Sea. Deep Sea Res 24:1195–1202

Mied RP, Lindemann GJ (1979) The propagation and evolution of cyclonic Gulf Stream rings. J Phys Ocenogr 9:1183–1206

Milder DM (1969) Ray and wave invariants for SOFAR channel propagation. J. Acoust Soc Am 46:1259–1263

Mills CA, Rhines P (1979) The deep western boundary current at the Blake-Bahama Outer Ridge – current meter and temperature observations 1977–78. WHOI Tech Rep WHOI-79-85, 77 pp

Miller R, Robinson AR, Haidvogel DB (1982) A baroclinic quasi-geostrophic open ocean model. J Comput Phys (in press)

Mintz Y (1979) On the simulation of the oceanic general circulation. Proceedings of the JOC Study Conference on climate models, Washington, DC, April 1978. GARP publ ser 22, World Meteorolog Organ, Geneva, 1979, Volume II, pp 607–687

Mizenko D, Chamberlin JL (1979) Anticyclonic Gulf Stream eddies off the northeastern United States during 1976. In: Goulet JR Jr, Haynes ED (eds) Ocean Variability in the US Fishery Conservation Zone, 1976. NOAA Tech Rep NMFS Circ 427, June 1979, pp 259–280

MODE Group (1978) The mid-ocean dynamics experiment. Deep Sea Res 25:859–910

MODE-1 Atlas Group (1977) Atlas of the mid-ocean dynamics experiment (MODE-1) Lee V, Wunsch C (ed), MIT, Cambridge, MA, 274 pp

MODE-I Intercomparison Group (1974) Instrument description and intercomparison report. POLYMODE Office, MIT, Cambridge, MA (unpublished manuscript)

Moller DA (1976) A computer program for the design and static analysis of single-point subsurface mooring systems; NOYFB. WHOI Ref No 76–59, Woods Hole Oceanogr Inst

Molinari RL (1970) Cyclonic ring spin-down in the North Atlantic. PhD Thesis, Texas A & M Univ, 106 pp

Molinari R, Kirwan AD Jr (1975) Calculations of differential kinematic properties from Lagrangian observations in the western Caribbean Sea. J Phys Oceanogr 5:483–491

Moore DW, Philander SGH (1977) Modeling the tropical oceanic circulation. In: Goldberg ED (ed) The Sea, Vol 6. Wiley & Sons, New York, pp 319–361

Morel P, Larchevec M (1974) Relative of constant level balloons in the 200-mb general circulation. J Atmos Sci 31:2189–2196

Morey R (1973) Evalution of long term deep sea effects on mooring line components. Charles Stark Draper Lab E-2748, Cambridge, MA

Morgan CW, Bishop JM (1977) An example of Gulf Stream eddy-induced water exchange in the mid-Atlantic Bight. J Phys Oceanogr 7:472–479

Morgan GW (1956) On the wind-driven ocean circulation. Tellus 8:301–320

Mosley W et al (1979) Summary of Freddex Developments prior to at-sea phase. Freddex Rpt. #1, Nav Res Lab, Washington, DC

Mountain DG, Shuhy JL (1980) Circulation near the Newfoundland Ridge. J Mar Res, 38:205–213

Mühry A (1864) Die Meeresströmungen an der Südspitze Afrikas. Petermanns geogr Mitteilungen 35–36

Müller P, Frankignoul C (1981) Direct atmospheric forcing of geostrophic eddies. J Phys Oceanogr 11:287–308

Müller TJ (1981) Current and temperature measurements in the north-east Atlantic during NEADS. Berichte aus dem Inst für Meereskunde Ch-Albrechts-Univ, Kiel, 90, 100 pp

Munk WH (1950) On the wind-driven ocean circulation. J Meteorol 7:79–93

Munk WH (1980) Horizontal deflection of acoustic paths by mesoscale eddies. J Phys Oceanogr 10:596–604

Munk WH, Palmén E (1951) Note on the dynamics of the Antarctic circumpolar current. Tellus 3:53–55

Munk W, Wunsch C (1979) Ocean Acoustic Tomography; a scheme for large scale monitoring. Deep Sea Res 26:123–161

Mysak L, Schott F (1977) Evidence for baroclinic instability of the Norwegian Current. JGR 82 (15):2087–2095

Nath JH (1977) Laboratory validation of numerical model drifting buoy-tethering-droguing system. Final Rep NOAA Data Buoy Office, NSTL Station, MS, Contract No 03-6-038-128

Needell GJ (1980) The distribution of dissolved silica in the deep western North Atlantic Ocean. Deep Sea Res 27:941–950

Neilson PB, Jacobsen TS (1980) An Intercomparison of acoustic, electromagnetic, and laser Doppler current meters at Stareso 1975. Inst Phys Oceanogr Rep No 41 Copenhagen Univ

Nelepo BA, Bulsakov NP, Timtichenko IE et al (1980) Synoptic Eddies in the Ocean. Naukova Dumka, Kiev, 286 pp (Russian)

Nelepo BA, Vasilyev AS, Korotayev GK (1979) Variability of the equatorial subsurface countercurrent in the Indian Ocean. Okeanologia 19 (1):5–11 (in Russian, English translation in Oceanol 19 (1):1–4)

Neumann G, Pierson WJ (1966) Principles of physical oceanography. Prentice-Hall Inc, NJ, USA, 545 pp

Newton CW (1961) Estimates of vertical motions and meridional heat exchange in Gulf Stream eddies and a comparison with atmospheric disturbances. J Geophys Res 66 (3):853–870

Newton JL, Aagaard K, Coachman LK (1974) Baroclinic eddies in the Arctic Ocean. Deep Sea Res 21:707–719

Niiler PP (1975) Variability in western boundary currents. In: Numerical Models of Ocean Circulation, Proc Nat Acad Sci Conf, October 1972

Niiler PP, (1976) Observations of low-frequency currents on the West Florida continental shelf. Mem Soc Roy Sci Liege 6:331–358

Niiler PP Mysak LA (1971) Barotropic waves along an eastern continental shelf. Geophys Fluid Dyn 2:273–278

Niiler PP, Richardson WS (1973) Seasonal variability of the Florida Current. J Mar Res 31:144–167

Niiler PP, Robinson AR (1967) The theory of free inertial jets. II. A numerical experiment for the path of the Gulf Stream. Tellus 19:601–619

Nilsson CS, Cresswell GR (1980) The formation and evolution of East Australian Current warm core eddies. Prog Oceanogr 9:133–184

Nilsson CS, Andrews JC, Scully-Power P (1977) Observations of eddy formation off East Australia. J Phys Oceanogr 7:659–669

Nishida H, White WB (1982) On the role of eddy processes in the momentum and kinetic energy balance of the Kuroshio Extension. J Phys Oceanogr 12 (2):160–170

Nowlin WD, Hubertz JM, Reid RO (1968) A detached eddy in the Gulf of Mexico. J Mar Res 26:185–186

Nowlin WD Jr, Whitworth T III, Pillsbury RD (1977) Structure and transport of the Antarctic Circumpolar Current at Drake Passage from short-term measurements. J Phys Oceanogr 7:788–802

Nowlin WD Jr, Pillsbury RD, Bottero J (1981) Observations of kinetic energy levels in the Antarctic Circumpolar Current at Drake Passage. Deep Sea Res 28:1–17

Nyson PA, Scully-Power P (1978) Sound propagation through an East Australian Current eddy. J Acoust Soc Am 65:1381–1388

O'Gara RM, Rossby HT, Spain DL (1982) SOFAR float pilot studies in the western North Atlantic 1975–1980, Data Report. Univ of Rhode Island, Graduate School of Oceanogr, Tech Rep

Oliger J, Sündstrom A (1978) Theoretical and practical aspects of some initial boundary value problems in fluid dynamics. SIAM J Appl Math 35:419–446

Olson DB (1967) Baroclinic instability and the spindown of a Gulf Stream. MS Thesis, Texas A&M Univ 82 pp

Olson DB (1980) The physical oceanography of two rings observed by the cyclonic ring experiment. Part II: Dynamics. J Phys Oceanogr 10:514–528

Olson DB, Spence TW (1978) Asymmetric disturbances in the frontal zone of a Gulf Stream ring. J Geophys Res 83 (9):4691–4695

Olson DB, Watts DR (1982) Ring-Gulf Stream interactions. J Geophys Res (submitted)

Oort AH (1964) Computations of eddy heat and density transports across the Gulf Stream. Tellus, 16:55–63

Oort AH (1971) The observed annual cycle in the meridional circulation of atmospheric energy. J Atmos Sci 28:325–339

Orlanski I (1969) The influence of bottom topography on the stability of jets in a baroclinic fluid. J Atmos Sci 26:1216–1232

Orlanski I, Cox MD (1973) Baroclinic instability in ocean currents. Geophys Fluid Dyn 4:297–332

Ortner PB (1978) Investigations into the seasonal deep chlorophyll maximum in the western North Atlantic, and its possible significance to regional food chain relationships. PhD Thesis, Woods Hole Oceanogr Inst 303 pp

Ortner PB, Wiebe PH, Haury L, Boyd S (1978) Variability in zooplankton biomass distribution in the northern Sargasso Sea: the contribution of Gulf Stream cold core rings. Fishery Bull 76 (2):323–334

Ortner PB, Hulburt EM, Wiebe PH (1979) Phytohydrography, Gulf Stream rings, and herbivore habitat contrasts. J Exp Mar Biol Ecol 39:101–124

Orszag SA (1971) Numerical simulation of incompressible flows within simple boundaries: Accuracy. J Fluid Mech 49:75

Ou HW (1980) On The propagation of free topographic Rossby waves near continental margins. Part 1: Analytical model for a wedge, J Phys Oceanogr 10:1051–1060

Ou HW, Beardsley RC (1980) On the propagation of free topographic Rossby waves near continental margins. Part 2: Numerical model. J Phys Oceanogr 10:1323–1339

Owen RW (1980) Eddies of the California Current system: physical and ecological characteristics. In: Power D (ed) The California Island: proceedings of a multidisciplinary symposium. Santa Barbara Museum Nat His, Cal, p 787

Owens WB (1979) Simulated dynamic balances for mid-ocean mesocale eddies. J Phys Oceanogr, 9:337–359

Owens WB, Bretherton FP (1978) A numerical study of mid-ocean mesoscale eddies. Deep Sea Res, 25:1–14

Owens WB, Luyten JR, Bryden HL (1982) Moored velocity measurements on the edge of the Gulf Stream recirculation. J Mar Res (in press)

Ozmidov RV, Belyayev VS, Yampolsky AD (1970) Some characteristics of turbulent energy transport and transformation in the ocean. Izvestia Akad Nauk SSSR Phys Atmos Okean 6 (3):285–291 (in Russian, English translation in Izvestiya Atm Ocean Phys 6 (3):160–163)

Paramonov AN, Kushnir BM, Zaburdaev VI (1979) Modern methods and measurement techniques for hydrological parameters in the ocean. Mar Hydrophys Inst, Acad Sci Sevastapol, USSR, Chapter 5, Section 5.2 (in Russian)

Parker CE (1971) Gulf Stream rings in the Sargasso Sea. Deep Sea Res, 18 (10):981–993

Parr AE (1935) Hydrographic relations between the so-called Gulf Stream and the Gulf of Mexico. Trans Am Geophys Union, pp 246–250

Parr AE (1937) Report on hydrographic observations at a series of anchor stations across the Straits of Florida. Bull Bingham Oceanogr Coll 6:3, 61 pp

Patterson S (1977) A cyclonic ring north of the polar front in Drake Passage. POLYMODE News 21 (unpublished manuscript)

Patzert WC, Bernstein RL (1976) Eddy structure in the central South Pacific Ocean. J Phys Oceanogr 6:392–394

Patzert WC, McNally GJ (1980) NORPAX drifting buoy program. Tropical Ocean-Atmosphere Newsletter No 3, NOAA Pac Mar Environ Lab, Seattle, WA (unpublished manuscript)

Patzert WC, Barnett TP, Sessions MH, Kilonsky B (1978) AXBT observations of tropical Pacific Ocean thermal structures during the NORPAX Hawaii-Tahiti Shuttle Experiment, November 1977 to February 1978. Ref 78–24, Scripps Inst Oceanogr, La Jolla, CA, 61 pp

Pearce AF (1975) Ode to an eddy. Nature 258:486

Pearce AF (1977) Some features of the upper 500 m of the Agulhas Current. J Mar Res 35:731–753

Pearcy WG, Krygier EE, Mesecar R, Ramsey F (1977) Vertical distribution and migration of oceanic micronekton off Oregon. Deep Sea Res 24:223–245

Pedlosky J (1964a) The stability of currents in the atmosphere and the ocean. Part I. J Atmos Sci 21:201–219

Pedlosky J (1964b) The stability of currents in the atmosphere and the ocean. Part II. J Atmos Sci 21:342–353

Pedlosky J (1977) On the radiation of mesoscale energy in the mid-ocean. Deep Sea Res 24:591–600

Pedlosky J (1979) Geophysical Fluid Dynamics. Springer-Verlag, New York, 624 pp

Peterson RG, Nowlin WD, Whitworth T III (1981) Generation and evolution of a cyclonic ring at Drake Passage in early 1979. J Phys Oceanogr (submitted)

Petrie B, Smith PC (1977) Low-frequency motions on the Scotian Shelf and Slope. Atmosphere 15:117–140

Philander SGH (1979) Variability of the tropical oceans. Dyn Atmos Oceans 3:191–208

Philander G, Düing W (1980) The oceanic circulation of the tropical Atlantic, and its variability, during GATE. Deep Sea Res, 26:1–27 (GATE suppl 2)

Phillips BF, Rimmer WD, Reid DD (1978) Ecological investigations of the late stage phyllosoma and puerulus larvae of the western rock lobster Panulirus longipes cygnus. Mar Biol 45:347–357

Phillips NA (1951) A simple three-dimensional model for the study of large-scale extratropical flow patterns. J Meteorol 8:381–394

Phillips NA (1954) Energy transformations and meridional circulations associated with simple baroclinic waves in a two-level quasi-geostrophic model. Tellus 6:273–286

Pierce AD (1965) Extension of the method of normal modes to sound propagation in an almost stratified medium. J Acoust Soc Am 37:19–27

Pillsbury RD, Bottero JS, Still RE (1977) A compilation of observations from moored current meters. Vol X. Currents, temperature and pressure in the Drake Passage during FDRAKE 75, February 1975 – February 1976. Oregon State Univ, School of Oceanogr, Data Rep, 67 pp

Pillsbury RD, Whitworth T III, Nowlin WD Jr, Sciremammano F (1979) Currents and temperatures as observed in Drake Passage during 1975. J Phys Oceanogr 9, 469–482

Pingree RD (1972) Mixing in the deep stratified ocean. Deep Sea Res 19:549–562

Pingree RD (1977) Mixing and stabilization of phytoplankton distributions on the Northwest European continental shelf. In: Steele TH (ed) Spatial Pattern in Plankton Communities. Plenum Press, New York, pp 181–220

Piton B, Poulain JF (1974) Resultat des mesures de courant superficiels au GEK effectuees avec le N.O. "Vauban" dans le sud-ouest de l'ocean Indien, 1973–74. ORSTOM Centre de Nosy-Bé Doc Sci No 47, 77 pp

Pochapsky TE (1975) The instrumented neutrally-buoyant float program at Columbia University. Lamont-Doherty Geol Observat Tech Rep No 3, CU-3-75, August

Pochapsky TE, Malone FD (1972) A vertical profile of deep horizontal current near Cape Lookout, North Carolina. J Mar Res 30:163–167

Pollard R (1973) Interpretation of near-surface current meter observations. Deep-Sea Res 20:261–268

Pollard RT (1982) Mesoscale (50–100 km) circulations revealed by inverse and classical analysis of the JASIN hydrographical data. J Phys Oceanogr (in press)

Polloni C, Mariano A, Rossby T (1981) Streamfunction maps of the Soviet Polymode current meter array. U Rhode Is, Grad Sch Oceanogr, Tech Rept 81–1:131 pp

Pounder ER (1980) Physical oceanography in the central Arctic. Transactions, Am Geophys Union 61 (17) (abstract)

Prangsma GJ (1977) Eddies and weathership records: a cautionary note. Deep Sea Res, 25:271–274

Price JF (1982) Diffusion statistics computed from SOFAR float trajectories in the western North Atlantic. (in press)

Price JF, Rossby HT (1982) Observations of a barotropic planetary wave in the western North Atlantic. J Mar Res, Suppl Vol 40 (in press)

Rao DP (1977) A comparative study of some physical processes governing the potential productivity of the Bay of Bengal and Arabian Sea. Thesis, Andhra Univ, Waltair, India

Regier L (1982) Mesoscale current fields observed with a shopboard profiling acoustic current meter. J Phys Oceanogr (submitted)

Regier L, Stommel H (1976) Trajectories of INDEX surface drifters. INDEX Occasional Notes No 5, Ocean Sci Center, Nova Univ, USA (unpublished manuscript)

Reid JL (1965) Physical oceanography of the region near Point Arguello. Univ of California, Inst Mar Res Ref 65–19

Reid JL Jr (1978) On the mid-depth circulation and salinity field in the North Atlantic Ocean. J Geophys Res 83:5063–5067

Reid JL, Nowlin WD Jr (1971) Transport of water through the Drake Passage. Deep Sea Res 18:51–64

Reid JL Jr, Schwatzlose RA, Brown DM (1963) Direct measurements of a small surface eddy off northern Baja California. J Marine Res 21:205–218

Reid RO, Elliott BA, Olson DB (1981) Available potential energy: A clarification. J Phys Oceanogr 11:15–29

Rennell J (1832) Investigation of the currents. The South African Current and Countercurrent from the Atlantic to the Indian Ocean. Chart engraved by J&C Walker, London

Rhines PB (1969) Slow oscillations in an ocean of variable depth; Part I. Abrupt topography, J Fluid Mech 37:161–189

Rhines PB (1970) Edge-, bottom-, and Rossby waves in a rotating stratified fluid. Geophys Fluid Dyn 1:273–302

Rhines PB (1971a) A note on long-period motions at Site D. Deep Sea Res 18:21–26

Rhines PB (1971b) A comment on the Aries observations. Philos Trans R Soc London, Ser A 270:461–463

Rhines PB (1973) Observations of the energy-containing oceanic eddies, and theoretical models of waves and turbulence. Boundary Layer Meteorol 4:345–360

Rhines PB (1975) Waves and turbulence on a β-plane. J Fluid Mech 69:417–443

Rhines PB (1977) The dynamics of unsteady currents. In: Goldberg ED, McCave IN, O'Brien JJ, Steele JH (eds) The Sea, Vol 6, Wiley & Sons, New York

Rhines PB (1979) Geostrophic turbulence. Ann Rev Fluid Mech 11:401–441

Rhines PB, Holland WR (1979) A theoretical discussion of eddy-driven mean flows. Dyn Atmos Oceans 3:289–325

Rhines PB, Young WR (1982) Homogenization of potential vorticity in planetary gyres. J Fluid Mech (in press)

Richards FA, Redfield AC (1955) Oxygen-density relationships in the western North Atlantic. Deep Sea Res 2:182–199

Richardson PL (1980) Gulf Stream ring trajectories. J Phys Oceanogr 10:90–104

Richardson PL (1981) Gulf Stream trajectories measured with free-drifting buoys. J Phys Oceanogr 11:999–1010

Richardson PL (1982a) Gulf Stream paths measured with free-drifting buoys. J Phys Oceanogr 11:999–1010

Richardson PL (1982b) Eddy kinetic energy in the North Atlantic from surface drifters. J Geo-
    phys Res (submitted)
Richardson PL, Knauss JA (1971) Gulf Stream and western boundary undercurrent observa-
    tions at Cape Hatteras. Deep Sea Res 18:1089–1109
Richardson PL, Mooney K (1975) The Mediterranean outflow – a simple advection-diffusion
    model. J Phys Oceanogr 5:476–483
Richardson PL, Strong AE, Knauss JA (1973) Gulf Stream eddies: recent observations in the
    western Sargasso Sea. J Phys Oceanog 3:297–301
Richardson PL, Cheney RE, Mantini LA (1977) Tracking a Gulf Stream ring with a free-drift-
    ing surface buoy. J Phys Oceanogr 7:580–590
Richardson PL, Cheney RE, Worthington LV (1978) A census of Gulf Stream rings, Spring
    1975. J Geophys Res 83 (C12):6136–6143
Richardson PL, Maillard C, Sanford TB (1979) The physical structure and life history of cy-
    clonic Gulf Stream ring Allen. J Geophys Res 84 (C12):7727–7741
Richardson PL, Wheat JJ, Bennett D (1979) Free-drifting buoy trajectories in the Gulf Stream
    system (1975–1978). A Data Report. WHOI Ref No 79–4, Woods Hole Oceanogr Inst
Richardson PL, Price JF, Owens WB, Schmitz WJ, Rossby HT, Bradley AM, Valdes JR, Webb
    DC (1981) North Atlantic Subtropical Gyre: SOFAR floats tracked by moored listening
    stations. Sci 213:435–437
Richardson WS, Schmitz WJ Jr (1965) A technique for the direct measurement of transport
    with applications to the Straits of Florida. J Mar Res 23:172–185
Richardson WS, Schmitz WJ Jr, Niiler PP (1969) The velocity structure of the Florida Current
    from the Straits of Florida to Cape Fear. Frederick C. Fuglister Sixtieth Anniversary Vol,
    Deep Sea Res 16:225–231 (suppl)
Richardson WS, White HJ Jr, Nemeth L (1972) A technique for the direct measurment of
    ocean currents from aircraft. J Mar Res 30:259–268
Richman JG, Wunsch C, Hogg NG (1977) Space and time scales of mesoscale motion in the
    western North Atlantic. Rev Geophys Space Phys 15:385–420
Ring Group (1981) Gulf Stream cold-core rings: their physics, chemistry and biology. Sci
    212:1091–1100
Riley GA (1951) Oxygen, phosphate, and nitrate in the Atlantic Ocean. Bull Bingham Ocea-
    nogr Collect 13:1–128
Riser SC (1982a) On the quasi-Lagrangian nature of SOFAR floats. Deep Sea Res (in press)
Riser SC (1982b) Quasi-Lagrangian signatures of mid-ocean dynamical processes (submit-
    ted)
Riser SC, Rossby HT (1983) Quasi-Lagrangian structure and variability of the subtropical
    western North Atlantic circulation. J Mar Res, 41 (1), 127–162 pp
Riser SC, Freeland HJ, Rossby HT (1978) Mesoscale motions near the deep western boundary
    of the North Atlantic. Deep Sea Res 25:1179–1191
Robinson AR (1965) A three-dimensional model of inertial currents in a variable density
    ocean. J Fluid Mech 21:211–223
Robinson AR (1975) The Variability of Ocean Currents. Reviews of Geophysics and Space
    Physics 13 (3):598–602
Robinson AR (1980) Dynamics of ocean currents and circulation: Results of POLYMODE
    and related investigations. U. S. POLYMODE Organizing Committee, Mass Inst of Tech-
    nol Cambridge, MA USA, 31 pp
Robinson AR, (1982) Dynamics of Ocean Currents and Circulation: Results of POLYMODE
    and Related Investigations, Topics in Ocean Physics. A Osborne and PM Rizzoli (editors)
    Soc. Italiana di Fisica, Bologna (Esevier, New York)
Robinson AR, Haidvogel DB (1980) Dynamical forecast experiments with a barotropic open
    ocean model. J Phys Oceanogr 10:1909–1928
Robinson AR, Niiler PP (1967) The theory of free inertial jets. I: Path and Structure. Tellus
    19:269–291
Robinson AR, Taft BA (1972) A numerical experiment for the path of the Kuroshio. J Mar
    Res 30:65–101
Robinson AR, Luyten JR, Fuglister FC (1974) Transient Gulf Stream meandering. Part I: An
    observational experiment. J Phys Oceanogr 4:237–255

Robinson AR, Luyten JR, Flierl G (1975) On the theory of thin rotating jets: a quasi-geostrophic time-dependent model. Geophys Fluid Dyn 6:211–244

Robinson AR, Harrison DE, Mintz Y, Semtner AJ (1977) Eddies and the general circulation of an idealized oceanic gyre: a wind and thermally driven primitive equation numerical experiment. J Phys Oceanogr 7:182–207

Robinson AR, Harrison DE, Haidvogel DB (1979) Mesoscale eddies and general ocean circulation models. Dyn Atmos Oceans 3:143–180

Robinson MK, Bauer RA, Schroeder EH (1979) Atlas of North Atlantic – Indian Ocean monthly mean temperatures and mean salinities of the surface layer. Nav Oceanogr Office Ref Public 18, Dep of the Navy, Washington, DC

Rochford DJ (1961) Hydrology of the Indian Ocean 1: The water masses in the intermediate depths of the south east Indian Ocean. Aust J Mar Freshwater Res 12:129–149

Rochford DJ (1962) Hydrology of the Indian Ocean 2. The surface waters of the south east Indian Ocean and Arafura Sea in the Spring and Summer. Aust J Mar Freshwater Res 13:226–251

Rochford DJ (1969) Seasonal variations in the Indian Ocean along 110°E 1: Hydrological structure of the upper 500 m. Aust J Mar Freshwater Res 20:1–50

Rochford DJ (1972) Nutrient enrichment of east Australian Coastal waters. I Evans Head upwelling. CSIRO Aust Div Fish Oceanogr Tech Pap 33, 17 pp

Roden GI (1970) Aspects of the mid-Pacific transition zone. J Geophys Res 75:1097–1109

Roden GI (1972) Temperature and salinity fronts at the boundaries of the subarctic and subtropical transition zone in the western Pacific. J Geophys Res 77:7155–7187

Roden GI (1977) On long-wave disturbances of dynamic height in the North Pacific. J Phys Oceanogr 7:41–49

Roden GI (1979) The depth variability of meridional gradients of temperature, salinity, and sound velocity in the western North Pacific. J Phys Oceanogr 9:756–767

Roden GI (1980) Mesothermohaline, sound velocity and baroclinic flow structure of the Pacific subtropical front during the winter of 1980. J Phys Oceanogr 11:658–675

Roe HSJ, Shale DM (1979) A new multiple rectangular midwater trawl (RMT 1+8M) and some modifications to the Institute of Oceanographic Sciences' RMT 1+8. Mar Biol 50:283–288

Roe HSJ, Angel MV, Babcock Y, Domanek P, James PT, Pugh PR, Thurston MH (1982) The Diel migrations and distributions of a mesopelagic community in the North-East Atlantic. Introduction and sampling procedures. Progr Oceanogr (in press)

Rooth CGH (1974) Tritium-oxygen correlations in subtropical thermoclines. IUGG Gen Assem/Grenoble, 1974 (abstract)

Rooth CGH, Ostlund HG (1974) Penetration of tritium into the Atlantic thermocline. Deep Sea Res 19:481–482

Ross CK, Needler GT (1976) Spectral analysis of the longterm oceanographic time series at Ocean Weather Station "P". JFRB 33 (10):2203–2212

Rossby CG (1936) Dynamics of steady ocean currents in the light of experimental fluid mechanics. Pap Phys Oceanogr Meteorol 5:1–43

Rossby HT (1982) Eddies and the general ocean circulation. Proc Conf Future of Oceanography, Woods Hole Oceanogr Inst

Rossby HT, Sanford TB (1976) A study of velocity profiles through the main thermocline. J Phys Oceanogr 6:766–774

Rossby HT, Webb D (1970) Observing abyssal motions by tracking Swallow floats in the SOFAR channel. Deep Sea Res 17:359–365

Rossby HT, Voorhis AD, Webb D (1975) A quasi-Lagrangian study of mid-ocean variability using long-range SOFAR floats. J Mar Res 33:355–382

Rossby HT, Price JF, Webb D (1982) The spatial and temporal evolution of a cluster of SOFAR floats in the Polymode Local Dynamics Experiment (LDE) (submitted)

Royer TC (1978) Ocean eddies generated by seamounts in the North Pacific. Sci 199:1063–1064

Royer TC, Hansen DV, Pashinski DJ (1979) Coastal flow in the northern Gulf of Alaska as observed by dynamic topography and satellite-tracked bouys. JPO 9:785–801

Saffman PG (1971) On the spectrum and decay of random two-dimensional vorticity distributions at large Reynolds number. Stud Appl Math 50:377–383

Salmon R (1978) Two-layer quasi-geostrophic turbulence in a simple special case. Geophys Astrophys Fluid Dyn, 10:25–52

Salmon R (1980) Baroclinic instability and geostrophic turbulence. Geophys Astrophys Fluid Dyn 15:167–211

Salmon R, Holloway G, Hendershott MC (1976) The equilibrium statistical mechanics of simple quasi-geostrophic models. J Fluid Mech 75:691–703

Sameoto DD, Jaroszynski LO, Fraser WB (1977) A multiple opening and closing plankton sampler based on the MOCNESS and NIO Nets. J Fish Res Bd Canada 34:1230–1235

Sanford TB, Drever RG, Dunlap JH (1974) The design and performance of a free-fall electromagnetic velocity profiler (EMVP). WHOI Ref No 74-46, Woods Hole Oceanogr Inst

Sanford TB, Drever RG, Dunlap JH (1978) A velocity profiler based on the principles of geomagnetic induction. Deep Sea Res 25:183–210

Sarmiento J, Bryan K (1982) A robust diagnostic model. J Phys Oceanogr (in press)

Sarmiento J, Rooth CGH, Roether W (1982) The North Atlantic tritium distribution in 1972. J Geophys Res (in press)

Saunders PM (1971) Anticyclonic eddies formed from shoreward meanders of the Gulf Stream. Deep Sea Res 18:1207–1219

Saunders PM (1973) The instability of a baroclinic vortex. J Phys Oceanogr 3:61–65

Saunders PM (1976) Near-surface current measurements. Deep Sea Res 23:249–258

Saunders PM (1980) CTD Data obtained during Discovery cruise 81. IOS Data Report No 17, Inst Oceanogr Sci Wormley Surrey UK 16 pp figures and tables

Saunders PM (1982) Circulation in the eastern North Atlantic. J Mar Res 40 (suppl)

Savchenko VG, Emery WJ, Vladimirov OA (1978) A cyclonic eddy in the Antarctic circumpolar current south of Australia: Results of Soviet-American observations aboard R/V Professor Zubov. J Phys Oceanogr 8:825–837

Savonius SJ (1931) The S-rotor and its applications. Mech Eng 53:333–338

Scarlet RI (1973) STD's in MODE – A grab bag of calibration problems. Proc STD Conf (suppl), Plessey Environmental Systems, San Diego, Cal

Scarlet RI (1975) A data processing method for salinity, temperature, depth profiles. Deep Sea Res 22:509–515

Schmitt RW, Evans DL (1978) An estimate of the vertical mixing due to salt fingers based on observations in the North Atlantic central water. J Geophys Res 83:2913–2919

Schmitz JE, Vastano AC (1975) Entrainment and diffusion in a Gulf Stream cyclonic ring. J Phys Oceanogr 5 (1):93–97

Schmitz JE, Vastano AC (1977) Decay of a shoaling Gulf Stream cyclonic ring. J Phys Oceanogr 7 (3):479–481

Schmitz WJ Jr (1974) Observations of low-frequency current fluctuations on the continental slope and rise near site D. J Mar Res 32:233–251

Schmitz WJ Jr (1976) Eddy kinetic energy in the deep western North Atlantic. J Geophys Res 81:4980–4982

Schmitz WJ Jr (1977) On the deep general circulation of the western North Atlantic. J Mar Res 35:21–28

Schmitz WJ Jr (1978) Observations of the vertical distribution of low frequency kinetic energy in the western North Atlantic. J Mar Res 36:295–310

Schmitz WJ Jr (1980) Weakly depth-dependent segments of the North Atlantic circulation. J Mar Res 38:111–133

Schmitz WJ Jr (1981a) Observations of eddies in the Newfoundland Basin. Deep Sea Res. 28:1414–1421

Schmitz WJ Jr (1981b) A comparison of the mid-latitude eddy fields in the western North Atlantic and North Pacific Oceans. J Phys Oceanogr 12:208–210

Schmitz WJ Jr, Hogg NG (1978) Observations of energetic low-frequency current fluctuations in the Charlie-Gibbs Fracture Zone. J Mar Res 36:725–734

Schmitz WJ Jr, Holland WR (1982) Numerical eddy resolving general circulation experiments: preliminary comparison with observation. J Mar Res 40 (1):15–117

Schmitz WJ Jr, Niiler PP (1969) A note on the kinetic energy exchange between fluctuations and mean flow in the surface layer of the Florida Current. Tellus 21:814–819

Schmitz WJ Jr, Owens WB (1979) Observed and numerically simulated kinetic energies for MODE eddies. J Phys Oceanogr 9:1294–1297

Schmitz WJ Jr, Price JF, Richardson PL, Owens WB, Webb DC, Cheney RE, Rossby HT (1981) A preliminary exploration of the Gulf Stream system with SOFAR floats. J Phys Oceanogr 11:1194–1204

Schott F, Bock M (1980) Determination of energy interaction terms and horizontal wavelengths for low-frequency fluctuations in the Norwegian Current. JGR 85:4007–4014

Schott F, Düing W (1976) Continental shelf waves in the Florida Straits. J Phys Oceanogr 6:451–460

Schumann EH (1976) High waves in the Agulhas Current. Mariners Weather Log 20:1–5

Sciremammano F Jr (1979) Observations of Antarctic polar front motions in a deep water expression. J Phys Oceanogr 9:221–226

Sciremammano F Jr (1980) The nature of the poleward heat flux due to low-frequency current fluctuations in Drake Passage. J Phys Oceanogr 10:843–852

Sciremammano F Jr, Pillsbury RD, Bottero JS, Still RE (1978) A compilation of observations from moored current meters. Vol XI. Currents, temperature and pressure in the Drake Passage during FDRAKE 76, February 1976 – January 1977. Oregon State Univ, Sch Oceanogr, Data Rep 68 pp

Sciremammano F Jr, Pillsbury RD, Nowlin WD Jr, Whitworth T III (1980) Spatial scales of temperature and flow at Drake Passage. J Geophys Res 85:4015–4028

SCOR Working Group 21 (1969) An intercomparison of some current meters. UNESCO Tech Pap Mar Sci 11, UNESCO, Paris, 70 pp

SCOR Working Group 21 (1974) An intercomparison of some current meters, II. UNESCO Tech Pap Mar Sci 17, UNESCO, Paris, 116 pp

SCOR Working Group 21 (1975) An intercomparison of some current meters, III. UNESCO Tech Pap Mar Sci 23, UNESCO, Paris, 42 pp

Scott BD (1978) Hydrological features of a warm core eddy and their biological implications. CSIRO Div Fish Oceanogr Rep 100:1–6

Scully-Power P et al (1975) A multisystem technique for the detection and measurement of warm core ocean eddies. Proc IEEE Ocean '75:761–766

Seaver G (1975) Two XBT sections in the North Atlantic. MODE Hot-Line News, No 84 (unpublished manuscript)

Seaver G, Kuleshov S (1979) XBT Accuracy. POLYMODE News No 72 (unpublished manuscript)

Semtner AJ, Holland WR (1978) Intercomparison of quasigeostrophic simulations of the western North Atlantic circulation with primitive equation results. J Phys Oceanogr 8:735–754

Semtner AJ, Holland WR (1980) Numerical simulation of equatorial ocean circulation. Part I. A basic case in turbulent equilibrium. J Phys Oceanogr 10:667–693

Semtner AJ, Mintz Y (1977) Numerical simulation of the Gulf Stream and mid-ocean eddies. J Phys Oceanog 7:208–230

Sessions MH (1979) AXBT's for oceanic measurements. Exposure 7, 5, November (unpublished manuscript)

Sessions MH, Ryan WR, Barnett TP (1974) AXBT Calibration and operation for NORPAX POLE experiment. Scripps Inst Oceanogr Ref Ser, No 74–31, 19 pp

Shapiro R (1971) The use of linear filtering as a parameterization of atmospheric diffusion. J Atmos Sci 28:523–531

Shulenberger F (1978) The deep chlorophyll maximum and mesoscale environmental heterogeneity in the western half of the North Pacific central gyre. Deep Sea Res 25 (12):1193–1208

Siedler G, Woods JD (1980) Introduction to a collection of papers on GATE oceanography and surface layer meteorology. Deep Sea Res 26:1–8 (GATE suppl 1)

Siedler G, Philander SGH (1982) Physics of the upper tropical ocean. (GATE Monograph W M O, Geneva, Garp Publ Ser No 25, pp 219–235

Simpson JP (1979) The effects of fouling on electromagnetic current meters. Exposure 7, 2, May (unpublished manuscript)

Smith PC (1976) Baroclinic instability in the Denmark Strait Overflow. JPO 6:355–371

Smith PC (1978) Low-frequency fluxes of momentum, heat, salt and nutrients at the edge of the Scotian Shelf. J Geophys Res 83:4079–4096

Smith PC, Petrie BD (1982) Low-frequency circulation at the edge of the Scotian Shelf. J Phys Oceanogr 12:28–46

Smith PC, Petrie B, Mann CR (1978) Circulation, variability and dynamics on the Scotian Shelf and Slope. J Fish Res Board Can 35:1067–1083

Smith R (1971) The ray paths of topographic Rossby waves. Deep Sea Res 18:477–483

Snodgrass RE (1968) Deep sea instrument capsule. Sci 162:78–87

Snodgrass F, Brown W, Munk W (1974) MODE: IGPP measurements of bottom pressure and temperature. J Phys Oceanogr 5:63–74

Solomon H, Ahlnas K (1978) Eddies in the Kamchatka Current. Deep Sea Res 25:403–410

Spain DL, O'Gara RM, Rossby HT (1980) SOFAR float data report of the Polymode Local Dynamics Experiment. Univ, Graduate Sch Oceanogr, Tech Rep 80-1, 200 pp

Spence TW, Legeckis R (1981) Satellite and hydrographic observations of low-frequency wave motions associated with a cold-core Gulf Stream ring. J Geophys Res 86:1945–1953

Spiegel SL, Robinson AR (1968) On the existence and structure of inertial boundary currents in a stratified ocean. J Fluid Mech 32:569–607

Spiesberger JL, Spindel RC, Metzger K (1980) Stability and identification of ocean acoustic multipaths. J Acoust Soc Am 67:2011–2017

Spindel RC (1979) An underwater acoustic pulse compression system. IEEE Trans Acoust Speech and Sig Proc, ASSP-27:723–728

Spindel RC, Spiesberger JL (1981) Multipath variability due to the Gulf Stream. J Acoust Soc Am 69:982–988

Stanton BR (1976) An oceanic frontal jet near the Norfolk Ridge northwest of New Zealand. Deep Sea Res 23:1207–1219

Stanton BR (1981) An oceanographic survey of the Tasman Front. NZ J Mar Freshwat Res 15:289–297

Stavropoulos CC, Duncan CP (1974) A satellite-tracked buoy in the Agulhas Current. J Geophys Res 79:2744–2746

Steele JH, Henderson EW (1979) Spatial patterns in North Sea plankton. Deep Sea Res 26, (8A):955–964

Stefannson U, Atkinson LP, Bumpus DF (1971) Hydrographic properties and circulation of the North Carolina shelf and slope waters. Deep Sea Res 18:383–420

Steinberg JC, Birdsall TG (1966) Underwater sound propagation in the Straits of Florida. J Acoust Soc Am 39:301–315

Stern ME (1961) The stability of thermocline jets. Tellus 8:503–508

Stern ME (1975) Ocean circulation physics. Acad Press, New York,

Stern ME, Turner JS (1969) Salt fingers and convecting layers. Deep Sea Res 16:497–511

Stockman WB, Koshlyakov MN, Ozmidov RV, Fomin LM, Yampolsky AD (1969) Long-term measurements of the physical field variability on oceanic polygons, as a new stage in the ocean research. Dokl Akad Nauk SSSR 186 (5):1070–1073 (in Russian, English version in Nat Inst Oceanogr translation T134)

Stokes AN (1977) On the types of moving front in quasilinear diffusion. Math Biosc, 31:307–315

Stommel H (1948) The western intensification of wind-driven ocean currents. Trans Am Geophys Union 29:202–206

Stommel H (1957) A survey of ocean current theory. Deep Sea Res 4:149–184

Stommel H (1965) The Gulf Stream: A Physical and Dynamical Description, 2nd edn Univ California Press, Berkeley, 248 pp (1st edn 1958, 202 pp)

Stommel H (1970) Future Prospects for Physical Oceanography. Science 168:1531–1537

Stommel H (1980) How the ratio of meridional flux of fresh-water to flux of heat fixes the latitude where low salinity intermediate water sinks. Tellus, 32:562–566

Stommel H, Stroup ED, Reid JL, Warren BA (1973) Transpacific hydrographic sections at Lats. 43°S and 28°S: the SCORPIO expedition – I Preface. Deep Sea Res 20:1–8

Stramma L (1981) Die Bestimmung der dynamischen Topographie aus Temperaturdaten aus dem Nordostatlantik. Berichte aus dem Inst für Meereskunde an der Christian-Albrechts-Univ Kiel Nr 84:66 pp

Stumpf HG, Legeckis RV (1977) Satellite observations of mesoscale eddy dynamics in the eastern tropical Pacific Oeean. J Phys Oceanogr 7:648–658

Suarez AA (1971) The propagation and generation of topographic oscillations in the ocean. PhD Thesis, MIT-WHOI, J Prog Oceanogr, 130 pp

Sverdrup HU (1933) On vertical circulation in the ocean due to the action of the wind with application to conditions within the Antarctic Circumpolar Current. Discovery Rep 7:139–170

Sverdrup HU (1947) Wind-driven currents in a baroclinic ocean with application to the equatorial currents of the eastern Pacific. Proc N A Sci US 33:318–326

Sverdrup HU, Johnson MW, Fleming RH (1942) The Oceans. Prentice-Hall, Englewood Cliffs NJ 1087 pp

Swallow JC (1955) A neutral-buoyancy float for measuring deep currents. Deep Sea Res 3:74–81

Swallow JC (1969) A deep eddy off Cape St Vincent. Deep Sea Res Suppl 16:285–296

Swallow JC (1971) The Aries current measurements in the western North Atlantic. Philos Trans Royal Soc London Ser A 270:451–461

Swallow JC (1976) Variable currents in the mid-ocean. Oceans 19:(3) 18–25

Swallow JC, Bruce JG (1966) Current measurements off the Somali coast during the southwest monsoon of 1964. Deep Sea Res 13:861–888

Swallow JC, McCartney BS, Millard NW (1974) The Minimode float tracking system. Deep Sea Res 21:573–595

Szabo D, Weatherly GL (1979) Energetics of the Kuroshio south of Japan. J Mar Res 37:531–556

Tabata S (1976) Evidence of a westward-flowing "subarctic countercurrent" in the north Pacific Ocean. JFRB 33:2168–2196

Taft BA (1972) Characteristics of the flow of the Kuroshio south of Japan. In: Stommel H, Yoshida K (eds) Kuroshio, physical aspects of the Japan Current. Univ Washington Press, pp 165–214

Taft BA (1978) Structure of the Kuroshio south of Japan. J Mar Res 36:77–117

Taft BA, Ramp SR, Dworski JG, Holloway G (1981) Measurements of deep currents in the central North Pacific. J Geophys Res 86:1955–1968

Taira K. Direct observations of current in the Kuroshio around the Izu Ridge (unpublished manuscript)

Taira K, Teramoto T (1981) Velocity fluctuations of the Kuroshio near the Uzu Ridge and their relationship to current path. Deep Sea Res 28:1187–1197

Tang CL (1979) Development of radiation fields and baroclinic eddies in a β-plane. J Fluid Mech 93:379–400

Tarbell SA (1980) A compilation of moored current meter data and associated oceanographic observations, Vol XXII (POLYMODE Array III Clusters A, B and Site Moorings). WHOI Ref No 80–40, Woods Hole Oceanogr Inst

Tarbell S, Spencer A, Payne RE (1978) A compilation of moored current meter data and associated oceanographic observations, XVII (POLYMODE Array II data). WHOI Tech Rep 78–49, Woods Hole Oceanogr Inst

Tareev BA (1965) Unstable Rossby waves and the instability of oceanic currents. Atmos Oceanic Phys Ser 1:426–438

Taylor GI (1921) Diffusion by continuous movements. Proc London Math Soc 2 (XX):196–211

Thompson RE, Stewart RW (1977) The balance and redistribution of potential vorticity within the ocean. Dyn Atmos Oceans 1:299–321

Thomson RORY (1971) Topographic Rossby waves at a site north of the Gulf Stream. Deep Sea Res, 18:1–19

Thompson RORY (1977) Observations of Rossby waves near site D. Prog Oceanog 7 (4):135–162

Thompson RORY, Luyten JR (1976) Evidence for bottom-trapped topographic Rossby waves from single moorings. Deep Sea Res 23:629–635

Thompson RORY, Veronis G (1980) Transport calculations in the Tasman and Coral Sea. Deep Sea Res 27:303–323

Tomosada A (1978) Oceanographic characteristics of a warm eddy detached from the Kuroshio east of Honsu, Japan. Bull Tokai Reg Fish Res Lab 94:59–103

Toole JM (1981) Intrusion characteristics in the Antarctic polar front. J Phys Oceanogr 11:780–793

Tranter DJ, Parker RR, Vaudrey DJ (1979) In vivo chlorophyll a fluorescence in the vicinity of warm core eddies off the coast of New South Wales 1. September 1978. CSIRO Division Fish Oceanogr 105:1–29

Tranter DJ, Parker RR, Cresswell GR (1980) Are warm-core eddies unproductive? Nature 284:540–542

Trenberth K (1979) Mean annual poleward energy transports by the oceans in the Southern Hemisphere. Dyn Ats Oceans 4:57–64

Tucholke BE, Wright WR, Hollister CD (1973) Abyssal circulation over the Greater Antilles Outer Ridge. Deep Dea Res 20:973–995

Ugincius P (1972) Ray Acoustics and Fermat's principle in a moving inhomogeneous medium. J Acoust Soc Am 51:1759–1763

US Navy Oceanographic Office (1973) Sound velocity structure of the Northeast Atlantic. NAVOCEANO Tech Note 930-1-73:38 pp

Vachon WA (1978) Present status and future directions of drifting buoy developments. Nat Tech Inf Serv, Springfield, VA, 32 pp

Van Leer JC, Düing W, Erath R, Kennelly E, Speidel A (1974) The cyclesonde: an unattended vertical profiler for scalar and vector quantities in the upper ocean. Deep Sea Res 21:385–400

Van Leer JC, Ross AE, Rohr L (1980) Velocity data observed by cyclesonde during POLYMODE. Univ Miami RSMAS Tech Rep Ref No UM-RSMAS-80005

Vastano AC, Hagan DE (1977) Observational evidence for transformation of tropospheric waters within cyclonic rings. J Phys Oceanogr 7:938–943

Vastano AC, Owens GE (1973) On the Acoustic Characteristics of a Gulf Stream cyclonic ring. J Phys Oceano 3:470–478

Vastano AC, Schmitz JE, Hagan DE (1980) The physical oceanography of two rings observed by the cyclonic ring experiment, Part I Physical structures. J Phys Oceanog 10:493–513

Vastano AC, Schmitz JE, Hagan DE (1982) The physical oceanography of two rings observed by the cyclonic ring experiment. Part II Core water relationships and ring anomalies. J Phys Oceanogr (submitted)

Venrick EL (1979) The lateral extent and characteristics of the North Pacific Central environment at 35°N. Deep Sea Res 26 (10A):1153–1178

Veronis G (1966) Wind-driven ocean circulation – Part 2 Numerical solutions of the non-linear problem. Deep Sea Res 13:31–55

Veronis G (1973) Model of world ocean circulation: I Wind-driven, twolayer. J Mar Res 31:223–288

Veronis G (1981) Dynamics of large-scale ocean circulation. In: Warren BA, Wunsch C (eds) Evolution of Physical Oceanography, MIT Press, Cambridge, MA pp 140–183

Volkman G (1975) Some XBT sections in the western North Atlantic. MODE Hot-Line News No 84 (unpublished manuscript)

Vonder Haar TH, Oort AH (1973) New estimate of annual poleward energy transport by northern hemisphere oceans. J Phys Oceanog 3:169–172

Voorheis GM, Aagaard K, Coachman LK (1973) Circulation patterns near the tail of the Grand Banks. J Phys Oceanogr 3 (4):397–405

Voorhis AD, Webb D (1973) Data summary and review of Shipboard SOFAR Float Program from September 1972 through July 1973. WHOI Ref No 73-74, Woods Hole Oceanogr Inst

Vukovich FM (1979) Some aspects of the oceanography of the Gulf of Mexico using satellite and in situ data. J Geophys Res 84:7749–7768

Vukovich FM, Crissman BW, Bushnell M, King WJ (1979) Gulf Stream boundary eddies off the east coast of Florida. J Phys Oceanogr 9:1214–1222

Walden RG, Panicker NN (1973) Performance analysis of Woods Hole taut moorings. WHOI Ref 73–31, Woods Hole Oceanogr Inst

Walden RG, Berteaux HO, Striffler F (1973) The design, logistics, and installation of a SOFAR float tracking station at Grand Turk Island, BWI WHOI Ref No 73–73, Woods Hole Oceanogr Inst

Walen RS (1976) Shore-based receivers used in POLYMODE program. WHOI Ref No 76–72, Woods Hole Oceanogr Inst

Warm Core Rings Executive Committee (1982) Multidisciplinary Program to Study Warm Core Rings. EOS 63(44):834–836

Warren BA (1963) Topographic influences on the path of the Gulf Stream. Tellus 15:167–183

Warren BA (1967) Notes on translatory movement of rings of current with application to Gulf Stream eddies. Deep Sea Res 14:505–524

Warren BA (1970) General circulation of the South Pacific. In: Wooster WS (ed) Scientific Exploration of the South Pacific. N A Sci Washington, p 33

Warren BA, Volkmann GH (1968) Measurement of volume transport of the Gulf Stream south of New England. J Mar Res 26:110–126

Watts DR (1982) On the potential vorticity distribution in the Gulf Stream. Dyn Atmos Oceans (in press)

Watts DR, Johns WE (1982) Gulf Stream meanders: observations on propagation and growth. Submitted to: J Geophys Res

Watts DR, Olson DB (1978) Gulf Stream ring coalescence with the Gulf Stream off Cape Hatteras. Sci 202:971–972

Watts R, Rossby HT (1977) Measuring dynamic heights with inverted echo sounder: results from MODE. J Phys Oceanogr 7:345–358

Webb DC (1977) SOFAR floats for POLYMODE. Proc of Ocean '77 Conf Rec, Los Angeles, CA October 17–19, 1977, MTS-IEEE Vol 2, pp 44B-1–44B-5

Webster F (1961) A description of Gulf Stream meanders off Onslow Bay. Deep Sea Res 8:130–143

Webster I, Golding TJ, Dyson N (1979) Hydrological features of the near shelf waters off Freemantle, Western Australia, during 1974. CSIRO, Div Fish Oceanogr Rep 106, 30 pp

Weihs D (1973) Optimal fish cruising speed. Nature 245 (5419):48–50

Weinberg NL, Clark JG (1980) Horizontal acoustic refraction through ocean mesoscale eddies and fronts. J Acoust Soc Am. 68:703–706

Weinberg NL, Zabalgogeazcoa X (1977) Coherent ray propagation through a Gulf Stream ring. J Acoust Soc Am, 62:888–894

Weisberg RH, Horigan A, Colin C (1979) Equatorially trapped Rossby-gravity wave propagation in the Gulf of Guinea. J Mar Res, 37:67–86

Weisberg RH, Miller L, Horigan A, Knauss JA (1980) Velocity observations in the equatorial thermocline during GATE. Deep Sea Res, GATE Suppl 2 Vol 26:217–248

Weller RA (1978) Observations of horizontal velocity in the upper ocean mode with a new vector measuring current meter. PhD Thesis, Univ of California, San Diego, 169 pp

Weller RA, Davis RG (1980) A vector measuring current meter. Deep Sea Res 27A:565–582

Wennekens MP (1959) Water mass properties of the Straits and Florida and related waters. Bull Mar Sci Gulf and Caribbean 9:1–52

White WB, McCreary JP (1976) On the formation of the Kuroshio meander and its relationship to the large-scale ocean circulation. Deep Sea Res, 23:33–47

Whitehead J (1975) Mean flow driven by circulation on a β-plane. Tellus, 27:358–364

Whitham GB (1974) Linear and non-linear waves. Wiley & Sons, New York, 636 pp

Whittle P (1962) Topographic correlation, power-law covariance functions and diffusion. Biometrika 49:305–314

Wiebe PH (1976) Biology of cold-core rings. Oceanus 19:69–76

Wiebe PH, Boyd SH (1978) Limits of Nematoscelis megalops in the Northwestern Atlantic in relation to Gulf Stream core rings. I Horizontal and vertical distributions. J Mar Res 36:119–142

Wiebe PH, Burt KH, Boyd SH, Morton AW (1976) A multiple opening-closing net and environmental sensing system for sampling zooplankton. J Mar Res 34:313–326

Wiebe PH, Hulburt EM, Carpenter EJ, Jahn AE, Knapp GP, Boyd SH, Ortner PB, Cox JL (1976) Gulf Stream cold core rings: large-scale interaction sites for open ocean plankton communities. Deep Sea Res 23:695–710

Willebrand J, Meinke J (1980) Statistical model of mesoscale temperature and current fluctuations over the Iceland-Faroe Ridge slope. ICES CM 1980/C:12, 16 pp + 12 figs (mimeo)

Williams GP (1978) Planetary circulations: I, Barotropic representation of Jovian and terrestrial turbulence. J Atmos Sci 35:1399–1426

Williams J (1793) Memoir of Jonathan Williams on the use of the thermometer in discovering banks, soundings, etc. Trans Am Philos Soc, 3:82–96

Williamson DI (1967) Some recent advances and outstanding problems in the study of larval Crustacea. Proc Symp Crustacea Pt 11, NIO, India, pp 815–823

Willmott AJ, Mysak LA (1980) Atmospherically forced eddies in the northeast Pacific. Trans Am Geophys Union 61 (17)

Wilson WD (1960) Speed of sound in sea water as a function of temperature, pressure and salinity. J Acoust Soc Am 32:641 pp

Wilson WS, Dugan JP (1978) Mesoscale thermal variability in the vicinity of the Kuroshio extension. J Phys Oceanogr 8:537–540

Wollard G, Latham R, Budd R (1969) Oceanographic parameters for acoustical determination between latitudes 22°N and 30°N along the 157°50′W meridian. Rep HIG-69-11, Hawaii Inst Geophys 20 pp

Woodward W, Mooers CNK, Jensen K (1978) Proceedings of a working conference on current measurement. Tech Rep DEL-SG-3-78, Coll Mar Stud Univ of Delaware, Newark 372 pp

Worthington LV (1959) The 18° water in the Sargasso Sea. Deep Sea Res 4:297–305

Worthington LV (1976a) On the North Atlantic circulation. The Johns Hopkins Oceanogr Stud No 6, Johns Hopkins Univ Press, Baltimore 110 pp

Worthington LV (1976b) Further notes on Big Babies. POLYMODE News, 3 (1) (unpublished manuscript)

Worthington LV (1977) Intensification of the Gulf Stream after the winter of 1976–77. Nature 270:415–417

Worthington LV, Metcalf WG (1961) The relationship between potential temperature and salinity in deep Atlantic water. Rapp Procès-Verbaux des Réunions, Cons Permanent Int 'Expl Mer 149:122–128

Wright DG (1981) Baroclinic instability in Drake Passage. J Phys Oceanogr 11:231–246

Wright WR, Worthington LV (1970) The water masses of the North Atlantic ocean: a volumetric census of temperature and salinity. Serial Atlas Mar Environment 19, 8 pp

Wroblewski JS (1980) A simulation of the distribution of Acartia clausi during Oregon upwelling. August 1973. J Plank Res 2:43–68

Wunsch C (1977) Determining the general circulation of the oceans: a preliminary discussion. Sci 196:871–875

Wunsch C (1981) Low frequency variability of the sea. In: Warren BA, Wunsch C (eds) Evolution of Physical Oceanography, scientific surveys in honor of Henry Stommel. MIT Press, Cambridge, MA pp 342–374

Wunsch C, Dahlen J (1973) A moored temperature and pressure recorder. Deep Sea Res 21:145–154

Wunsch C, Gaposchkin EM (1980) On using satellite altimetry to determine the general circulation of the oceans with application to geoid improvement. Rev Geophys Space Phys 18:725–745

Wunsch C, Grant B (1982) Towards the general circulation of the North Atlantic Ocean. Prog Oceanogr 11:1–59

Wyrtki K (1962a) Geopotential topographics and associated circulation in the southeastern Indian Ocean. Aust J Mar Freshwater Res 13:1–17

Wyrtki K (1962b) Geopotential topographics and associated circulation in the western South Pacific Ocean. Aust J Mar Freshwater Rev 13:89–105

Wyrtki K (1967) The spectrum of ocean turbulence over distances between 40 and 1000 Kilometers. Deutsche Hydrogr Z 20:176–186

Wyrtki K (1971) Oceanographic atlas of the International Indian Ocean Expedition. N Sci Foundation, US Government Printing Office, Washington, 531 pp

Wyrtki KL (1978) Lateral oscillations of the Pacific Equatorial Countercurrent. J Phys Oceanogr 8:530–532

Wyrtki K, Meyers G (1975) The trade wind field over the Pacific Ocean, part I. The mean field and the mean annual variation. Rep HIG-75-1 p 26 Hawaii Inst Geophys, Honolulu

Wyrtki K, Magaard L, Hager J (1976) Eddy energy in the oceans. J Geophys Res 81:2641–2646

Wyrtki K, Firing E, Halpern D, Knox R, McNally GJ, Patzert WC, Stroup ED, Taft BA, Williams R (1981) The Hawaii to Tahiti Shuttle Experiment. Sci 211:22–28

Young WR, Rhines PB, Garrett CJR (1982) Shear flow dispersion, internal waves and horizontal mixing in the ocean. J Fluid Mech (in press)

Zoutendyk P (1970) Zooplankton density in the Southwest Indian Ocean. Proc Symp Oceanogr S Afr, Durban

Subject Index

Intense Atmospheric Vortices

Proceedings of the Joint Symposium (IUTAM/IUGG)
held at Reading (United Kingdom) July 14–17, 1981

Editors: **L. Bengtsson, Sir J. Lighthill**

1982. 195 figures. XII, 326 pages
(Topics in Atmospheric and Oceanographic Sciences)
ISBN 3-540-11657-5

With contributions by *F. K. Browand, M. Challa,
D. R. Davies, R. P. Davies-Jones, Du Xing-yuan,
M. P. Escudier, H. P. Evans, A. E. Gill, W. M. Gray,
R. Hide, G. J. Holland, E. J. Hopfinger, J. B. Klemp,
Y. Kurihara, L. M. Leslie, D. K. Lilly, B. A. Lugovtsov,
T. Maxworthy, A. D. McEwan, K. V. Ooyama, R. P. Pearce,
R. L. Pfeffer, R. Rotunno, R. C. Sheets, J. Simpson,
R. K. Smith, J. T. Snow, R. E. Tuleya, Xiao Wen-jun, Xie
An, Zhou Zi-Dong*

The concept of vorticity is of central importance in fluid mechanics, and the change and variability of atmospheric flow is dominated by transient vortices of different time and space scales. Or particular importance are the most intense vortices, such as hurricanes, typhoons and tornadoes, which are associated with extreme and hazardous weather events of great concern to society.

This book examines the different mechanisms for vorticity intensification that operate in two different kinds of meteorological phenomena of great importance, namely the tropical cyclone and the tornado. The understanding of these phenomena has grown in recent years due to increased and improved surveillance by satellites and aircraft, as well as by numerical modelling and simulation, theoretical studies and laboratory experiments. The book summarizes these recent works with contributions from observation studies (from radio sonde data, aircraft, satellites and radars) and from studies concerning the physical mechanism of these vortices by means of theoretical, numerical or laboratory models. The book contains articles by the leading world experts on the meteorological processes and on the fundamental fluid dynamics mechanism, for vorticity intensification.

Springer-Verlag
Berlin
Heidelberg
New York
Tokyo

J. Pedlosky

Geophysical Fluid Dynamics

1982. 180 figures. XII, 624 pages
Springer Study Edition
ISBN 3-540-90745-9
(Springer study edition based on corrected second
printing of the original edition)

Contents: Preliminaries. – Fundamentals. – Inviscid
Shallow-Water Theory. – Friction and Viscous Flow. –
Homogeneous Models of the Wind-Driven Oceanic
Circulation. – Quasigeostrophic Motion of a Stratified
Fluid on a Sphere. – Instability Theory. – Ageostro-
phic Motion. – Selected Bibliography. – Index.

From the reviews:
"An excellent high-level physical and mathematical
treatment of the dynamics of the oceans and atmosphere. Pedlosky's volume is directed to graduate students and scientists pursuing research in oceanography and meteorology... Topics fully discussed include the equations of motion in a rotating coordinate
frame, the concept of potential vorticity, geostrophic
motion, homogeneous models of oceanic circulation,
shallow water dynamics, Rossby waves, and baroclinic
and barotropic instability..."
Choice

"...everyone interested in the major motions of our
atmosphere and ocean (should) buy and study... this
outstanding book. Former students and colleagues
who have heard lectures and seminars by Joseph
Pedlosky will recognize again the clarity of logic and
illuminating perception that he brings to his subject."
Bulletin American Meteorological Society

"... this is an excellent book well worth a prominent
position on the bookshelf or desk of my serious
students of G. D. F. We all find a great deal of useful
information between its covers, information that
forms the basis for the subject and is unlikely to be
supplanted very quickly."
Earth-Science Reviews

Springer-Verlag
Berlin
Heidelberg
New York
Tokyo